CELL BIOLOGY AND PATHOLOGY OF MYELIN

Evolving Biological Concepts and
Therapeutic Approaches

ALTSCHUL SYMPOSIA SERIES

Series Editors: Sergey Fedoroff and Gary D. Burkholder

CELL BIOLOGY AND PATHOLOGY OF MYELIN

Evolving Biological Concepts and
Therapeutic Approaches

Edited by

Bernhard H. J. Juurlink
Richard M. Devon
J. Ronald Doucette
Adil J. Nazarali
David J. Schreyer
and
Valerie M. K. Verge

University of Saskatchewan
Saskatoon, Saskatchewan, Canada

SPRINGER SCIENCE+BUSINESS MEDIA, LLC

Library of Congress Cataloging-in-Publication Data

Cell biology and pathology of myelin : evolving biological concepts
  and therapeutic approaches / edited by Bernhard H.J. Juurlink, ...
  [et al.].
      p.   cm. -- (Alschul symposia series ; v. 4)
  Includes bibliographical references and index.

      1. Myelin sheath--Pathophysiology--Congresses.   2. Demyelination-
  -Congresses.   3. Myelination--Congresses.   4. Myelin proteins-
  -Congresses.   I. Juurlink, B.H.J., 1946-   . II. Series.
  RC366.C45 1997
  616.8'047--dc21   97-8136
                                                                CIP

ISBN 978-1-4613-7726-9   ISBN 978-1-4615-5949-8 (eBook)
DOI 10.1007/978-1-4615-5949-8

Proceedings of the Fourth International Altschul Symposium on Cell Biology and Pathology of Myelin:
Evolving Biological Concepts and Therapeutic Approaches, held June 27–29, 1996, in Saskatoon,
Saskatchewan, Canada

© 1997 Springer Science+Business Media New York
Originally published by Plenum Press New York in 1997

http://www.plenum.com

10 9 8 7 6 5 4 3 2 1

PREFACE

This volume contains papers presented at the Fourth International Altschul Symposium, held June 27–29, 1996, at the University of Saskatchewan in Saskatoon, Saskatchewan, Canada. The Altschul Symposia Series are held in memory of Rudolf Altschul, a graduate of the University of Prague and a pioneer in the fields of vascular and nervous system biology. Dr. Altschul was head of the Department of Anatomy at the University of Saskatchewan from 1955 to 1963. The Altschul Symposia are made possible by an endowment left by Anni Altschul, Dr. Altschul's wife, and by other contributions given by the sponsors listed at the end of this Preface.

The objective of the Fourth Altschul Symposium, entitled *Cell Biology and Pathology of Myelin: Evolving Biological Concepts and Therapeutic Approaches*, was to facilitate the transfer of ideas on the biology and pathology of myelin from the research laboratory to the clinic by providing a forum for discussing the evolving biological concepts regarding myelin function in health and disease.

Myelin facilitates the temporal resolution of information flow within the nervous system through its role in increasing the conduction velocity of action potentials. Many of the problems encountered in demyelinating diseases such as multiple sclerosis or dysmyelinating diseases such as cerebral palsy are due to deficiencies in information flow that arise when parts of an axon are not myelinated. Ideally, treating such diseases requires the development of therapeutic approaches designed either to prevent demyelination from occurring in the first place or to facilitate myelination. In the case of multiple sclerosis, for example, such a therapeutic approach demands an intricate understanding of the manner in which the immune system interacts with the macromolecules comprising the myelin sheath. An additional challenge is how best to promote remyelination in patients afflicted with a demyelinating or dysmyelinating disease. The evolution of therapies to promote remyelination has occurred in tandem with our obtaining a better understanding of the factors involved in controlling the proliferation, migration, and differentiation of myelinating cells. Myelin also appears to be a major player in events that prevent axonal regeneration in the central nervous system. This problem is the opposite of that encountered in demyelinating or dysmyelinating diseases, in that the mere presence of myelin macromolecules impedes the ability of the nervous system to properly repair itself after injury.

A theme that ran throughout the symposium was that myelin functions as more than a simple insulating sheath. It is fitting, therefore, that the first part of this volume deals with myelin as a dynamic structure, concentrating on the molecular interaction with the molecules comprising the myelin sheath. The process of myelination also involves a dynamic interaction among several cell populations. Thus, the second part of this volume addresses the cellular and molecular control over a glial cell's expression of the myelinating phenotype. The final part of the volume discusses evolving therapeutic approaches in the treatment of demyelinating/dysmyelinating diseases and for promotion of functional recovery after spinal cord injury.

After reading the chapters in this volume, the reader will agree that over the last decade we have come a long way in unraveling the cell–cell interactions that govern gene expression involved in myelination and axonal growth. Understanding the mechanisms that govern these cell–cell interactions is opening the path to the development of rational and effective neurotherapies in the area of demyelinating/dysmyelinating diseases and spinal cord injuries.

The symposium, using seed monies from the Altschul Trust Fund, could not have been held without the generous financial support of the following funding agencies and corporations: College of Medicine, University of Saskatchewan; Heart and Stroke Foundation of Saskatchewan; March of Dimes Birth Defects Foundation; Medical Research Council of Canada; Multiple Sclerosis Society of Canada; National Multiple Sclerosis Society (U.S.A.); Paralyzed Veterans of America's Spinal Cord Research Foundation; Saskatoon District Health Board, University of Saskatchewan (President's Office); Cameco Corporation; Cedarlane Laboratories; Ciba-Geigy Canada; Janssen-Ortho Inc.; Mandel Scientific Co. Ltd; Serotec Ltd; Summit Biotechnology; and Waverly International.

The participation of many members of the Department of Anatomy and Cell Biology was integral in ensuring the success of the symposium, and we are indebted to them all, particularly Dr. Gary D. Burkholder, Head of the Department of Anatomy and Cell Biology, Don Gurnsey, Gregg Parchomchuk, Irene Partridge, and Shirley West.

Bernhard H. J. Juurlink
Richard M. Devon
J. Ronald Doucette
Adil J. Nazarali
David J. Schreyer
Valerie M. K. Verge

CONTENTS

MYELIN AS A DYNAMIC STRUCTURE

FACTORS CONTROLLING THE EXPRESSION OF THE MYELINATING PHENOTYPE

PROMOTING FUNCTIONAL RECOVERY IN WHITE MATTER

STRUCTURE AND FUNCTION OF MYELIN, AN EXTENDED AND BIOCHEMICALLY MODIFIED CELL SURFACE MEMBRANE

R. H. Quarles, R. G. Farrer, and S. H. Yim

Myelin and Brain Development Section
Laboratory of Molecular and Cellular Neurobiology
NINDS
Bethesda, Maryland 20892

INTRODUCTION

The formation of myelin involves the extension and biochemical modification of plasma membranes of oligodendrocytes or Schwann cells (Morell, et al., 1994). Although there is a continuous surface membrane from the perikaryon of the myelin-forming cell to compact myelin, there are many specializations within the membrane which exhibit variability in terms of structure, function and biochemical composition. In addition to the compact myelin itself, the specialized domains include the surface membranes of the cell body and processes, surface and periaxonal membranes of the myelin sheath, paranodal loops, and incisures. When considering function or pathogenic mechanisms involving myelin sheaths, it is important to distinguish the constituents of compact myelin from those of these other myelin-related membranes. Many of the myelin-related membranes are adjacent to cytoplasmic pockets in the sheaths where dynamic processes such as transmembrane signaling and ion flux are likely to occur. The first part of this article consists of a brief overview the principal proteins in the compact myelin and myelin-related membranes, respectively. Since glycoproteins are important components of plasma membranes in general, it is not surprising that many of the proteins in both the compact myelin and the myelin-related membranes are glycosylated. This article will consider plasma membrane proteins which are specific for myelin-forming cells or have been studied primarily in the context of myelination, but it should be remembered that these myelin-related membranes probably contain many other components which are common to plasma membranes of most cells in the body. Glycosphingolipids are another characteristic component of cell surface membranes, and the second part of this article will describe recent research in our laboratory concerning the expression and possible function of gangliosides in differentiating oligodendrocytes.

Cell Biology and Pathology of Myelin, edited by Juurlink *et al.*
Plenum Press, New York, 1997

OVERVIEW OF PROTEINS IN COMPACT MYELIN AND MYELIN-RELATED MEMBRANES OF THE CNS AND PNS

Compact Myelin

Schematic representations of the proteins of compact CNS and PNS myelin showing their orientations with respect to the lipid bilayer of the membrane are shown in Fig. 1. Proteolipid protein (PLP) and PO protein are the major integral membrane proteins of compact CNS myelin and PNS myelin, respectively. Whereas PLP is not glycosylated, PO protein and the quantitatively less prominent, peripheral myelin protein-22 (PMP-22) of compact PNS myelin each contain a single oligosaccharide moiety. Both CNS and PNS myelin contain myelin basic protein (MBP, P1) and P2 protein which associate with the cytoplasmic side of the layered membranes.

PLP and MBP have generally been assumed to stabilize the intraperiod and major dense lines of compact CNS myelin, respectively (Morell, et al., 1994), but their contributions to the structure of compact myelin may not be as straightforward as previously thought (see articles in this volume by Moscarello, Klugmann, et al, and Macklin and Duchala). Furthermore, PLP and especially its DM-20 isoform are thought to have another function in developing oligodendrocytes in addition to a structural role in compact myelin (Skoff and Knapp, 1992).

PO glycoprotein has a single extracellular immunoglobulin-like domain and is believed to stabilize the intraperiod line of PNS myelin by homophilic interactions (Uyemura, et al., 1992). The carbohydrate moiety of PO and its cytoplasmic domain are required for these interactions (Filbin and Tennekoon, 1991; Filbin and Tennekoon, 1993; Filbin et al., this volume). PMP-22 is a quantitatively less prominent component of compact PNS myelin whose function is not known (Suter and Snipes, 1995; Suter, this vol-

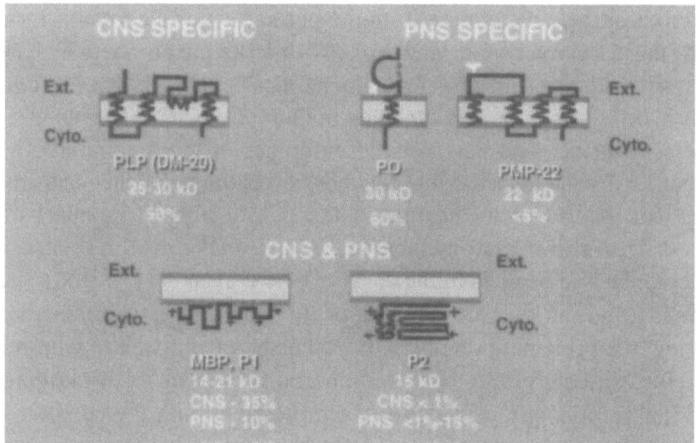

Figure 1. Schematic diagrams of the proteins of compact CNS and PNS myelin. The drawings show the apparent relationships of each protein to the lipid bilayer of the myelin membrane. Ext., extracellular side of membrane; Cyto., cytoplasmic side of membrane. Ig-like domains are represented by disulfide bonded loops, and oligosaccharides are indicated by the white antennae. PLP (DM-20), proteolipid protein and its smaller DM-20 isoform ; PO, PO glycoprotein; PMP-22, peripheral myelin protein-22; MBP, myelin basic protein (also called P1 protein in the terminology for PNS myelin); P2, P2 protein. The figures under the abbreviations indicate the molecular weights of the proteins (or the range for isoforms) and the percentage of the total protein of isolated myelin that each represents.

ume). Since it accounts for only a few percent of the total protein of PNS myelin, it seems unlikely to serve a major structural role like P0. However, its importance for myelin formation and maintenance is clear in view of the severe dysmyelination associated with abnormalities of its gene occurring in trembler mutant mice and Charcot-Marie Tooth disease, type 1a. The fact that oligosaccharide structures of PMP-22 include the adhesion-related HNK-1 epitope suggests that its function could involve cell-cell or membrane-membrane interactions (Hammer, et al., 1993; Snipes, et al., 1993). The overall structure of PMP-22, with 4 putative transmembrane domains, is similar to that of the major proteolipid protein (PLP) and related DM-20 of CNS myelin, and it has been speculated that PMP-22 could play a role for the PNS similar to one of the functions of PLP/DM-20 in the CNS. The four putative transmembrane domains in PMP-22 and PLP raise the possibility that these proteins could function by forming pores or channels to facilitate the transport of nutrients, ions, or other substances between myelin layers. PNS myelin also contains MBP, and both MBP and the positively charged cytoplasmic domain of P0 appear to be important for stabilization of the major dense line (Morell, et al., 1994; Martini, et al., 1995). Furthermore, another positively charged protein called P2 is present in significant amounts at the cytoplasmic surface of PNS myelin membranes in some species. P2 could play a structural role in the major dense line, although it also may function as a fatty acid binding protein (Martenson and Uyemura, 1992; Tennekoon, this volume). P2 is also present in small amounts in CNS myelin of some species.

Myelin-Related Membranes

The structures of several proteins that are associated with membranes of CNS myelin sheaths that are distinct from compact myelin are schematically represented in Fig. 2. Several are quantitatively minor glycoproteins, such as myelin-associated glycoprotein (MAG), myelin-oligodendrocyte glycoprotein (MOG), and oligodendrocyte-myelin glycoprotein (OMgp), and may be involved in recognition, adhesion and cellular signaling. The enzyme 2',3'-cyclic nucleotide 3'-phosphodiesterase (CNP) is another myelin-related pro-

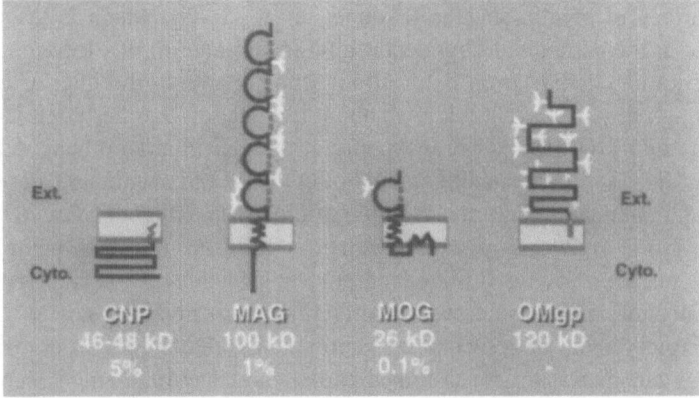

Figure 2. Schematic diagrams of the myelin-related proteins in the CNS. The drawings show the apparent relationships of each protein to the lipid bilayer of the membrane. Ext., extracellular side of membrane; Cyto., cytoplasmic side of membrane. Ig-like domains are represented by disulfide bonded loops, and oligosaccharides are indicated by the white antennae. CNP, 2',3'-cyclic nucleotide 3'-phosphodiesterase; MAG, myelin-associated glycoprotein; MOG, myelin/oligodendrocyte glycoprotein; OMgp, oligodendrocyte-myelin glycoprotein. The figures under the abbreviations indicate the molecular weights of the proteins (or the range for isoforms) and the percentage of the total protein of isolated myelin that each represents.

tein that is localized in oligodendroglial cytoplasm and associates with many of the myelin-related membranes described above (Trapp et al, 1988). CNP is also a good candidate to function in the dynamic aspects of myelin formation and maintenance, although its function remains enigmatic (Tsukada and Kurihara, 1992; Braun et al., this volume). Recent findings suggest that it could have a cytoskeletal function during process extension by oligodendrocytes (Braun et al., this volume).

MAG is a transmembrane glycoprotein in the immunoglobulin (Ig) superfamily with 5 extracellular Ig-like domains and a cytoplasmic domain that can be modified by phosphorylation and alternative mRNA splicing (Quarles, et al., 1992). It is selectively localized in the periaxonal oligodendroglial and Schwann cell membranes (Trapp, et al., 1989) where it is believed to function in glia-axon interactions. MAG is in the I-type lectin subgroup of the Ig-superfamily (Kelm, et al., 1994) and probably interacts with sialic acid-containing oligosaccharides on a glycoprotein or glycolipid ligand of the axolemma. In the PNS, MAG is also localized in mesaxons, lateral loops, and Schmidt-Lanterman incisures (Trapp, et al., 1989) where it may have an additional function in the interactions of the adjacent Schwann cell membranes at these sites. Recent reports showing that MAG interacts with fyn tyrosine kinase, FAK kinase and phospholipase Cγ (Jaramillo, et al., 1994; Umemori, et al., 1994; Umemori et al., this volume) suggest that binding to its ligand could activate second messenger cascades within myelin-forming cells and thereby transmit signals from neurons to oligodendrocytes or Schwann cells. Furthermore, recent findings that MAG affects neurite outgrowth (McKerracher, et al., 1994; Mukhopadhyay, et al., 1994; McKerracher, this volume) and that MAG-null mice develop axonal abnormalities (Fruttiger, et al., 1995; Yin et al, this volume) suggest that this glycoprotein might also be involved in transmitting signals in the reverse direction from myelin-forming cells to neurons

MOG is another member of the Ig superfamily associated with CNS myelin sheaths. It has one Ig-like domain and two hydrophobic domains only one of which appears to pass completely through the bilayer (Gardinier, et al., 1992; Phamdinh, et al., 1994; Hilton, et al., 1995; Gardinier et al., this volume). Quantitatively, it accounts for an even smaller percentage of total myelin proteins than MAG. Interestingly, it is highly concentrated on the outside surface of myelin sheaths (Brunner, et al., 1989), which is the converse of MAG which is in the periaxonal oligodendroglial membrane. In this location, it could be involved transmitting signals from extracellular components to the interior of oligodendrocytes, although little is currently known about its function. Furthermore, its surface location suggests that MOG could be an important antigen in autoimmune demyelinating diseases of the CNS, and a substantial amount of evidence has accumulated to support this hypothesis both in animal models and multiple sclerosis. The oligodendrocyte-myelin glycoprotein (OMgp) is a 120 kD glycoprotein that is enriched in human white matter, expressed on the surface of cultured oligodendrocytes, is anchored to membranes through a phosphatidyl inositol linkage, binds peanut agglutinin, and contains 9 potential attachment sites for asparagine-linked oligosaccharides and a serine/threonine rich domain that contains probable attachment sites for O-linked carbohydrates (Mikol and Stefansson, 1988; Mikol, et al., 1990; Mikol, et al., 1993). It contains a short cysteine-rich motif at the N-terminus and a series of tandem leucine-rich repeats similar to those which have been described in other proteins implicated in adhesion.

MAG and CNP are also expressed in the PNS by myelin-forming Schwann cells, although at lower concentrations than in the CNS. Although MOG and OMgp are specific for CNS myelin sheaths as their names imply, there are other myelin-related proteins associated with PNS myelin. For example, there is a 170 kD glycoprotein (P170k) that was first detected in PNS myelin and Schwann cells by lectin binding (Shuman, et al., 1986) and was recently further characterized and called Schwann cell membrane glycoprotein or SAG (Dieperink, et al., 1992). P170k/SAG appears to be in regions of the myelin sheaths distinct from compact myelin based on preliminary immunocytochemical observations and was hypothesized to play a role in Schwann cell-axon interactions. Another myelin-related protein in PNS myelin sheaths is the gap-junction protein, connexin-32, (Bergoffen, et al., 1993; Bruzzone, et al., 1994; Balice-Gordon et al., this volume). Although not traditionally thought to be one of the common myelin-related proteins, its importance for PNS myelination is clear since mutations of connexin-32 cause the X-linked form of Charcot-Marie Tooth disease. It is localized primarily in Schmidt-Lanterman incisures and lateral loops, and may function to allow small molecules and ions to diffuse radially between adjacent layers of myelinating Schwann cells.

GANGLIOSIDES OF MYELIN AND MYELIN-RELATED MEMBRANES

The preceding overview concentrates on the proteins of myelin and myelin-related membranes as do most of the other articles in this volume. However, it is well known that compact myelin contains 70–80% lipid, including some that are characteristic of myelin and myelin-forming cells such as galactocerebroside (GalC) and ethanolamine plasmalogen (Morell, et al., 1994). Furthermore, as much as half the dry weight of plasma membranes in general is accounted for by lipids. Although the lipid bilayer serves as a permeability barrier in the cell surface membrane into which the various integral membrane proteins are inserted, there are also numerous examples in which particular lipids have been shown to modulate the biological activities of membrane-associated enzymes or other membrane proteins. Of particular interest with regard to myelination, GalC has been shown to be part of a signaling pathway that affects calcium influx and cytoskeleton organization in cultured oligodendrocytes (Dyer and Benjamins, 1990; Dyer, this volume).

Gangliosides are a class of complex sialic acid-containing glycosphingolipids which are characteristic components of plasma membranes and have been implicated in modulating the activity of membrane proteins as mentioned above (Ledeen 1989; Saqr et al., 1993). They are present in high concentration in the surface membranes of neurons, but smaller amounts are present in plasma membranes of essentially all cells of the body. The major gangliosides in CNS tissue are a monosialoganglioside (GM1), two disialogangliosides (GD1a and GD1b), and a trisialoganglioside (GT1b), but there are also a large number of quantitatively less prominent gangliosides. The ganglioside composition of purified CNS myelin is quite different from neuronal membranes in that GM1 is the major component, and it contains relatively little of the polysialogangliosides (Morell et al., 1994). CNS myelin from some species, including humans, also contains an unusual ganglioside that is formed by adding sialic acid to GalC (GM4). The ganglioside composition of purified PNS myelin is different from CNS myelin. The major PNS myelin ganglioside in some species, including rats and humans, is a glucosamine-containing monosialoganglioside (LM1) (Morell et al., 1994).

Expression of Gangliosides in Cultured Oligodendrocytes

Cell surface staining with antibodies to gangliosides and other glycosphingolipids has been used to identify various stages of differentiation of cultured rat cells in the oligodendrocyte-type 2 astrocyte (O-2A) lineage (Dubois-Dalcq and Armstrong, 1992). In particular, anti-GD3 antibodies and the A2B5 monoclonal antibody have been used to immunostain cells at the O-2A progenitor stage. A2B5 has been reported to react with a number of polysialogangliosides, especially GT1c (Eisenbarth, et al., 1982; Kasai and Yu, 1983; Kundu, et al., 1983). However, a biochemical analysis of the gangliosides of cultured rat cells in the O-2A lineage (McCarthy and de Vellis, 1980) had not been done until recently (Yim et al., 1994; Yim et al., 1995a). These studies revealed that these differentiating oligodendrocytes express little of the major gangliosides of CNS tissue described above and that the most prominent gangliosides are GM3 and GD3. The structures of these two gangliosides are shown in Fig. 3 and differ from each other only by a single sialic acid moiety. The autoradiogram in Fig. 3 shows that GD3 is the major ganglioside synthesized in oligodendrocyte cultures early in the process of differentiation when most are at the O-2A progenitor stage (Yim, et al., 1994). As the cells, differentiate to mature oligodendroctyes GD3 continues to be synthesized, and the synthesis of GM3 ganglioside increases substantially. Resorcinol staining of the gangliosides of differentiated oligoden-

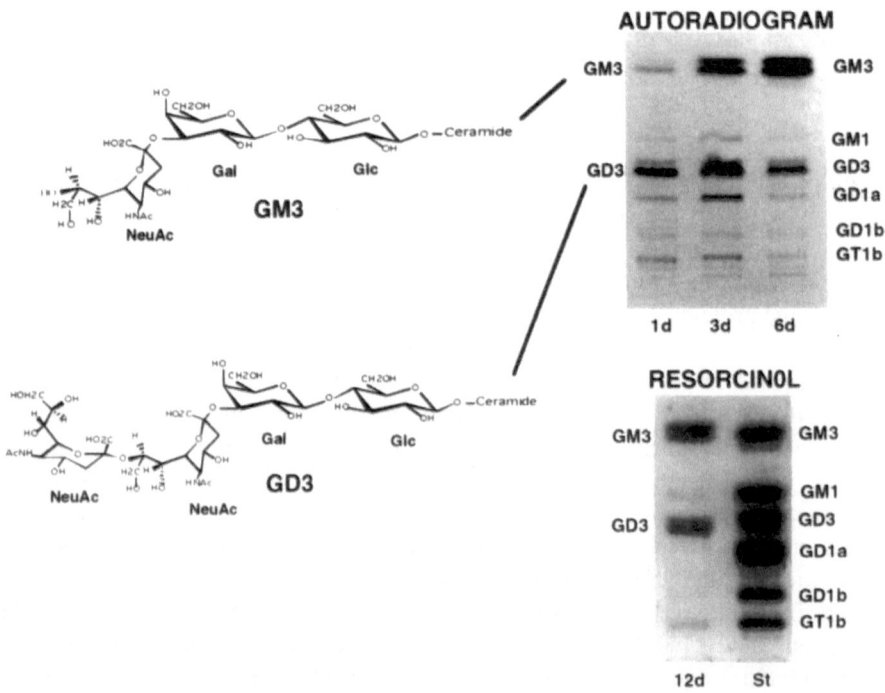

Figure 3. Gangliosides of differentiating oligodendrocytes. Upper right - Metabolic labeling. O-2A lineage cells which were maintained in culture for 1, 3 and 6 days after plating on polylysine coated dishes were incubated with [^{14}C]galactose for 20 h. The gangliosides were isolated, chromatographed and detected by autoradiography. GD3 was the major labeled ganglioside when the cells were first cultured, but the amount of GM3 synthesized by the cells increased greatly with differentiation. Reproduced from Yim et al. (1994) with permission. Lower right - Resorcinol staining of gangliosides in differentiated oligodendrocytes. The ganglioside fraction was isolated from O-2A lineage cells that had differentiated for 12 days after plating on polylysine coated dishes, chromatographed on a TLC plate, and stained for sialic acid with resorcinol. St - known ganglioside standards. Reproduced from Yim et al. (1995a) with permission. The results show that GM3 and GD3 are the major gangliosides expressed in O-2A lineage cells, and the structures of these two gangliosides are shown in the left part of the figure.

drocytes on TLC plates (Fig. 3) revealed that GD3 and GM3 are quantitatively the major components (Yim, et al., 1995a), as suggested by the metabolic labeling. GM3 and GD3 are also enriched over whole brain in bulk isolated rat oligodendrocytes (Yu et al., 1989), but the isolated cells contained a much higher content of more complex gangliosides than the cultured oligodendrocytes. This difference may be due to *in vitro* versus *in vivo* conditions, loss of most processes from oligodendrocytes during bulk isolation, a difference in the extent of differentiation, or other factors.

These biochemical findings on cultured O-2A lineage cells are consistent with the fact that oligodendrocyte progenitors are immunostained by anti-GD3 antibodies. However, they are more difficult to reconcile with their immunostaining by A2B5, since these cells in the O-2A lineage express very little of the polysialogangliosides that have been reported to react most strongly with this antibody. Therefore, experiments involving immunostaining of TLC plates similar to those shown in Fig. 3 were undertaken to identify the glycolipid antigen in these cells that reacted most strongly with A2B5. The results showed that there is a quantitatively minor glycolipid in the cells that reacts very strongly with A2B5 and does not correspond to one of the glycolipids revealed by resorcinol or orcinol staining nor to one of the well known ganglioside standards expressed in neural tissue. Furthermore, the amount of this glycolipid antigen decreases as the cells differentiate to mature oligodendrocytes, suggesting that it is the A2B5-reactive antigen that has been used widely to identify O-2A progenitors in these cultures. Further experiments revealed that these progenitors express a number of sulfated gangliosides (Farrer and Quarles, 1997). These findings are of particular interest not only because they show that the A2B5 antigen in these cells is a novel ganglioside, but also because sulfated gangliosides have not previously been described in neural tissue. Their identification in differentiating oligodendrocytes raises many questions concerning their possible functions in these cells in particular and the nervous system in general.

Function of GM3 Ganglioside in Oligodendrocyte Differentiation

The findings that the expression of GM3 ganglioside increases dramatically as progenitors differentiate into more mature oligodendrocytes (Fig. 3) suggests that this ganglioside could play a key role in the differentiation process. To explore this possibility, experiments were undertaken in which the effects of adding exogenous GM3 to these cells were investigated (Yim, et al., 1994). There is an extensive literature indicating that exogenous gangliosides have neuronotrophic and neuritogenic effects on cultured neurons (Ledeen, 1989), but the effects of exogenous gangliosides on cultured glia have received relatively little attention. In the case of neurons, the effects of exogenous gangliosides are believed to be due at least in part to their insertion into surface membranes thereby modulating the function of various membrane components (Saqr, et al., 1993). One of the best studied examples of the functional insertion of a ganglioside into plasma membranes involves the role of GM1 as the receptor for cholera toxin (Fishman, et al., 1993). The addition of exogenous GM1 to cells not expressing this ganglioside and insensitive to cholera toxin renders them sensitive to the action of the toxin.

Addition of GM3 ganglioside to primary cultures of rat oligodendrocytes resulted in the elaboration of thicker and more highly branched processes in comparison to untreated control cultures (Yim, et al., 1994). The processes and membranous arrays of the treated cells immunostained intensely with the O4 and O1 antibodies. The more extensive processes and membranes in the GM3-treated cells in comparison to control cells are most obvious in low power immunostained micrographs (Fig. 4, a and b). The greater differentiation of GM3 treated oligodendrocytes could also be demonstrated biochemically by increased incorporation of radioactive precursors into GalC, sulfatide and MAG (Fig. 5). However, GM3 did not appreciably affect the synthesis of MBP. Thus the GM3 effect seems to be most pronounced on oligodendroglial components such as GalC and MAG that have been implicated in the early aspects of myelination.

Another difference between GM3-treated and control cultures was many fewer contaminating GFAP-positive astrocytes in the treated cultures (Fig. 4, c and d). A likely explanation for this is that some progenitors which had begun to develop toward type 2 astrocytes reversed their direction of differentiation and became oligodendrocytes. This explanation was supported by the finding that cells that were immunopositive for both GalC and GFAP were detected frequently in GM3-treated cultures but seldom in control cultures (Yim, et al., 1994). Thus, GM3 appears to promote plasticity in the differentiation process.

These results have shown that exogenous GM3 ganglioside enhances the differentiation of oligodendrocytes in the direction of myelin formation. The results are reminiscent of the well known neuritogenic effects of exogenous gangliosides on cultured neurons, but they differ in a very significant way. Whereas the effects of gangliosides on neuronal cultures generally exhibit very little specificity with regard to what ganglioside(s) is used (Ledeen, 1989), the effect on oligodendrocytes appeared to be quite specific for GM3. No other ganglioside tested, including GM1, GD3 or GD1a, affected the differentiation of oligodendrocytes. Thus it seems likely that the exogenous GM3 augments the role of the endogenous GM3 which is normally synthesized in increasing amounts as the oligodendrocyte progenitors differentiate. A similar specific effect of GM3 ganglioside

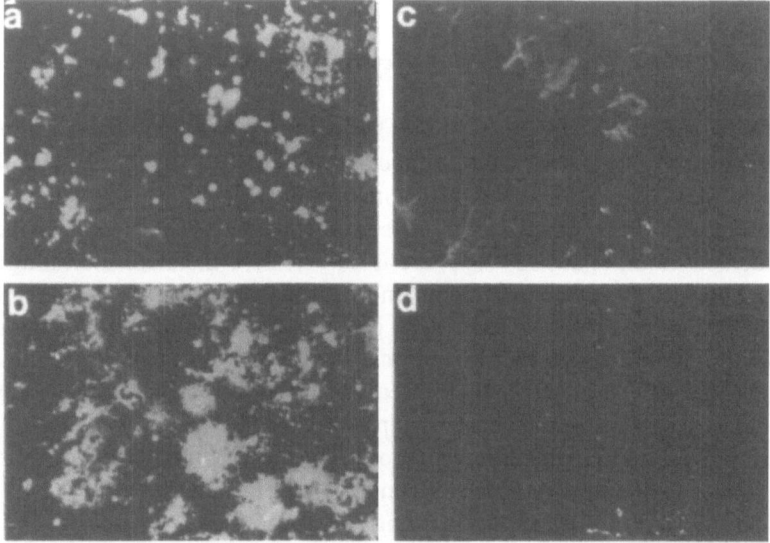

Figure 4. Low power micrographs of control and GM3 treated oligodendrocyte cultures doubly immunostained for GalC and GFAP. *Cultures of O-2A lineage cells were examined 9 days after plating on polylysine.* a, c: Control cultures; b, d: Cultures treated with GM3 (50 µg/ml) for 8 days before examination. a ,b: Immunostaining with the O1 antibody and rhodamine-labeled second antibody; c, d: Immunostaining for GFAP and fluorescein-labeled second antibody. Note the much more extensive O1-positive membranous arrays and the decreased number of GFAP-positive cells in the GM3-treated cells. Bar = 50 µm. Reproduced from Yim et al. (1994) with permission.

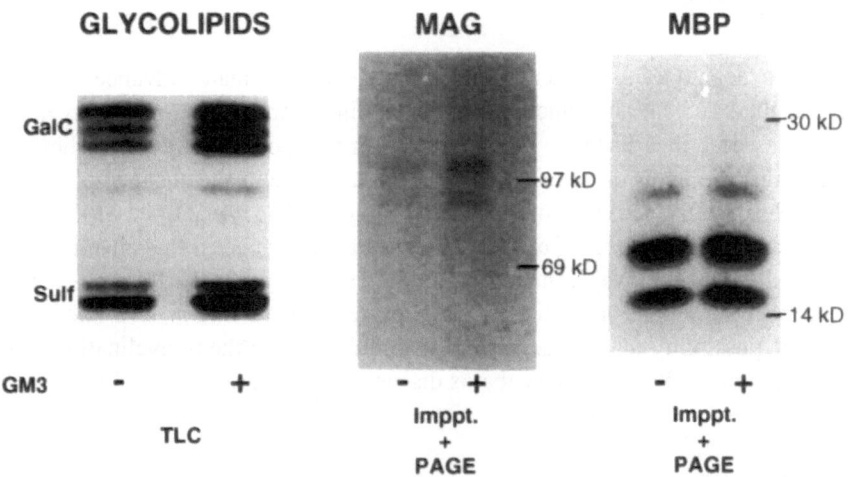

Figure 5. Effects of GM3 ganglioside on the incorporation of radioactive precursors into myelin or myelin-related components. O-2A lineage cells that had been in culture for several days in the presence (+) or absence (-) of 50 μg/ml GM3 ganglioside were incubated overnight with a radioactive precursor and the metabolically labeled constituents were detected by autoradioagraphy. The left panel is an autoradiogram of a TLC plate on which glycosphingolipids labeled with radioactive galactose were separated. GalC, galactocerebroside; Sulf, sulfatide. The right two panels are autoradiograms of immunoprecipitated (Imppt.) MAG and MBP, respectively, that had been separated by SDS polyacrylamide gel electrophoresis (PAGE) after metabolic labeling with a radioactive amino acid. GM3 increased the synthesis of GalC, sulfatide and MAG, but had little effect on MBP synthesis. Reproduced from Yim et al. (1994) with permission.

has been demonstrated in enhancing the differentiation of human leukemic cell lines toward the monocyte/macrophage lineage (Saito, 1993). Similarly to the oligodendrocytes, GM3 expression increases during normal differentiation to monocytes and is enhanced by exogenous GM3, so it may be that similar molecular mechanisms are operating in the two systems.

Current research in our laboratory is directed toward elucidating the biochemical mechanisms by which GM3 promotes oligodendrocyte differentiation. Since it is well known that a variety of growth factors affect oligodendrocyte differentiation (Dubois-Dalcq and Armstrong, 1992) and exogenous gangliosides have been shown to modulate the activities of growth factor receptors in a variety of cells (Hakomori and Igarashi, 1993; Saqr, et al., 1993; Yates, et al., 1995), modulation of an oligodendroglial growth factor receoptor is one possibility that is under investigation. Exogenous gangliosides have also been shown to modulate the activities of protein kinases, and it was demonstrated that the phosphorylation of proteins in GM3-treated oligodendrocytes is generally down regulated (Yim, et al., 1994). Despite the general decreased protein phosphorylation, however, the phosphorylation of MAG was increased by GM3 treatment. Whether this simply relates to the increased expression of MAG alluded to above or reflects activation of a MAG-mediated signaling system in the cells remains to be established. It is interesting that protein kinase C is a key enzyme that phosphorylates MAG in cultured oligodendrocytes (Bambrick and Braun, 1991; Kirchhoff, et al., 1993; Yim, et al., 1995b), and protein kinase C has been implicated in differentiation and process formation by oligodendrocytes (Vartanian, et al., 1986; Althaus, et al., 1991; Yong, et al., 1994; Yong, this volume).

CONCLUSIONS

Modern techniques of molecular biology are leading to major advances in our under-standing of the structure and function of proteins in compact myelin and myelin-related membranes as described in many of the articles in this book. However, it should not be overlooked that the lipids of myelin-forming cells and myelin sheaths may play important roles in modulating the activities of the various proteins. The capacity of GM3 ganglioside to enhance the differentiation of oligodendrocytes as described in this chapter is just one example of this. In some ways, this observation is similar to the well established neurito-genic and neuronotrophic properties of gangliosides *in vitro* and *in vivo* (Ledeen, 1989) and raise the possibility that exogenous GM3 could promote the remyelinating aspects of neural regeneration following injury that is discussed elsewhere in this volume.

REFERENCES

Althaus HH, Schroter J, Spoerri P, Schwartz P, Kloppner S, Rohmann A, Neuhoff V (1991): Protein kinase C stimulation enhances the process formation of adult oligodendrocytes and induces proliferation. J Neurosci Res 29:481–489.

Bambrick LL, Braun PE (1991): Phosphorylation of myelin-associated glycoprotein in cultured oligodendrocytes. Dev Neurosci 13:412–416.

Bergoffen J, Scherer SS, Wang S, Scott MO, Bone LJ, Paul DL, Chen K, Lensch MW, Chance PF, Fischbeck KH (1993): Connexin mutations in X-linked Charcot-Marie-Tooth disease. Science 262:2039–2042.

Brunner C, Lassmann H, Waehneldt TV, Matthieu JM, Linington C (1989): Differential ultrastructural localization of myelin basic protein, myelin/oligodendroglial glycoprotein, and 2',3'-cyclic nucleotide 3'-phosphodi-esterase in the CNS of adult rats. J Neurochem 52:296–304.

Bruzzone R, White TW, Scherer SS, Fischbeck KH, Paul DL (1994): Null mutations of connexin32 in patients with X-linked Charcot-Marie-Tooth disease. Neuron 13:1253–1260.

Dieperink ME, O'Neill A, Magnoni G, Wollmann RL, Heinrikson RL, Zucher NH, Stefansson K (1992): SAG: a Schwann cell membrane glycoprotein. J Neurosci 12:2177–2185.

Dubois-Dalcq M, Armstrong RC (1992): The oligodendrocyte lineage during myelination and remyelination. In Martenson RE (eds): Myelin: Biology and Chemistry. Boca Raton: CRC Press pp 81–122.

Dyer CA, Benjamins JA (1990): Glycolipids and transmembrane signaling: antibodies to galactocerebroside cause an influx of calcium in oligodendrocytes. J Cell Biol 111:625–633.

Eisenbarth GS, Shimizu K, Bowring MA, Wells S (1982): Expression of receptors for tetanus toxin and mono-clonal antibody A2B5 by pancreatic islet cells. Proc Natl Acad Sci U S A 79:5066–5070.

Farrer RG, Quarles RH (1997): Expression of sulfated gangliosides in the central nervous system. J Neurochem 68: 878–881.

Filbin MT and Tennekoon GI (1991): The role of complex carbohydrates in adhesion of the myelin protein, P0. Neuron 7:845–855.

Filbin MT, Tennekoon GI (1993): Homophilic adhesion of the myelin Po protein requires glycosylation of both molecules in the homophilic pair. J Cell Biol 122:451–459.

Fishman PH, Pacuszka T, Orlandi PA (1993): Gangliosides as receptors for bacterial enterotoxins. Adv Lipid Res 25:165–187.

Fruttiger M, Montag D, Schachner M, Martini R (1995): Crucial role for the myelin-associated glycoprotein in the maintenance of axon-myelin integrity. Eur J Neurosci 7:511–515.

Gardinier MV, Amiguet P, Linington C, Matthieu JM (1992): Myelin/oligodendrocyte glycoprotein is a unique member of the immunoglobulin superfamily. J Neurosci Res 33:177–187.

Hakomori S, Igarashi Y (1993): Gangliosides and glycosphingolipids as modulators of cell growth, adhesion, and transmembrane signaling. Adv Lipid Res 25:147–162.

Hammer JA, O'Shannessy DJ, De LM, Gould R, Zand D, Daune G, Quarles RH (1993): Immunoreactivity of PMP-22, P0, and other 19 to 28 kDa glycoproteins in peripheral nerve myelin of mammals and fish with HNK1 and related antibodies. J Neurosci Res 35:546–558.

Hilton AA, Slavin AJ, Hilton DJ, Bernard CC (1995): Characterization of cDNA and genomic clones encoding hu-man myelin oligodendrocyte glycoprotein. J Neurochem 65:309–318.

Jaramillo ML, Afar DE, Almazan G, Bell JC (1994): Identification of tyrosine 620 as the major phosphorylation site of myelin-associated glycoprotein and its implication in interacting with signaling molecules. J Biol Chem 269:27240–27245.

Kasai N, Yu RK (1983): The monoclonal antibody A2B5 is specific to ganglioside GQ1c. Brain Res 277:155–158.

Kelm S, Pelz A, Schauer R, Filbin MT, Tang S, de BM, Schnaar RL, Mahoney JA, Hartnell A, Bradfield P, Crocker PR (1994): Sialoadhesin, myelin-associated glycoprotein and CD22 define a new family of sialic acid-dependent adhesion molecules of the immunoglobulin superfamily. Curr Biol 4:965–92.

Kirchhoff F, Hofer HW, Schachner M (1993): Myelin-associated glycoprotein is phosphorylated by protein kinase C. J Neurosci Res 36:368–381.

Kundu SK, Pleatman MA, Redwine WA, Boyd AE, Marcus DM (1983): Binding of monoclonal antibody A2B5 to gangliosides. Biochem Biophys Res Commun 116:836–842.

Ledeen RW (1989): Biosynthesis, metabolism and biological effects of gangliosides. In Margolis RU and Margolis RK (eds): Neurobiology of Glycoconjugates. New York: Plenum Press pp 43–83.

Martenson R, Uyemura K (1992): Myelin P2, a neuritogenic member of the family of cytoplasmic lipid binding proteins. In Martenson R (eds): Myelin: Biology and Chemistry. Boca Raton: CRC Press pp 509–530.

Martini R, Mohajeri MH, Kasper S, Giese KP, Schachner M (1995): Mice doubly deficient in the genes for P0 and myelin basic protein show that both proteins contribute to the formation of the major dense line in peripheral nerve myelin. J Neurosci 15:4488–4495.

McCarthy KD, de Vellis J (1980): Preparation of separate astroglial and oligodendroglial cell cultures from rat cerebral tissue. J Cell Biol 85:890–902.

McKerracher L, David S, Jackson DL, Kottis V, Dunn RJ, Braun PE (1994): Identification of myelin-associated glycoprotein as a major myelin-derived inhibitor of neurite growth. Neuron 13:805–811.

Mikol DD, Gulcher JR, Stefansson K (1990): The oligodendrocyte-myelin glycoprotein belongs to a distinct family of proteins and contains the HNK-1 carbohydrate. J Cell Biol 110:471–479.

Mikol DD, Rongnoparut P, Allwardt BA, Marton LS, Stefansson K (1993): The oligodendrocyte-myelin glycoprotein of mouse: primary structure and gene structure. Genomics 17:604–610.

Mikol DD, Stefansson K (1988): A phosphatidylinositol-linked peanut agglutinin-binding glycoprotein in central nervous system myelin and on oligodendrocytes. J Cell Biol 106:1273–1279.

Morell P, Quarles RH, Norton WT (1994): Myelin formation, structure and biochemistry. In Siegel GJ (eds): Basic Neurochemistry: Molecular, Cellular, and Medical Aspects. New York: Raven Press Ltd. pp 117 - 144.

Mukhopadhyay G, Doherty P, Walsh FS, Crocker PR, Filbin MT (1994): A novel role for myelin-associated glycoprotein as an inhibitor of axonal regeneration. Neuron 13:757–767.

Phamdinh D, Allinquant B, Ruberg M, Dellagaspera B, Nussbaum JL, Dautigny A (1994): Characterization and expression of the cDNA coding for the human myelin/oligodendrocyte glycoprotein. J Neurochem 63:2353–2356.

Quarles R, Colman D, Salzer J, Trapp B (1992): Myelin-associated glycoprotein: Structure-function relationships and involvement in neurological diseases. In Martenson R (eds): Myelin: Biology and Chemistry. Boca Raton: CRC Press pp 4413–4448.

Saito M (1993): Bioactive gangliosides: differentiation inducers for hematopoietic cells and their mechanism(s) of actions. Adv Lipid Res 25:303–327.

Saqr HE, Pearl DK, Yates AJ, Van Brocklyn J, Bremer EG, Yates AJ (1993): A review and predictive models of ganglioside uptake by biological membranes J Neurochem 61:395–411.

Shuman S, Hardy M, Pleasure D (1986): Immunochemical characterization of peripheral nervous system myelin 170,000-Mr glycoprotein. J Neurochem 47:811–818.

Skoff RP, Knapp PE (1992): Phenotypic expression of X-linked genetic defects affecting myelination. In Martenson RE (eds): Myelin: Biology and Chemistry. Boca Raton FL: CRC Press pp 653–676.

Snipes GJ, Suter U, Shooter EM (1993): Human peripheral myelin protein-22 carries the L2/HNK-1 carbohydrate adhesion epitope. J Neurochem 61:1961–1964.

Suter U, Snipes CJ (1995): Peripheral myelin protein-22: Facts and hypotheses. J Neurosci Res 40:145–151.

Trapp BD, Andrews SB, Cootauco C, Quarles R (1989): The myelin-associated glycoprotein is enriched in multivesicular bodies and periaxonal membranes of actively myelinating oligodendrocytes. J Cell Biol 109:2417–2426.

Trapp BD, Bernier L, Andrews SB, Colman DR (1988): Cellular and subcellular distribution of 2′,3′-cyclic nucleotide 3′-phosphodiesterase and its mRNA in the rat central nervous system. J. Neurochem. 51:859–869.

Tsukada Y, Kurihara T (1992): 2′,3′-Cyclic nucleotide 3′-phosphodiesterase: Molecular charcterizartion and possible functional significance. In Martenson R (eds): Myelin: Biology and Chemistry. Boca Raton: CRC Press pp 449–480.

Umemori H, Sato S, Yagi T, Aizawa S, Yamamoto T (1994): Initial events of myelination involve Fyn tyrosine kinase signalling. Nature 367:572–576.

Uyemura K, Kitamura K, Miura M (1992): Structure and molecular biology of P0 glycoprotein. In Martenson RE (eds): Myelin: Chemistry and Biology. Boca Raton, FL: CRC Press pp 481–508.

Vartanian T, Szuchet S, Dawson G, Campagnoni AT (1986): Oligodendrocyte adhesion activates protein kinase C-mediated phosphorylation of myelin basic protein. Science 234:1395–1398.

Yates AJ, Saqr HE, Van Brocklyn J (1995): Ganglioside modulation of the PDGF receptor. A model for ganglioside functions. J Neurooncol 24:65–73.

Yim SH, Farrer RG, Hammer JA, Yavin E, Quarles RH (1994): Differentiation of oligodendrocytes cultured from developing rat brain is enhanced by exogenous GM3 ganglioside. J Neurosci Res 38:268–281.

Yim SH, Farrer RG, Quarles RH (1995a): Expression of glycolipids and myelin-associated glycoprotein during the differentiation of oligodendrocytes: comparison of the CG-4 glial cell line to primary cultures. Dev Neurosci 17:171–180.

Yim SH, Toda K, Goda S, Quarles RH (1995b): Comparison of the phosphorylation of myelin-associated glycoprotein in cultured oligodendrocytes and Schwann cells. J. Mol. Neurosci. 6:63–74.

Yong VW, Dooley NP, Noble PG (1994): Protein kinase C in cultured adult human oligodendrocytes: a potential role for isoform alpha as a mediator of process outgrowth. J Neurosci Res 39:83–96.

Yu RK, Macala LJ, Farooq M, Sbaschnig-Agler M, Norton WT, Ledeen RW (1989) Ganglioside and lipid composition of bulk isolated rat and bovine oligodendroglia. J Neurosci Res 23:136–141.

2

MYELIN BASIC PROTEIN, THE "EXECUTIVE" MOLECULE OF THE MYELIN MEMBRANE

Mario A. Moscarello

Department of Biochemistry Research
Hospital for Sick Children
Toronto, Ontario
Canada M5G 1X8

INTRODUCTION

Myelin basic protein (MBP) has been known since the mid 1960's. Its primary structure was reported by two groups independently, Eylar et al. reported on the bovine MBP while Carnegie reported on the human MBP (Eylar et al., 1971; Carnegie, 1971). The molecular weight of this protein was computed to be 18500. It is devoid of cysteinyl residues, is highly basic (12 lysyl and 19 arginyl residues) and has 11 prolines, 3 of which occur as a triproline sequence.

The human MBP gene was located on chromosome 18 (Kalmholz et al., 1987), consisting of seven exons (Kalmholz et al., 1988; Streicher and Stoffel, 1989). More recently it has been shown to be part of a larger MBP gene, golli MBP, containing additional exons (Pribyl et al., 1993). The significance of this larger gene is not understood at this time. Alternative splicing of the primary transcripts of the original gene gives rise to various isoforms in the 14–21 kDa range in rodents but less variation has been found in humans (17–21.5 kDa), in which the 18.5 kDa variant is the principle isoform (Fig. 1). To add further to the complexity each isoform exists as several components or "charge isomers". A separation of 8–10 charge isomers of the 18.5 kDa isoform was reported (Deibler et al., 1975; Chou et al., 1976) by chromatography on CM52 columns at pH 10.6. Therefore for each isoform of MBP, 8–10 isomers may exist giving a total of 30–40 members of the MBP family. This fact must be taken into consideration when the function and interactions of MBP are being studied. Since it is unlikely that nature has generated these various forms of MBP for no reason, it must be assumed that each has a specific role or function in the myelin sheath. If this were not the case, nature would have made only a single MBP molecule. Most of the modifications that have been studied are enzymatic in nature and therefore energy requiring reactions e.g. phosphorylation. If no specific role were assigned to the phosphorylated MBP molecules, the process would be wasteful of large amounts of metabolic energy. The purpose of my presentation is to convince you that different roles

Cell Biology and Pathology of Myelin, edited by Juurlink *et al.*
Plenum Press, New York, 1997

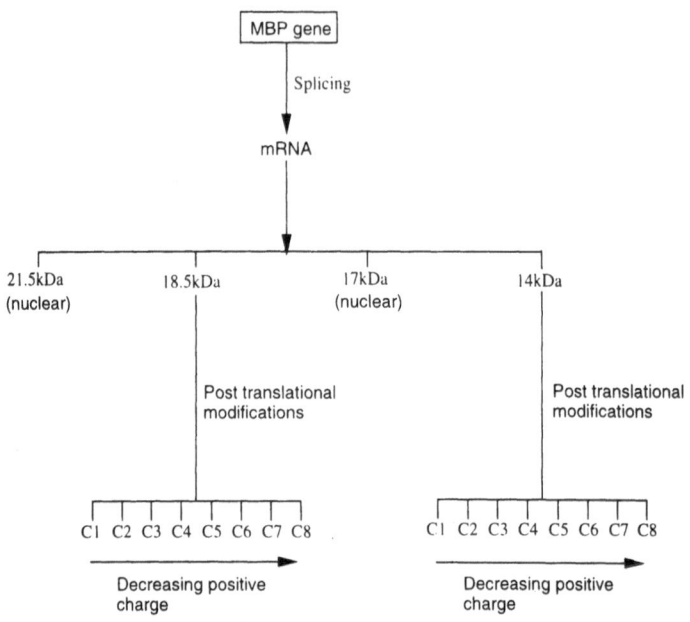

Figure 1. The human MBP family consisting of isoforms and charge isomers.

can be assigned to some members of the MBP family already and that this area of investigation should be highly rewarding.

A structural role for MBP has been accepted since 1971, when the sequence became known. The important structural role of MBP was demonstrated in an electron microscopic immunocytochemical study (Schwob et al., 1985) in which it was shown that MBP was incorporated into myelin before PLP. In fact PLP was only found in myelin after several layers of compact myelin had surrounded the axon. This role was to bring together the cytoplasmic faces of the oligodendrocyte to form the major dense line (MDL) observed by electron microscopy. Although MBP has been found at the MDL, several reports suggest it is also found at the intraperiod line (IPL). In particular we have localized the citrullinated MBP at this location (McLaurin et al., 1993). In a hybrid oligodendrocyte line (MO3–13), the citrullinated MBP has been found on the external surface of the cell, where it is expected to be, to form the IPL (Ursell et al., 1995). Phosphorylation at Thr 98 in the MAP kinase site is well accommodated into the folded structure, shielding the phosphate. This MBP was localized only at the MDL (Yon et al., 1996). These examples which will be discussed in more detail later, demonstrated the division of labour amongst MBP molecules involved in maintaining the structure and metabolic activity of myelin.

An active metabolic role, especially involved in reactions related to signal transduction, is now well established. Thus MBP can bind GTP, can be ADP-ribosylated, contains a MAP kinase site and stimulates the activity of a phosphatidylinositol-specific phospholipase C. Again not all members of the MBP family are active in these reactions e.g. the citrullinated MBP cannot be ADP-ribosylated and does not affect the activity of phospholipase C. The fact that an antibody, specific for phosphorylated Thr 98 in the MAP kinase site only labelled the MDL, suggests that the citrullinated MBP is not a substrate for the kinase, even though the MAP kinase sequence is intact in "C-8". The nuclear localization of the 17 and 21.5 kDa isoforms but not the 14 or 18 kDa isoforms in non-glial cells suggest that these isoforms may have regulating functions in myelinogenesis (Staugaitis et al., 1996). It is quite clear that, understanding the structure of myelin is dependent on unravelling the various roles of the MBP family. It is for this reason that I have referred to "MBP" as the "executive" molecule of the myelin sheath - it is responsible for many interactions (Fig. 2).

METABOLIC INTERACTIONS INVOLVING MBP

MBP has been shown to be a substrate for a number of enzymic as well as non enzymic reactions. A list of reactions occurring at various residues is shown in Table I. Most of the modifications decrease the positive charge on the molecule either by the addition of a negatively charged group such as phosphate or the removal of positive charge such as the deimination of arginyl residues which are converted to citrulline. Increase in negative charge may affect protein interactions with lipids. When the increase in negative charge results from phosphorylation, a reversible modification, destabilization will be local. Reversal of the modification returns the membrane structure to its "resting" state. On the other hand, non reversible modifications such as deimination of arginyl residues may result in destabilization for prolonged periods which may culminate in myelin breakdown.

INVOLVEMENT OF MBP IN REACTIONS RELATED TO THE TRANSDUCTION OF SIGNALS

Contrary to common belief that myelin functions as an electrical insulator for the axon, the presence of a number of enzymes associated with myelin suggests that many metabolic reactions take place in the membrane system. We are primarily concerned with those in which MBP is involved. A short discussion of some of these will be given here.

MBP is a substrate for protein kinases, several of which have been described in myelin. These include cAMP-dependent protein kinases; calcium, phospholipid dependent

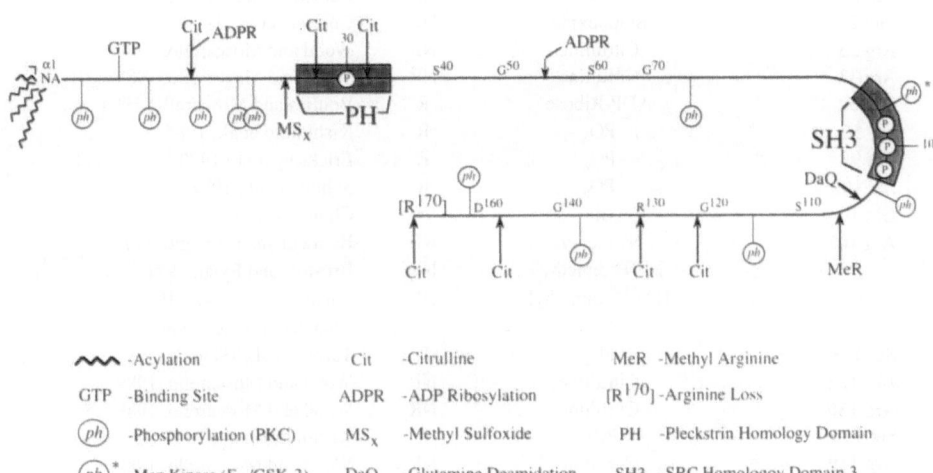

Figure 2. Sites of post translational modifications of human MBP (18.5 kDa) and the src homology 3 (SH3) and Pleckstrin homology (PH) domains.

protein kinases (protein kinase C) and calmodulin dependent protein kinases (reviewed by Ullmer, 1988). Because of the presence of a large number of kinases in myelin the potential to phosphorylate a variety of sites, thereby generating a large number of charge isomers transiently, would allow MBP to alter its interactions with the lipid bilayer for a short time in the immediate environment permitting the completion of an enzymic reaction, or permitting exposure of an immunologically active site subsequently returning it to its "resting" state by reversal of the reaction.

Of the different sites phosphorylated, two are noteworthy. When Ser 7 was phosphorylated by protein kinase C, it was refractory to hydrolysis by acid phosphatase. The Ser 7 was postulated to interact with Arg 5 and Arg 9 stabilizing a β-structure (Ramwani and Moscarello, 1990). From NMR studies of the structure of peptide 1–14, residues 7–12 have been postulated to form an antigenic site (Mendz et al., 1995). Phosphorylation at Ser 7 would have important effects on antibody recognition of this site, by altering the charge and possibly the conformation. Threonine 98 (human sequence) occurs in a mitogen activated protein kinase site (MAP kinase) in the sequence Thr-Pro-Arg-Thr[98]-Pro-Pro-Pro (Erickson et al., 1990). Using an antibody which reacts specifically with phosphorylated Thr 98, we have localized this phosphorylated MBP to the major dense line of myelin in cryosections by immunogold electron microscopy (Yon et al., 1996). This unexpected result can be explained by stabilization of the β-structure by interaction of the phosphate group with Lys 105 of one strand with Arg 33 of an adjacent strand or with other Arg residues nearby in the Stoner structure (Stoner, 1984). A newer model proposed by Harauz et al. (personal communication, see Abstract, Beniak et al. in this vol-

Table I. Modifications of MBP

Residue modified	Modifying group	Reversibility	Reference
Ala 1	acetyl-decyl	NR	Moscarello et al., 1992
Gln 3	GTP	R	Chan et al., 1988
Ser 7	PO$_4$	R (?)	Ramwani and Moscarello, 1990
Arg 9	Citrulline	NR	Wood and Moscarello, 1989
	ADP-Ribose	R	Boulias and Moscarello, 1994
Ser 12	PO$_4$	R	Ramwani and Moscarello, 1990
Met 21	Sulphoxide	NR	Cheifetz et al., 1994
Arg 25	Citrulline	NR	Wood and Moscarello, 1989
Arg 33	Citrulline	NR	Wood and Moscarello, 1989
Arg 54	ADP-Ribose	R	Boulias and Moscarello, 1994
Ser 56	PO$_4$	R	Kishimoto et al., 1985
Thr 98	PO$_4$	R	Erickson et al., 1990
Ser 102	PO$_4$	R	Schultz et al., 1988
Gln 103	Glu	?	Chou et al., 1976
Arg 107	N^w methyl	NR	Baldwin and Carnegie, 1971
	N^{w}N^w dimethyl	NR	Brostoff and Eylar, 1971
	N^{w}N$^{|w|}$ dimethyl	NR	Small and Carnegie, 1982
			Chanderkar et al., 1986
Ser 115	PO$_4$	R	Turner et al., 1984
Arg 122	Citrulline	NR	Wood and Moscarello, 1989
Arg 130	Citrulline	NR	Wood and Moscarello, 1989
Ser 136	PO4	R	Kishimoto et al., 1985
Arg 159	Citrulline	NR	Wood and Moscarello, 1989
Arg 170	Citrulline	NR	Wood and Moscarello, 1989

ume) may suggest interaction with Arg 122. In any case the phosphorylated Thr 98 site, although basic at the level of primary structure becomes apolar with phosphorylation allowing the site to be "buried" in the folded structure. Several MAP kinases have been detected in isolated rat brain myelin (Bhat and Zhang, 1996) and oligodendrocytes. Using anti pan ERK antibodies, the presence of ERK immunoreactive material in addition to $p42^{mapK}$ and $p44^{mapK}$ have been detected with the in-gel kinase assay. Localization of the phosphorylated substrate to the MDL suggests that the enzyme should be nearby (Yon et al., 1996).

The presence of a ligand for SH3 domain is found in the triproline sequence of MBP (residues 95–101). Preliminary results in our laboratory have shown that MBP binds GST-fusion proteins with established SH3 domains such as spectrin, PLC γ, GAP and p85 but not to src (unpublished data). The presence of arginine at position 97 (human sequence) suggests that MBP SH3 ligand site is class I type (Chen and Schreiber, 1996). Since this same region of MBP has been demonstrated to be a substrate for MAP kinase (Erickson et al., 1990), the regulation of binding to an SH3 domain probably involves a phosphorylation/dephosphorylation mechanism.

A second messenger adenylate cyclase system has been suggested by the studies from Ledeen's laboratory (Larocca et al., 1987). They demonstrated muscarinic cholinergic subtype M_1 and M_2 receptors were present in isolated rat brain myelin; myelin adenylate cyclase was inhibited by carbachol but activated by forskolin; a GTP analogue, GppNHp decreased the affinity for carbachol. Shortly thereafter, MBP was found to bind GTP at a single site (Gln 3) with a stoichiometry of one mole/mole MBP (Chan et al., 1988).

The GTP binding study prompted a search for a GTP-binding domain. Halliday 1984, proposed that all G proteins possessed four regions which made up the GTP binding domain. Residues 124–130 of human MBP, Gly-Phe-Gly-Tyr-Gly-Gly-Arg corresponds to Halliday region Gly-X-X-X-X-Gly-Lys; residues Asp-Ser-Ile-Gly (39–42) of MBP correspond to Halliday region Asp-X-X-Gly; residues 80–84 of MBP Gln-Asp-Glu-Asn correspond to Asn-X-X-Asp. These sequences can come together to form the GTP binding domain if the model proposed by Stoner 1984, is used. With the newer model of Beniak et al. (personal communication), electron microscopic data have been used to modify the Stoner structure. We expect to obtain better definition of the GTP binding domain. The presence of other domains i.e. SH3 and Pleckstrin (PH), suggest that the existence of several domains in MBP (see above) are consistent with an important signalling role for this molecule.

Mono-(ADP-ribosylation) of proteins is a characteristic of proteins involved in signal transduction cascades. ADP-ribose is transferred from NAD^+ to an acceptor amino acid in a protein, e.g. with cholera toxin as the source of ADP-ribosyl-transferase, ADP-ribose was transferred to G-proteins such as the α-subunit of transducin (Van Dop et al., 1984) and ADP-ribosylation factors (Tsia et al., 1988). Arginyl residues were identified as the acceptor sites. We demonstrated that MBP (component 1, C-1) accepted two moles of ADP-ribose/mole of protein in the presence of cholera toxin (Boulias and Moscarello, 1994). However "C-8", the citrullinated MBP (discussed later), was not an acceptor for ADP-ribose. Since arginyl residues 9 and 54 were identified as the ADP-ribosylation sites the presence of citrulline at Arg 9 and conformational factors around Arg 54, as demonstrated for G-actin (Vandekerckhove et al., 1987), accounted for the failure of C-8 to accept ADP-ribose.

Although cholera toxin is widely used as a source of ADP-ribosyltransferase activity, the demonstration that isolated myelin membranes contain an endogenous ADP-ribosyltransferase, which ADP-ribosylated MBP, strengthened our case that MBP was involved in a signal transduction cascade in myelin. The enzyme has been partially purified and its properties are being studied (Boulias et al., 1995; Yoon and Moscarello, 1996).

The isolation and characterization of a phosphatidyl inositol specific phospholipase C (PI-PLC) from myelin, which was stimulated by C-1 but not by the citrullinated MBP demonstrated that this pathway was active in myelin (Tompkins and Moscarello, 1991 and 1993). The activation was shown to be dependent on specific arginyl residues in MBP. Modification of 13 arginyl residues of C-1 by 1, 2-cyclohexanedione decreased the ability of C-1 to stimulate PI-PLC by 50%. A synthetic peptide corresponding to residues 24–33, with 3 arginyl residues at positions 25, 31 and 33 stimulated PI-PLC almost as much as C-1. Replacement of arginyl residues 25 and 31 by citrullinyl residues abolished stimulation. Thus Arg residues 25 and 31 are considered essential arginines (Tompkins and Moscarello, 1993). This peptide site in MBP has homology with known PH domains (Pleckstrin Homology), suggesting that the stimulation of phospholipase C may involve interaction with this PH domain (Pawson, 1995). The role of the PH domain, whether to bind substrate such as phosphatidylinositol-4, 5-bisphosphate or to facilitate interaction between the signalling molecule and membrane is not understood at this time (Pawson, 1995).

Diacylglycerol generated by the hydrolysis of phosphatidylinositol, 4, 5, bisphosphate by PI-PLC, activates a Ca^{++}, phosphatidylserine dependent, protein kinase C which uses MBP as a major substrate (Turner et al., 1982). Phosphorylation of MBP components in vitro with a purified protein kinase C resulted in phosphorylation to different extents for the various components. Phosphorylation of C-1, 2, 3 and 4 showed that C-1 incorporated the least amount of phosphate while C-4 incorporated the greatest amount (Ramwani and Moscarello, 1990). Phosphorylation of MBP decreased the ability to aggregate lipid vesicles (Cheifeitz and Moscarello, 1985), a measure of its ability to interact with the lipid bilayer and increased bilayer permeability to entrapped spin labels (Cheifetz et al., 1985). Consistent with the suggestion made earlier, phosphorylation at particular sites can decrease protein interactions with the bilayer resulting in local permeability changes which would be reversed with dephosphorylation.

MBP AS A STRUCTURAL MOLECULE

In an early study Napolitano et al. (1967) reported that the myelin sheath of rat sciatic nerve remained essentially intact after extraction of the lipids as determined by the removal of 98% of the fatty acids. They concluded that the lamellar structure of myelin was maintained in the absence of lipids. Unfortunately no study of the myelin proteins was made. Since the extraction procedure involved several steps including one with acidified chloroform-methanol, it is probably safe to assume that most of the proteolipids were removed leaving the MBPs behind to maintain the structure.

Using a model membrane system consisting of phosphatidylglycerol and MBP, we demonstrated that increasing the concentration of MBP from 0–35% in the vesicles produced sharp reflections in X-ray diffraction spectra, similar to the reflections found in natural myelin (Brady et al., 1981). A more detailed study, using the various components of MBP instead of the mixture of MBP components demonstrated that component 1 (C-1), the most cationic was the most effective of the MBP molecules at inducing multilayer compaction, whereas component-8 ("C-8"), the citrullinated MBP was unable to induce multilayer arrangement of the lipid vesicles.

Similar model membrane studies using phosphatidylcholine vesicles and the major human proteolipid resulted in the formation of small unilamellar vesicles as the concentration of protein was increased in the vesicles (Brady et al., 1979). Therefore the proteolipids had the effect of fragmenting large multilayered vesicles, the opposite effect to that observed with MBP. We concluded at that time that MBP was responsible for inducing and maintaining the unique multilayered arrangement of myelin.

More recent work with genetically engineered and natural mutant mice supports our conclusion. Two gene deletion mice ("knock-outs") have been reported recently. The one was a "knock-out" of the gene for myelin associated glycoprotein (MAG) (Li et al., 1994). The other was a "knock-out" of the gene for proteolipid protein (PLP) (Boison and Stoffel, 1994). No dramatic pathologies were observed in either case. In the MAG "knock-out", the myelin sheath was similar to normal with a normal repeat distance as determined by X-ray diffraction. In the PLP "knock-out" significant changes were observed in the myelin structure by electron microscopy. However in both "knock-outs", the major dense lines were intact and the animals lived and reproduced normally. Behavioural defects were detected in both "knock-outs" by specialized tests, but the defects were not life threatening. We posited, that the MBPs in both "knock-outs" were unaltered. In fact we showed subsequently that all the isoforms of mouse MBP and the complete complement of charge isomers (components) were present (unpublished data).

On the other hand, the naturally occurring shiverer mutant which contains a deletion in most of the MBP gene, and is therefore essentially an MBP "knock-out", fails to myelinate significantly and survives only 2–3 months. Correction of the phenotype by supplying the MBP gene was partially successful. With 20% of the normal amount of MBP, the animals were able to produce myelin and survive for 6 months (Readhead et al., 1987). Although all the isoforms of MBP were demonstrated in the "rescued" mice, a study of the microheterogeneity revealed that none of the cationic isomers were present, or were present in amounts too low to be detected. We concluded that for complete rescue, normal myelin can only be formed if all the isoforms and isomers of MBP are present.

LOCALIZATION OF MBP

Localization of MBP to the major dense line (MDL) of myelin has been supported by several studies, e.g. MBP was shown to be inaccessible to anti MBP antibodies unless the membrane was disrupted by mechanical or chemical means (Korngurth and Anderson 1965; Herndon et al., 1973; Guarnieri et al., 1974; Sternberg et al., 1978; Omlin et al., 1982). Labelling of intact cat spinal cord suggested that MBP was inaccessible to chemical labelling reagents (Poduslo and Braun 1975; Golds and Braun, 1978) from which the authors concluded that MBP was at the MDL, corresponding to the cytoplasmic surfaces of the oligodendrocyte.

Other studies suggested that MBP was located at the intraperiod line (IPL) of myelin. An X-ray diffraction study of freshly prepared and partially autolysed myelin placed MBP at the IPL since this space collapsed after autolysis (Dickinson et al., 1970). In an immunohistochemical study MBP was localized at the IPL in the peripheral nervous system (Mendell and Whitaker, 1978), but was not localized in the central nervous system.

A resolution to these conflicting positions has become possible with the study of localization of the components or charge isomers. The citrullinated MBP offered the possibility of raising a specific antibody to the ureido group of citrulline (McLaurin and Moscarello, 1990). With this antibody, the citrullinated MBP, "C-8", was found at the IPL predominantly, whereas with an antibody which recognized all the components, most of the labelling was found at the MDL. Therefore the least cationic component was preferentially localized to the IPL which is a 25Å space while the more tightly compacted MDL is 17Å in width. Since these studies were carried out in cryosections of brain biopsies (taken for intractable epilepsy), the deiminated form of arginine could not have arisen from post mortem autolysis.

PATHOLOGICAL IMPLICATIONS OF THE CITRULLINATED MBP

Increased levels of the citrullinated MBP ("C-8") have been reported in MBP isolated from victims of multiple sclerosis (Moscarello et al., 1994). The increase of about 3 fold, was not found in MBP isolated from several other chronic neurological diseases such as Alzheimer's, Parkinson's, Huntington's, amyotropic lateral sclerosis and motor neuron disease. High levels of "C-8" were found in infants less than two years old. In fact, 90% of the MBP was of the C-8 type. The high amount of "C-8" correlated with a poorly compacted myelin sheath as measured by the low phase transition temperature of the membrane. These studies prompted us to suggest that the myelin membrane in MS was developmentally immature, and as a result more susceptible to degradation (Moscarello et al., 1994).

Since the above-mentioned studies were done on autopsy material it was impossible to rule out a peculiar post mortem change in MS which had not occurred in other neurological diseases. When a transgenic animal model of demyelination became available (Simons-Johnson et al., 1995), we were able to determine that changes in MBP preceded demyelination (Mastronardi et al., 1993), and the changes in MBP were similar to the changes observed in MS (Mastronardi et al., 1996). Although clinical signs of demyelination appeared at 3 months of age, changes in MBP were found at 2 months accompanied by a small increase in the number of astrocytes. This transgenic model containing 70 copies of the cDNA for DM20 demyelinated spontaneously i.e. without injection of antigen, addresses the early changes in myelin stability. No changes in MBP were observed in the EAE model, whether acute or chronic (Mastronardi et al., 1996). We concluded that long standing brain pathology could not account for the changes observed in the MBP isolated from transgenic mice and by implication the increase in C-8 in MS cases was unlikely to be due to long standing brain pathology.

In our report on the decreased cationic nature of MBP in MS, we showed that the proportion of "C-8" was increased by about 3 fold over normal. However none of the "C-8" which we studied contained more than 6–7 moles of citrulline/mole of MBP. In a recent report on a single case of fulminating MS (Marburg's disease) of 6 weeks duration in a 26 year old woman, 18 of the 19 arginyl residues were deiminated and recovered as citrulline (Wood et al., 1996). The "C-8" MBP was increased approximately 7 fold over normal. Therefore the severity of MS correlated with the extent of increase of the relative proportion of "C-8" and with the number of arginyl residues deiminated. The greater the proportion of "C-8" and the greater the number of arginyl residues deiminated, the greater the severity of the disease.

MECHANISM OF DEMYELINATION

The mechanism by which changes in MBP initiate a series of reactions culminating in demyelinating disease is not known. The loss of positive charge resulting from the deimination of arginyl residues has been shown to decrease the ability of MBP to interact with the lipid bilayer (Wood and Moscarello, 1989). Whereas the cationic charge isomers (C-1) aggregate vesicles consisting of phosphatidylcholine/phosphatidylserine, "C-8" was unable to aggregate the vesicles, demonstrating that the loss of positive charge decreased the ability of the protein to interact with the lipid. The X-ray diffraction studies in which "C-8" was unable to compact the multilayered vesicles (Brady et al., 1981) have been mentioned already.

In addition to effects of decreased positive charge on protein-lipid interactions, citrulline itself can have destabilizing effects on the membrane. If citrulline is considered as a mono-substituted urea it can be expected to maintain some of the properties of urea. Urea is known to have a rigid planar structure with no rotations. It possesses strong electron rich clouds and is a strong chelator (Ochoa et al., 1980). Although there is no evidence of an interaction between urea and hydrocarbon chains, a mixed clathrate-like structure forms around hydrocarbon molecules resulting in increased solubility of the hydrocarbon due to smaller free energy of dissociation caused by the replacement of water by urea in the solvation region (Mizutani et al., 1989). Therefore the consequence of deimination of arginyl residues in MBP will affect electrostatic interactions with the lipid and the presence of the ureido group may disrupt the solvation shell around the hydrocarbons of membrane lipids. The resultant of these two changes will be to disrupt lipid protein interactions. In addition the strong chelating properties of urea which could bind essential metal ions such as Zn^{++} will further destabilize the structure of myelin.

Zinc represents a good metal candidate for chelation to the ureido group of citrulline since it was shown to be in higher concentration than other metals in brain (Bourre et al., 1987). In an X-ray diffraction study Zn^{++} was more effective than other divalent cations at inhibiting the swelling of extracellular spaces of myelin (IPL) (Inoue and Kirschner, 1984). A specific binding of Zn^{++} to MBP in solution has been reported by fluorescence studies (Cavatorta et al., 1994). Other divalent cations, Ca^{++}, Co^{++}, Mn^{++} did not affect the fluorescence spectra. Although evidence is not available, they postulate that Zn^{++} binding may occur at residues 23–26 of MBP (His Ala Arg His). All studies were done with unfractionated MBP. Similar studies with the various charge isomers may be rewarding especially since Arg 25 is deiminated in C-8. In model membrane studies, complexes between galactosylceramide (Gal C) and sulphatide with metals, Zn^{++} had the greatest effect (Stewart and Boggs, 1993). The Zn^{++} was shown to complex with the hydroxyl groups of Gal C. This interaction has been implicated in membrane adhesion (Kojima and Hakamori, 1989; Eggens et al., 1989; Kojima and Hakemori, 1991). Chelation of Zn^{++} by the ureido group of citrulline would decrease membrane adhesion, further destabilizing the membrane. The model of MBP presented at this symposium in which all ureido groups of citrulline are external and available to the aqueous media in the IPL supports the abovementioned theory of demyelination (Beniak et al., 1996).

CONCLUDING REMARKS

Molecular biological technologies have made the identification of new genes routine. Every week a new gene is reported, some of which are associated with disease processes. While the finding of new genes remains important, especially when a candidate protein can be identified, it does little to solve the mechanism by which various pathologies arise. Recent advances in our understanding of RNA-editing, a process by which the mRNA can be modified by nucleotide replacement or modification of existing bases changes specific codons (Simpson and Emeson, 1996). The protein translated will not be the one encoded by the DNA sequence. Therefore it is important to understand the structure of the translated protein in order that its role or function, be it structural or catalytic be understood. Since it is the protein and not the nucleic acid which represents the constituent which is used in membrane assembly, the properties of the protein become of utmost importance.

With this view in mind we have presented our studies on MBP which we have referred to as the "executive" molecule of the myelin sheath. Functions other than structural for MBP have recently been postulated (Staugaitis et al., 1996). They proposed that different isoforms, i.e. those containing exon 2 of MBP may be localized to the nucleus during myelinogenesis, where they may function as transcription factors. The specific binding of Zn^{++} mentioned in the previous section, speculated to involve residues 23–26 of MBP, represents a HisXXHis motif of zinc fingers (Coleman, 1992). Only by studying the various members of the MBP family i.e. the isoforms and charge isoforms, has it become clear that some members may be structural while others may be involved in complex metabolic reactions in myelin especially those dealing with the transduction of signals. Our studies are still at an early stage because only a few members of the extended MBP family have been studied. Completion of these studies is essential in order that the structure of the normal myelin sheath be understood. Parallel with these studies the elucidation of changes in MBP in diseases such as multiple sclerosis will lead to an understanding of the mechanism of disease from which specific therapeutic models should arise spontaneously.

REFERENCES

Baldwin GS, Carnegie PR (1971): Specific enzyme methylation of an arginine in the experimental allergic encephalomyelitis protein from human myelin. Science 171:579–581.

Bhat NR, Zhang P (1996): Activation of mitogen activated protein kinases in oligodendrocytes. J Neurochem 66:1986–1994.

Boison D, Stoffel W (1994): Disruption of the compact myelin sheath of axons of the central nervous system in proteolipid protein deficient mice. Proc Natl Acad Sci (USA) 91:11709–11713.

Boulias C, Moscarello MA (1994): ADP-ribosylation of human myelin basic protein. J Neurochem 60:351–359.

Boulias C, Mastronardi FG, Moscarello MA (1995): ADP-ribosyltransferase activity in myelin membranes isolated from human brain. Neurochem Res 20:1269–1277.

Bourre JM, Cloez I, Galliot M, Buisine A, Dumont O, Piciotti M, Prouillet F, Bourdon R (1987): Occurrence of manganese, copper and zinc in myelin. Alterations in the peripheral nervous system of dysmyelinating trembler mutant are at variance with brain mutants (quaking and shiverer). Neurochem Int 10:281–286.

Brady GW, Birnbaum PS, Moscarello MA, Papahadjopoulos D (1979): Liquid diffraction analysis of the model membrane system, egg lecithin and myelin protein (N-2). Biophys J 25:23–42.

Brady GW, Murthy NS, Fein DB, Wood DD, Moscarello MA (1981): The effect of basic myelin protein on multilayer membrane formation. Biophys J 34:345–360.

Carnegie PR (1971): Amino acid sequence of the encephalitogenic protein of human myelin. Biochem J 123:57–67.

Cavatorta P, Giovanelli S, Bobba A, Riccio P, Szabo AG, Quagliariello E (1994): Myelin basic protein interaction with zinc and phosphate. Fluorescence studies on the water-soluble form of the protein. Biophys J 66:1174–1179.

Chan CK, Ramwani J, Moscarello MA (1988): Myelin basic protein binds GTP at a single site in the N-terminus. Biochem Biophys Res Comm 152:1468–1473.

Chanderkar LP, Park, WK, Kim, S (1986): Studies on myelin-basic-protein methylation during mouse brain development. Biochem J 240: 471–479.

Cheifetz S, Moscarello MA (1985): Effect of basic protein charge microheterogeneity on protein-induced aggregation of lipid vesicles containing a mixture of acidic and neutral phospholipids. Biochemistry 24:1909–1914.

Cheifetz S, Boggs JM, Moscarello MA (1985): Increase in vesicle permeability mediated by myelin basic protein: effect of phosphorylation of basic protein. Biochemistry 24:5170–5175.

Cheifetz S, Moscarello MA, Deber CM (1984): NMR investigation of the charge isomers of bovine myelin basic protein. Arch Biochem Biophys 233:151–160.

Chen JK, Schreiber SL (1996): Combinatorial synthesis and multidimensional NMR-spectroscopy: an approach to understanding protein-ligand interactions. Agnew Chemie 34:953–969.

Chou FC-H, Chou JC-H, Shapira R, Kibler RF (1976): Basis of microheterogeneity of myelin basic protein.

Coleman JE (1992): Zinc proteins: enzymes, storage, proteins, transcription factors and replication proteins. Ann Rev Biochem 61:897–946.

Deibler GE, Martenson RE, Kramer AJ, Kies MW, Miyamoto E (1975): The contribution of phosphorylation and loss of carboxyterminal arginine to the microheterogeneity of myelin basic protein. J Biol Chem 250:7931–7938.

Dickinson JP, Jones KM, Aparicio SR, Lumsden CE (1970): Localization of encephalitogenic basic protein in the intraperiod line of lamellar myelin. Nature (Lond) 227:1133–1134.

Eggens I, Fenderson B, Toyokuni T, Dean B, Stroud M, Hakomori S (1989): Specific interaction between Lex and Lex determinants. A possible basis for cell recognition in preimplantation embryos and in embryonal carcinoma cells. J Biol Chem 264:9476–84.

Erickson AK, Payne DM, Martino PA, Rossomondo AJ, Shabanowitz J, Weber MJ, Hunt DF, Sturgill TW (1990): Identification by mass spectrometry of threonine 97 in bovine myelin basic protein as a specific phosphorylation site for mitogen-activated protein kinase. J Biol Chem 265:19728–19735.

Eylar EH, Brostoff S, Hashim G, Caccam J, Burnet P (1971): Basic A1 protein of the myelin membrane, the complete amino acid sequence. J Biol Chem 246:5770–5784.

Golds E, Braun PE (1978): Cross-linking studies on the conformation and dimerization of myelin basic protein. J Biol Chem 253:8171–8177.

Guarnieri M, Himmelstein J, McKhann G (1974): Isolated myelin quantitatively adsorbs antibody to basic protein. Brain Res 72:172–176.

Halliday K (1984): Regional homology in GTP binding proto oncogene products and elongation factors. J Cyc Nuc Res 9:435–448.

Herndon RW, Rauch HC, Einstein ER (1973): Immunoelectron microscopic localization of the encephalitogenic basic protein in myelin. Immunol Commun 2:163–172.

Inouye H, Kirschner DA (1984): Effects of ZnCl$_2$ on membrane interactions in myelin of normal and shiverer mice. Biochim Biophys Acta 776:197–208.

Kalmholz J, de Ferra F, Prickett C, Lazzarini RA (1986): Identification of three forms of human myelin basic protein by cDNA cloning. Proc Natl Acad Sci (USA) 83:4962–4966.

Kalmholz J, Spielman R, Goglin K (1987): The human myelin basic protein gene: chromosomal localization and RFLP analysis. Am J Hum Genetics 40:365–373.

Kalmholz J, Toffeneti, Lazzarini RA (1988): Organization and expression of the human myelin basic protein gene. J Neurosci Res 21:62–70.

Kojima N, Hakamori S (1991): Cell adhesion, spreading, and motility of GM3-expressing cells based on glycolipid-glycolipid interaction. J Biol Chem 266:17552–17558.

Kojima N, Hakemori S (1989): Specific interaction between gangliotriaosylceramide and sialosyllactosylceramide (GM3) as a basis for specific cellular recognition between lymphoma and melanoma cells. J Biol Chem 264: 20159–20162.

Korngurth S, Anderson L (1965): Localization of a basic protein of various species with the aid of fluorescence and electron microscopy. J Cell Biol 26:157–166.

Larocca JN, Ledeen RW, Dworkin B, Makaran MH (1987) Muscarinic receptor binding and muscarinic receptor-mediated inhibition of adenylate cyclase in rat brain myelin. J Neurosci 7:3869–3876.

Li C, Tropak MB, Gerlai R, Clapoff S, Abramow-Newerly W, Trapp B, Paterson A, Roder J (1994): Myelination in the absence of myelin associated glycoprotein. Nature 369:747–750.

Mastronardi FG, Ackerley CA, Arsenault L, Roots BJ, Moscarello MA (1993): Demyelination in a transgenic mouse: a model for multiple sclerosis. J Neurosci Res 36:315–324.

Mastronardi FG, Al-Sabbagh A, Nelson PA, Rego J, Roots BI, Moscarello MA (1996): Myelin basic protein is not affected at the post translational level in chronic EAE mouse brain. J Neurosci Res 44:344–349.

Mastronardi FG, Mak B, Ackerley CA, Roots BI, Moscarello MA (1996): Myelin basic protein changes in DM20 transgenic mice are similar to changes in MBP in multiple sclerosis. J Clin Invest 97:349–358.

McLaurin J, Moscarello MA (1990): The preparation of antibodies reactive against the citrulline-containing charge isomers of myelin basic protein but not against the arginine-containing charge isomer. Anal Biochem 191:272–277.

McLaurin J, Ackerley CA, Moscarello MA (1993): Localization of basic protein in human myelin. J Neurosci Res 35:618–628.

Mendell JR, Whitaker JN (1978): Immunocytochemical localization studies of myelin basic protein. J Cell Biol 76:502–511.

Mendz JL, Barden JA, Martenson RE (1995): Conformation of a tetrapeptide epitope of myelin basic protein. Eur J Biochem 231:659–666.

Mizutani Y, Kamagawa K, Nakanishi K (1989): Effect of urea on hydrophobic interaction: Raman difference spectroscopy on C-H stretching vibration of acetone and C-N stretching vibration of urea. J Phys Chem 93:5650–5654.

Moscarello MA, Pang H, Pace-Asciak CR, Wood DD (1992): The N-terminus of human myelin basic protein consists of C_2, C_4, C_6 and C_8 alkycarboxylic acids. J Biol Chem 267:9779–9782.

Moscarello MA, Wood DD, Ackerley C, Boulias C (1994): Myelin in multiple sclerosis is developmentally immature. J Clin Invest 94:146–154.

Napolitano L, LeBaron F, Scaletti J (1967): Preservation of myelin lamellar structure in the absence of lipid. Biophys J 34:817–826.

Ochoa JL, Porath J, Kempf J, Egly JM (1980): Electron donor-acceptor properties of urea and its role in charge-transfer chromatography.

Omlin FX, Webster H de F, Palkovits CE, Cohen SR (1982): Immunocytochemical localization of basic protein in major dense line regions of central and peripheral myelin. J Cell Biol 95:242–248.

Pawson T (1995): Protein modules and signalling networks. Nature 373:573–580.

Poduslo JF, Braun PE (1975): Topographical arrangement of membrane proteins in the intact myelin sheath: Lactoperoxidase incorporation of iodine into myelin surface protein. J Biol Chem 250:1099–1105.

Pribyl TM, Campagnoni CW, Kampf K, Kashima T, Handley VW, McMahon J, Campagnoni AT (1993): The human myelin basic protein gene is included within a 179 kilobase transcription unit: expression in the immune and central nervous system. Proc Natl Acad Sci (USA) 90:10695–10699.

Ramwani J, Moscarello MA (1990): Phosphorylation of charge isomers (components) of human myelin basic protein: identification of phosphorylated sites. J Neurochem 55:1703–1710.

Readhead C, Popko B, Takahashi H, Shine HD, Saavedra RA, Sidman RL, Hood L (1987): Expression of myelin basic protein gene in transgenic shiverer mice: correction of the dysmyelinating phenotype. Cell 48:703–712.

Roth HS, Kronquist KE, Kelero de Rosbo N, Crandall BF, Campagnoni AT (1987): Evidence for the expression of four myelin basic protein variants in the developing human spinal cord through cDNA cloning. J Neurosci Res 17:321–328.

Schwob VS, Clark HB, Agrawal D, Agrawal HC (1985): Electron microscopic immunocytochemical localization of myelin proteolipid protein and myelin basic protein to oligodendrocytes in rat brain during myelination. J Neurochem 45:559–571.

Simons-Johnson R, Roder JC, Riordan JR (1995): Overexpression of the DM20 myelin proteolipids causes central nervous system demyelination in transgenic mice. J Neurochem 64:967–976.

Simpson L, Emeson RB (1996): RNA editing. Annu Rev Neurosci 19:27–52.

Small DH, Carnegie PR (1982): In vivo methylation of an arginine in chicken myelin basic protein. J Neurochem 38:184–190.

Staugaitis SM, Colman DR, Pedraza L (1996): Membrane adhesion and other functions of the myelin basic proteins. Bioessays 18:13–18.

Sternberger NH, Iyotoma Y, Kies MW, Webster H de F (1978): Immunocytochemical method to identify basic protein in myelin forming oligodendrocytes of newborn rat CNS. J Neurocytol 7:251–263.

Stewart R, Boggs JM (1993): A carbohydrate-carbohydrate interaction between galactosylceramide-containing liposomes and cerebroside sulfate-containing liposomes: dependence on the glycolipid ceramide composition. Biochem 32:10666–10674.

Stoner GL (1984): Pedicted folding of β-structure in myelin basic protein. J Neurochem 43:433–447.

Streicher R, Stoffel W (1989): The organization of the human myelin basic protein gene: comparison with the mouse. Biol Chem Hoppe Seyler 370:503–510.

Tompkins TA, Moscarello MA (1991): A 57 kDa phosphatidylinositol-specific phospholipase C from bovine brain. J Biol Chem 266:4228–4236.

Tompkins TA, Moscarello MA (1993): Stimulation of bovine brain phospholipase C activity by myelin basic protein requires arginyl residues in peptide linkage. Arch Biochem Biophys 302:476–483.

Tsai SC, Noda M, Adamik R, Chang PP, Chen HC, Moss J, Vaughan M (1988): Stimulation of choleragen enzymatic activities by GTP and two soluble proteins purified from bovine brain. J Biol Chem 263:1768–1772.

Turner RS, Chou C-H J, Kibler RF, Kuo JF (1982): Basic protein in brain myelin is phosphorylated by endogenous phospholipid sensitive calcium-dependent protein kinase. J Neurochem 39:1397–1404.

Ullmer JB (1988): The phosphorylation of myelin proteins. Prog Neurobiol 31:241–259.

Ursell MRM, McLaurin J, Wood DD, Ackerley CA, Moscarello MA (1995): Localization and partial characterization of a 60 kDa citrulline-containing transport form of myelin basic protein from MO3–13 cells and human white matter. J Neurosci Res 42:41–53.

Van Dop C, Tsubokawa M, Bourne HR, Ramachandran J (1984): Amino acid sequence of retinal transducin at the site ADP-ribosylated by cholera toxin. J Biol Chem 259:696–696.

Vandekerckhove J, Schering B, Barmann M, Aktories K (1987): Clostridium perfringens iota toxin ADP-ribosylates skeletal muscle actin in Arg 177. Febs Letts 225:48–52.

Wood DD, Moscarello MA (1989): The isolation, characterization and lipid aggregating properties of a citrulline-containing myelin basic protein. J Biol Chem 264:5121–5127.

Yon SM, Ackerley CA, Mastronardi FG, Groome N, Moscarello MA (1996): Identification of a mitogen-activated protein kinase site in human myelin basic protein in situ. J Neuroimmunol 65:55–59.

Yoon I-S, Moscarello MA (1996): ADP-ribosyltransferase activity of myelin. M.Sc. Thesis, University of Toronto, Toronto, Canada.

3

MYELIN-ASSOCIATED GLYCOPROTEIN

Ligand and/or Receptor

Xinghua Yin and Bruce D. Trapp

Department of Neurosciences
Research Institute
The Cleveland Clinic Foundation
9500 Euclid Avenue
Cleveland, Ohio 44195

MYELIN-ASSOCIATED GLYCOPROTEIN

The myelin-associated glycoprotein (MAG) is quantitatively a minor constituent of the total protein found in myelin isolated from the central ($\approx 1.0\%$) and peripheral ($\approx 0.1\%$) nervous systems (Quarles et al. 1973a; Figlewicz et al. 1981). MAG has an apparent molecular weight of 100kD, of which 30% is carbohydrate. Although its precise function is unknown, it is generally accepted that as a member of the immunoglobulin gene superfamily, MAG functions in membrane-membrane interactions. Based on its biochemical properties and enrichment in periaxonal membranes, MAG was initially considered important to the initiation and progression of myelination. The production of MAG-deficient mice, however, established that myelination can occur in the absence of MAG (Montag et al. 1994; Li et al. 1994). Recently several laboratories (Mukhopadhyay et al. 1994; McKerracher et al. 1994; Schafer et al. 1996) have provided evidence that MAG can inhibit axonal regeneration. These data raise the possibility that MAG functions as a ligand that regulates axonal properties (Filbin, 1995).

Biochemistry of MAG

The amino acid sequence of CNS MAG deduced from cDNA clones (Lai et al. 1987; Arquint et al. 1987; Salzer et al. 1987) predicts a single transmembrane domain, a large extracellular domain that contains five immunoglobulin-like regions and eight potential N-linked glycosylation sites, and one of two possible cytoplasmic domains that contain putative phosphorylation sites (Fig 1). Two developmentally regulated MAG polypeptides (72-kD and 67-kD) are generated by alternate splicing of a single gene. These peptides have identical extracellular and transmembrane domains. The cytoplasmic

Cell Biology and Pathology of Myelin, edited by Juurlink *et al.*
Plenum Press, New York, 1997

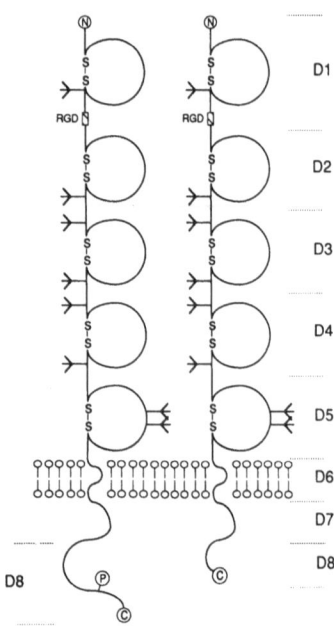

Figure 1. Models of the 72-kD and 67-kD polypeptides. Five immuno-globulin-like domains are indicated as disulfide-bonded (S-S) loops and the oligosaccharide side chains are indicated by branched structures. Exon 6 represents the transmembrane domain. The two forms are generated by alternative splicing and differ in C-terminal cytoplasmic domain D8 or D8*. A potential site for tyrosine kinase phosphorylation on the 72-kD form is indicated by the P. Reproduced with permission from Lai et al., 1987.

domain of the 67-kD polypeptide has 10 different amino acids and lacks 54 amino acids that are present in the 72-kD polypeptide. The 72-kD polypeptide is quantitatively the predominant form during early and active stages of CNS myelination, whereas the two iso-forms are present at equal amount in the mature CNS (Quarles et al. 1973b; Frail et al. 1985). In the peripheral nervous system, the 67-kD polypeptide represents at least 95% of total MAG at all stages of development (Frail et al. 1985; Tropak et al. 1988).

Location of MAG in Mature CNS and PNS Internodes

To date, all evidence indicates that MAG is expressed only by myelin forming cells. In the CNS (Fig. 2A), MAG is enriched exclusively in the periaxonal membrane of myelin internodes (Sternberger et al. 1979). In adult PNS myelin internodes (Fig. 2B and C), MAG is enriched in periaxonal, paranodal, Schmidt-Lanterman incisure, and inner and outer mesaxon membranes (Trapp et al. 1989a; Trapp and Quarles, 1982; Martini and Schachner, 1986). A unifying ultrastructural feature of all MAG-containing membranes (see Figure 3A) is that they appose other membranes by a 12–14-nm gap (Trapp, 1988; Trapp and Quarles, 1982). The bulk and polarity of the extracellular domain of MAG are sufficient to spatially maintain this distance. Amino acid homologies between extracellular domains of MAG and other immunoglobulin-like molecules such as N-CAM (Cunningham et al. 1987) support its potential role in membrane-membrane interaction. It is therefore likely that MAG functions as a membrane spacer, receptor and/or ligand. Confirmation of the potential receptor/ligand functions of MAG awaits identification of axolemmal molecules that interact with MAG. It is also possible that MAG interacts homotypically in mesaxon membranes, Schmidt-Lanterman incisures, and paranodal loops of PNS myelin internodes (Trapp et al. 1984b; Trapp, 1988). Since MAG is not present in axonal membranes, MAG-axonal interactions would have to occur by heterotypic interactions that may involve other members of the immunoglobulin gene superfamily that are present in the axolemma.

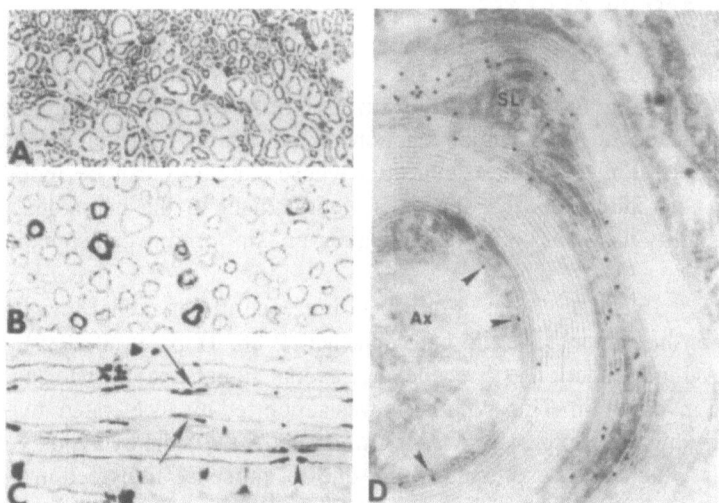

Figure 2. Distribution of MAG in 1-μm-thick Epon sections from adult rat spinal cord (A) and sciatic nerve (B and C), and in an ultrathin cryosection from rat peripheral nerve (D). MAG is enriched in periaxonal regions of myelin fibers in both CNS (A) and PNS (B). Some fibers in the PNS have darker and thicker bands of staining (B). In longitudinal orientation these are resolved as paranodal regions (C, arrowheads) or Schmidt-Lanterman incisures (C, arrows). In ultrathin cryosections of peripheral nerve (D), MAG immunogold labeling is enriched in Schmidt-Lanterman incisures (SL) and periaxonal membranes (D, arrowheads), but not in compact myelin. Scale bars A-C = 20 μm; (D) = 0.25 μm. Reproduced with permission from Trapp et al., 1989a.

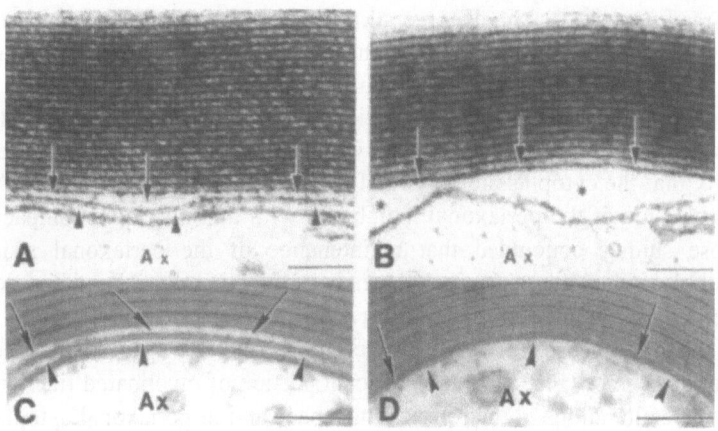

Figure 3. Electron micrographs of periaxonal regions of myelinated fibers from control (A,C), Quaking (B) and MAG-deficient (D) mice. In control fibers a 12–14 nm periaxonal space (A,C, arrowheads) separates a MAG-containing Schwann cell periaxonal membrane from the extracellular leaflet of the axolemma, and a Schwann cell cytoplasmic collar (A,C, arrows) separates the cytoplasmic leaflet of the periaxonal membrane and inner compact myelin lamellae. In regions of MAG-deficient Quaking fibers (B) and in MAG knock out mice the periaxonal space is dilated (B,asterisks) or compressed (D,arrowhead) and the cytoplasmic collar absent (C,D, arrow). Ax=axon. Scale bar = 0.1 μm. Reproduced with permission from Trapp et al., 1984b; Li et al., 1994.

Potential Functions of MAG

The function of a protein is mediated, in part, by its location and its biochemical composition. Correlating the subcellular localization of MAG with its biochemical properties can provide clues to its potential function. MAG has the potential to be multifunctional because (1) it is a complex molecule whose functions can be modulated both extracellularly and intracellularly, and (2) its enrichment in the periaxonal membrane raises the possibility that MAG is involved in axon-myelin forming cell signaling.

Initiation of Myelination. Based on MAG presence in periaxonal membrane during initial axonal ensheathment (Martini and Schacher, 1986; Trapp et al. 1989), it was proposed that MAG may function as a receptor for axonal signals that initiate and sustain the progression of myelination. This hypothesis was supported by accelerated myelination by Schwann cells which overexpressed MAG (Owens et al. 1990) and by hypomyelination by Schwann cells which underexpressed MAG due to antisense transfection (Owens and Bunge, 1991). While these hypotheses were suggestive and reasonable, normal myelination in MAG deficient mice established that MAG is not required for myelination to begin or proceed *in vivo* (Montag et al., 1994; Li et al. 1994).

Membrane Spacer. Based on the correlation between MAG's predicted biochemical properties and the ultrastructure of MAG containing membranes, it was proposed (Trapp and Quarles, 1982) that MAG may function as a membrane spacer; a simple concept that MAG was a structural protein that maintained a 12–14 nm space between apposing membranes. While this correlation was established in normal myelin internodes, a role for MAG as a spacer in the periaxonal membrane has been supported by a number of studies which demonstrated that the presence of MAG and a normal (12–14 nm) periaxonal space is rarely dissociated. This includes pathological conditions which result in axonal swelling, axonal shrinkage, swelling of the Schwann cell periaxonal cytoplasmic collar, and invagination of the Schwann cell periaxonal membrane and axolemma into swollen axoplasm (Trapp, Quarles, 1984; Trapp et al. 1984a). Another possible function of periaxonal MAG was raised by investigation of the hypomyelinated fibers in mutant Quaking mice. Many fibers in the ventral root of adult Quaking mice undergo a slowly progressive demyelination and remyelination and alterations in myelin which included dilation of the periaxonal space and loss of the periaxonal cytoplasmic collar (Fig. 3B). These data raised the possibility that the cytoplasmic part of MAG played a role in preventing the fusion of the cytoplasmic side of the periaxonal membrane with the innermost compact myelin lamellae. These studies concluded that maintenance of the periaxonal space and the Schwann cell periaxonal cytoplasmic collar may be functionally related and dependent on the presence of MAG in the Schwann cell periaxonal membrane.

Of all the early theories of MAG's function, its role as a spacer is best supported by the MAG-deficient mice (Fig. 3D). A significant portion of myelinated fibers in MAG-deficient mice have alterations in their periaxonal space and/or periaxonal cytoplasmic collar (Li et al. 1994). It should be stressed, however, that these changes were not found in all fibers. Thus, MAG may be part of the spacer mechanism in the periaxonal membrane.

Membrane Motility. Another function proposed for MAG was membrane movement. This hypothesis was reasonable since all MAG-containing membranes have the ability to move. Most obvious is the periaxonal membrane during axonal ensheathment and mesaxon membrane during spiral growth of the myelin membranes (Bunge et al. 1989). If

MAG was a membrane spacer and possibly an adhesion molecule connecting its C-terminal to a microfilament based motility system could easily link MAG to this essential function of Schwann cells. In support of this, the microfilament components, F-actin, spectrin and ankyrin were shown to colocalize with MAG in the PNS myelin internodes (Trapp et al. 1989b). Normal myelination in MAG-deficient mice makes it unlikely that MAG has a direct role in the movement of Schwann cell membranes. It is possible, however, that MAG may interact with the Schwann cell cytoskeleton.

Endocytic Pathway. The developmental change in the C-terminal of MAG suggests that it has more than one function within the periaxonal membrane of CNS myelin internodes. Although light microscopic immunocytochemical studies reveal no significant difference in L- and S-MAG distribution in CNS myelin internodes (Bo et al. 1995), recent studies have linked L-MAG but not S-MAG to an endocytic pathway in actively myelinating oligodendrocytes. The first association of MAG with endosomes was demonstrated by studies which determined the location of MAG in ultrathin cryosections at times when the 72-kD polypeptide (7 days) and 67-kD polypeptide (adult) were quantitatively the abundant form (Trapp et al. 1989a). Whereas MAG is confined to the periaxonal membrane of CNS myelin internodes in both 7-day-old and adult animals, its distribution within oligodendrocyte cytoplasm differed. Specifically, during early stages of myelination, MAG is enriched in endosomes (Fig. 4) that were located in oligodendrocyte perinuclear cytoplasm, processes extending to myelin internodes, the outer tongue process, paranodal loops, and inner tongue process of the myelin internode. These data indicate that during early stages of CNS myelination, MAG is associated with an endocytic pathway that originates in the periaxonal membrane of CNS myelin internodes (Fig. 4A) and terminates in oligodendrocyte perinuclear cytoplasm (Fig. 4C). Recent studies of Quaking mice (Bo et al. 1995) have confirmed and extended these earlier observations and show that L-MAG, but not S-MAG, is selectively removed from periaxonal membrane of CNS-myelinated fi-

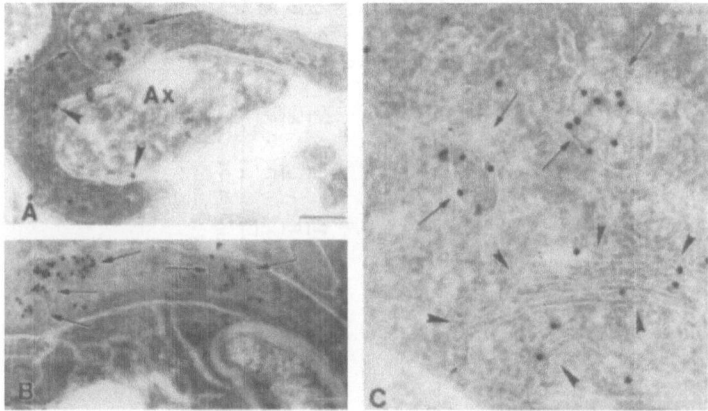

Figure 4. Ultrathin cryosections of 7-day-old rat spinal cord stained with MAG antibodies and colloidal gold. MAG is detected in periaxonal membranes (A, arrowheads) in endosomes (A, arrows) during early stages of axonal ensheathment. During active stages of myelination, MAG is enriched in endosomes (B, arrows), and distributed diffusely in oligodendrocyte processes (B,arrowheads) which extend to myelin internodes. In oligodendrocyte perinuclear cytoplasm, MAG is also enriched in endosomes (C, arrows) and Golgi membranes (C, arrowhead). Scale bar = 0.2 μm. Reproduced with permission from Trapp et al., 1989a.

Table 1. Identification of tyrosine internalization signals in the cytoplasmic doman of L-MAG

	Amino acid sequence (Prefer random coil)												Tyrosine position	Length of tail	Amino acid sequence (Polar or positively charged)											
	-8	-7	-6	-5	-4	-3	-2	-1	0	+1	+2	+3			-8	-7	-6	-5	-4	-3	-2	-1	0	+1	+2	+3
L-MAG	R	I	S	G	A	P	D	K	Y	E	S	E	35	90	R	I	S	G	A	P	D	K	Y	E	S	E
S-MAG	R	I	S	G	A	P	D	K	Y	E	S	R	35	45	R	I	S	G	A	P	D	K	Y	E	S	R
L-MAG	K	R	R	P	T	K	D	S	Y	T	L	T	65	90	G	K	R	P	T	K	D	S	Y	T	L	T

Sequences are oriented from NH_2 to COOH terminus. L-MAG contains two motifs at tyrosine position 35 and 65. S-MAG contains one motif at tyrosine 35. The endocytosis motif consists of a tyrosine residue which is flanked by amino acids that prefer a random coil conformation and/or that are polar or positively charged. Amino acids which meet these criteria are underlined. Reprinted from (Bo et al. 1995).

bers by receptor-mediated endocytosis. The loss of L-MAG from Quaking periaxonal membranes results from increased endocytosis of L-MAG and possibly a decrease in L-MAG production (Bo et al. 1995).

Receptor-mediated endocytosis is a ligand and energy-dependent process which is regulated, in part, by specific amino acid sequence signals. Endocytic signals usually reside in the cytoplasmic domain of the receptor. A tyrosine internalization signal, which is necessary and sufficient for endocytosis through clathrin-coated pits, has been identified on a number of transmembrane glycoproteins, including the polyimmunoglobulin receptor and the low density lipoprotein (LDL) receptor (Ktistakis et al. 1990). This signal consists of a motif of 8 to 10 amino acids with a central tyrosine that is flanked by amino acids which are basic or polar or that prefer a random coil confirmation. L-MAG contains two such motifs at tyrosine 35 (Y35) and tyrosine 65 (Y65). S-MAG contains one motif at Y35 (Table I). Since tyrosine 35 is at the splice site which distinguished L- and S-MAG, the Y35 motif is not identical in L- and S-MAG. Whether endocytosis of L-MAG requires the motif at Y35, Y65, or both, remains to be established. Receptor-mediated endocytosis has also been linked with tyrosine phosphorylation (Sibley et al. 1987). Fyn, a nonreceptor-type tyrosine kinase of the src family that is activated during early stages of myelination, can be coimmunoprecipitated with L-MAG. In addition, antibodies binding to the extracellular domain of L-MAG, but not S-MAG, resulted in rapid increases in fyn tyrosine activity (Umemori et al. 1994). These observations potentially link L-MAG to intercellular signaling pathways, and raises the possible connection between Fyn tyrosine phosphorylation and endocytosis of L-MAG. A possible role for Fyn in CNS myelination is indicated by defective myelination in Fyn-deficient mice (Umemori et al. 1994).

The physiological consequences of L-MAG endocytosis are presently unknown. Endocytosis is a mechanism for ligand internalization, transcytosis or protein sorting. Since MAG is only enriched in the periaxonal membrane of CNS myelin internodes, it is unlikely that its presence in endosomes indicates transcytosis or a mechanism of protein targeting. Elucidation of the possibility that the endocytosis of L-MAG is involved in axon-oligodendrocyte signalling awaits identification of MAG ligands.

Maintenance and Maturation of Myelinated Fibers. Although MAG-deficient mice form relatively normal myelin internodes, abnormalities of myelin can be detected as the mice age. In addition to alterations in the periaxonal space and cytoplasmic collar, some axons are surrounded by multiple myelin sheaths and redundant myelin lamellae (Montag et al. 1994) and unmyelinated fibers can be surrounded by compact myelin (Fig. 5). Myelin pathology continues to increase with age and in mice older than 8 months the maintenance of some myelinated fibers is compromised, resulting in Wallerian-like degeneration (Fruttiger et al. 1995). These observations raise the possibility that MAG has a crucial role in maintenance of myelinated fiber integrity (Fruttiger et al. 1995).

In vitro studies have implicated MAG in the promotion of neurite outgrowth from neonatal dorsal root ganglion cells (Johnson et al. 1989). The relevance of these data to developing neurons *in vivo* is unclear as growing neurites are not likely to encounter MAG during brain development. In contrast to the axonal growth promoting properties mentioned above, recent studies have indicated that MAG can inhibit neurite outgrowth from more mature neurons maintained *in vitro* (Mukhopadhyay et al. 1994; McKerracher et al. 1994). Myelin is a known inhibitor of axonal regeneration and Schwabb and colleagues (Schwab, 1990) have identified molecules other than MAG which inhibits axonal regeneration *in vitro*. Immunodepletion of MAG from myelin reduced the inhibitory effect or axonal regeneration by approximately 60% (McKerracher et al. 1994). In addition, re-

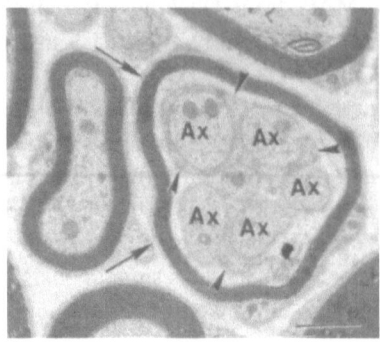

Figure 5. Electron micrograph of sciatic nerve from an adult MAG-deficient mouse. Myelin sheath (arrows) surrounds multiple axons which are partially ensheathed by Schwann cell processes (arrowheads). Ax=axons. Scale bar = 1 µm.

cent studies have demonstrated that MAG coated beads can cause growth cone collapse *in vitro* (Li et al. 1996). While these *in vitro* observations are consistent with a role for MAG in inhibiting axonal regeneration, the relevance of these observations needs to be established *in vivo*. This is particularly important to axonal regeneration in the CNS because removal of myelin debris following axonal transection is a prolonged process. An important question is whether regenerating axons in the CNS ever encounter MAG. MAG is located at the inner aspect of the myelin internode and tends to be located at the center of degenerating myelin debris. MAG-enriched membranes are surrounded by multiple membranes which are MAG-negative. It is possible that the extracellular domain of MAG may be cleaved during Wallerian degeneration and as a soluble product can reach and affect regenerating neurites.

Recent evidence indicates that MAG may inhibit axonal regeneration in peripheral nerve *in vivo* (Schafer et al. 1996). In these studies, MAG-deficient mice were crossed with C57BL/Wld[s] mice that have delayed Wallerian degeneration following axonal injury. In crushed nerves of C57BL/Wld[s] mice expressing MAG, 16% of myelin sheaths were associated with regrowing axons, while this number was doubled in MAG-deficient C57BL/Wld[s] mice. These observations suggest that the absence of MAG may contribute to the improved axonal regrowth. Since myelin is efficiently and rapidly removed following axotomy in the PNS, the physiological relevance of these observations to axonal regeneration in the normal peripheral nervous system is unclear following nerve transection. It is possible, however, that MAG may inhibit regeneration of axon sprouts along surviving fibers in partial nerve transection (Filbin, 1995).

Developmentally, myelination in the PNS is initiated by a signal from the axon to Schwann cell (Aguayo et al. 1976; Weinberg and Spencer, 1976). Myelination also has trophic effects on axons, causing larger axonal caliber (Windebank et al. 1985), increased neurofilament spacing (deWaegh et al. 1992; Sanchez et al. 1996), and greater neurofilament numbers (Sanchez et al. 1996). Although little is known about the molecular interactions which regulate myelin's trophic effect on the maturation of axons, MAG is an attractive candidate. Normal myelination in MAG-deficient mice (Montag et al., 1994; Li et al. 1994) and evidence that MAG can inhibit axonal regeneration (Mukhopadhyay et al. 1994; McKerracher et al. 1994; Schäfer et al. 1996) raise the possibility that MAG functions as a ligand that affects axonal properties. In this regard, we have recently detected significant alterations in axons of myelinated fibers in MAG-deficient mice, suggesting that MAG functions as a ligand that influences the organization of axonal cytoskeleton (Yin et al., submitted for publication).

SUMMARY

This chapter has reviewed much of the present literature regarding the possible function of MAG. Normal myelination in MAG-deficient mice raised serious doubts that MAG functions in the initiation and/or progression of myelination. As the MAG knockout mice were described, MAG research was taking a new direction focused on the role of MAG as a ligand which modulates axonal properties. It should be stressed that these data are no more convincing than earlier studies which indicated that MAG was a receptor for axonal signals that initiated myelination (Owens and Bunge, 1991). The possibility that MAG functions as a ligand, however, is attractive and the next few years should establish if this new outlook on MAG is correct. In either case, it appears that MAG may function more in maturation or maintenance of myelin-axon units rather than during early stages of nerve development.

REFERENCES

Aguayo AJ, Epps J, Charron L, Bray GM (1976) Multipotentiality of Schwann cells in cross anastomosed and grafted myelinated and unmyelinated nerves: Quantitative microscopy and radioautography. Brain Res. 104:1–20

Arquint M, Roder J, Chia L-S, Down J, Wilkinson O, Bayley H, Braun P, Dunn R (1987) Molecular cloning and primary structure of myelin-associated glycoproteins. Proc. Natl. Acad. Sci. USA 84:600–604

Bo L, Quarles RH, Fujita N, Bartoszewicz Z, Sato S, Trapp BD (1995) Endocytic depletion of L-MAG from CNS myelin in quaking mice. J. Cell. Biol. 131(6):1811–1820

Bunge RP, Bunge MB, Bates M (1989) Movements of the Schwann cell nucleus implicate progression of the inner (axon-related) Schwann cell process during myelination. J. Cell. Biol. 109:273–284

Cunningham BA, Hemperly JJ, Murray BA, Prediger EA, Brackenberry R, Edelman GM (1987) Neural adhesion molecule: structure, immunoglobulin-like domains, cell surface modulation and alternative RNA splicing. Science 236:799–806

deWaegh SM, Lee VM-Y, Brady ST (1992) Local modulation of neurofilament phosphorylation, axonal caliber, and slow axonal transport by myelinating Schwann cells. Cell 68:451–463

Figlewicz DA, Quarles RH, Johnson D, Barbarash GR, Sternberger NH (1981) Biochemical demonstration of the myelin-associated glycoprotein in the peripheral nervous system. J. Neurochem. 37:749–758

Filbin MT (1995) Myelin-associated glycoprotein: a role in myelination and in the inhibition of axonal regeneration. Curr. Opin. Neurobiol. 5:588–595

Frail DE, Webster Hd, Braun PE (1985) Developmental expression of the myelin-associated glycoprotein in the peripheral nervous system is different from that in the central nervous system. J. Neurochem. 45:1308–1310

Fruttiger M, Montag D, Schachner M, Martini R (1995) Crucial role for the myelin-associated glycoprotein in the maintenance of axon-myelin integrity. Eur. J. Neurosci. 7:511–515

Johnson PW, Abramow-Newerly W, Seilheimer B, Sadoul R, Tropak MB, Arquint M, Dunn RJ, Schachner M, Roder JC (1989) Recombinant myelin-associated glycoprotein confers neural adhesion and neurite outgrowth function. Neuron 3:377–385

Ktistakis NT, Thomas D, Roth MG (1990) Characteristics of the tyrosine recognition signal for internalization of transmembrane surface glycoproteins. J Cell. Biol. 111:1393–1407

Lai C, Brow MA, Nave K-A, Noronha AB, Quarles RH, Bloom FE, Milner RJ, Sutcliffe JG (1987) Two forms of 1B236/myelin-associated glycoprotein (MAG), a cell adhesion molecule for postnatal neural development, are produced by alternative splicing. Proc. Natl. Acad. Sci. USA 84:4337–4341

Li, Shibata, Braun, McKerracher, Roder, Kater, David (1996) Myelin-associated glycoprotein inhibits neurite/axon growth and causes growth cone collapse. J. Neurosci. Res. in press:

Li C, Tropak MB, Gerlai R, Clapoff S, Abramow-Newerly W, Trapp B, Peterson A, Roder J (1994) Myelination in the absence of MAG. Nature 369:747–750

Martini R, Schachner M (1986) Immunoelectron microscopic localization of neural cell adhesion molecules (L1, N-CAM, and MAG) and their shared carbohydrate epitope and myelin basic protein in developing sciatic nerve. J. Cell. Biol. 103:2439–2448

McKerracher L, David S, Jackson DL, Kottis V, Dunn RJ, Braun PE (1994) Identification of myelin-associated glycoprotein as a major myelin-derived inhibitor of neurite growth. Neuron 13:805–811

Montag D, Giese KP, Bartsch U, Martini R, Land Y, Blüthmann H, Karthigasan J, Kirschner DA, Wintergerst ES, Nave K-A, Zielasek J, Toyka KV, Lipp H, Schachner M (1994) Mice deficient for the myelin-associated glycoprotein show subtle abnormalities in myelin. Neuron 13:229–246

Mukhopadhyay G, Doherty P, Walsh FS, Crocker PR, Filbin MT (1994) A novel role for myelin-associated glyco-protein as an inhibitor of axonal regeneration. Neuron 13:757–767

Owens GC, Boyd CJ, Bunge RP, Salzer JL (1990) Expression of recombinant myelin-associated glycoprotein in primary Schwann cells promotes the initial investment of axons by myelinating Schwann cells. J. Cell Biol. 111:1171–1182

Owens GC, Bunge RP (1991) Schwann cells infected with a recombinant retrovirus expressing myelin-associated glycoprotein antisense RNA do not form myelin. Neuron 7:565–575

Quarles RH, Everly JL, Brady RO (1973a) Evidence for the close association of a glycoprotein with myelin. J. Neurochem. 21:1177–1191

Quarles RH, Everly JL, Brady RO (1973b) Myelin-associated glycoprotein: a developmental change. Brain Res. 58:506–509

Salzer JL, Holmes WP, Colman DR (1987) The amino acid sequences of the myelin-associated glycoproteins: ho-mology to the immunoglobulin gene superfamily. J Cell Biol. 104:957–965

Sanchez I, Hassinger L, Paskevich PA, Shine HD, Nixon RA (1996) Oligodendroglia regulate the regional expan-sion of axon caliber and local accumulation of neurofilaments during development independently of mye-lin formation. J Neurosci. 16:5095–5105

Schafer M, Fruttiger M, Montag D, Schachner M, Martini R (1996) Disruption of the gene for the myelin-associ-ated glycoproteins improves axonal regrowth along myelin in C57BL/Wlds mice. Neuron 16:1107–1113

Schäfer M, Fruttiger M, Montag D, Schachner M, Martini R (1996) Disruption of the gene for the myelin-associ-ated glycoprotein improves axonal regrowth along myelin in C57BL/Wld[S] mice. Neuron 16:1107–1113

Schwab ME (1990) Myelin-associated inhibitors of neurite growth and regeneration in the CNS. TINS 13:452–456

Sibley DR, Benovic JL, Caron MG, Lefkowitz RJ (1987) Regulation of transmembrane signaling by receptor phosphorylation. [Review]. Cell 48:913–922

Sternberger NH, Quarles RH, Itoyama Y, Webster Hd (1979) Myelin-associated glycoprotein demonstrated immu-nocytochemically in myelin and myelin-forming cells of developing rats. Proc. Natl. Acad. Sci. USA 76:1510–1514

Trapp BD, Quarles RH, Griffin JW (1984a) Myelin-associated glycoprotein and myelinating Schwann cell-axon interaction in chronic beta,beta'-iminodipropionitrile neuropathy. J. Cell Biol. 98:1272–1278

Trapp BD, Quarles RH, Suzuki K (1984b) Immunocytochemical studies of quaking mice support a role for the myelin-associated glycoprotein in forming and maintaining the periaxonal space and periaxonal cytoplas-mic collar of myelinating Schwann cells. J. Cell Biol. 99:594–606

Trapp BD (1988) Distribution of the myelin-associated glycoprotein and P_0 protein during myelin compaction in Quaking mouse peripheral nerve. J. Cell Biol. 107:675–685

Trapp BD, Andrews SB, Cootauco C, Quarles RH (1989a) The myelin-associated glycoprotein is enriched in mul-tivesicular bodies and periaxonal membranes of actively myelinating oligodendrocytes. J. Cell Biol. 109:2417–2426

Trapp BD, Andrews SB, Wong A, O'Connell M, Griffin JW (1989b) Co-localization of the myelin-associated gly-coprotein and the microfilament components f-actin and spectrin in Schwann cells of myelinated fibers. J. Neurocytol. 18:47–60

Trapp BD, Quarles RH (1982) Presence of the myelin-associated glycoprotein correlates with alterations in the pe-riodicity of peripheral myelin. J. Cell Biol. 92:877–882

Trapp BD, Quarles RH (1984) Immunocytochemical localization of the myelin-associated glycoprotein: Fact or artifact? J. Neuroimmunol. 6:231–249

Tropak MB, Johnson PW, Dunn RJ, Roder JC (1988) Differential splicing of MAG transcripts during CNS and PNS development. Mol. Brain Res. 4:143–155

Umemori H, Sato S, Yagi T, Aizawa S, Yamamoto T (1994) Initial events of myelination involve Fyn tyrosine ki-nase signalling. Nature 367:572–576

Weinberg HJ, Spencer PS (1976) Studies on the control of myelinogenesis. II. Evidence for neuronal regulation of myelin production. Brain Res. 113:363–378

Windebank AJ, Word P, Bunge RP, Dyck PJ (1985) Myelination determines the caliber of dorsal root ganglion neurons in culture. J. Neurosci. 6:1563–1567

4

MYELIN/OLIGODENDROCYTE GLYCOPROTEIN

A Molecular Analysis

Minnetta V. Gardinier,[1,2] Pauline A. Ballenthin,[1] John F. Kroepfl,[1] and Laura R. Viise[1]

[1]Department of Pathology
[2]Institute for Neuroscience
Northwestern University Medical School
303 East Chicago Avenue
Chicago, Illinois 60611

INTRODUCTION

Myelinogenesis is an essential, highly regulated event that occurs during early mammalian development whereby internodal axonal segments are enveloped by compacted multilamellar membranes. A single oligodendrocyte in the central nervous system (CNS) has the capacity to extend numerous processes, which develop into elaborate specialized membranes that ensheathe multiple larger diameter axonal segments. Historically, CNS myelin had been characterized as a relatively inert membrane with a relatively simple profile of proteins, including proteolipid proteins (PLP, DM20), myelin basic proteins (MBP), myelin-associated glycoprotein (MAG), and 2',3'-cyclic nucleotide 3'-phosphodiesterase (CNP). However it is now clear that myelin is an extremely dynamic structure with many quantitatively minor proteins that clearly play significant roles in myelination.

Myelin/oligodendrocyte glycoprotein (MOG) is a CNS-specific integral membrane protein identified by the mouse monoclonal antibody, 8–18C5 that is directed against rat cerebellar glycoproteins (Linington et al., 1984). M_2, initially described as a CNS-specific autoantigen, was subsequently shown to be immunoreactive with MOG antibody and identical to MOG (Lebar et al., 1986). Immunization of susceptible species (i.e., SJL mice, Lewis rats) with CNS tissue (or purified myelin) induces an immune-mediated inflammatory CNS demyelinating disease, experimental allergic encephalomyelitis (EAE). EAE has been used extensively as an animal model for multiple sclerosis (MS), and this model has demonstrated MOG's importance in the immunopathogenesis of demyelination (Schluesener et al., 1987; Lassmann et al., 1988; Linington et al., 1988). We have shown that purified recombinant MOG (residues 1–125) induces both histologic and clinical EAE in

Cell Biology and Pathology of Myelin, edited by Juurlink *et al.*
Plenum Press, New York, 1997

mice (Amor et al., 1994). MOG has also been implicated as a target autoantigen in the pathogenesis of MS (Xiao et al., 1991). Indeed, MOG's localization to oligodendrocyte cell bodies, oligodendrocyte processes, and the outermost surface of myelin sheaths (Linington et al., 1988; Brunner et al., 1989) makes it a readily accessible target for attack during immune-mediated demyelination.

While MOG is clearly involved in demyelinating disease pathology, its normal biological function has yet to be defined. It has been described as a late marker of oligodendrocyte differentiation and is associated with *myelinating* oligodendrocytes (Scolding et al., 1989; Matthieu and Amiguet, 1990; Solly et al., 1996). Purified MOG migrates as a 26–28 kDa doublet band on denaturing SDS acrylamide gels, and deglycosylation yields a single 25 kDa band (Matthieu and Amiguet, 1990; Amiguet et al., 1992). Rat MOG's peptide sequence was deduced from a full length cDNA clone, and it shows that MOG is a member of the immunoglobulin (Ig) superfamily with a single Ig-like domain (Gardinier et al., 1992). Based upon its similarity with Ig superfamily members, we proposed that MOG serves as an adhesion or receptor molecule. A 1.6 kb MOG mRNA is expressed only in CNS tissue and at maximal levels during active myelination. We also uncovered two highly unusual features for MOG. First, MOG transcripts utilize a series of overlapping, rare polyadenylation signals, each of which has been shown in other models to be very inefficient at mRNA cleavage and polyadenylation, suggesting that MOG mRNA may be inherently unstable. Second, this Ig superfamily member contains two extremely hydrophobic domains and was the first Ig-like molecule identified with more than a single potential membrane spanning domain.

We have recently determined that the C-terminal hydrophobic region is not truly membrane spanning but is more likely associated with the cytoplasmic face of the plasma membrane, perhaps forming a "hairpin" loop (Kroepfl et al., 1996). This membrane topology has interesting ramifications relative to another surprising observation we made regarding alternative splice variants of MOG. A striking dichotomy in MOG gene expression was revealed in a comparative study of human and mouse MOG mRNAs. This developmental analysis revealed that human MOG transcripts exhibit an extensive array of alternatively spliced mRNAs, while no splicing was found in murine CNS tissues (Ballenthin and Gardinier, 1996). Paradoxically, alternative splicing of human MOG would result in variations restricted to MOG's C-terminal cytoplasmic domains that are otherwise 100% conserved between humans and rodents (mouse, rat). This review explores how some of these splice variants may have arisen in humans and discusses their proposed disposition in the oligodendroglial plasma membrane.

MOG PROTEIN IS HIGHLY CONSERVED AMONG MAMMALS

MOG cDNA sequences have been reported for four mammalian species - rat (Gardinier et al., 1992), mouse (Gardinier and Matthieu, 1993; Pham-Dinh et al., 1993), human (Gardinier and Matthieu, 1993; Pham-Dinh et al., 1994; Hilton et al., 1995) and cow (Pham-Dinh et al., 1993). Among these species, peptide sequence identity ranges between 89% and 96% (Figure 1); no residue additions or deletions occur. Upon closer inspection of amino acid differences, we noted that approximately two-thirds of these substitutions are conservative changes (Figure 1B). While several residue substitutions occur within MOG's first membrane soluble domain, they are conservative, and hydrophobicity within this region is entirely retained. In contrast, MOG's second hydrophobic region is nearly 100% conserved with only a single amino acid change found in bovine

A

```
rat     GQFRVIGPGH PIRALVGDEA ELPCRISPGK NATGMEVGWY RSPFSRVVHL  50
mouse   .........Y .......... .......... .......... ..........
human   ........R. .........V .......... .......... .P........
cow     .......... .........V .......... .......... .P........

rat     YRNGKDQDAE QAPEYRGRTE LLKESIGEGK VALRIQNVRF SDEGGYTCFF 100
mouse   .......... .......... ....T.S... ..T....... ..........
human   .......GD. .......... ....DA.... ..T..R.... ....F....
cow     .......E.. .........Q ....T..... ..T..R.... ....F....

rat     RDHSYQEEAA VELKVEDPFY WINPGVLALI ALVPHLLQV SVGLVFLFLQ 150
mouse   .......... M......... .V....T... ....TI.... ..........
human   .......... M......... .VS...V.L .VL.V....I T.....C..
cow     .......... M......... .V.. .VL.V....I T.....C..

rat     HRLRGKLRAE VENLHRTFDP HFLRVPCWKI TLFVIVPVLG PLVALIICYN 200
mouse   .......... ......I... .......... .......... ..........
human   Y......... ......I... .......... .......... ..........
cow     R.....W... ......I... ....M..... .......... ..........

rat     WLHRRLAGQF LEELRNPF                                   218
mouse   .......... ........
human   .......... ........
cow     .......... ........
```

B

		% Conservative Substitutions			
		Rat	Mouse	Human	Cow
% Identity	Rat		78% (7/9)	78% (18/23)	75% (15/20)
	Mouse	96% (209/218)		70% (16/23)	67% (14/21)
	Human	89% (195/218)	89% (195/218)		67% (8/12)
	Cow	91% (198/218)	90% (197/218)	94% (206/218)	

Figure 1. Sequence alignment of MOG peptide sequences. *Panel A:* Predicted peptide sequences from rat, mouse, human, and bovine cDNAs (Genbank numbers M99485, U64572, U64564, and L21757, respectively) are aligned with sequence differences noted below the corresponding site in rat MOG. Amino acid identities are indicated by a period (.). Two Cys residues (●) are indicated that form the disulfide linkage for MOG's single Ig-like domain. MOG's N-linked carbohydrate attachment site is shown at Asn-31 (○). Two potential membrane spanning domains are highlighted by double lines above and below the relevant sequences. Key residues that might maintain MOG's unique membrane topology are indicated within MOG's second hydrophobic domain - cysteine (☆), proline (▮). *Panel B:* Pairwise sequence comparisons are summarized for MOG among the four species that have been sequenced to date. Percent identity is shown in the lower left diagonal half, and a substitution analysis is presented in the upper right diagonal half. Amino acid substitutions are analyzed using a PAM250 (percent accepted mutation) scoring matrix, and percent conservative substitutions are reported. Percentages are given with residue totals included in parentheses.

MOG. Indeed MOG's C-terminal third, including the second hydrophobic domain, is strikingly conserved. This observation suggests that it represents a significant functional element within MOG.

MOG'S TOPOLOGY IN THE PLASMA MEMBRANE

Ig-like molecules typically transduce signals across the lipid bilayer via receptor or adhesion mechanisms, and our studies are directed at ascertaining whether MOG plays a similar role in CNS oligodendrocytes. However, our sequence analysis of rat MOG cDNA predicted an *atypical* member of the Ig superfamily with *two* potential membrane spanning domains (Gardinier et al., 1992). All previously described Ig-like molecules had only a single or no transmembrane domain (Williams and Barclay, 1988). Subsequently only one other Ig family member, integrin-associated protein/ovarian antigen 3 (IAP/OA3), has been predicted to have multiple membrane spanning domains (Campbell et al., 1992; Lindberg et al., 1993). These observations for both MOG and IAP/OA3 set precedents for further topological diversity within the Ig superfamily.

As described above, MOG's second hydrophobic domain with its flanking hydrophilic residues (MOG152-218) is nearly 100% identical among four mammalian species. The strong conservation of this unique area within MOG led us to investigate models of its membrane topology. While we initially proposed a topological model with two membrane spanning domains, positioning both the N- and C-termini at the extracellular membrane surface, the possibility of a single membrane spanning domain was also acknowledged (Figure 2). A full length mouse MOG cDNA was stably transfected into human embryonic kidney carcinoma (HEK293) cells, and G418-resistant cell lines were examined for MOG expression by immunoblot analysis of cell lysates (Kroepfl et al., 1996). MOG was not detected in HEKNEO (vector only) or untransfected HEK293 cells. The 26–28 kDa doublet in HEKMOG cells comigrates with MOG found in normal rat brain myelin, also indicative that MOG is appropriately glycosylated in the heterologous cells.

Three specific MOG antisera were generated to ascertain on which side of the plasma membrane MOG's three hydrophilic domains resided. MOG1-125 Ab specifies MOG's single Ig-like domain (Figure 2; refer to residue numbers at membrane bilayer interfaces). The short hydrophilic stretch linking MOG's two hydrophobic domains is identified by MOG154-169 Ab. MOG's C-terminal hydrophilic tail is immunoreactive with MOG198-218 Ab. Unpermeabilized cells were fixed with 4% paraformaldehyde. Follow-

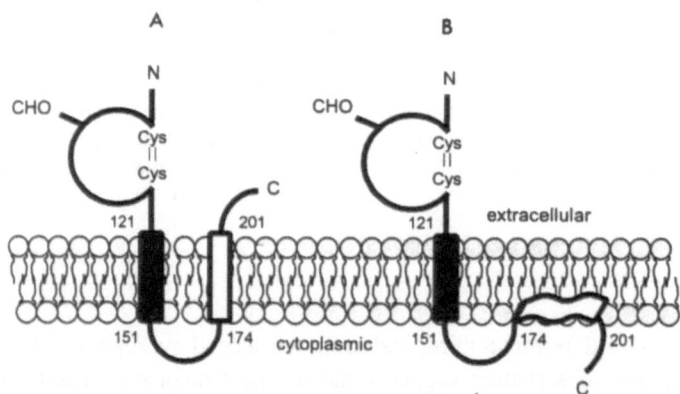

Figure 2. Two models for MOG's disposition in the plasma membrane. In both models, MOG's glycosylated (CHO) N-terminal Ig-like domain is found on the extracellular side of the lipid bilayer. *Model A:* Both of MOG's hydrophobic domains (black and white boxes) span the plasma membrane, and the C-terminus is on the extracellular side. *Model B:* Only MOG's N-terminal hydrophobic domain (black box) is transmembrane, while the C-terminal hydrophobic domain (white box) is likely to associate, but not cross, the lipid bilayer. As reference points, residue numbers are provided at the juxtamembrane hydrophilic-hydrophobic interfaces. (Adapted from Kroepfl et al., 1996)

ing fixation, cells could be permeabilized with 0.2% saponin. MOG-expressing heterologous cells (HEKMOG) were tested under these conditions with β-tubulin Ab, which requires cell permeabilization for its access to β-tubulin, a cytoplasmic protein (Figure 3, panels d and h). Each antiserum was then used under these conditions for permeabilized vs. unpermeabilized cells.

MOG's glycosylated Ig-like domain was expected to show localization to the extracellular surface of the plasma membrane, and thus it would be accessible to MOG1-125

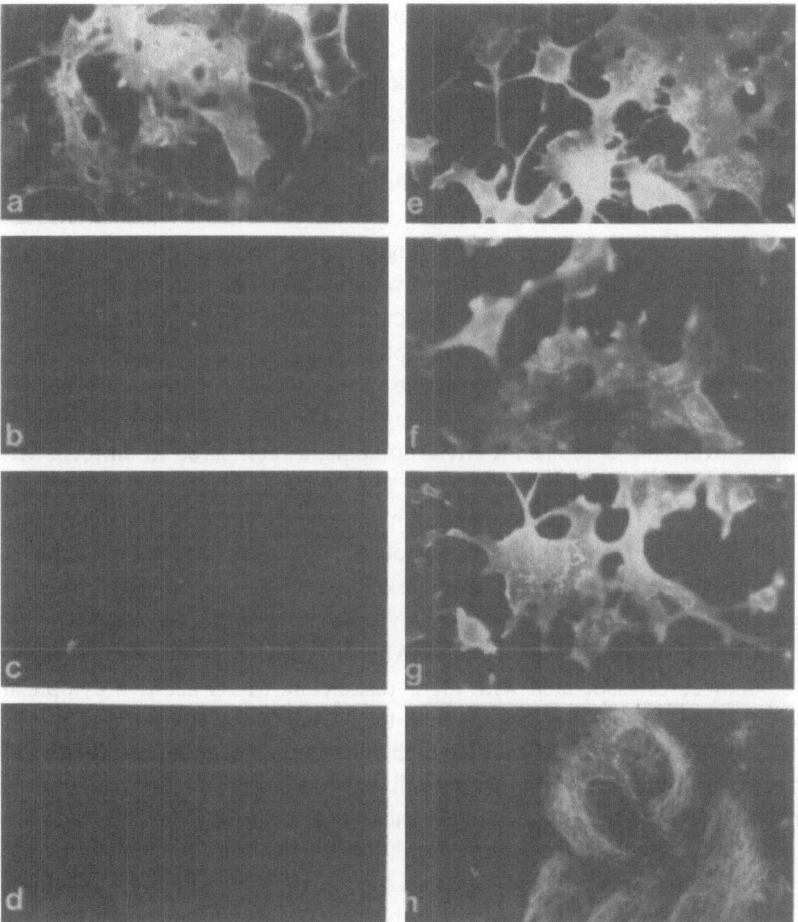

Figure 3. Mouse MOG topology in the cell membrane. HEKMOG cells were immunostained with MOG1-125 (a, e), MOG154-169 (b, f), MOG198-218 (c, g) or β-tubulin antisera (d, h). Cells were fixed with paraformaldehyde but *un*permeabilized (a-d) or permeabilized with 0.2% saponin (d-h). Unless otherwise indicated, primary antisera were diluted 1:50, and cells were incubated overnight at 4°C. MOG epitopes were detected with Texas Red-conjugated goat anti-rabbit IgG (1:200), and β-tubulin was identified with fluorescein-conjugated goat anti-mouse IgG (1:100)(37°C, 3 hr). Anti-MOG1-125 staining gave strong reactivity for surface staining with both *un*permeabilized and permeabilized HEKMOG cells (a, b, respectively). In contrast, intense staining with MOG154-169, MOG198-218, or β-tubulin antisera (1:100) required saponin permeabilization of HEKMOG cells (f, g, h, respectively); no surface reactivity was found with *un*permeabilized HEKMOG cells (b, c, d). HEKNEO cells showed no immunoreactivity for these domain-specific MOG antisera (data not shown). A Zeiss Axiophot epifluorescence microscope was used for photography with Kodak Ektachrome 400 film (×630 magnification).

Ab under either condition - permeabilized or unpermeabilized. Indeed this Ab stains both unpermeabilized and saponin permeabilized HEKMOG cells (Figure 3, panels a and e). Either model shows that residues linking MOG's two hydrophobic domains would be found intracellularly (see Figure 2). Thus MOG154-169 Ab would require permeabilization of the HEKMOG cells for access to this region. Unpermeabilized HEKMOG cells were not immunoreactive, suggesting that this domain is indeed intracellular and inaccessible to MOG154-169 Ab (Figure 3, panel b). Saponin permeabilization rendered these epitopes accessible to MOG154-169 Ab (Figure 3, panel f). Alternative models evolved when we considered where MOG's C-terminal tail might lie relative to the plasma membrane (see Figure 2). Strong immunofluorescence with MOG198-218 Ab was observed only upon membrane permeabilization; unpermeabilized HEKMOG cells showed no immunoreactivity with this Ab (Figure 3, panels c and g). We concluded that the single Ig-like domain is MOG's only extracellular domain and that only MOG's N-terminal hydrophobic domain crosses the plasma membrane (Figure 2B).

These data clearly support our second model (Figure 2B) with only a single membrane spanning domain. Specific regions within MOG protein may now be more accurately targeted for involvement in possible signaling events. While MOG has been characterized as a marker for myelinating oligodendrocytes, the cellular distribution of this glycoprotein has been described in areas *excluded* from compact myelin. As an Ig-like molecule, MOG may serve as a receptor molecule that transmits an extracellular signal to the interior of oligodendrocytes. Ligand/receptor interactions can be mimicked (or disrupted) with antibodies against the membrane protein. This paradigm has been used to treat oligodendrocytes with MOG mAb resulting in a striking redistribution of MOG to areas specifically overlying cytoplasmic MBP domains, as compared to uniform MOG distribution in untreated oligodendrocytes (Dyer and Matthieu, 1994). Prolonged treatment with MOG mAb led to a widespread disruption and depolymerization of fine microtubule networks within the membrane sheets. These'observations imply that MOG may communicate intracellularly via oligodendroglial cytoskeleton.

This membrane topology study targets cytoplasmic domains within MOG that may interact directly with cytoskeletal elements. Alternatively, MOG may interact indirectly with "linker" molecules associated with the cytoskeleton. MOG has two cytoplasmic hydrophilic domains, the MOG151-175 loop and the MOG201-218 C-terminal tail, that may be accessible to these structures. MOG's highly conserved C-terminal hydrophobic domain may be responsible for fixing the three-dimensional conformation that enables these interactions to occur. We believe it is likely that this hydropathic domain must be tucked into the lipid bilayer. Cys177 and Cys198 might be stabilized in juxtamembrane positions through a disulfide linkage (compare Figures 1A and 2B). Alternatively, cysteine residues near such transition zones at the plasma membrane have also been found to be acylated. Inducible palmitoylation is another avenue for signal transduction mechanisms (for reviews - Casey, 1995; Ponimaskin and Schmidt, 1995). Reversible acylation around this hydrophobic domain within MOG might affect the strength of its association with the lipid bilayer and/or its interaction with other membrane-associated proteins (i.e., cytoskeletal proteins, MBP, CNP).

MOG is the only Ig-like family member with a second prominent hydrophobic domain that is probably inserted into the lipid bilayer as a reentrant hairpin loop. Similarly, studies have recently identified two classes of integral membrane proteins, K^+ and ionotropic glutamate receptors, with a pore-forming membrane-associated domain that is predicted to form a hairpin structure on one surface of the plasma membrane (for review - Dani and Mayer, 1995; Wood et al., 1995). Three proline residues within MOG's second hydrophobic domain might also result in significant bending of this region, allowing it to exit and reenter on the cytoplasmic side (compare Figures 1A and 2B). The strong genetic conservation of MOG's second hydrophobic domain and its flanking hydrophilic residues further suggests the functional significance of these MOG domains at or near the cytoplasmic face of the oligodendrocyte plasma membrane.

ALTERNATIVE SPLICING OF HUMAN MOG TRANSCRIPTS

One human MOG clone, pMOG25.6, among those isolated during two cDNA library screens suggested a possible alternative splicing event within MOG's terminal exon. An alternative 3' splice acceptor site within this exon was found that resulted in loss of the 5' 250 nucleotides of this exon. Translation of this transcript would replace the last four amino acids with another nine residues (RNPF → LFHLEALSG), resulting in a slightly lengthened peptide of 25.6 kDa (as compared to an expected 25.1 kDa for MOG). This observation prompted us to widen our investigation for the identification and isolation of MOG splice variants. An RT-PCR analysis was done using total cellular RNA isolated from both adult and fetal human CNS tissues at different stages of development. For maximal sensitivity and specificity, a MOG-specific 3' antisense primer was used for cDNA synthesis with reverse transcriptase, followed by two rounds of PCR amplification using two distinct pairs of nested MOG-specific oligonucleotides flanking MOG's coding region (PCR conditions: annealed at 60°C, 30 sec; extended at 72°C, 30 sec; denatured at 95°C, 30 sec; 30 cycles). We not only confirmed the expected MOG25.1 (1190 nt) and MOG25.6 (940 nt) transcripts, but surprisingly, we also observed several additional MOG-specific amplification bands from CNS mRNA only (Figure 4A). These RT-PCR fragments were isolated and subcloned for sequence analysis to identify other potential splice variants (Ballenthin and Gardinier, 1996).

Sequence analysis of these clones revealed a more complex splicing pattern than anticipated, and we implemented a nomenclature that indicates the predicted peptide size encoded by each splice variant (i.e., MOG25.6 would encode a 25.6 kDa peptide). These

Figure 4. RT-PCR analysis of human and mouse MOG transcripts. *Panel A:* MOG-specific transcripts from RT-PCR analysis are obtained from human CNS tissue. Total cellular RNA (1 µg) from spinal cord (16, 17, and 19 weeks fetal gestation) and brain (21 weeks fetal gestation; A, adult) were used for cDNA synthesis. Plasmid cDNA, pMOG25.1, is used for a positive control (P). *Panel B:* Alternatively spliced MOG transcripts in mouse brainstem were not observed. A single MOG-specific transcript from RT-PCR analysis was obtained from mouse brainstem total RNA (1 µg). RNA was isolated at embryonic day 19/20 (E), day of birth (0), and 5–60 days as indicated. A full length mouse MOG cDNA, pMOG39, was used for a positive control (P). In both panels, negative controls included both no template (NT) and liver (L) RNA. *Hind III*-digested λ DNA and *Hae III*-digested φX174 DNA were used as markers (M); marker sizes are indicated as nucleotides on the left side. DNA bands were visualized by ethidium bromide staining of 1% agarose TAE gels. (Adapted from Ballenthin and Gardinier, 1996)

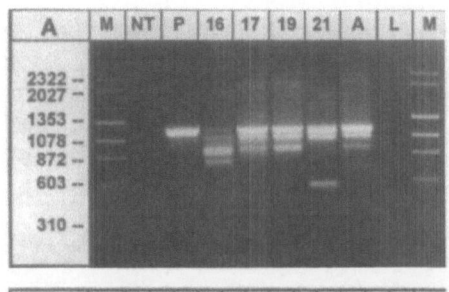

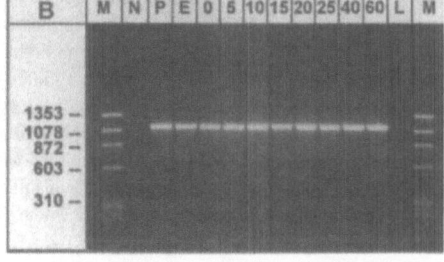

splicing events involved intron/exon junctions as previously identified for the human MOG gene (Hilton et al., 1995; Roth et al., 1995). Besides the internal 3' splice acceptor site within MOG's terminal exon, we found two previously unidentified exons, requiring reorganization of the human MOG gene structure, which now comprises 10 exons (Figure 5). Indeed, both of these new exons with appropriate flanking consensus splice sequences (Stamm et al., 1994) were confirmed and found to lie within previously reported introns of the human MOG gene (Roth et al., 1995).

All MOG splice variants possessed exons that encoded MOG's extracellular Ig-like domain (exon 2), its single transmembrane domain (exon 4), and the short, adjacent hydrophilic cytoplasmic region (exons 5 and 6). A complex splicing pattern prominently arises in cytoplasmic peptide regions encoded by exons 7 through 10. The internal 3' splice acceptor site that was found in MOG25.6 originally was also found in four other clones (MOG22.7, MOG21.0, MOG20.2, and MOG16.3b). Only three clones (MOG25.1, MOG20.5, MOG16.3a) contained a fully intact exon 10. As mentioned above, this alternative splicing within exon 10 would give rise to two distinct C-termini for MOG. Confirmation of these splicing events were obtained with a second RT-PCR strategy using a splice-specific oligonucleotide that flanks the exon 9/10b splice junction (Ballenthin and Gardinier, 1996).

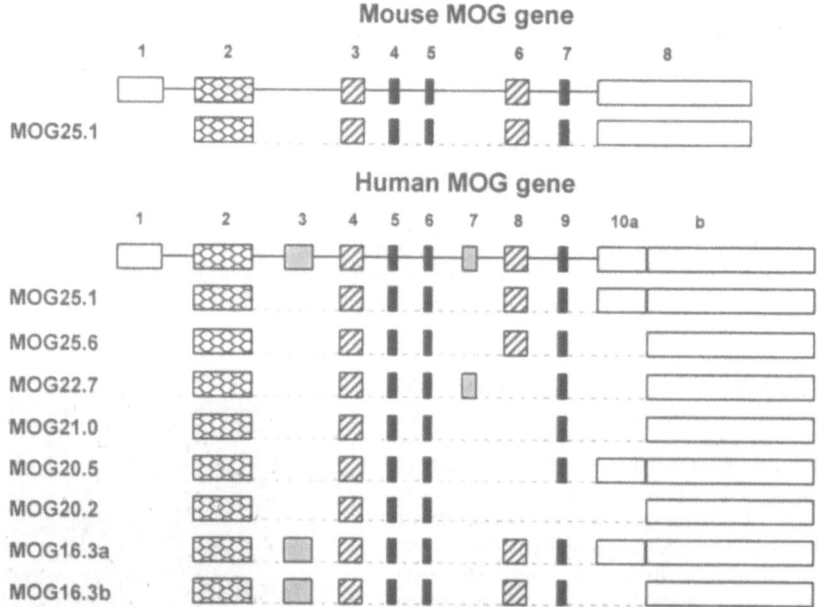

Figure 5. Mouse and human MOG gene structure with alternative exon usage identified among human MOG gene transcripts. The mouse MOG gene structure is shown with its single mRNA transcript represented below. The current human MOG gene structure is given with eight alternatively spliced transcripts represented below the gene. Untranslated regions at the 5' and 3' ends are indicated by white boxes. MOG's single Ig-like domain is encoded by exon 2 (gray mesh box). Exons encoding MOG's two hydrophobic domains are shown as diagonally striped boxes. Exons for MOG's cytoplasmic hydrophilic domains denoted by black boxes. Newly identified exons 3 and 7 identified in the human MOG gene are depicted with gray boxes. Intron regions (not drawn to scale) are shown as a thin solid black line in both mouse and human genes. Sequences for each of these MOG cDNAs are available through Genbank accession numbers U64572 (mouse) and U64564-U64571 (human).

Exon 8, which encodes MOG's unique second hydrophobic domain, was absent in four clones - MOG22.7, MOG21.0, MOG20.5, and MOG20.2. While our membrane topology data demonstrate that this hydrophobic domain does not span the lipid bilayer (Figures 2B and 3), we predict that it is highly likely to be associated with the plasma membrane and responsible for maintaining the three-dimensional conformation of MOG's cytoplasmic domain. While both MOG20.5 and MOG21.0 lack exon 8, the latter clone shows an additional splicing event with the removal of exon 10a also. Exon 6 is spliced directly to exon 10b in MOG20.2, which results in a severely truncated C-terminal tail. One clone, MOG22.7, contains a newly identified exon 7, which would encode a *hydrophilic* domain that would replace MOG's second hydrophobic domain (encoded by exon 8). Upon loss of exon 8 encoding MOG's second hydrophobic domain, all four of these splice variants would predict a "classical" Ig family member possessing only a single hydrophobic membrane-spanning domain. Independently, another laboratory has recently reported four of these seven human MOG variants, including MOG25.6, MOG22.7, MOG20.5, and MOG20.2 (Pham-Dinh et al., 1995).

Exon 3, a second novel exon, was found among two MOG-specific transcripts, and translation of these MOG mRNAs would result in MOG peptides with a striking structural alteration. These *soluble* peptides would contain only a single Ig-like domain with *no* transmembrane domain due to translation termination within exon 3, which would occur prior to translation of MOG's transmembrane domain encoded by exon 4. Again, confirmation of these transcripts was obtained by an alternative RT-PCR strategy using an exon 2/3 splice-specific oligonucleotide (Ballenthin and Gardinier, 1996). Both MOG16.3a and MOG16.3b show inclusion of exon 3; MOG16.3b also excludes exon 10a. Multiple in-frame termination sites occur within exon 3.

Given the complex pattern of splicing evident in the human MOG gene, we undertook an extensive developmental study in mice (embryonic day 19/20 through postnatal day 60). We opted to investigate total cellular RNA isolated from brainstem, an area enriched in myelin. As above, we used a MOG-specific antisense oligonucleotide and amplified the cDNA through two rounds of PCR with nested MOG-specific oligonucleotide primer pairs. Unequivocally, the mouse MOG gene shows no evidence of alternative splicing with only a single MOG transcript evident at each time point tested (Figure 4B). Thus in stark contrast to the wealth of human MOG splice variants, none are observed in mouse.

PROPOSED MEMBRANE TOPOLOGY OF HUMAN MOG VARIANTS

The study described above summarizes our data on eight MOG-specific transcripts that show an intriguing pattern of alternative splicing that is restricted to MOG's cytoplasmic domains occurring during human CNS development. Sequence analysis from several rat and mouse MOG cDNA clones never revealed any hint of alternative splicing, and our recent RT-PCR analysis of mouse brainstem RNA at various stages in development independently confirmed this observation. Likewise, MOG25.1 and MOG25.6 were both identified from library screenings and RT-PCR. These MOG variants are evident at all ages tested and appear to be the most abundant MOG variants (Figure 4A). Based upon numbers of bands observed, the earliest developmental stages in our study (16–19 weeks gestation) revealed the most complex splicing pattern, as compared to that from later ages (21 weeks gestation through adult). We propose that those earlier, less abundant alternatively spliced forms of MOG may play a critical role during initial stages of human MOG gene expression. Such developmental shifts in gene expression have also been noted for proteins translated from other myelin-specific transcripts, including PLP/DM20 (Gardinier and Macklin, 1988), MBP and MAG mRNAs (for review - Campagnoni, 1988).

The predicted peptide sequences of these splice variants would suggest striking variability that we believe would be restricted to MOG's cytoplasmic domains (Figure 6). It is likely that all variants would be glycosylated at Asn-31, as this site occurs within MOG's Ig-like domain and is evident in all of our clones. Except for the two MOG16.3 clones, the other MOG mRNAs are identical through the region encoded by exon 6, encompassing MOG's extracellular Ig-like domain, its transmembrane domain, and the short 14-residue cytoplasmic loop encoded by exons 5 and 6. In the context of our model from the membrane topology study, these MOG peptides predict a highly polymorphic C-terminal cytoplasmic region as a result of the observed splicing events among MOG mRNAs.

Platelet endothelial cell adhesion molecule-1 (PECAM-1), another Ig-like molecule, has five or more splice variants. These PECAM-related translation products have an impact on various properties within their respective cytoplasmic domains (i.e., phosphorylation, cytoskeletal association, and adhesion)(Kirschbaum et al., 1994). Parallel findings

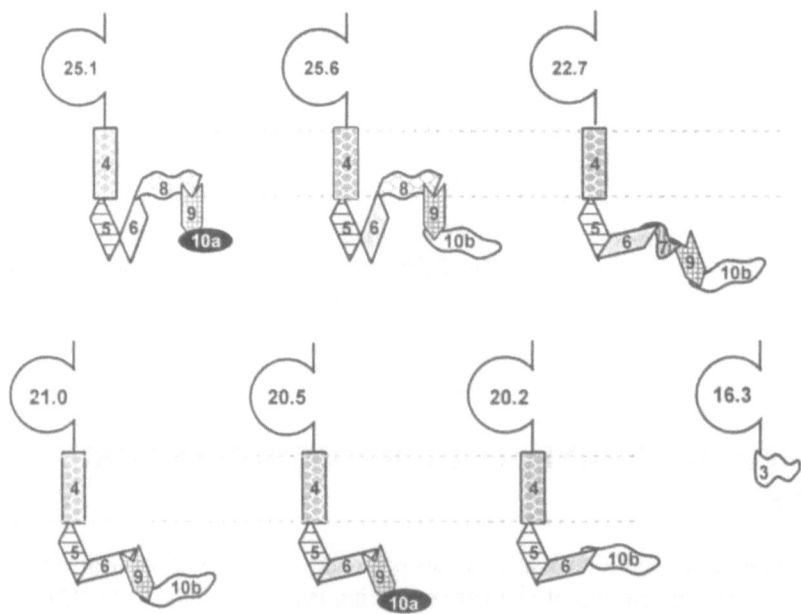

Figure 6. Hypothetical models of MOG's splice variants in humans. MOG's single Ig-like loop is the only extracellular element and is encoded by exon 2. Except for MOG16.3, translation products of all MOG variants would be integral membrane proteins. MOG16.3 peptide would terminate before MOG's first hydrophobic domain and is predicted to be soluble. Splicing events based upon differential use of exons 7–10 would result in a variety of potential secondary structures within the intracellular compartment. Only two of eight variants (MOG 25.1, MOG25.6) express domains for both hydrophobic domains (boxes 4 and 8). Our current data indicate that MOG's second hydrophobic domain does not span the lipid bilayer (dashed lines). Differential splicing of exons 3 and 10 would result in three distinct C-termini (3, 10a, and 10b). Numbers in each box indicate the exon encoding that element.

have been reported for other classical molecules involved in adhesion and receptor signaling molecules, including Ig-like proteins (Siever and Verderame, 1994; Lin et al., 1995; Friesel and Maciag, 1995) and integrins (Song et al., 1993; Fornaro et al., 1995; van der Flier et al., 1995). It is believed that this variability within cytoplasmic domains facilitates a flexibility and wider repertoire for intracellular interactions. Perhaps the diversity found among the cytoplasmic domains of human MOG's variants accommodates different activities within the intracellular compartment, and this interplay may be affected by dynamic events that occur during oligodendrocyte differentiation.

The most conspicuous MOG variants are the MOG16.3 transcripts that would encode a *soluble* form of MOG. This MOG peptide would be generated from a transcript containing the newly identified exon 3 that is spliced in after the exon encoding MOG's single Ig-like domain. In these transcripts, the translation products would terminate prematurely due to multiple in-frame stop codons (Ballentin and Gardinier, 1996). Thus termination would occur *without* translation of MOG's only transmembrane domain and result in a significantly shortened and soluble MOG peptide. Roth and colleagues reported that nine of 15 *Alu* retroposons in the human MOG gene are localized to a 6.5 kb intron following exon 2. We observed that exon 3 is embedded within the tenth *Alu* element, which is inserted in an antisense orientation. This orientation yields introduction of novel splice sites that can lead to production of dysfunctional proteins, premature peptide truncations, or formation of soluble (vs. membrane bound) peptide variants (for review - Makałowski et al., 1994). We are currently investigating if this soluble form of MOG protein is expressed in human CNS tissue. Since *Alu* sequences are unique to humans and higher primates, we believe that expression of exon 3 is most likely restricted to these species. The exon 3 splice products are the only MOG variants that are directly linked to *Alu* repetitive elements.

While human MOG is extensively spliced throughout development, not a single mouse MOG splice variant is detectable at any age studied. This striking dichotomy of expression between human and mouse MOG genes remains a puzzle, except for the one *Alu* insertion giving rise to exon 3. This pattern of MOG variants is also paradoxical in that all of the splicing occurs within exons encoding MOG's cytoplasmic elements, which are nearly 100% conserved with only two conservative amino acid substitutions between rat and human CNS tissue.

FUTURE DIRECTIONS

Now that we know how MOG25.1 is inserted into the plasma membrane of oligodendrocytes, one of our goals is to identify MOG's functional elements - both extracellular and intracellular. Adhesion and binding assays may help us isolate a specific ligand that interacts with MOG's extracellular Ig-like domain. In addition, MOG's cytoplasmic domains may have an interplay with other intracellular molecules involved in cell signaling events. While we have identified several human alternatively spliced MOG transcripts, we still need to demonstrate that their peptide counterparts exist in vivo. These studies will necessitate the preparation of highly specific peptide antisera. The two newly identified exons, 3 and 7, are excellent candidate antigens for this purpose. Moreover, splice-specific peptide antisera could be generated from peptides encoded by the exon 9/10b splice junction and exon 6/10b splice junction. It will be intriguing to discern the species-specific role(s) played by human MOG variants during development.

ACKNOWLEDGMENTS

This research was supported by grants from the National Multiple Sclerosis Society (RG 2638-A-1) and Northwestern University (93-1014-64). J.F.K. and L.R.V. are supported by the Cellular and Molecular Basis of Disease Training Grant (NIH T32-GM08061).

REFERENCES

Amiguet P, Gardinier MV, Zanetta J-P, Matthieu J-M (1992): Purification and partial structural and functional characterization of mouse myelin/oligodendrocyte glycoprotein. J Neurochem 58:1676–1682.

Amor S, Groome N, Linington C, Morris MM, Dornmair K, Gardinier MV, Matthieu J-M, Baker D (1994): Identification of epitopes of myelin oligodendrocyte glycoprotein for the induction of experimental allergic encephalomyelitis in SJL and Biozzi AB/H mice. J Immunol 153:4349–4356.

Ballenthin PA, Gardinier MV (1996): Myelin/oligodendrocyte glycoprotein is alternatively spliced in humans but not mice. J Neurosci Res 46: 271–286.

Brunner C, Lassmann H, Waehneldt TV, Matthieu J-M, Linington C (1989): Differential ultrastructural localization of myelin basic protein, myelin/oligodendroglial glycoprotein, and 2',3'-cyclic nucleotide 3'-phosphodiesterase in the CNS of adult rats. J Neurochem 52:296–304.

Campagnoni AT (1988): Molecular biology of myelin proteins from the central nervous system. J Neurochem 51:1–14.

Campbell IG, Freemont PS, Foulkes W, Trowsdale J (1992): An ovarian tumor marker with homology to vaccinia virus contains an Ig-like region and multiple transmembrane domains. Cancer Res 52:5416–5420.

Casey PJ (1995): Protein lipidation in cell signaling. Science 268:221–225.

Dani JA, Mayer MI (1995): Structure and function of glutamate and nicotinic acetylcholine receptors. Curr Opin Neurobiol 5:310–317.

Dyer CA, Matthieu J-M (1994): Antibodies to myelin/oligodendrocyte-specific protein and myelin/oligodendrocyte glycoprotein signal distinct changes in the organization of cultured oligodendroglial membrane sheets. J Neurochem 62:777–787.

Fornaro M, Zheng DQ, Languino LR (1995): The novel structural motif Gln795-Gln802 in the integrin beta 1C cytoplasmic domain regulates cell proliferation. J Biol Chem 270:24666–24669.

Friesel RE, Maciag T (1995): Molecular mechanisms of angiogenesis: fibroblast growth factor signal transduction. FASEB J 9:919–925.

Gardinier MV, Amiguet P, Linington C, Matthieu J-M (1992): Cloning and sequence analysis of myelin/oligodendrocyte glycoprotein cDNAs: a novel member of the immunoglobulin gene superfamily. J Neurosci Res 33:177–187.

Gardinier MV, Macklin WB (1988): Myelin proteolipid protein gene expression in jimpy and jimpy[msd] mice. J Neurochem 51:360–369.

Gardinier MV, Matthieu J-M (1993): Murine and human MOG are highly conserved: cDNA analysis. Trans Amer Soc Neurochem 24:234.

Hilton AA, Slavin AJ, Hilton DJ, Bernard CCA (1995): Characterization of cDNA and genomic clones encoding human myelin oligodendrocyte glycoprotein. J Neurochem 65:309–318.

Kirschbaum NE, Gumina RJ, Newman PJ (1994): Organization of the gene for human platelet/endothelial cell adhesion molecule-1 shows alternatively spliced isoforms and a functionally complex cytoplasmic domain. Blood 84:4028–4037.

Kroepfl JF, Viise LR, Charron AJ, Linington C, Gardinier MV (1996): Investigation of myelin/oligodendrocyte glycoprotein membrane topology. J Neurochem 67:2219–2222.

Lassmann H, Brunner C, Bradl M, Linington C (1988): Experimental allergic encephalomyelitis: the balance between encephalitogenic T lymphocytes and demyelinating antibodies determines size and structure of demyelinated lesions. Acta Neuropathol 75:566–576.

Lebar R, Lubetzki C, Vincent C, Lombrail P, Boutry J-M (1986): The M_2 autoantigen of central nervous system myelin, a glycoprotein present in oligodendrocyte membrane. Clin Exp Immunol 66:423–443.

Lin SH, Luo W, Earley K, Cheung P, Hixson DC (1995): Structure and function of C-CAM1: effects of the cytoplasmic domain on cell aggregation. Biochem J 311:239–245.

Lindberg FP, Gresham HD, Schwarz E, Brown EJ (1993): Molecular cloning of integrin-associated protein: an immunoglobulin family member with multiple membrane-spanning domains implicated in $\alpha_v\beta_3$-dependent ligand binding. J Cell Biol 123:485–496.

Linington C, Webb M, Woodhams PL (1984): A novel myelin-associated glycoprotein defined by a mouse monoclonal antibody. J Neuroimmunol 6:387–396.

Linington C, Bradl M, Lassmann H, Brunner C, Vass K (1988): Augmentation of demyelination in rat acute allergic encephalomyelitis by circulating mouse monoclonal antibodies directed against a myelin/oligodendrocyte glycoprotein. Amer J Pathol 130:443–454.

Makałowski W, Mitchell GA, Labuda D (1994): Alu sequences in the coding regions of mRNA: a source of protein variability. Trends Genet 10:188–193.

Matthieu J-M, Amiguet P (1990): Myelin/oligodendrocyte glycoprotein expression during development in normal and myelin-deficient mice. Dev Neurosci 12:293–302.

Pham-Dinh D, Allinquant B, Ruberg M, Gaspera BD, Nussbaum J-L, Dautigny A (1994): Characterization and expression of the cDNA coding for the human myelin/oligodendrocyte glycoprotein. J Neurochem 63:2353–2356.

Pham-Dinh D, Gaspera BD, Kerlero de Rosbo N, Dautigny A (1995): Structure of the human myelin/oligodendrocyte glycoprotein gene and multiple alternative spliced isoforms. Genomics 29:345–352.

Pham-Dinh D, Mattei M-G, Nussbaum J-L, Roussel G, Pontarotti P, Roeckel N, Mather IH, Artzt K, Lindahl KF, Dautigny (1993): Myelin/oligodendrocyte glycoprotein is a member of a subset of the immunoglobulin superfamily encoded within the major histocompatibility complex. Proc Natl Acad Sci USA 90:7990–7994.

Ponimaskin E, Schmidt MFG (1995): Acylation of viral glycoproteins: structural requirements for palmitoylation of transmembrane proteins. Biochem. Soc. Trans. 23:565–568.

Roth MP, Malfroy L, Offer C, Sevin J, Enault G, Borot N, Po P (1995): The human myelin oligodendrocyte glycoprotein gene: complete nucleotide sequence and structural characterization. Genomics 28:241–250.

Schluesener HJ, Sobel RA, Linington C, Weiner HL (1987): A monoclonal antibody against a myelin oligodendrocyte glycoprotein induces relapses and demyelination in central nervous system autoimmune disease. J Immunol 139:4016–4021.

Scolding NJ, Frith S, Linington C, Morgan BP, Campbell AK, Compston DAS (1989): Myelin/oligodendrocyte glycoprotein (MOG) is a surface marker of oligodendrocyte maturation. J Neuroimmunol 22:169–176.

Siever DA, Verderame MF (1994): Identification of a complete Cek7 receptor protein tyrosine kinase coding sequence and cDNAs of alternatively spliced transcripts. Gene 148:219–226.

Solly SK, Thomas J-L, Monge M, Demerens C, Lubetzki C, Gardinier MV, Matthieu J-M, Zalc B (1996): Myelin/oligodendrocyte glycoprotein (MOG) expression is associated with myelin deposition. Glia 18:39–48.

Song WK, Wang W, Sato H, Bielser DA, Kaufman SJ (1993): Expression of alpha 7 integrin cytoplasmic domains during skeletal muscle development: alternate forms, conformational change, and homologies with serine/threonine kinases and tyrosine phosphatases. J Cell Sci 106:1139–1152.

Stamm S, Zhang MQ, Marr TG, Helfman DM (1994): A sequence compilation and comparison of exons that are alternatively spliced in neurons. Nucl Acids Res 22:1515–1526.

van der Flier A, Kuikman I, Baudoin C, van der Neut R, Sonnenberg A (1995): A novel beta 1 integrin isoform produced by alternative splicing: unique expression in cardiac and skeletal muscle. FEBS Letter 369:340–344.

Williams AF, Barclay NA (1988): The immunoglobulin superfamily-domains for cell surface recognition. Annu Rev Immunol 6:381–405.

Wood MW, VanDongen HMA, VanDongen AMJ (1995): Structural conservation of ion conduction pathways in K channels and glutamate receptors. Proc Natl Acad Sci USA 92:4882–4886.

Xiao B-G, Linington C, Link H (1991): Antibodies to myelin-oligodendrocyte glycoprotein in cerebrospinal fluid from patients with multiple sclerosis and controls. J Neuroimmunol. 31:91–96.

5

MOLECULAR BIOLOGY OF HEREDITARY MOTOR AND SENSORY NEUROPATHIES

U. Suter

Department of Biology
Institute of Cell Biology
Swiss Federal Institute of Technology
ETH-Hönggerberg
CH-8093 Zürich, Switzerland

INTRODUCTION

Myelin is mainly composed of lipids with a quantitatively minor, but functionally important contribution of proteins. During the development of the PNS and in adulthood, the continuous bi-directional dialogue between axon and Schwann cell regulates a temporally and spatially precise pattern of myelin protein expression in the complex myelin structure as a prerequisite for the correct function of the various myelin proteins (reviewed by Snipes and Suter, 1995a).

In addition to the long known and well characterized PNS myelin proteins Protein zero (P0), myelin basic protein (MBP), Protein two (P2) and myelin-associated glycoprotein (MAG; reviewed by Lemke, 1993), a number of novel myelin-associated proteins have recently been described including peripheral myelin protein 22 (PMP22; Snipes et al., 1992), connexin 32 (Cx32; Bergoffen et al., 1993; Scherer et al., 1995a), periaxin (Gillespie et al., 1994; Scherer et al., 1995b), plasmolipin (Fischer and Sapirstein, 1994; Gillen et al., 1996), E-cadherin (Fannon et al., 1995), and the myelin and lymphocyte protein (MAL; Schaeren-Wiemers et al., 1995; Kim et al., 1995). Mutations in three of the PNS myelin proteins, PMP22, Cx32 and P0, have been discovered in human, linking their respective genes to various mutations in hereditary motor and sensory neuropathies (HMSN; reviewed by Snipes and Suter, 1995b; Suter and Snipes, 1995a). This finding has sparked renewed interest in the field of not only geneticists and clinicians, but also neuroscientists and cell biologists. Indeed, the analysis of the nature of the different mutations and their correlation to the observed dysmyelinating and/or demyelinating phenotypes has already contributed considerably to our knowledge of the development and maintenance of the PNS (reviewed by Suter and Snipes, 1995a). Furthermore, the discovery of mutated myelin protein genes as culprits in HMSN provides an illustrative example how seemingly quite disparate research avenues involving various approaches aimed at a) the elucidation of the genetic basis of the HMSN subtype Charcot-Marie-Tooth disease (CMT), b) the careful description of the variable neurological CMT phenotypes, c) the understanding of the molecular mechanisms of nerve degeneration and regeneration after injury, and d) the description of the spontaneous mouse mutants *Trembler (Tr)* and *Trembler-J (Tr-J)* have synergistically contributed to a scientific breakthrough (reviewed by Suter et al., 1993).

CLINICAL PHENOTYPES OF HMSN

Charcot-Marie-Tooth Disease

Dominantly inherited CMT has been first described in the last century by two French physicians, Charcot and Marie (Charcot and Marie, 1886), and the English physician Tooth (Tooth, 1886). The onset of CMT is usually in the second decade of life, and patients are affected by a slowly progressing disease which is manifested mainly by muscular atrophy that impairs the function of the feet and lower legs and, less frequently, the hands and forearms. Foot drop walking gaits, high arches, hammer toes, loss of reflexes and scoliosis are common, but also sensory disturbances are observed (Lupski et al., 1991; Dyck et al., 1993). However, there is a large variability of symptoms within the same family. Even between identical twins, the severity of disabilities ranges from only mild disturbances to being confined to the wheel chair at early age (Garcia et al., 1995).

Generally, two major types of CMT can be distinguished based on electrophysiological examination and nerve pathology. The most frequent form, CMT type 1 (CMT1; HMSN-I), is characterized by significantly reduced nerve conduction velocity (NCV) due to detoriation of both sensory and motor nerve fibers, with segmental demyelination and onion bulb formation by supernumerary Schwann cells as prominent pathological features (reviewed by Snipes and Suter, 1995b). In contrast, NCV is not reduced in CMT type 2 (CMT2; HMSN-II) and morphological hallmarks are the reduction in the number of large diameter myelinated axons without significant signs of demyelination and remyelination (Dyck et al., 1993).

Dejerine-Sottas Syndrome (DSS)

The classically defined congenital DSS has been considered as a separate HMSN entity (HMSN-III) based on its recessive inheritance and its severe phenotype which is characterized by major loss of muscle and sensory functions as well as limb ataxia (Dyck et al., 1993). However, recent findings in molecular genetics have questioned the separate classification of DSS (see below).

Hereditary Neuropathy with Liability to Pressure Palsies (HNPP)

The autosomal dominant HNPP has traditionally not been included in HMSN, a concept which is currently being revised based on genetic evidence (see below). Phenotypically, HNPP is a mild inherited neuropathy which is characterized by recurrent episodes of sensory and motor nerve palsies that are usually precipitated by minor trauma (e.g. pressure-caused) to peripheral nerves (Windebank, 1993). Sural nerve biopsies of HNPP patients reveal multiple sausage-like myelin thickenings (tomacula) along individual nerve fibers as a non-exclusive fingerprint of the disease. In elderly patients, the course of the disease is often progressive and resembles the typical features of a CMT1-like neuropathy.

GENETICS OF HMSN

CMT Type 1A (CMT1A) and HNPP

Recent progress in molecular genetics revealed that 70% of spontaneous and hereditary CMT cases are associated with an intrachromosomal duplication of approximately 1.5 megabases on chromosome 17p11.2 (classified as CMT1A; Lupski et al., 1991). The gene encoding the peripheral myelin protein PMP22 which has been previously shown to carry point mutations in the naturally occurring demyelinating *Tr* and *Tr-J* mouse mutants (Suter et al., 1992a,b) was mapped to the duplicated segment suggesting a PMP22 gene-dosage effect as the potential genetic basis of CMT1A (Patel et al., 1992). In support of this hypothesis, rare non-conservative missense point mutations within the putative trans-membrane regions of PMP22 have been found to be associated with non-duplication CMT1A (Roa et al., 1993a; Roa et al., 1993b; Valentijn et al., 1992) and DSS (Roa et al., 1993c). The CMT1A duplication appears to be generated by a specific recombination event which is mediated by repetitive sequences that contain a *mariner*-like transposon element (CMT1A-REP; Reiter et al., 1996), and the predicted reciprocal allele carrying the deletion corresponding to the chromosomal segment duplicated in CMT1A is linked to HNPP (Chance et al., 1993). In agreement with these results, a frame-shift mutation causing a premature translation stop of the PMP22 protein has been identified in a non-deletion HNPP family (Nicholson et al., 1994).

CMT Subtype 1B (CMT1B)

Based on the finding of a defect affecting a myelin protein gene as the culprit in CMT1A, other genes encoding components of myelin became likely candidates to be mutated in hereditary demyelinating CMT1. Indeed, several point mutations in the gene encoding the major myelin protein P0 (*MPZ*) have been identified and classified as CMT1B (reviewed in Suter and Snipes, 1995a). In addition, rare P0 mutations have been found in patients diagnosed with DSS (reviewed by Roa et al., 1996).

X-Linked CMT1 (CMTX)

Extensive efforts aimed at the mapping of the gene responsible for X-linked forms of demyelinating CMT lead to the identification of the gap junction protein Cx32 as a likely candidate based on the localization of the Cx32 (*GJB*) gene in the critical region of the X-chromosome. Northern blot and immunohistochemical analyses revealed that Cx32 is expressed in peripheral nerves in association with myelin and molecular analyses confirmed that mutations affecting the Cx32 gene are found in CMTX (Bergoffen et al., 1993; reviewed by Spray and Dermietzel, 1995).

PROTEINS AFFECTED IN CMT

Peripheral Myelin Protein-22 (PMP22)

PMP22 is predominately localized in compact PNS myelin (Snipes et al., 1992; Snipes and Suter, 1995a, Haney et al., 1996). The *PMP22* gene is a member of a family of hydrophobic proteins that contain four potential membrane-associated domains (Welcher et al., 1991; Spreyer et al., 1991; Pareek et al., 1993; Taylor et al., 1995; Taylor and Suter, 1996). Structurally, PMP22 resembles proteolipid protein (PLP), the major CNS myelin protein, and intriguing parallels are shared by the genetics of PMP22 and PLP in that both, gene dosage and point mutations are linked to myelin deficiencies (reviewed by Suter and Patel, 1994; Nave and Boespflug-Tanguy, 1996). Similar to PLP, the biological function of PMP22 remains enigmatic. PMP22 might be involved in adhesion processes since it carries the carbohydrate epitope L2/HNK1 (Snipes et al., 1993), or it may function as an adhesive pore or channel protein based on structural considerations (reviewed by Suter and Snipes, 1995b). A more general role of PMP22 in cell biology is indicated by *in vitro* data demonstrating that overexpression of PMP22 after gene transfer leads to a significant prolongation of the G1 phase in Schwann cells (Zoidl et al., 1995) and induces signs of apoptosis in NIH3T3 cells (Fabbretti et al., 1995). This hypothesis is indirectly supported by a) the widespread expression of PMP22 in embryonic mouse development and in adulthood (Baechner et al., 1995; Parmatier et al., 1995), b) PMP22 gene regulation by a complex system involving a myelinating Schwann cell-specific and a more ubiquitously expressed promoter (Suter et al., 1994), and c) the finding of a large PMP22 gene family with high expression in several tissues outside of the PNS (Marvin et al., 1995; Taylor et al., 1995; Taylor and Suter, 1996).

Protein Zero (P0)

The compact PNS myelin component P0 accounts for approximately 50% of total myelin protein and its expression is restricted to Schwann cells (reviewed by Mirsky and Jessen, 1996). Structurally, P0 carries a single immunoglobulin-like domain which characterizes this protein as a prototypic member of a large family of transmembrane glycoproteins which are involved in cellular recognition (reviewed by Martini, 1994). In support of this hypothesis, *in vitro* and *in vivo* approaches have shown that the extracellular domain of P0 is involved in adhesion processes (reviewed by Lemke, 1993). Furthermore, the positively charged cytoplasmic domain of P0 appears to be required for the integrity of the major dense line, possibly by interaction with negatively charged membrane phospholipids (Martini et al., 1994).

Connexin32 (Cx32)

Cx32 has been first identified as a structural component of gap junctions in the liver (reviewed by Kumar and Gilula, 1996). However, the biological function of Cx32 in the PNS remains unclear since the classical morphology of gap junctions has not been observed in this tissue. It has been proposed based on the localization of Cx32 in uncompacted portions of myelin, the Schmidt-Lantermann incisures and paranodal loops, that Cx32 is a component of 'reflexive' gap junctions which link apposed layers of the same Schwann cell (Scherer et al., 1995a). It is conceivable that in this way, an efficient

Schwann cell cytoplasmic channel network is maintained that provides a 'short cut' for small molecules to cross through myelin.

ANIMAL MODELS FOR CMT

Animals which are affected by mutations in the PMP22, P0 and Cx32 genes are not only of interest as models for CMT to explore potential disease mechanisms and to test possible therapeutic interventions, such mutants are also likely to yield novel information about biological functions of the proteins involved.

PMP22 Animal Mutants

As previously indicated, the spontaneous mouse mutants *Tr* and its allelic variant *Tr-J* provided the key to the elucidation of PMP22 as the mutated gene in the most common forms of CMT (CMT1A; reviewed by Suter et al. 1993). In fact, the same PMP22 point mutation associated with *Tr-J* was also found in a human family diagnosed with CMT1 (Valentjin et al., 1992). However, the data obtained from human genetic studies favored a gene-dosage effect (e.g. CMT1A duplication) as the most frequent cause of CMT. This hypothesis is supported by the demyelinating phenotype of transgenic mice carrying approx. eight copies of a 560 kilobases-long yeast artificial chromosome (YAC) that contains also the human *PMP22* gene (Huxley et al., 1996). Another series of transgenic mice were generated with approximately 15 and 30 additional copies of exclusively the mouse *pmp22* gene (derived from a cosmid; Magyar et al., 1996). The predominant dysmyelinating phenotype of these animals provided the formal proof that increased PMP22 gene dosage *per se* is sufficient to cause severe PNS myelin deficiencies. Finally, a rat transgenic model for CMT1A was constructed that contains approximately three copies of the mouse *pmp22* gene (same cosmid as used by Magyar et al., 1996: Sereda et al.. 1996). This rat strain closely mimics the CMT1 phenotype including PNS hypomyelination, reduced NCV, muscle weakness and Schwann cell 'onion bulb' formation as a fingerprint of a continuous process of demyelination and remyelination. Motor nerve fibers appear to be more affected than sensory fibers in these animals but the cellular and molecular mechanisms involved in this intriguing phenomenon remain to be determined. If the PMP22 gene dosage is further increased by breeding PMP22-transgenic rats to homozygosity, the animals are affected by complete dysmylination, in agreement with the results obtained in the mouse (Magyar et al., 1996). In summary, the combined data suggest that impaired Schwann cell development and maintenance is a likely disease mechanism in CMT1A.

In a complementary set of experiments, the *pmp22* gene was inactivated using homologous recombination in embryonic stem cells. Heterozygous mice that have retained only one functional *pmp22* gene copy show focal myelin thickenings and sausage-like structures (e.g. tomacula) in peripheral nerves comparable to the pathology observed in HNPP patients (Adlkofer et al., 1995). Thus, loss of one PMP22 gene copy is responsible for the hereditary neuropathy HNPP. Homozygous animals which are completely devoid of PMP22 expression are characterized by a delay in the onset of myelination, extensive tomacula formation at young age which progresses to a CMT1-like demyelination, mild axonal loss and reduced NCV in the adult organism (Adlkofer et al.. 1995). These results demonstrate conclusively that PMP22 is required for the correct development of peripheral nerves, the determination of myelin thickness and stability, and the maintenance of myelin and axons.

P0-Deficient Mice

The generation of homozygous P0-deficient mice confirmed the role of this protein in myelin compaction, in particular in the maintenance of the myelin intraperiod line (Giese et al., 1992). Demyelination and dysmyelination are common feature in young P0-deficient animals and resemble to some degree the DSS phenotype. The genetic alteration of Schwann cells also affects the initiation of myelin spiral formation as evidenced by retarded myelination. In addition, axons show morphological changes including significantly reduced calibers. Thus, normal P0 expression is also involved in determining the phenotype of the axonal partner, suggesting an important role for this protein in axon-Schwann cell interactions. Heterozygous P0-deficient mice which have retained one functional P0 allele show normal myelination but develop signs of demyelination including 'onion bulb' formation at older age reminiscent of CMT1B (Martini et al. 1995). Thus, since all P0 mutations detected in CMT1B patients so far are present in a heterozygous state (reviewed by Roa et al., 1996), some of these mutations may constitute null alleles.

Cx32-Deficient Mice

Mice carrying a null mutation in the Cx32 gene have been generated by gene targeting techniques to examine the functional role of Cx32 and to prove its causative role in hereditary peripheral neuropathies (Nelles E, Willecke K, personnel communication). As anticipated, peripheral nerves of mice deficient in Cx32 exhibit signs of myelin degeneration and regeneration (onion bulb formation) reminiscent of CMTX. The observed hypomyelination displays a late onset by being rare at the age of four weeks but becoming prominent at the age of four months and in older animals (Anzini P, Neuberg DH-H, Nelles E, Willecke K, Schachner M, Suter U, Martini R, unpublished observations). In addition, Cx32-deficient mice show often prominent periaxonal collars which appear to be increased in volume. This swelling and disorganization may be interpreted as a potential indicator of a toxic effect due to deficient communication between the Schwann cell cytoplasmic inner and outer compartments.

FUTURE PERSPECTIVES

We have now a comprehensive set of rodent mutants at hand which will enable us to study disease processes in HMSN in detail. It is anticipated that we will also gain considerable further insights into the function of the proteins involved and their interaction during Schwann cell development and myelinogenesis from the analysis and cross-breeding of these animals.

ACKNOWLEDGMENT

This work was supported by a grant of the Swiss National Science Foundation (to U.S.).

REFERENCES

Adlkofer K, Martini R, Aguzzi A, Zielasek J, Toyka KV, Suter U (1995): Hypermyelination and demyelinating peripheral neuropathy in PMP22-deficient mice. Nature Genet 11:274–280.

Baechner D, Liehr T, Hameister H, Altenberger H, Grehl H, Suter U, Rautenstrauss B (1995): Widespread expression of the peripheral myelin protein-22 gene (PMP22) in the neural and non-neural tissues during murine development. J Neurosci Res 42:735–741.

Bergoffen J, Scherer SS, Wang S, Oronzi Scott M, Bone LJ, Paul DL, Chen K, Lensch MW, Chance P, Fischbek K (1993): Connexin mutations in X-linked Charcot-Marie-Tooth disease. Science 262:2039–2042.

Chance PF, Alderson MK, Leppig KA, Lensch MW, Matsunami N, Smith B, Swanson PD, Odelberg SJ, Disteche CM, Bird TD (1993). DNA deletion associated with hereditary neuropathy with liability to pressure palsies. Cell 72:143–51.

Charcot J-M, Marie P (1886): Sur une forme particulière d'atrophie musculaire progressive souvent familiale debutant par les pieds et les jambes et atteignant plus dard les mains. Rev Méd (Paris) 6:97–138.

Dyck PJ, Chance P, Lebo R, Carney JA (1993): Hereditary motor and sensory neuropathies. In Dyck PJ, Thomas PK, Griffin JW, Low PA, Poduslo JF (eds): "Peripheral neuropathy" Philadelphia: Saunders WB, pp 194–1136.

Fabbretti E, Edomi P, Brancolini C, Schneider C (1995): Apoptotic phenotype induced by overexpression of wild-type gas3/PMP22: its relation to the demyelinating peripheral neuropathy CMT1A. Genes & Devel 9:1846–1856.

Fannon AM, Sherman DL, Ilyina-Gragerova G, Brophy PJ, Friedrich VLJ, Colman DR (1995). Novel E-cadherin-mediated adhesion in peripheral nerve: Schwann cell architecture is stabilized by autotypic adherens junctions. J Cell Biol 129:189–202.

Fischer I, and Sapirstein VS (1994): Molecular cloning of plasmolipin; characterization of a novel proteolipid restricted to brain and kidney. J Biol Chem 269:24912–24919.

Garcia CA, Malamut RE, England JD, Parry GS, Liu P, Lupski JR (1995): Clinical variability in two pairs of identical twins with the Charcot-Marie-Tooth disease type 1A duplication. Neurology 45:2090–2093.

Giese KP, Martini R, Lemke G, Soriano P, Schachner M (1992): Mouse P0 gene disruption leads to abnormal expression of recognition molecules and degeneration of myelin and axons. Cell 71:565–576.

Gillen C, Gleichmann M, Greiner-Petter R, Zoidl G, Kupfer S, Bosse F, Auer J, Mueller. HW (1996): Full-length cloning, expression and cellular localization of rat plasmolipin mRNA, a proteolipid of PNS and CNS. Eur J Neurosci 8:405–414.

Gillespie CS, Sherman DL, Blair GE, Brophy PJ (1994): Periaxin, a novel protein of myelinating Schwann cells with a possible role in axonal ensheathment. Neuron 12:497–508.

Haney C, Snipes GJ, Shooter EM, Suter U, Garcia C, Griffin JW, Trapp BD (1996): Ultrastructural distribution of PMP22 in Charcot-Marie-Tooth disease type 1A. J Neuropath Exp Neurol 55:290–299.

Huxley C, Passage E, Manson A, Putzu G, Figarella-Branger D, Pelissier JF, Fontes M (1996): Construction of a mouse model of Charcot-Marie-Tooth disease type 1A by pronuclear injection of human YAC DNA. Hum Mol Genet 5:563–569.

Kim T, Fiedler K, Madison DL, Krueger WH, Pfeiffer SE (1995): Cloning and characterization of MVP17: A developmentally regulated myelin protein in oligodendrocytes. J Neurosci Res 42:413–422.

Kumar NM, Gilula NB (1996): The gap junction communication channel. Cell 84:381–388.

Lemke G (1993): The molecular genetics of myelination: an update. Glia 7:263–271.

Lupski JR, Garcia CA, Parry GJ, Patel PI (1991): Charcot-Marie-Tooth polyneuropathy syndrome: clinical, elecrophysiological, and genetic aspects. In Apel S (ed): "Current Neurology" Chicago: Mosby-Yearbook, pp 1–25.

Lupski JR, de Oca Luna RM, Slaugenhaupt S, Pentao L, Guzzetta V, Trask BJ, Saucedo-Cardenas O, Barker DF, Killian JM, Garcia CA, Chakravarti A, Patel PI (1991): DNA duplication associated with Charcot-Marie-Tooth disease type 1A. Cell 66:219–32.

Magyar JP, Martini R, Ruelicke T, Aguzzi A, Adlkofer K, Dembic Z, Zielasek J, Toyka KV, Suter U (1996): Impaired differentiation of Schwann Cells in transgenic mice with increased PMP22 gene dosage. J Neurosci In press.

Martini R (1994). Expression and functional roles of neural cell surface molecules and extracellular matrix components during development and regeneration of peripheral nerves. J Neurocytol 23:1–28.

Martini R, Mohajeri HM, Kasper S, Giese KP, Schachner M (1994): Mice doubly deficient in the genes for P0 and myelin basic protein show that both proteins contribute to the formation of the major dense line in peripheral nerve myelin. J Neurosci 15:4488–4495.

Martini R, Zielasek R, Toyka KV, Giese KP, Schachner M (1995): Protein zero (P0)-deficient mice show myelin degeneration in peripheral nerves characteristic of human inherited neuropathies. Nature Genet 11:281286.

Marvin KW, Wataru F, Jetten, AM (1995): Identification and characterization of a novel squamous cell-associated gene related to PMP22. J Biol Chem 270: 28910–28916.

Mirsky R, Jessen KR (1996): Schwann cell development, differentiation and myelination. Cur Opin Neurobiol 6:89–96.

Nave K, Boespflug-Tanguy O (1996): X-linked developmental defects of myelination: From mouse mutants to human genetic diseases. The Neuroscientist 1:33–43.

Nicholson GA, Valentijn LJ, Cherryson AK, Kennerson ML, Bragg TL, DeKroon RM, Ross DA, Pollard JD, McLeod JG, Bolhuis PA, Baas F (1994): A frame shift mutation in the PMP22 gene in hereditary neuropathy with liability to pressure palsies. Nature Genet 6:263–266.

Pareek S, Suter U, Snipes GJ, Welcher AA, Shooter EM, Murphy RA (1993): Detection and processing of peripheral myelin protein PMP22 in cultured Schwann cells. J Biol Chem 268:10372–10379.

Parmantier E, Cabon F, Braun C, D'Urso D, Mueller HW, Zalc B (1995): Peripheral myelin protein-22 is expressed in rat and mouse brain and spinal cord motoneurons. Eur J Neurosci 7:1080–1088.

Patel PI, Roa BB, Welcher AA, Schoener-Scott R, Trask BJ, Pentao L, Snipes GJ, Garcia CA, Francke U, Shooter EM, Lupski JR, Suter U (1992). The gene for the peripheral myelin protein PMP-22 is a candidate for Charcot-Marie-Tooth disease type 1A. Nature Genet 1:159–165.

Reiter LT, Murakami T, Koeuth T, Pentao L, Muzni DM, Gibbs RA, Lupski JR (1996): A recomination hotspot responsible for two inherited peripheral neuropathies is located near a *mariner* transposon-like element. Nature Genet 12:288–297.

Roa BB, Garcia CA, Pentao L, Killian JM, Trask BJ, Suter U, Snipes GJ, Shooter EM, Patel PI, Lupski JR (1993a): Evidence for a recessive *PMP22* point mutation in Charcot-Marie-Tooth disease type 1A. Nature Genet 5:189–194.

Roa BB, Garcia CA, Suter U, Kulpa DA, Wise CA, Mueller J, Welcher AA, Snipes GJ, Shooter EM, Patel PI, Lupski JR (1993b): Charcot-Marie-Tooth disease type 1A: Association with a spontaneous point mutation in the PMP22 gene. N Engl J Med 329:96–101.

Roa BB, Dyck PJ, Marks HG, Chance PF, Lupski JR (1993c): Dejerine-Sottas syndrome associated with point mutation in the peripheral myelin protein 22 *(PMP22)* gene. Nature Genet 5:269–272.

Roa BB, Warner LE, Garcia CA, Russo D, Lovelance R, Chance PF, Lupski JR (1996): Myelin protein zero (*MPZ*) gene mutations in nonduplication type 1 Charcot-Marie-Tooth disease Hum Mutat 7:36–45.

Schaeren-Wiemers N, Valenzuela DM, Frank M, Schwab ME (1995): Characterization of a rat gene, rMAL, encoding a protein with four hydrophobic domains in central and peripheral myelin. J Neurosci 15:5753–5764.

Scherer SS, Deschenes SM, Xu Y-t, Grinspan JB, Fischbeck KH, Paul DL (1995a): Connexin32 is a myelin-related protein in the PNS and the CNS. J Neurosci 15:8281–8294.

Scherer SS, Xu Y-T, Bannerman PGC, Sherman DL, Brophy PJ (1995b): Periaxin expression in myelinating Schwann cells: modulation by axon-glial interactions and polarized localization during development. Development 121:4265–4273.

Sereda M, Griffiths I, Pühlhofer A, Stewart H, Rossner MJ, Zimmermann F, Magyar JP, Schneider A, Hund E, Meinck H-M, Suter U, Nave KA (1996): A rat transgenic model for Charcot-Marie-Tooth disease. Neuron In press.

Snipes GJ, Suter U, Welcher AA, Shooter EM (1992): Characterization of a novel peripheral nervous system myelin protein (PMP22/SR13). J Cell Biol 117:225–238.

Snipes GJ, Suter U, Shooter EM (1993): Human peripheral myelin protein-22 carries the L2/HNK-1 carbohydrate epitope. J Neurochem 61:1961–1964.

Snipes GJ, Suter U (1995a): Molecular anatomy and genetics of myelin proteins in the peripheral nervous system. J Anat 186:483–494.

Snipes GJ, Suter U (1995b). Molecular basis of common hereditary motor and sensory neuropathies in humans and in mouse models. Brain Path 5:233–247.

Spray DC, Dermietzel R (1995): X-linked Charcot-Marie-Tooth disease and other potential gap-junction diseases of the nervous system. Trends Neurosci 18:256–262.

Spreyer P, Kuhn G, Hanemann CO, Gillen C, Schaal H, Kuhn R, Lemke G, Muller, HW (1991): Axon-regulated expression of a Schwann cell transcript that is homologous to a 'growth arrest-specific' gene. EMBO J 10:3661–3668.

Suter U, Welcher AA, Ozcelik T, Snipes GJ, Kosaras B, Francke U, Billings GS, Sidman RL, Shooter EM (1992a). Trembler mouse carries a point mutation in a myelin gene. Nature 356:241–244.

Suter U, Moskow JJ, Welcher AA, Snipes GJ, Kosaras B, Sidman RL, Buchberg AM, Shooter EM (1992b): A leucine-to-proline mutation in the putative first transmembrane domain of the 22-kDa peripheral myelin protein in the trembler-J mouse. Proc Natl Acad Sci USA 89:4382–4386.

Suter U, Welcher AA, Snipes GJ (1993): Progress in the molecular understanding of hereditary peripheral neuropathies reveals new insights into the biology of the peripheral nervous system. Trends Neurosci 16:50–56.

Suter U, Snipes GJ, Schoener-Scott R, Welcher AA, Pareek S, Lupski JR, Murphy RA, Shooter EM, Patel PI (1994): Regulation of tissue-specific expression of alternative peripheral meyelin protein-22 (*PMP22*) gene transcripts by two promoters. J Biol Chem 269:25795–25808.

Suter U, Patel, PI (1994): Genetic basis of inherited peripheral neuropathies. Hum Mutat 3:95–102.

Suter U, Snipes GJ (1995a): Biology and genetics of hereditary motor and sensory neuropathies. Ann Rev Neurosci 18:45–75.

Suter U, Snipes GJ (1995b): Peripheral myelin protein 22: Facts and hypotheses. J Neurosci Res 40:145–151.

Taylor V, Welcher AA, Amgen Est Programm, Suter U (1995): Epithelial membrane protein-1, peripheral myelin protein 22 and lens membrane protein 20 define a novel gene family. J Biol Chem 270:28824–28833.

Taylor V, Suter U (1996): Epithelial membrane protein-2 and epithelial membrane protein-3: Two novel members of the peripheral myelin protein 22 gene family. Gene In the press.

Tooth HH (1886): The peroneal type of progressive muscular atrophy. London England: HK Lewis.

Valentijn LJ, Baas F, Wolterman RA, Hoogendijk JE, Bosch NHA, Zorn I, Gabreels-Festen AAWM, deVisser M, Bolhuis PA (1992): Identical point mutations of PMP-22 in Trembler-J mouse and Charcot-Marie-Tooth disease type 1a. Nature Genet 2:288–291.

Welcher AA, Suter U, De Leon M, Snipes GJ, Shooter EM (1991): A myelin protein is encoded by the homologue of a growth arrest-specific gene. Proc Natl Acad Sci USA 88:7195–7199.

Windebank AJ (1993): Inherited recurrent focal neuropathies. In Dyck PJ, Thomas PK, Griffin JW, Low PA, Poduslo JF (eds): "Peripheral neuropathy" Philadelphia: Saunders WB, pp 194–1136.

Zoidl G, Blass-Kampmann SD, 'Urso D, Schmalenbach C, Müller HW (1995): Retroviral-mediated gene transfer of the peripheral myelin protein PMP22 in Schwann cells: modulation of cell growth. EMBO J 14:1122–1128.

6

MUTATIONS OF THE PROTEOLIPID PROTEIN GENE

A Molecular Mechanism of CNS Dysmyelination

M. Klugmann,[1] M. H. Schwab,[1] M. Jung,[1] A. Pühlhofer,[1] A. Schneider,[1]
F. Zimmermann,[1] I. R. Griffiths,[2] and K.-A. Nave[1]

[1]Zentrum für Molekulare Biologie (ZMBH)
University of Heidelberg, Germany
[2]Department of Veterinary Clinical Studies
University of Glasgow, United Kingdom

INTRODUCTION

Proteolipid protein (PLP) is the most abundant integral membrane protein of myelin in central nervous system (CNS), but the normal function of this myelin component has been difficult to define. The primary structure of PLP has been determined by direct protein sequencing and through cDNA cloning: PLP comprises 276 residues and its primary translation product lacks a N-terminal signal peptide (Milner et al., 1985 and references herein). Four highly hydrophobic stretches are likely to constitute membrane-spanning domains and the most widely accepted topological model of PLP assumes that both the N- and C-terminus are located at the cytoplasmic membrane surface. Such a model of PLP as a "4 helix bundle" protein is supported by both theoretical considerations and experimental data (Popot et al., 1991; Weimbs and Stoffel, 1992). By alternative mRNA splicing a second PLP isoform, termed DM20, is generated which lacks 35 residues from the intracellular loop region (Nave et al., 1987a). DM20 is less abundant than PLP in compacted myelin but constitutes the more prevalent isoform in early CNS development (Dickinson et al., 1996) and in some non-glial cells (the function of proteolipids outside the nervous system is not known). The PLP/DM20 primary structure is highly conserved in evolution and is 100% identical in mouse and man, suggesting that PLP engages in multiple protein-protein interactions with little tolerance to evolutionary changes. More recently, DM20 has emerged as the prototype of a new protein family that includes two neuronal membrane proteins, M6A (EMA) and M6B (Yan et al., 1993).

Cell Biology and Pathology of Myelin, edited by Juurlink et al.
Plenum Press, New York, 1997

PLP GENETICS

The interest in the function of PLP/DM20 has risen when mutations were identified in the murine PLP gene that cause CNS dysmyelination and abnormal motor development. We have identified two of these mutants which differ in phenotype depending on the nature of the specific molecular defect. In the *rumpshaker* mouse, a point mutation (Ile186->Thr) causes hypomyelination of the CNS without a reduction in the number of myelin-forming oligodendrocytes (Schneider et al., 1992), and the overall phenotype of this mutant is mild. In the more severely affected *jimpy* mouse, a point mutation at the end of intron 4 causes a defect of PLP mRNA splicing, the abnormal loss of exon 5, and a frameshift of translation (Nave et al., 1986; Hudson et al., 1987; Nave et al., 1987b). *Jimpy* mice are nearly myelin-deficient as most oligodendrocytes die prematurely with morphological features of apoptosis (Skoff, 1995 for review). In *jimpy*, less than 5% of the normal number of CNS axons become electrically insulated, and all "myelinated" axons are thinly wrapped and myelin membranes are ultrastructurally abnormal (Duncan et al., 1989). Whereas *rumpshaker* mice are viable, *jimpy* animals die in the fourth or fifth week of age, usually with severe seizures.

PLP mutations in the mouse cover a wide spectrum of phenotypical changes and provide a good model for Pelizaeus-Merzbacher disease (PMD; MIM No. 260600) in humans, likewise associated with mutations of the PLP gene (reviewed in Hodes et al., 1994; Nave and Boespflug-Tanguy, 1996). By clinical criteria, PMD has been further classified into the severe *connatal* form (early onset, rapid progression) and a less severe *classical* form (slow progression), as well as "transitional forms". Recent genetic evidence suggests that even X-linked spastic paraplegia type-2 (SPG-2), which is clinically distinct, constitutes a relatively mild form of PMD, associated with specific PLP mutations. In fact, the index family of this disease carries the same point mutation and amino acid substitution that defines *rumpshaker* in mice (Kobayashi et al., 1994). A mild dysmyelination was also reported for a new mouse mutant, plp^{neo}, defined by a targeted manipulation of the PLP gene (here the neomycin resistance gene is located in antisense orientation in intron 3 of the PLP gene). Plp^{neo} mice have clear motor deficits which are presumably caused by dysmyelination of small caliber axons and abnormal myelin compaction of large fibers (Boison and Stoffel, 1994; Boison et al., 1995).

Taken together, mutations of the PLP gene result in a myelin pathology of varying severity, probably in a continuum of disease expression. To establish a clear genotype-phenotype correlation has been difficult, because even conservative substitutions (Gencic and Hudson, 1990) can have the same severe consequences on oligodendrocyte survival and myelination as the *jimpy* mutation which results in a severely truncated myelin protein.

To genetically rescue PLP mutant mice by transgenic complementation, we have reintroduced the wild-type PLP gene into the mouse germ line (Schneider et al., 1995). These experiments were carried out to determine whether heterozygous oligodendrocytes are affected and to test for possible gain-of-function effects (in the CNS of heterozygous *jimpy* females only one PLP gene is expressed due to random X chromosome inactivation). Expression of an autosomal PLP transgene improved significantly the behavioural phenotype of *rumpshaker* mice (Schneider et al., in preparation), but there was no phenotypical improvement of PLP-transgenic *jimpy* mice, despite the fact that the PLP transgene was properly expressed at the RNA and protein level (Schneider et al., 1995). This provided indirect evidence that dysmyelination in *jimpy* is not simply a loss of PLP function effect and that mutant PLP is possibly by itself partly responsible for the *jimpy* phenotype.

In the course of these transgenic complementation studies it was further observed that (in the absence of any PLP mutation) the 2–3 fold transcriptional overexpression of a wildtype PLP gene is sufficient to cause a dysmyelinated phenotype in mice, quite similar to the spontaneous mouse mutants (Readhead et al., 1994; Kagawa et al., 1994). Also two-fold overexpression of human PLP/DM20 turns the wildtype gene into a disease gene, as PLP gene duplications have now been found in about 50% of all familial cases (reviewed in Nave and Boespflug-Tanguy, 1996). Thus, glial cells are highly sensitive to the structural alterations of PLP and its expression level, and there is a genetically-dominant effect of protein misfolding on apoptotic oligodendrocyte death and dysmyelination.

A Complete *Null* Allele of the Mouse PLP Gene

Transgenic studies have suggested that the most severe forms of dysmyelination in mutant mice could be a combination of gain- and loss-of-function effects. We have tested our working hypothesis that mutations in the PLP gene cause abnormal oligodendrocyte death secondary to the expression of misfolded proteins. By homologous recombination in mouse embryonic stem cells, we have generated transgenic mice which completely lack expression of PLP or related polypeptides (*plp^{null}* allele). To this end, we constructed a gene targeting vector which eliminates, after homologous recombination, the translation start site for PLP and its alternatively-spliced DM20 isoform. The initiation codon, located on exon 1 (Ikenaka et al., 1988), and a 2.5 kb fragment of the first intron were replaced by the neomycine (neo) resistance gene as shown in Fig 1. Unexpectedly, these mice reveal a normal phenotype, on gross inspection and show none of the behavioural characteristics of naturally occuring PLP mutants. Hemizygous (plp^{null}/Y) mutants reproduce normally and

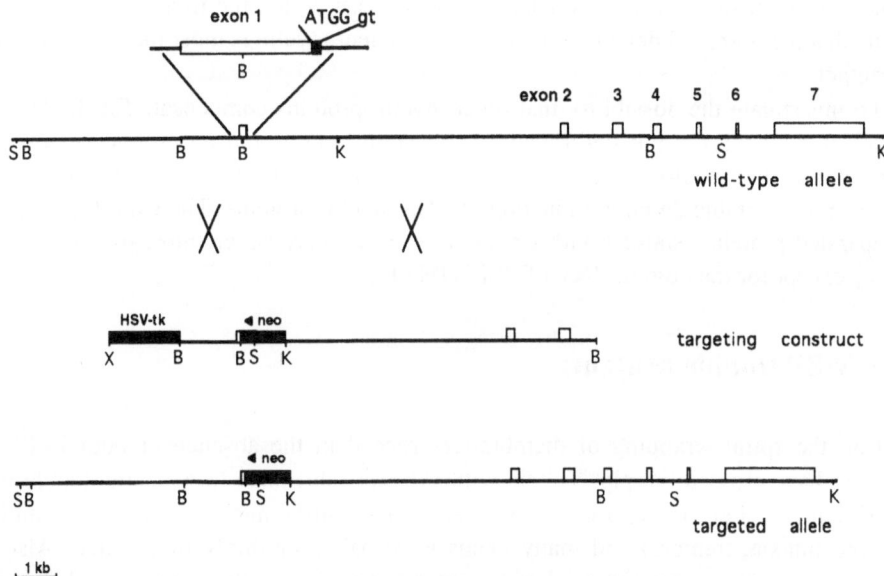

Figure 1. Recombination schema for the complete inactivation of the mouse proteolipid protein gene. The wildtype gene structure is shown on top, the targeting construct and the targeted allele below. Homologous recombination deletes the translational initiation codon (ATG) located adjacent to the first exon-intron boundary (see enlargement on top). The neomycin resistance gene (neo) is transcribed in antisense orientation relative to PLP. Restriction sites are B (BamHI), K (KpnI), S (SphI), and X (XhoI). Modified from Klugmann et al. (1997).

are long-lived (Klugmann et al., 1997). Even more specialized locomotor tests, such as a rotarod test, have failed to show a deficit of motor performance.

There are no signs of abnormal cell death in the oligodendrocyte lineage and, contrary to our expectation, CNS myelin is assembled in the PLP null allele as a multilamellar compacted structure (Klugmann et al., 1997), questioning the view that PLP is required to obtain membrane adhesion and myelin compaction. To analyze possible morphological consequences of PLP-deficiency, we also studied heterozygous females, because random X chromosome-inactivation allows to compare side-by-side PLP-deficient and wild-type myelin, processed under identical conditions for immunostaining and electronmicroscopy. In these chimeras, 50% of the myelinated fibers were PLP-positive, the theoretical value after random X-inactivation, demonstrating that PLP-deficient oligodendrocytes survive and ensheath normally in the competitive situation of the chimera. By immunostaining, PLP deficiency did not alter the staining of other myelin proteins, specifically the periaxonal localization of MAG was intact, indicating that this adhesion molecule has not substituted for PLP in compacted myelin (Klugmann et al., 1997). Also astrocytosis was not a feature. At the ultrastructural level, myelin was normally compacted in the absence of PLP/DM20 and the G-ratio of axon to fiber diameter was not altered. However, in many areas the intraperiod line was barely distinguishable from the major dense line, reminiscent of ultrastructural abnormalities in the *md rat* and in *jimpy* mice (Duncan et al., 1989).

Interestingly, in the absence of PLP and DM20, the overall stability of compacted myelin appeared reduced. Perfusion-fixation of tissues may not strictly prevent the artefactual delamination of myelin. We noticed in the analysis of chimeric mice that myelinated fibers which lacked PLP staining were more likely to show delamination artefacts than wildtype fibers. This observation was most striking when brains were poorly fixed, suggesting that PLP and/or DM20 contribute to the physical stability of compacted myelin. There were no signs of in vivo delamination, at least within the first year of life. This suggests that the observed delamination of PLP-deficient myelin is most likely a post-fixation artefact.

To investigate the possibility that other myelin proteins compensate for the lack of PLP/DM20, we analyzed purified myelin by Western blotting but no protein emerged at the abundance that would be sufficient to compensate for the lack of PLP/DM20, and there was no detectable de novo induction of PNS myelin proteins. The overall pattern of size-separated proteins, stained with Coommasie Blue or by silver impregnation, is also the same except for the obvious lack of PLP/ DM20.

PLP * MBP Double Mutants

Can the spiral wrapping of membranes proceed in the absence of both PLP and MBP? We have crossed the plp^{null} allele with *shiverer* (shi) mice to generate double-mutants (Klugmann et al., 1997). Both shi/shi and shi/shi*plp-/Y mice showed a very similar phenotype (ataxia, tremors) and many axons were naked or thinly myelinated. Also in common with *shiverer*, myelin of double mutants was frequently decompacted, lacking the major dense line. However, the intraperiod line was identifiable and often increased in density. All double mutants were long-lived (>6 months) whereas shi/shi mice died at about three months of age. The morphological basis for this finding awaits to be identified.

Ectopic Expression of Myelin Proteins

The cellular mechanisms which underlie dysmyelination and premature oligodendrocyte cell death, when triggered by mutations or overexpression of the PLP/DM20 gene, appear to be complex. We have therefore studied some aspects of PLP expression in a "simple" cell culture system. When cloned cDNAs for PLP or DM20 are overexpressed in transiently transfected COS-7 cells, using a strong viral promoter, the encoded proteins can be immunostained on the cell surface and do not appear to immediately kill the host cell (Jung et al., 1996). PLP and DM20 polypeptides derived from murine PLP mutants or from PMD patients are largely retained in the endoplasmic reticulum of the cell and fail to be surface expressed (Gow et al., 1994). Using the conformation-sensitive monoclonal antibody O10, we have obtained direct immunological evidence that the overall structure of PLP and DM20 in mutant mice is significantly altered (Jung et al., 1996). Recently, it has been suggested that the ability of mutant DM20, but not PLP, to exit the endoplasmic reticulum of transfected COS7 cells and to reach the cell surface correlates with the clinical severity of the corresponding mutation in human PMD (Gow and Lazzarini, 1996). Our *in vivo* data suggest that the lack of functional DM20 from the oligodendrocyte cell surface is unlikely to be by itself be the cause of dysmyelination. More likely, proteolipids that are retained inside the cell interfere with other critical functions necessary for oligodendrocyte survival and normal myelin assembly.

DISCUSSION

It has become increasingly clear that the pleiotropic defects of myelination in *jimpy* mice and other rodent mutants (modelling Pelizaeus-Merzbacher disease) result from the complicated interaction of loss-of-function and gain-of-function effects, but the relative contribution of each has been difficult to assess (Nave, 1995; Griffiths, 1996). We have

Table 1. Comparison of PLP mouse mutants, grouped by increasing clinical severity and the absence or presence of protein coding capacity

PLP mouse mutant	Polypeptide encoded	Abnormal motor development	Dysmyelination of CNS axons	Myelin ultrastructure
plp[null]	none	not detectable	absent	compact (unstable)
plp[neo]	159aa[1]	mild	small caliber axons	loose / compact[4]
plp[rsh]	276aa[2]	moderate	small and large caliber axons	compact
plp[jp]	242aa[1]	severe (ataxia and seizures)	most axons and oligodendrocyte death	compact (very thin)
plp[msd]	276aa[3]			
plp-transgenic	276aa			compact

The number of predicted amino acids (aa) is indicated and can deviate from full-length PLP (276aa) as a result of abnormal RNA splicing[1]. Single amino acid substitutions[2,3] are Ile[186]->Thr and Ala[242]->Val, respectively. PLP is coexpressed with DM20 (not indicated), except in the isoform-specific plp[neo] allele. All steady-state PLP mRNA levels are reduced except for PLP-transgenic mice. Dysmyelination is defined as a missing or very thin axonal ensheathment. For plp[neo] mice different results[4] were reported by Boison et al. (1995) and Rosenbluth et al. (1996). For additional references see text and Nave (1995). Modified from Klugmann et al. (1997)

now uncoupled these two disease processes by completely eliminating expression of a targeted PLP gene in mice, and have compared this new mutant with previously identified alleles. Clearly, mice which have no coding capacity for PLP or PLP-related polypeptides differ from other mutants at the molecular, cellular and behavioural level. Most important, with respect to PLP function, has been the observation that PLP/DM20-deficient oligodendrocytes differentiate completely and spirally enwrap all myelination-competent axons, irrespective of their caliber (Klugmann et al., 1997). Our findings thus question the previously held view that PLP is required to assemble compacted myelin, a model that was self-suggestive given the abundance of this protein. PLP-deficient myelin is compacted in vivo but extra-sensitive to the osmotic stress during perfusion-fixation, yielding numerous "fixation artifacts" and suggesting a stabilizing function of PLP in the myelin architecture. Proteolipids may engage in homophilic interactions and form "struts" that expand the width of the IPL and bridge the extracelluar gap. Such specialized interactions may stabilize compact myelin but require the prior association of the two membranes in order to engage. We propose that these primary junctions are stabilized by secondary PLP/DM20-dependent bonds, a mechanism that has similarities with a "zipper".

Two previously described PLP mutants, plp^{neo} and $dm20^{neo}$, differ from plp^{null} mice at the genetic, cellular and behavioural level (Boison and Stoffel, 1994; Boison et al., 1995). All available PLP mutant mice, spontaneous and engineered ones, suggest that even low amounts of truncated (but PLP-related) polypeptides act as effective inhibitors of myelination. This was independently suggested by the complementation experiments in which the *jimpy* allele had emerged as genetically dominant over an autosomal wild-type PLP transgene (Nadon et al., 1994; Schneider et al., 1995). Comparing the phenotype of mutant alleles (Table 1) with the ability of each to encode a gene product, we conclude that dysmyelination in mice, as a developmental disorder affecting myelin synthesis, is dependent on the expression of PLP or PLP-related polypeptides. In the order of increasing pathological severity, the consequences of abnormal PLP gene expression are (1) the inability of oligodendrocytes to recognize and myelinate small-caliber axons, (2) the premature arrest of myelination with abnormally thin sheaths, (3) the failure to recognize and myelinate large-caliber axons, and (4) oligodendrocyte death with features of apoptosis.

The phenotype of PLP-deficent mice appears to contradict corresponding findings in humans. Raskind et al. (1991) have reported a case with Pelizaeus-Merzbacher disease (PMD) and a deletion of the entire PLP gene, which demonstrates that PLP-deficiency is not tolerated in man. However, symptoms of the disease were not obvious after birth (as in connatal PMD), and the patient has survived for over forty years (T. Bird, personal communication). A similar case of a human PLP null mutation has been studied by O. Boespflug-Tanguy et al. (pers. communication). In humans, the lack of PLP is not lethal but leads to a "mild" form of PMD with later onset and slower progression. We hypothesise that in the absence of proteolipids, myelin is at a high risk of premature break-down and degenerates within a few years. Such a disease mechanism has clearly features of a developmental disorder in humans, because it interferes with normal motor development in children. In contrast, the plp^{null} allele of the mouse has no functional consequences within the reproductive age, but may become manifest as a degenerative disease in aged animals. It is anticipated that this issue of species-specific disease progression will be relevant also for other mouse models of human genetic disorders.

ACKNOWLEDGMENTS

Research in the authors' lab was supported by grants from the Deutsche For-schungsgemeinschaft (SFB317), and the European Biomed-2 Program. Address for corre-spondence: Dr. Klaus-Armin Nave, University of Heidelberg, ZMBH, Im Neuenheimer Feld 282, D-69120 Heidelberg, Germany (e-mail: nave@sun0.urz.uni-heidelberg.de). We thank O. Boespflug-Tanguy for helpful discussions.

REFERENCES

Boison D and Stoffel W (1994): Disruption of the compacted myelin sheath of axons of the central nervous system in proteolipid protein-deficient mice. Proc. Natl. Acad. Sci. (USA) 91: 11709–11713.

Boison D Büssow H D'Urso D Müller HW and Stoffel W (1995): Adhesive properties of proteolipid protein are responsible for the compaction of CNS myelin sheaths. J. Neurosci. 15: 5502–5513.

Dickinson PJ Fanarraga ML Griffiths IR Barrie JA Kyriakides E and Montague P (1996): Oligodendrocyte pro-genitors in the embryonic spinal cord express DM-20. Neuropath. Appl. Neurobiol. 22:188–198.

Duncan ID Hammang JP Goda S and Quarles RH (1989): Myelination in the jimpy mouse in the absence of pro-teolipid protein. Glia 2: 148–154.

Gencic S and Hudson LD (1990): Conservative amino acid substitution in the myelin proteolipid protein of jim-py^md mice. J. Neurosci. 10: 117–124.

Gow A Friedrich VL and Lazzarini RA (1994): Many naturally occuring mutations of myelin proteolipid protein impair its intracellular transport. J. Neurosci. Res. 37: 574–583.

Gow A and Lazzarini, RA (1996): A cellular mechanism governing the severity of Pelizaeus-Merzbacher disease. Nat. Genet. 13: 422–428.

Griffiths IR (1996): Myelin mutants: model systems for the study of normal and abnormal myelination. Bioessays 18: 789–797.

Hodes ME Pratt VM and Dlouhy SR (1994): Genetics of Pelizaeus-Merzbacher disease. Dev. Neurosci. 15: 383–394.

Hudson LD Berndt JA Puckett C Kozak CA and Lazzarini RA (1987): Aberrant splicing of proteolipid protein mRNA in the dysmyelinating jimpy mutant mouse. Proc. Natl. Acad. Sci. (USA) 84: 1454–1458.

Ikenaka K Furuichi T Iwasaki Y Moriguchi A Okano H and Mikoshiba K (1988): Myelin proteolipid protein gene structure and its regulation of expression in normal and *jimpy* mutant mice. J. Mol. Biol. 199: 587–596.

Jung M Sommer I Schachner M and Nave K-A (1996): Monoclonal antibody O10 defines a conformationally sen-sitive cell-surface epitope of proteolipid protein (PLP): evidence that PLP misfolding underlies dysmyeli-nation in mutant mice. J. Neurosci. 24: 7920–7929.

Kagawa T Ikenaka K Inoue Y Kuriyama S Tsujii T Nakao J Nakajima K Aruga J Okano H and Mikoshiba K (1994): Glial cell degeneration and hypo- myelination caused by overexpression of myelin proteolipid pro-tein gene. Neuron 13: 427–442.

Klugmann M Schwab MH Pühlhofer A Schneider A Zimmermann F Griffiths IR and Nave K-A (1997): Assembly of CNS myelin in the absence of proteolipid protein. Neuron (in press).

Kobayashi H Hoffman EP and Marks HG (1994): The *rumpshaker* mutation in spastic paraplegia. Nat. Genet. 7: 351–352.

Milner RJ Lai C Nave K-A Lenoir D Ogata J and Sutcliffe JG (1985): Nucleotide sequences of two mRNAs for rat brain myelin proteolipid protein. Cell 42: 931–939.

Nadon N Arnheiter H and Hudson LD (1994): A combination of PLP and DM20 transgenes promotes partial myelination in the jimpy mouse. J. Neurochem. 63: 822–833.

Nave K-A Lai C Bloom FE and Milner RJ (1986): Jimpy mutant mouse: a 74-base deletion in the mRNA for mye-lin proteolipid protein and evidence for a primary defect in RNA splicing. Proc. Natl. Acad. Sci. (USA) 83: 9264–9268.

Nave K-A Lai C Bloom FE and Milner RJ (1987a): Splice site selection in the proteolipid protein (PLP) gene tran-script and primary structure of the DM-20 protein of central nervous system myelin. Proc. Natl. Acad. Sci. (USA) 84: 5665–5669.

Nave K-A Bloom FE and Milner RJ (1987b): A single nucleotide difference in the gene for myelin proteolipid pro-tein defines the jimpy mutation in mouse. J. Neurochem.49: 1873–1877.

Nave K-A (1995): Neurological mouse mutants: a molecular genetic analysis ofmyelin proteins. in: Neuroglia
    (Eds.: H Kettenmann and B Ransom),
Oxford University Press, New York, pp. 571–586.
Nave K-A and Boespflug-Tanguy O (1996): Developmental defects of myelin formation: from X-linked mutations
    to human dysmyelinating diseases. The Neuroscientist 2: 33–43.
Popot J Pham-Dinh D and Dautigny A (1991): Major myelin proteolipid: the 4-alpha- helix topology. J. Membr.
    Biol. 120: 233–246.
Raskind WH Williams CA Hudson LD and Bird TD (1991): Complete deletion of the proteolipid protein gene
    (PLP) in a family with X-linked Pelizaeus-Merzbacher disease. Am. J. Hum. Genet. 49: 1355–1360.
Readhead C Schneider A Griffiths IR and Nave K-A (1994): Premature arrest of myelin formation in transgenic
    mice with increased proteolipid protein gene dosage. Neuron 12: 583–595.
Rosenbluth J Stoffel W and Schiff R (1996): Myelin structure in proteolipid protein (PLP)-null mouse spinal cord.
    J. Comp. Neurol. 371: 336–344.
Schneider A Montague P Griffiths I Fanarraga M Kennedy P Brophy P and Nave K-A (1992): Uncoupling of hy-
    pomyelination and glial cell death by a mutation in the proteolipid protein gene. Nature 358: 758–761.
Schneider A Readhead C Griffiths IR and Nave K-A (1995): Dominant-negative action of the *jimpy* mutation in
    mice complemented with an autosomal transgene for myelin proteolipid protein. Proc. Natl. Acad. Sci.
    (USA) 92: 4447–4451.
Skoff RP (1995): Programmed cell death in the dysmyelinating mutants. Brain Path. 5: 283–288.
Weimbs T and Stoffel W (1992): Proteolipid protein (PLP) of CNS myelin: positions of free, disulfide-bonded,
    and fatty acid thioester-linked cysteine residues and implications for the membrane topology of PLP. Bio-
    chemistry 31: 12289- 12296.
Yan Y Lagenaur C and Narayanan V (1993): Molecular cloning of M6: identification of a PLP/DM-20 gene fam-
    ily. Neuron 11: 423–431.

7

MYELIN PROTEINS AS MEDIATORS OF SIGNAL TRANSDUCTION

Charissa A. Dyer

The Children's Hospital of Philadelphia
Department of Neurology
Abramson Research Center
34th and Civic Center Blvd.
Philadelphia, Pennsylvania 19104

INTRODUCTION

Myelin is a specialized membrane that wraps multiple times around an axon and is essential for the proper conduction of action potentials. In the central nervous system (CNS), myelin is produced by oligodendrocytes. Because of the difficulty in studying myelin assembly in vivo, the elaboration of the myelin sheath is poorly understood. Current knowledge of the cytoarchitecture of oligodendrocyte membrane sheets has been obtained largely by studying oligodendrocytes in culture. The focus of this review is to discuss specific signaling pathways that may play key regulatory roles in the development and maintenance of the myelin membrane sheet.

During the initial stages of myelination in vivo, oligodendrocytes extend multiple processes, and from the distal tip of each process, a membrane sheet is produced which wraps around an axon (for review, see Pfeiffer et al., 1994). The membrane sheet is regulated in size and shape, and therefore, the assembly of a specialized cytoskeleton is expected to be essential for the production of the sheet. As myelination proceeds, much of the cytoplasm between the intracellular faces of the membrane sheet is extruded, resulting in a compact, multilamellar structure. The removal of cytoplasm is likely to be an energy dependent process regulated by signaling systems, since compaction occurs following axonal contact. Compact CNS myelin contains a network of interlamellar junctions called radial component which extend from the outer most wrap to the inner most wrap of membrane (Kosaras and Kirschner, 1990). Radial component has been isolated and shown to be enriched in proteins that are also found in the cytoskeletal veins of cultured oligodendrocyte membrane sheets, i.e. tubulin, actin, 2′,3′-cyclic nucleotide 3′-phosphohydrolase (CNPase) and certain isoforms of myelin basic protein (MBP) (Karthigasan et al., 1994). These data suggest that radial component in myelin serves as a cytoskeletal scaffolding structure. Furthermore, these data indicate that the study of the organization of oligoden-

Cell Biology and Pathology of Myelin, edited by Juurlink *et al.*
Plenum Press, New York, 1997

drocyte membrane sheets in culture may be beneficial for understanding how myelin is assembled.

CYTOARCHITECTURE OF CULTURED OLIGODENDROCYTE
MEMBRANE SHEETS

Cultured oligodendrocytes provide an excellent system in which to study the cytoarchitecture of membrane sheets. Oligodendrocytes appear to develop in the same manner in
vivo and in vitro. In vitro, oligodendrocytes activate the same sets of genes during their
various stages of development in the same time frame as they do in vivo (Pfeiffer et al.,
1994). Oligodendrocytes also produce membrane sheets in culture; this process does not
depend on the presence of neurons and their axons (Mirsky et al., 1980; Sarlieve et al.,
1980; Szuchet et al., 1980; Barbarese and Pfeiffer, 1981; Knapp et al., 1987; Rome et al.,
1986). These sheets are flat and large, with diameters as great as 200 μm.

Fully assembled sheets consist of a membrane stretched over a unique cytoskeleton.
The membrane surface components galactocerebroside (GalC) and myelin/oligodendro-
cyte specific protein (MOSP) are normally expressed uniformly over the entire surface of
the sheets (Figure 1a) (Dyer and Benjamins, 1988a; Dyer and Matthieu, 1994). Large cytoplasmic microtubular vein-like structures extend from the cell body into the membrane
sheet where they branch into a lacy network of smaller veins (Figure 1b) (Dyer and Benjamins, 1989a). Colocalized along these microtubular structures are actin filaments and
CNPase (Figure 1c,d) (Dyer and Benjamins, 1989a). The microtubule/F-actin/CNPase
vein-like structures surround areas, or domains, of MBP (Figure 1g,h) (Dyer and Benjamins, 1988b).

Studies of MBP-deficient shiverer oligodendrocytes suggest that MBP is required
for the normal assembly of the membrane sheet (Dyer et al., 1995). Although microtubules are present in shiverer oligodendrocyte membrane sheets, they are abberently organized. Also, these studies showed that MBP is essential for the assembly of actin into
filaments and for the normal colocalization of CNPase along microtubules (Dyer et al.,
1997).

SIGNALING SYSTEMS THAT REGULATE CYTOSKELETON IN
CULTURED OLIGODENDROCYTE MEMBRANE SHEETS

Specific antibodies reactive with myelin surface membrane markers, hypothesized
to mimic the action of endogenous ligands in the brain, have been used in oligodendrocyte
binding studies to obtain an understanding of possible in vivo regulatory mechanisms controlling cytoskeleton assembly. Such binding studies have revealed the existence of two
signaling pathways that bring about distinct changes in cytoskeleton in cultured oligodendrocyte membrane sheets (for review, see Dyer, 1993).

The interaction of specific antibodies with GalC, a membrane surface glycolipid,
triggers a signaling pathway that is mediated by MBP (Dyer et al., 1994); a cascade of
events subsequently occurs. Within minutes after antibody binding, a sustained influx of
calcium occurs, and oligodendrocytes acquire an increase of about 250 nM in intracellular
calcium (Dyer and Benjamins, 1990). The increase in intracellular calcium appears to be
maintained for as long as the antibody is bound to the cell. The membrane surface anti-

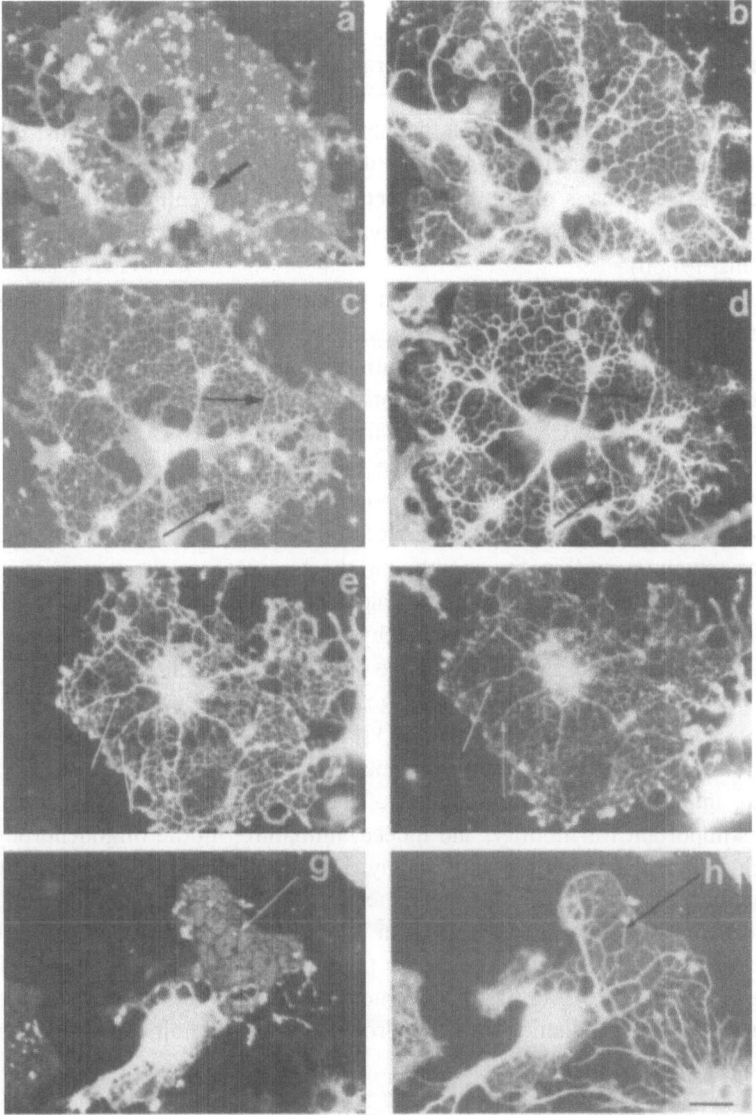

Figure 1. Cytoarchitecture of cultured oligodendrocyte membrane sheet. a) Membrane surface GalC staining (arrow at cell body). b) Same cell shown in a double stained for polymerized tubulin; note the lacy network of microtubular structures throughout the membrane sheets. c) Oligodendrocyte immunostained for internal CNPase. d) Same cell shown in c double stained for polymerized tubulin. Note that CNPase is colocalized along microtubular structures (arrows). e) Oligodendrocytes immunostained for CNPase. f) Same cell in f double stained for filamentous actin (F-actin). Note that F-actin and CNPase are colocalized in vein-like structures (arrows). g) Oligodendrocyte immunostained for cytoplasmic MBP. h) Same cell in g double stained for polymerized tubulin. Note that microtubular veins (arrows) surround domains of MBP (arrows). Bar = 20 μm. (Reproduced with permission from Dyer and Benjamins, 1989a,b and Dyer et al., 1988b)

body:GalC complexes redistribute into areas that directly overlie cytoplasmic MBP domains within the sheets (Dyer and Benjamins, 1989a,b). This is followed by calcium-dependent depolymerization of microtubules and fusion of MBP domains (Dyer and
Benjamins, 1990). Actin filaments do not disassemble during this process (Dyer, unpublished observations). At this point, the membrane sheets in vitro resemble unrolled compact myelin sheets in vivo, since cytoplasmic channels that normally surround MBP
domains in the cultured oligodendrocyte membrane sheets are apparently lost. These data
suggest that activation of the GalC/MBP pathway may be important for *in vivo* compaction of myelin. After 3 days of continuous activation of this pathway, membrane sheets
contract; this process is reversible upon removal of the antibody from the medium (Dyer
and Benjamins, 1988b).

In addition to the GalC/MBP pathway, a second signaling pathway has been characterized in cultured wild type oligodendrocytes. The MOSP/CNPase pathway, when activated, also results in a cascade of events. Antibody binding triggers a transient influx of
calcium (Dyer, 1993). Membrane surface anti-MOSP:MOSP complexes redistribute directly over internal F-actin/CNPase/microtubular structures (Dyer and Matthieu, 1994),
and subsequently microfilaments disassemble throughout the oligodendrocytes (Dyer et
al., 1997). However, microtubules do not depolymerize and membrane sheets remain extended. After 3 days of continuous activation of this pathway, the width and number of the
microtubular structures within the membrane sheets increases dramatically (Dyer and
Matthieu, 1994), and the amount of unassembled actin diminishes (Dyer et al., 1997). Finally, the number of long processes produced by the oligodendrocytes also increases.

Thus, these two signaling pathways have opposite effects on the oligodendrocyte cytoskeleton, i.e. the GalC/MBP pathway leads to depolymerization of microtubules and stabilization of actin filaments, while the MOSP/CNPase pathway results in growth of
microtubules and disassembly of actin filaments. It is conceivable that these two signaling
pathways are important in the development of the oligodendrocytes. An actin based cytoskeleton is important for motility, and indeed, immature, motile oligodendrocytes have
an actin-rich cytoskeleton (Wilson and Brophy, 1989). As oligodendrocytes mature, actin
filament number appears to decrease and their cytoskeleton becomes microtubule-rich
(Wilson and Brophy, 1989). A critical step in oligodendrocyte development is the exten-
tion of processes, which is dependent on the polymerization of microtubules. If 1) these
two pathways play key roles in the development of myelin, and 2) they do not function
properly during myelin production, oligodendrocyte/myelin pathology would be expected.

THE ROLE OF CYTOSKELETON IN
OLIGODENDROCYTE/MYELIN PATHOLOGY

It should be possible to predict the source of a cytoskeletal defect in oligodendrocytes based upon our knowledge of the regulation of cytoskeletal assembly. For example,
the absence of MOSP or CNPase should produce an oligodendrocyte in which the
GalC/MBP pathway is dominant. These oligodendrocytes would assemble actin filaments,
but growth of microtubules would be impaired; the expectant phenotype would be enlarged cells that extend short processes and no membrane sheets. A similar phenotype
would be expected of oligodendrocytes that express MOSP and CNPase, but over-express
MBP, thus causing constitutive activation of the GalC/MBP pathway. On the other hand,
the absence of MBP would result in oligodendrocytes in which the MOSP/CNPase pathway was unopposed. In these oligodendrocytes, unchecked microtubule growth and actin

disassembly would be expected to result in cells that extend multiple lengthy processes and no membrane sheets. Another expected feature would be the presence of multiple nuclei within individual cells, since disassembly of actin filaments in dividing cells results in the inability to form the contractile ring, a structure essential for cytokinesis (Satterwhite and Pollard, 1992). A similar phenotype would be expected in oligodendrocytes which over-express CNPase.

Currently, no known mouse mutant has been identified that lacks MOSP or CNPase, or which overexpresses MBP, and therefore it is not possible to test our hypothesis that in these mice, oligodendrocytes would be unable to extend lengthy processes. However, mouse mutants have been identified which lack MBP, the mld and the shiverer mouse. We have examined the phenotype of oligodendrocytes from the shiverer mouse both in vivo and in vitro. The predominant phenotype is similar our predicted phenotype, i.e. cells able to extend processes but not membrane sheets (Dyer et al., 1995). Multinucleated cells were also identified (Dyer, unpublished observations).

Interestingly, although only one phenotype of MBP-deficient oligodendrocytes is expected, oligodendrocytes from shiverer mice exhibit many phenotypes (Dyer et al., 1995; Dyer, unpublished observations). For example, subpopulations of shiverer oligodendrocytes exist which have extreme phenotypes, i.e. they either have blebbed cell membranes and do not extend processes (Dyer, unpublished observations) or they produce membrane sheets that contain normally organized cytoskeleton (Dyer et al., 1995). The mechanism underlying the production of these phenotypes is unknown; however, one theory may account for these findings. It may be that subpopulations of oligodendrocytes differentially regulate gene expression of a protein(s) that can substitute for MBP. One likely substitute protein for MBP is golli-mbp, which can express exons 1 and/or 2 from the "classic" MBP gene (Roach et al., 1983; Takahashi et al., 1985; Molineaux et al., 1986; Campagnoni et al., 1993). Indeed, evidence suggests that golli-mbp is synthesized in some MBP-deficient shiverer oligodendrocytes in vivo (Campagnoni et al., 1993). Since the association of CNPase with microtubules in oligodendrocytes is dependent upon the presence of MBP domains (Dyer et al., 1995), the increased amount of substitute MBP may lead to increased association of CNPase with microtubules, and the subsequent increased activity of the MOSP/CNPase pathway. Therefore, the shiverer oligodendrocyte phenotype might be a direct result of how much substitute MBP is expressed. In other words, shiverer oligodendrocytes which do not express any substitute MBP protein would be expected not to be able to activate either the GalC/MBP or MOSP/CNPase pathways because they 1) do not express MBP and 2) can not associate CNPase with microtubules. These cells would be expected to have enlarged somas that bleb, but do not extend processes. Those shiverer oligodendrocytes which produce normal membrane sheets would be expected to have synthesized large amount of substitute MBP protein. Studies are currently ongoing to examine these possibilities.

REFERENCES

Barbarese E, Pfeiffer SE (1981) Developmental regulation of myelin basic protein in dispersed cultures. Proc Natl Acad Sci USA 78:1953–1957.

Campagnoni AT, Priby TM, Campagnoni CW, Kampf D, AMur-Umarjee S, Landry CF, Handley VW, Newman SL, Garbay B, Kiramura K (1993) Structure and developmental regulation of Golli-mbp, a 105-kilobase gene that encompasses the myelin basic protein gene and is expressed in cells in the oligodendrocyte lineage in the brain. J Biol Chem 268:4930–4938.

Dyer CA (1993) Novel oligodendrocyte transmembrane signaling systems. Investigations utilizing antibodies as ligands. Mol Neurobiol 7:1–22.

Dyer CA, Benjamins JA (1988a) Redistribution and internalization of antibodies to galactocerebroside by oligodendroglia. J Neurosci 8:883–891.

Dyer CA, Benjamins JA (1988b) Antibody to galactocerebroside alters organization of oligodendroglial membrane sheets in culture. J Neurosci 8:4307–4318.

Dyer CA, Benjamins JA (1989a) Organization of oligodendroglial membrane sheets. I: Association of myelin basic protein and 2',3'-cyclic nucleotide 3'-phosphohydrolase with cytoskeleton. J Neurosci Res 24:201–211.

Dyer CA, Benjamins JA (1989b) Organization of oligodendroglial membrane sheets:II. Galactocerebroside:antibody interactions signal changes in cytoskeleton and myelin basic protein. J Neurosci Res 24:212–221.

Dyer CA, Benjamins JA (1990) Glycolipids and transmembrane signaling: antibodies to galactocerebroside cause an influx of calcium in oligodendrocytes. J Cell Biol 111:625–633.

Dyer CA, Matthieu JM (1994) Antibodies to MOSP and MOG regulate cytoskeletal structure in cultured oligodendrocytes. J Neurochem 62:777–787.

Dyer CA, Philibotte TM, Wolf MK, Billings-Gagliardi S (1994) Myelin basic protein mediates extracellular signals that regulate microtubule stability in oligodendrocyte membrane sheets. J Neurosci Res 39:97–107.

Dyer CA, Philibotte TM, Billings-Gagliardi S, Wolf MK (1995) Cytoskeleton in myelin basic protein-deficient shiverer oligodendrocytes. Dev Neurosci 17:53–62.

Dyer CA, Philibotte TM, Wolf MK, Billings-Gagliardi S (1997) Regulation of cytoskeketon by myelin components: Studies on Shiverer oligodendrocytes carrying an MBP transgene. Dev Neurosci, in press.

Karthigasan J, Kosaras B, Nguyen J, Kirschner DA (1994) Protein and lipid composition of radial component-enriched CNS myelin. J Neurochem 62:1203–1213.

Knapp PE, Bartlett WP, Skoff RP (1987) Oligodendrocyte development in vitro; Double-label immunostaining for galactocerebroside and 2',3'-cyclic nucleotide 3'-phosphohydrolase. J Neurochem. 48S:S30.

Kosaras B, Kirschner DA (1990) Fine-structure and supramolecular organization of the radial component of CNS myelin. In Duncan ID, Skoff RP, Colman D (eds): "Myelination and Dysmyelination." New York:The New York Academy of Sciences, pp430–434.

Mirsky RJ, Winter J, Abney ER, Pruss RM, Gavrilovic J, Raff MC (1980) Myelin-specific proteins and glycolipids in rat Schwann cells and oligodendrocytes in culture. J Cell Biol 84:483–494.

Molineaux SM, Engh H, de Gerra F, Hudson L, Lazzarini RA (1986) Recombination within the myelin basic protein gene created the dysmyelinating shiverer mouse mutation. Proc Natl Acad Sci USA 83:7542–7546.

Patterson GML, Smith CD, Kimura LH, Britton BA, Carmeli S (1993) Action of tolytoxin on cell morphology, cytoskeletal organization, and actin polymerization. Cell Motility and the Cytoskeleton 24:39–48.

Pfeiffer SE, Warrington AE, Bansal R (1994) The oligodendrocyte and its many cellular processes. Trends Cell Biol 3:191–197.

Roach A, Boylan K, Horvath S, Prusiner SB, Hood LE (1983) Characterization of cloned cDNA representing rat myelin basic protein: Absence of expression in brain of shiverer mutant mice. Cell 34:799–806.

Rome LH, Bullock PN, Chiappelli R, Cardwell M, Adinolfi AM, Swanson D (1986) Synthesis of a myelin-like membrane by oligodendrocytes in culture. J Neurosci Res 15:49–65.

Sarlieve LL, Rao GS, Campbell GLEM, Pieringer RA (1980) Investigations on myelination in vitro: Biochemical and morphological changes in cultures of dissociated brain cells from embryonic mice. Brain Res 189:79–90.

Satterwhite LL, Pollard TD (1992) Cytokinesis. Curr Opin Cell Biol 4:43–52.

Szuchet S, Stefansson K, Wollman RL, Dawson G, Arnason BGW (1980) Maintenance of isolated oligodendrocytes in long-term culture. Brain Res 200:151–164.

Takahashi N, Roach A, Teplow DB, Prusiner SB, Hood L (1985) Cloning and characterization of the myelin basic protein gene from mouse; one gene can encode both 14-kD and 18.5-kD MBPs by the alternate use of exons. Cell 42:139–148.

Wilson l, Brophy PJ (1989) Role for the oligodendrocyte cytoskeleton in myelination. J Neurosci Res 22:439–448.

8

CNP IN MYELINATION

Overexpression Alters Oligodendrocyte Morphogenesis

Michel Gravel,[1] Bruce Trapp,[2] John Peterson,[2] and Peter E. Braun[1]

[1]Department of Biochemistry and Department of
    Neurology and Neurosurgery
McGill University
Montreal, Quebec, Canada H3G 1Y6
[2]Department of Neurosciences
Cleveland Clinic Foundation
Cleveland, Ohio 44195

INTRODUCTION

2′,3′-Cyclic nucleotide 3′-phosphodiesterase (CNP) is a critical component of the molecular machinery that mediates early events in myelinogenesis, in particular the elaboration of oligodendrocyte processes that sample the extracellular space and engage axons in order to initiate myelination. This protein is well known for its in vitro enzymatic activity, as yet of no known physiological importance, but which serves as a useful analytical marker for myelination-related membranes in the CNS even though it does not contribute to the structural lattice of myelin lamellae (Braun et al, 1988; Vogel and Thompson, 1988; Trapp et al, 1989; Sprinkle, 1989; Thompson, 1992; Tsukada and Kurihara, 1992; Gravel et al, 1994).

In the developing mammalian CNS the CNP gene is expressed (as mRNA) in cells that are believed to be oligodendrocyte precursors (Scherer et al, 1994). The CNP protein can be detected in immature oligodendrocytes, both in vitro (Pfeiffer et al, 1993) and in vivo (Braun et al, 1988) at the time of, or perhaps before, the appearance of galactocerebroside. Biochemical and cell biological studies have demonstrated several properties of CNP that are shared by protein components of signal transduction cascades. For example, CNP, like many G proteins, is isoprenylated at the C-terminus, a modification that is essential for membrane binding and is, in part, responsible for binding of CNP to the cytoskeletal matrix (Braun et al, 1991; DeAngelis et al, 1994; DeAngelis and Braun, 1994; DeAngelis and Braun, 1996a,b). All known members of the large family of isoprenylated proteins are components of cell signalling pathways.

Cell Biology and Pathology of Myelin, edited by Juurlink *et al.*
Plenum Press, New York, 1997

Additionally, these proteins share with CNP the common feature of C-terminal carboxylmethylation (Cox et al, 1994), a modification that itself may contribute to signal transduction (Stock, 1991; Clarke, 1992; Philips et al, 1993). These shared features only imply that CNP might also belong to this family of cell regulators, but other, more physiological data support our view that CNP participates in regulatory events of myelinogenesis. Although the exact nature of this role is not yet clear several observations allow us to conceptualize a model to assist in the design of experimental strategies and to interpret new evidence for the function of CNP.

We adopted several experimental approaches to address the question of CNP involvement in early stages of oligodendrocyte development. In preliminary experiments, we attempted to interfere with the normal synthesis of CNP with antisense oligonucleotides. Two fifteen-mer oligos, designed to tandemly span the initiation start site of the message for CNP were added to primary cultures of oligodendrocytes, at a stage just before cellular projections appear. Control oligonucleotides representing the sense orientation of the mRNA at the same site did not affect the normal elaboration and growth of filopodia and cellular processes, whereas the antisense oligomers almost completely prevented the oligodendrocyte processes from being extended, with the cells remaining viable and round in appearance (Bernier and Braun, 1991; DeAngelis and Braun, unpublished). Although further experimentation is required to consolidate and extend these findings, our suggestion of a role for CNP in process extension seems justified. Further support for this notion was derived from other investigations in which we transfected several cell types (unrelated to oligodendrocytes) with the cDNA for CNP1 and showed that CNP expressing cells underwent major changes of their normal morphology, and developed networks of filopodia and large processes (DeAngelis et al, 1994; DeAngelis and Braun, 1994). This profound influence of CNP on heterologous cells was dependent on modification of the C-terminus by isoprenylation, demonstrating that the targeting of the protein to specific sites on or near the plasma membrane was a prerequisite for morphological plasticity to occur. In related experiments we showed that CNP is associated intracellularly with the actin-based cytoskeleton (DeAngelis and Braun, 1996a,b), extending earlier observations by us and by others on its intracellular localization (Pereyra et al, 1988; Wilson and Brophy, 1989; Dyer and Benjamins, 1989; Gillespie et al, 1989; Braun et al, 1990).

In an independent approach to the question of how oligodendrocyte process outgrowth is regulated, Yong et al (1994) showed that protein kinase C (PKC) is a mediator of the underlying events, but the protein substrates in the signal cascade were not identified. Since CNP isoform 2 (CNP2) is phosphorylated by PKC (Vartanian et al, 1986) it follows that there may be a linkage of our observations on CNP and those on the involvement of PKC in process outgrowth, strengthening the notion of CNP as a component of an important signal transduction cascade.

In light of the need for an in vivo system in which CNP gene expression could be modified, we elected to generate several lines of transgenic mice that express the human CNP transgene, thereby enabling us to examine the consequence of increased CNP gene dosage on cells and tissues in which CNP is deemed to be important. Our initial description of the effect of elevated CNP gene expression on the CNS (Gravel et al, 1996) supports the observations summarized above, and has suggested new directions in the pursuit of detailed mechanisms by which CNP operates in the whole scheme of myelinogenesis. We present here some recent observations and interpretations in the context of an evolving hypothesis for the function of CNP.

PHENOTYPIC CHARACTERISTICS OF CNP-OVEREXPRESSING MICE

Generation of three lines of mice that express the human CNP transgene, with levels of CNP up to 6-fold greater than the endogenous levels, has been described (Gravel et al, 1996). Western blots of brain homogenates show the relative overexpression of CNP (Figure 1). In this gel system, due to the enormous number of proteins present only mouse isoform 1 (CNP1) is visible; CNP2 is normally only one tenth as abundant in the mouse. Human CNP1 migrates faster than mouse CNP (Waehneldt and Malotka, 1980) and is visible here as a distinct band (lanes 2, 3 and 4), whereas human CNP2 co-migrates with mouse CNP1. The relatively abundant expression of CNP2 in humans is also evident when the transgene is expressed in each of the mouse lines. This pattern of human CNP production remains about the same, when mature animals (90d) are compared to those at 30 days, showing that oligodendrocytes remain generally metabolically robust. Similarly, the near-normal synthesis of other myelin proteins (Gravel et al, 1996) attests to the vitality of the oligodendrocytes.

In normal rodent brain, most of the CNP is tightly associated with myelin and myelin-related membranes; this avid binding is largely dependent on isoprenylation at the C-terminus (Braun et al, 1991; DeAngelis and Braun, 1996a; 1996b). We asked whether the extensive overproduction of CNP in L191 mice, as the human form, might exceed the cellular capacity to isoprenylate the whole pool. Accordingly, we fractionated brain homegenates into soluble and total membrane fractions, and assayed these for CNP enzymatic activity. We found that in both L191 and in normal mice about 98% of CNP could be accounted for in the membranes, demonstrating that even at 6-fold higher rates of synthesis, the pool of isoprenoid precursors is not limiting, and that isoprenyltransferases have the capacity to modify the CNP. Therefore, the developmental aberrations we describe below cannot be attributed to inadequate processing of CNP.

When we examined the expression of CNP, proteolipid protein (PLP) and myelin basic protein (MBP) during development we noted that the timing of expression differs from normal, in that the peak of expression of all three genes occurs about a week earlier than in normal mice (Figure 2). Although we do not yet fully understand this large alteration in the developmentally regulated expression of myelin genes it appears likely that

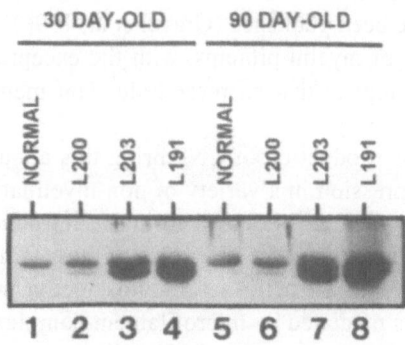

Figure 1. Western blot showing overexpression of CNP in normal and transgenic mice. Protein samples (50 μg) of total brain homogenates from 30 d and 90 d mice were separated by SDS-PAGE (12%), and CNP in electroblots was visualized by enhanced chemiluminescence after exposure to monoclonal antiCNP.

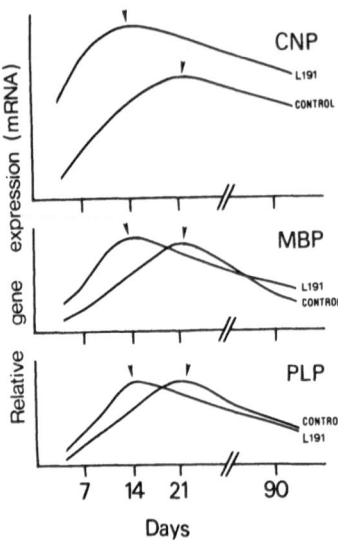

Figure 2. Myelin gene expression is altered in transgenic mice that overexpress CNP. Quantification of mRNA for CNP, MBP and PLP was achieved by densitometric scans of northern blots at different stages of brain development (Gravel et al, 1996), shown here as postnatal ages. Arrows indicate the approximate ages of maximum gene expression.

CNP is a determinant of oligodendrocyte maturation. Anticipating that this phenomenon might be reflected in the in vitro growth characteristics we cultured ODC from neonatal brains of line L191, mice that evinced the highest level of CNP overexpression and noted profound differences in morphology. Whereas normal mouse ODC at day 19 in culture are associated with abundant membrane sheets emanating from relatively few processes, L191 ODC produce a more abundant network of filopodia and cell processes and almost no membrane sheets (Figure 3).

Although the significance of this in the context of myelinogenesis cannot yet be assessed, some insights can be gleaned from other experiments. First, immunocytochemistry of brain sections has revealed massive intramyelinic vacuolation in CNP-overexpressing mice, with abundant evidence of redundant myelin formation (Gravel et al, 1996). Figure 4 shows the presence of typical vacuoles in sub-cortical white matter. Although they are abundant in L191 (homozygous mice that have 18 copies of the human CNP gene) there are fewer of these redundant membrane structures in L203 mice, that have only 6 copies of this gene. The vacuoles and their associated myelin sheaths are visualized by antibodies to both CNP (panels A,C,E) and proteolipid protein (panels B,D,F), demonstrating membrane continuity with myelin. Ultrastructural details of these, that illustrate their redundant myelin characteristics, have been published (Gravel et al, 1996). Given the essentially normal electrophoretic profile of myelin proteins, with the exception of as much as a 600% increase in CNP, it seems logical that all these redundant membrane loops are a consequence of excess CNP.

How might an excess production of CNP bring this about? We think our previous studies of ectopic CNP expression in a variety of non-myelinating cell types (DeAngelis and Braun, 1994) provides a clue. As described in the introduction, expression of CNP induced profound changes in cell surface dynamics of these cells. In light of the evidence for an association of CNP with the actin-based cytoskeleton, we conclude provisionally that one function of CNP is mediated by microfilament complexes near the plasma membrane that are responsible for outgrowth of filopodia and processes. We envision that CNP acts in a manner not unlike that of the widely distributed G protein RhoA, which serves to regulate actin stress fibres at focal adhesions, permitting the cytoskeleton to interact with appropriate integrins that in turn pull against the extracellular matrix to

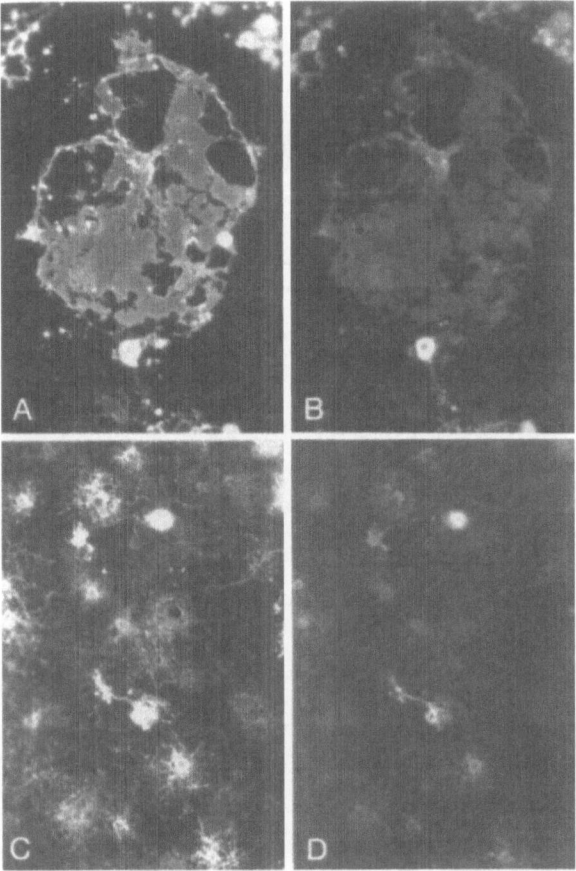

Figure 3. Overexpression of CNP alters the morphology of oligodendrocytes in culture. Primary mixed glial cultures were prepared according to Yong et al, 1991. Briefly, brains from normal mice (A and B) and L191 mice (CNP is 6X normal; C and D) were trypsinized and mechanically dissociated. Size selection by sequential filtration provided enrichment of oligodendrocytes. Cultured cells were observed by immunocytochemistry with anti-galactocerebroside (A and C) and monoclonal antiMBP (B and D).

achieve changes in cell morphology (Bussey, 1996). By extrapolation to myelinating cells in which CNP is constitutively expressed, we surmise that excess CNP results in excessive extensions of filopodia and processes, giving rise to redundant membrane configurations that are ultimately observed as intra-myelinic vacuoles. In vivo studies are underway to verify this.

It is clear that great morphological adaptation must occur in the process of myelination. The mechanisms by which CNP helps to mediate these essential events are not yet clear but we can tender a hypothesis, based on recent advances in the biology of cell surface plasticity and the participation of morphogenic switching mechanisms involving G-proteins. Figure 5 summarizes diagrammatically the known cell biological characteristics of CNP and suggests a scenario to account for the participation of CNP in the dynamics of ODC morphogenesis. Briefly, we know that isoprenylated CNP is targeted to discrete loci

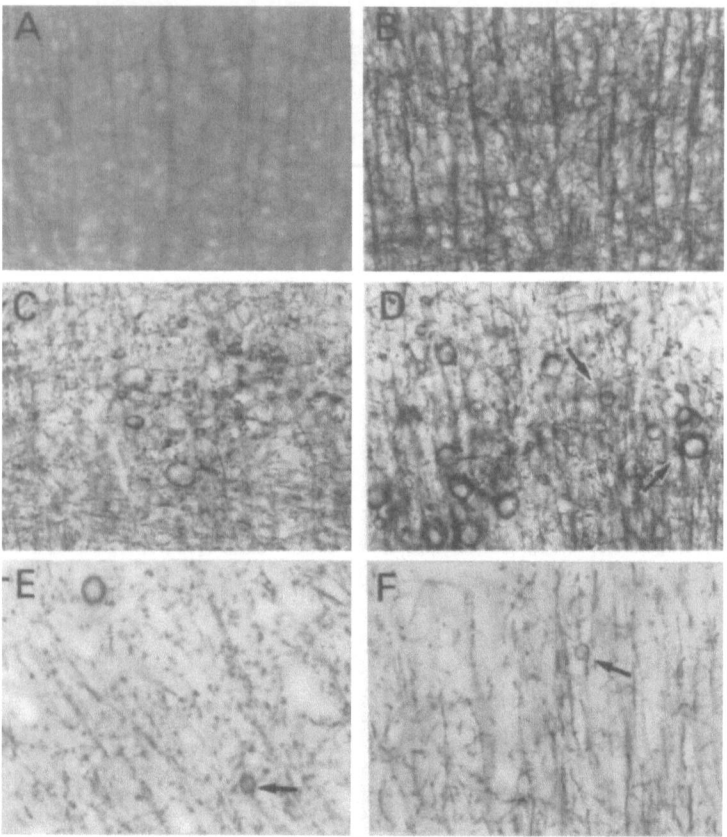

Figure 4. Abnormalities of myelinated axons are revealed by immunocytochemistry of transgenic mice. Sections (30 microns) from sub-cortical brain regions were immunostained with antiCNP (A,C,E) and antiPLP (B,D,F) and revealed by standard peroxidase reactivity. Vacuoles surrounded by PLP or CNP-containing membranes (arrows) are abundant in L191 mice (C,D) and less abundant in L203 mice (E,F) which have fewer copies of the human CNP gene. Detailed characteristics of similar vacuoles have been described (Gravel et al, 1996).

and that it binds to the membrane and to the cytoskeleton. From the known generic mechanisms for morphogenesis in which filopodial extensions are mediated by actin stress fibres recruited with the help of isoprenylated G proteins (e.g. RhoA) to focal adhesions that employ integrins (Takai et al, 1995) we conjecture that CNP contributes to similar, specialized oligodendrocyte complexes whose formation is a necessary prelude to engagement of axons and myelination. We further propose that this is a regulated event which may involve elements of a signal transduction cascade. In this regard it is worth nothing that one CNP isoform (CNP2) is phosphorylated on two serines (ser 9 and ser 22; O'Neill and Braun, to be published) and that both kinases PKA and PKC appear to be involved. When taken together with the well documented observations on the involvement of PKC in ODC process outgrowth (Yong et al, 1994) we believe that CNP is a strong candidate substrate for PKC, and that this phosphorylation may be part of a signal cascade that initiates the early stages of myelinogenesis. Various facets of this hypothesis are currently under investigation.

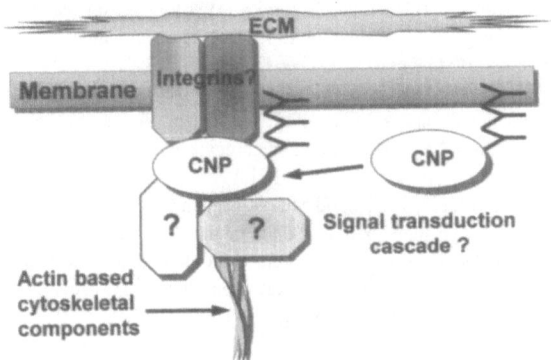

Figure 5. Diagrammatic representation of CNP binding to the membrane and cytoskeleton. The participation of CNP in a signal transduction pathway and its interaction with oligodendrocyte integrins is hypothetical. Protein binding partners (?) are not yet fully characterized.

ACKNOWLEDGMENTS

We thank Vicky Kottis for expert technical assistance, and J.P. Julien and D. Houle, of the McGill Center for Neuroscience Research for their advice and assistance in the production of transgenic mice. We also thank W. Wilcox for her technical help, and Enoch Gao for performing CNPase assays. We acknowledge with gratitude grant support from the MRC and the Multiple Sclerosis Society of Canada (P.E.B.) and the National Institute of Health (NNDS NS29818; B.T.).

REFERENCES

Bernier L and Braun PE (1991): Relationship of CNP to regulated events in myelinogenesis. Trans. Amer. Soc. Neurochem. 22:265.

Braun PE Bambrick LL Edwards AM and Bernier L (1990): 2',3'-cyclic nucleotide 3'-phosphodiesterase has characteristics of cytoskeletal proteins; a hypothesis for its function. Annals N.Y. Acad. Sci. 605:55–65.

Braun PE DeAngelis D Shtybel WW and Bernier L (1991): Isoprenoid modification permits 2'.3'-cyclic nucleotide 3'-phosphodiesterase to bind to membranes. J. Neurosci Res. 30:540–544.

Braun PE Sandillon D Edwards A Matthieu JM and Privat A (1988) Immunocytochemical localization by electron microscopy of 2,3'-cyclic nucleotide 3'-phosphodiesterase in developing oligodendrocytes of normal and mutant brain. J. Neurosci. 8: 3057–3066.

Bussey H (1996): Rho returns: its targets in focal adhesions. Science 273:203.

Clarke S (1992): Protein isoprenylation and methylation at carboxyl-terminal cysteine residues. Annu. Rev. Biochem. 61:355–386.

DeAngelis D and Braun PE (1996a): 2',3'-cyclic nucleotide 3'-phosphodiesterase binds to actin-based cytoskeletal elements in an isoprenylation independent manner. J. Neurochem, in press.

DeAngelis D and Braun PE (1996b): Binding of CNP to myelin: An in vitro study. J. Neurochem. 66:2523–2531.

DeAngelis D Cox M Gao E and Braun PE (1994): Cellular and molecular characteristics of CNP suggest regulatory mechanisms in myelinogenesis. In Salvati S (ed): "A Multidisciplinary Approach to Myelin Diseases II", New York: Plenum Press, pp. 49–58.

Dyer CA and Benjamins JA (1989): Organization of oligodendroglial membrane sheets. I. Association of myelin basic protein and CNP with cytoskeleton. J. Neurosci. Res. 22:201–211.

Gillespie CS Wilson R Davidson A and Brophy PJ (1989): Characterization of a cytoskeletal matrix associated with myelin from rat brain. Biochem. J. 260:689–696.

Gravel M DeAngelis D and Braun PE (1994): Molecular cloning and characterization of rat brain 2',3'- cyclic nucleotide 3'-phosphodiesterase isoform 2. J. Neurosci. Res. 38:243–247.

Gravel M Peterson J Yong VW Kottis V Trapp B and Braun PE (1996): Overexpression of 2',3'-cyclic nucleotide 3'-phosphodiesterase in transgenic mice alters oligodendrocyte development and produces aberrant myelination. Mol. & Cell Neurosci. 7:453–466.

Pereyra PM Horvath E and Braun PE (1988): Triton X-100 extractions of CNS myelin indicate a possible role for the minor myelin proteins in the stability of lamellae. Neurochem. Res. 13:583–595.

Pfeiffer SE Warrington AF and Bansal R (1993): The oligodendrocyte and its many cellular processes. Trends Cel Biol. 3:191–197.

Philips MR Pillinger MH Staud R Volker C Rosenfield MG Weismann G and Stock JB (1993): Carboxyl methylation of ras-related proteins during signal transduction in neutrophils. Science 259:977–980.

Sprinkle TJ (1989): 2',3'-cyclic nucleotide 3'-phosphodiesterase, an oligodendrocyte-Schwann cell and myelin-associated enzyme of the nervous system. CRC Crit. Rev. Neurobiol. 4:235–301.

Stock J (1991): Balancing effector outputs. Current Biology 1:154–156.

Takai Y Sasaki T Tamaka K and Nakanishi H (1995): Rho as a regulator of the cytoskeleton. Trends in Biochem. Sci. 20:227–231.

Thompson RJ (1992): 2',3'-cyclic nucleotide 3'-phosphodiesterase and signal transduction in CNS myelin. Biochem. Soc. Trans. 20:621–626.

Trapp BD Bernier L Andrews SB and Colman D (1988): Cellular and subcellular distribution of CNP and its mRNA in the rat CNS. J. Neurochem. 51:859–868.

Tsukada Y and Kurihara T (1992): 2',3'-cyclic nucleotide 3'-phosphodiesterase: Molecular characterization and possible functional significance. In Myelin: Biology and Chemistry (R.E. Martensen, Ed.), pp. 449–480. CRC Press, Boca Raton, FL.

Vogel US and Thompson RJ (1988): Molecular structure, localization and possible functions of the myelin associated enzyme 2',3'-cyclic nucleotide 3'-phosphodiesterase. J. Neurochem. 50:1667–1677.

Waehneldt TV and Malotka J (1980): Comparative electrophoretic study of the wolfgram proteins in myelin from several mammalia. Brain Res. 189:592–587.

Wilson R and Brophy PJ (1989): Role for the oligodendrocyte cytoskeleton in myelination. J. Neurosci. Res. 22:439–448.

Yong VW Cheung JCB Uhm JH and Kim SU (1991): Age dependent decrease of process formation by cultured oligodendrocytes is augmented by protein kinase C stimulation. J. Neurosci. Res. 29:87–99.

Yong VW Dooley NP and Nobel PG (1994): Protein kinase C in cultured adult human oligodendrocytes: A potential role for isoform alpha as a mediator of process outgrowth. J. Neurosci. Res. 39:83–96.

9

THE ROLE OF THE GAP JUNCTION PROTEIN CONNEXIN32 IN THE MYELIN SHEATH

Steven S. Scherer,[*][1] Linda J. Bone,[2] Suzanne M. Deschênes,[2]
Kenneth Fischbeck,[1] and Rita J. Balice-Gordon[3]

[1]Department of Neurology
[2]Department of Molecular and Cellular Biology
[3]Department of Neuroscience
The University of Pennsylvania Medical Center
Philadelphia, Pennsylvania 19104

SUMMARY

X-linked Charcot-Marie-Tooth disease (CMTX) is caused by mutations in the gap junction gene connexin32 (*Cx32*). To date, 89 different mutations have been found, including deletions, insertions, missense, and nonsense mutations, and at least two mutations in the non-coding region. Different mutations appear to have different effects on the synthesis and localization of Cx32 protein. Both myelinating Schwann cells and oligodendrocytes express Cx32 mRNA and protein, and the localization of Cx32 protein matches the location of putative gap junctions previously seen by freeze-fracture electron microscopy. In myelinating Schwann cells, Cx32 protein is found in incisures and paranodes. Preliminary data suggest that gap junctions form a radial channel that directly traverses the myelin sheath. It is plausible that mutations in *Cx32* disrupt the gap junctions in the myelin sheath, thereby abolishing this radial pathway and leading to peripheral neuropathy.

CHARCOT-MARIE-TOOTH DISEASE

Inherited neuropathies were first described in 1886 by Charcot and Marie and independently by Tooth, and are thus referred to as CMT disease (for a comprehensive discussion, see Dyck et al., 1993). CMT is common, affecting about 1 in 3000 people (Skre, 1974). In most kindreds, the sole manifestation of their disease is peripheral neuropathy; other tissues are unaffected.

---

* Corresponding author, Steven S. Scherer, M.D., Ph.D., Department of Neurology, 460 Stemmler Hall, 36th Street and Hamilton Walk, The University of Pennsylvania Medical Center, Philadelphia, Pennsylvania 19104

The pioneering work of Dyck and Lambert (Dyck and Lambert, 1968a; Dyck and Lambert, 1968b) clarified that there are both demyelinating/hypertrophic and neuronal/ax-onal forms, termed CMT type 1 and 2, respectively. CMT1 has an earlier onset, with variable but definite slowing of nerve conduction velocities. Affected individuals typically develop the symptoms and signs of a length-dependent peripheral neuropathy by their late adolescence, and the disease progresses slowly and symmetrically. The peripheral nerves may be enlarged (hence the term "hypertrophic") and contain "onion bulbs", which are concentric accumulations of supernumerary Schwann cells around remyelinated axons, believed to arise from repeated episodes of demyelination and remyelination (Webster et al., 1967; Gabreels-Festen et al., 1992). CMT2 is much less common, with an onset in middle age, and is slowly progressive. The nerve conduction velocities are not slowed, and the nerves are not enlarged and do not contain onion bulbs. On the basis of this evidence, one might anticipate that inherited defects affecting myelinating Schwann cells cause CMT1, whereas inherited defects affecting neurons, including their axons, cause CMT2.

While most kindreds affected with CMT disease show dominant autosomal inheritance, there have been sporadic reports of sex-linked inheritance, beginning with Herringham (Herringham, 1889). X-linked kindreds of CMT have not always been recognized (Harding and Thomas, 1980a; Harding and Thomas, 1980b), but they account for 10% to 16% of patients in other series (Skre, 1974; Ionasescu et al., 1993). In CMTX kindreds, there is no male-to-male transmission, and males are affected before their twenties (Allan, 1939; Erwin, 1944; Woratz, 1964; de Weerdt, 1978; Fryns and Van den Berghe, 1980; Phillips et al., 1985; Rozear et al., 1987; Hahn et al., 1990; Skre, 1974). In some studies, obligate female carriers were not reported to be affected, hence these kindreds were considered to have X-linked recessive CMT. In other kindreds, all obligate carriers were affected, but later and less severely than males, so that these kindreds were considered to have X-linked dominant CMT. The historical confusion regarding whether CMTX is dominant or recessive can be attributed to the lack of ascertainment. Obligate female carriers may not have clinical manifestations until much later than males, and electrophysiological testing may be required to determine whether an individual is affected. The likely reason that females are less severely affected than males of the same kindred is that the X chromosome is randomly inactivated (Lyon, 1963). In obligate female carriers, half of the Schwann cells would be expected to express the mutant CMTX allele, and the other half, the normal allele, thereby mitigating the severity of the disease. Finally, these kindreds with alleged recessive CTMX should not be confused with the rare kindreds with true recessive CMTX, in which affected individuals have early onset, axonal neuropathy and distinctive clinical features such as spasticity, hearing loss, and mental retardation (Cowchock et al., 1985; Ionasescu et al., 1991; 1992). The disease trait in these recessive kindreds maps to a different region of the X chromosome (Ionasescu et al., 1992).

There are clinical, electrophysiological, and pathological differences between CMTX and autosomal dominant forms of CMT. As emphasized by Hahn et al. (1990), there is pronounced loss of distal axons, with resultant atrophy of distal hand and foot muscles and diminished acral sensation. Nerve conduction velocities are "intermediate"; faster than that typically found in CMT1 patients, but slower than that in CMT2 patients (Nicholson and Nash, 1993; Timmerman et al., 1996). Nerve biopsies reveal more axonal loss and fewer onion bulbs than typically seen in CMT1 (Rozear et al., 1987; Hahn et al., 1990). For these reasons, CMTX has often been considered to be an "axonal" neuropathy, rather than a "demyelinating" neuropathy (see below). One CMTX kindred was even considered to have CMT2 until it was determined that there was a mutation in *Cx32* (Timmerman et al., 1996). This controversy underscores the lack of knowledge about the cellular and molecular pathogenesis of inherited neuropathies. Schwann cells clearly interact with axons through as yet undetermined mechanisms (de Waegh et al., 1992; Reynolds and Woolf, 1993), so that genetic abnormalities that primarily affect Schwann cells could also secondarily affect the axons they ensheathe or myelinate.

THE MOLECULAR BASIS OF CMT1 DISEASE

Linkage mapping in CMT1 kindreds quickly lead to the realization that CMT1 is genetically diverse. Most kindreds map to chromosome 17, and a few to chromosome 1; CMT was thus subdivided into CMT1A and CMT1B, respectively. One of the major advances in the cellular and molecular biology of myelin is the demonstration that mutations in the gene encoding peripheral myelin protein 22 kD (*PMP22*) cause CMT1A, and mutations in the gene encoding protein zero (P_0) cause CMT1B. The details of these discoveries go beyond the scope of this review, but this work has been well summarized in several recent reviews (Roa and Lupski, 1994; Suter and Snipes, 1995; Chance, 1997). It should be emphasized that both *PMP22* and P_0 genes are highly expressed by myelinating Schwann cells, and the corresponding proteins are prominent components of compact myelin. The expression of mutated *PMP22* or P_0 genes in myelinating Schwann cells is probably sufficient and necessary for the development of demyelinating neuropathy (Magyar et al., 1996; Sereda et al., 1996).

Beginning in the early 1980's, the position of CMTX trait on the X chromosome was investigated by several laboratories. Using linkage analysis, the earliest reports excluded the distal short and long arms of the X chromosome. CMTX was further localized to the proximal segment of the long arm using both restriction fragment length and short tandem repeat polymorphic markers (Gal et al., 1985; Beckett et al., 1986; Fischbeck et al., 1986; Goonewardena et al., 1988; Ionasescu et al., 1988; Haites et al., 1989; Fischbeck et al., 1990; Mostacciuolo et al., 1991; Bergoffen et al., 1993b; Fain et al., 1994; Le Guern et al., 1994; Hahn et al., 1990). Recombination analysis further refined CMTX to about a 1.5 megabase interval, which contained three candidate genes. cell cycle gene 1, the γ subunit of the interleukin-2 receptor, and *Cx32* (Bergoffen et al., 1993a; 1993b; Cochrane et al., 1994). Since the defective genes that cause CMT1A and CMT1B are expressed by myelinating Schwann cells, we screened each of these candidate genes for expression in peripheral nerve by northern blot analysis of rat tissues. Of the three candidate genes, only Cx32 mRNA was detectably expressed in nerve, at levels comparable to that in brain and spleen, but less than that in liver (Fig. 1). The widespread expression of Cx32 mRNA contrasts sharply with the phenotype of CMTX patients, whose only consistently reported clinical problem is peripheral neuropathy.

To determine whether CMTX patients have mutations in *Cx32*, we amplified the open reading frame of the human *Cx32* by PCR, and directly sequenced the PCR products (Bergoffen et al., 1993a). In our initial report, we found 7 different mutations in 8 different families. These results have been confirmed and extended in other kindreds from North America, Europe, and Australia (Cherryson et al., 1994; Fairweather et al., 1994; Ionasescu et al., 1994; 1996; Orth et al., 1994; Bone et al., 1995; Sorour and Upadhyaya, 1995; Nelis et al., 1996a,b; Oterino et al., 1996; Ressot et al., 1996; Schiavon et al., 1996; Tan et al., 1996; Timmerman et al., 1996), bringing the total number of mutations affecting the open reading frame to 87 (in 122 kindreds). These mutations are compiled in Table 1 and shown schematically in Fig. 2. The mutations include missense and nonsense muta-

Table 1. Connexin32 coding region mutations in CMTX patients

Codon affected	Nucleotide change	Amino acid change	Reference
3	TGG → TCG	Trp → Ser	(Ionasescu et al., 1996)
7	TAC → TGC	Tyr → Cys	(Schiavon et al., 1996)
12	GGC → AGC	Gly → Ser	(Bergoffen et al., 1993a)
13	GTG → TTG	Val → Leu	(Bone et al., 1995)
15	CGG → CAG	Arg → Gln	(Fairweather et al., 1994)
15	CGG → TGG	Arg → Trp	Janssen, personal communication
15	CGG → TGG	Arg → Trp	Sorour&Upadhyaya, personal communication
15	CGG → TGG	Arg → Trp	(Nelis et al., 1996b)
15	CGG → TGG	Arg → Trp	(Nelis et al., 1996b)
1st Transmembrane domain			
22	CGA → CAA	Arg → Gln	Stolle, personal communication
22	CGA → CAA	Arg → Gln	(Ionasescu et al., 1995)
22	CGA → CAA	Arg → Gln	(Ionasescu et al., 1996)
22	CGA → TGA	Arg → stop	(Ionasescu et al., 1994)
22	CGA → CAA	Arg → Gln	(Nelis et al., 1996b)
22	CGA → TGA	Arg → stop	Janssen, personal communication
22	CGA → TGA	Arg → stop	(Ressot et al., 1996)
22	CGA → CCA	Arg → Pro	(Ressot et al., 1996)
22	CGA → GGA	Arg → Gly	(Ressot et al., 1996)
25	CTC → TTC	Leu → Phe	(Nelis et al., 1997)
26	TCG → TTG	Ser → Leu	Athena Diagnostics Inc., personal communication
26	TCG → TTG	Ser → Leu	(Nelis et al., 1997)
28	ATC → AAC	Ile → Asn	Athena Diagnostics Inc., personal communication
30	ATC → AAC	Ile → Asn	(Bone et al., 1995)
34	ATG → ACG	Met → Thr	(Tan et al., 1996)
35	GTG → ATG	Val → Met	(Cherryson et al., 1994)
38	GTG → ATG	Val → Met	(Orth et al., 1994)
1st Extracellular loop			
40	GCA → GTA	Ala → Val	(Nelis et al., 1996b)
44	TGG → TTG	Trp → Leu	Athena Diagnostics Inc., personal communication
49	TCT → TAT	Ser → Tyr	(Timmerman et al., 1996)
56	CTC → TTC	Leu → Phe	(Latour et al., 1997)
60	TGC → TTC	Cys → Phe	(Fairweather et al., 1994)
63	GTT → ATT	Val → Ile	(Fairweather et al., 1994)
63	GTT → ATT	Val → Ile	Janssen, personal communication
65	TAT → TGT	Tyr → Cys	(Bone et al., 1995)
2nd Transmembrane domain			
73	CAT → AT	frameshift	(Fairweather et al., 1994)
75	CGG → CAG	Arg → Gln	(Tan et al., 1996)
75	CGG → TGG	Arg → Trp	Athena Diagnostics Inc., personal communication
77	UGG → UCG	Trp → Ser	(Ionasescu et al., 1996)
80	CAG → CGG	Gln → Arg	(Ionasescu et al., 1995)
85	TCC → TGC	Ser → Cys	Janssen, personal communication
86	ACC → GCC	Thr → Ala	(Sorour and Upadhyaya, 1995)
86	ACC → TCC	Thr → Ser	Athena Diagnostics Inc., personal communication
87	CCA → TCA	Pro → Ser	Janssen, personal communication
87	CCA → GCA	Pro → Ala	(Nelis et al., 1997)
89	CTC → CCC	Leu → Pro	Janssen, personal communication

Table 1. (*Continued*)

Codon affected	Nucleotide change	Amino acid change	Reference
Intracellular loop			
91/92	GTGGCC →GT- - CC	frameshift	Sorour&Upadhyaya, personal communication
92	GCC → G- -	frameshift	Sorour&Upadhyaya, personal communication
93	ATG → GTG	Met → Val	(Nelis et al., 1996b)
94	CAC → TAC	His → Tyr	Garbern, personal communication
95	GTG → ATG	Val → Met	Kant, personal communication
95	GTG → ATG	Val → Met	(Bone et al., 1995)
100	CAC → TAC	His → Tyr	Sorour&Upadhyaya, personal communication
100	CAC → TAC	His → Tyr	Athena Diagnostics Inc., personal communication
102	GAG → GGG	Glu → Gly	(Ionasescu et al., 1994)
102	GAG → GGG	Glu → Gly	(Ionasescu et al., 1994)
102	GAG → GGG	Glu → Gly	(Ionasescu et al., 1996)
102	GAG → GGG	Glu → Gly	(Ionasescu et al., 1996)
103	AAG → GAG	Lys → Glu	Athena Diagnostics Inc., personal communication
107	CGG → TGG	Arg → Trp	Tan et al., 1996)
107	CGG → TGG	Arg → Trp	Athena Diagnostics Inc., personal communication
111-116	18 bp deletion	6 aa deletion	(Cherryson et al., 1994)
111-116	18 bp deletion	6 aa deletion	(Ionasescu et al., 1995)
124	AAG → AG	frameshift	Athena Diagnostics Inc., personal communication
3rd Transmembrane domain			
133	TGG → CGG	Trp → Arg	(Bone et al., 1995)
133	TGG → TGA	Trp → stop	(Ressot et al., 1996)
137	ATC → AC	frameshift	(Bone et al., 1995)
139	GTG → ATG	Val → Met	(Bone et al., 1995)
139	GTG → ATG	Val → Met	(Bergoffen et al., 1993a)
139	GTG → ATG	Val → Met	(Bergoffen et al., 1993a)
139	GTG → ATG	Val → Met	(Nelis et al., 1996b)
142	CGG → TGG	Arg → Trp	(Bergoffen et al., 1993a)
142	CGG → TGG	Arg → Trp	(Ionasescu et al., 1994)
142	CGG → TGG	Arg → Trp	(Nelis et al., 1996b)
142	CGG → TGG	Arg → Trp	(Ionasescu et al., 1996)
142	CGG → CAG	Arg → Glu	Athena Diagnostics Inc., personal communication
143	TTG deletion	Leu deletion	(Fairweather et al., 1994)
2nd Extracellular loop			
154	TAT → TAA	Tyr → stop	Athena Diagnostics Inc., personal communication
156	CTC → CGC	Leu → Arg	(Bergoffen et al., 1993a)
156	CTC → CGC	Leu → Arg	(Bone et al., 1995)
157	TAC → TGC	Try → Cys	Athena Diagnostics Inc., personal communication
158	CCT → GCT	Pro → Ala	(Cherryson et al., 1994)
160	TAT → CAT	Tyr → His	Athena Diagnostics Inc., personal communication
164	CGG → TGG	Arg → Trp	(Ionasescu et al., 1995)
164	CGG → TGG	Arg → Trp	(Oterino et al., 1996)
164	CGG → TGG	Arg → Trp	(Ionasescu et al., 1996)
172	CCC → TCC	Pro → Ser	(Bergoffen et al., 1993a)
172	CCC → CTC	Pro → Leu	Athena Diagnostics Inc., personal communication
175	AAC → AAAC	frameshift	(Bergoffen et al., 1993a)
179	TGC → CGC	Cys → Arg	Athena Diagnostics Inc., personal communication
181	GTG → ATG	Val → Met	Athena Diagnostics Inc., personal communication
182	TCC → ACC	Ser → Thr	(Cherryson et al., 1994)
183	CGC → TGC	Arg → Cys	Athena Diagnostics Inc., personal communication
185	ACC deletion	Thr deletion	Athena Diagnostics Inc., personal communication
186	GAG → GAGG	frameshift	(Ionasescu et al., 1995)
186	GAG → AAG	Glu → Lys	(Bergoffen et al., 1993a)
186	GAG → TAG	Glu → stop	(Ionasescu et al., 1994)

(continued)

Table 1. (*Continued*)

Codon affected	Nucleotide change	Amino acid change	Reference
4th Transmembrane domain			
187	AAA → GAA	Lys → Glu	Athena Diagnostics Inc., personal communication
198	TCT → TTT	Ser → Phe	Athena Diagnostics Inc., personal communication
201	TGC → CGC	Cys → Arg	Dahl, personal communication
204	CTC → GTC	Leu → Val	Athena Diagnostics Inc., personal communication
205	AAT → AGT	Asn → Ser	Sorour&Upadhyaya, personal communication
205	AAT → AGT	Asn → Ser	Athena Diagnostics Inc., personal communication
Carboxy terminus			
208	GAG → AAG	Glu → Lys	(Fairweather et al., 1994)
211	TAC → TAA	Tyr → stop	(Tan et al., 1996)
215	CGG → TGG	Arg → Trp	(Fairweather et al., 1994)
215	CGG → TGG	Arg → Trp	(Ressot et al., 1996)
217	TGT → TGA	Cys → stop	(Ionasescu et al., 1994)
217	TGT → TGA	Cys → stop	(Ionasescu et al., 1996)
219	CGC → TGC	Arg → Cys	Athena Diagnostics Inc., personal communication
220	CGA → TGA	Arg → stop	(Bone et al., 1995)
220	CGA → TGA	Arg → stop	(Bone et al., 1995)
220	CGA → TGA	Arg → stop	(Fairweather et al., 1994)
220	CGA → TGA	Arg → stop	(Ionasescu et al., 1994)
230	CGC → TGC	Arg → Cys	Athena Diagnostics Inc., personal communication
235	TTC → TGC	Phe → Cys	Athena Diagnostics Inc., personal communication
238	CGC → CAC	Arg → His	(Nelis et al., 1997)
265-273	29 bp deletion	deletion/frameshift	(Ionasescu et al., 1995)
281	TCG → TAG	Ser → stop	(Nelis et al., 1997)

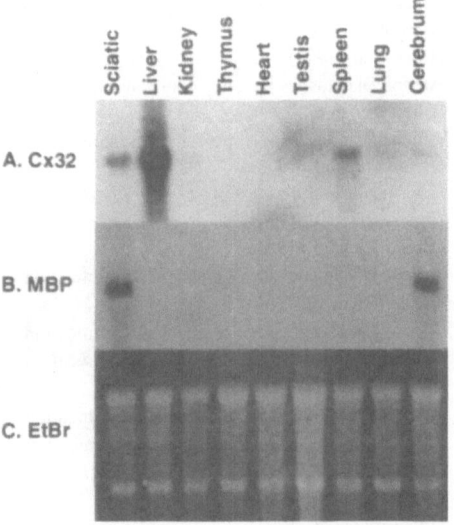

Figure 1. Northern blot analysis of Cx32 mRNA in various tissues from adult rats. Each lane contains 10 μg of total RNA. The blot was probed with a radiolabeled cDNA probe for Cx32 mRNA (panel A), then reprobed for myelin basic protein (MBP) mRNA (panel B), which is expressed by myelinating Schwann cells and oligodendrocytes; panel (C) shows the ethidium bromide staining of the gel before transfer.

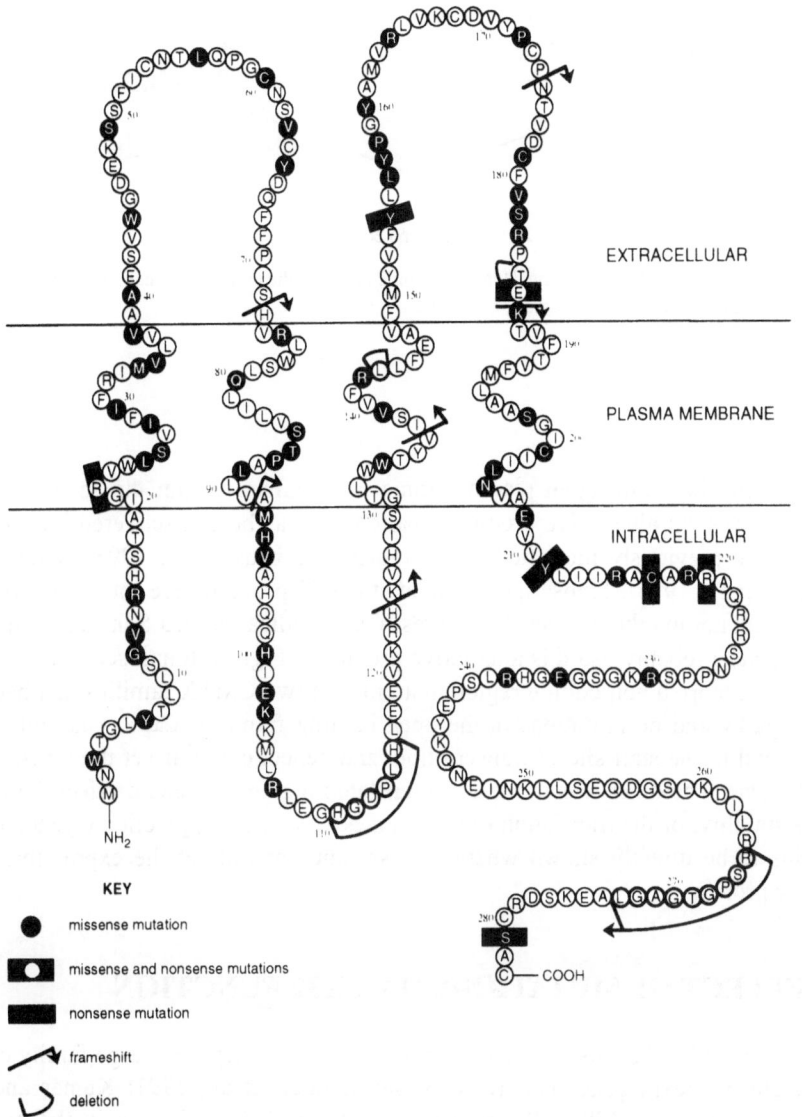

EXTRACELLULAR

PLASMA MEMBRANE

INTRACELLULAR

NH₂

KEY

● missense mutation

◨ missense and nonsense mutations

■ nonsense mutation

⤹ frameshift

⤷ deletion

Figure 2. Mutations in the open reading frame of *Cx32*. For details of the individual mutations, see Table 1.

tions, as well as deletions and insertions that may also cause a change in the open reading frame, and affect all parts of the Cx32 protein. Some mutations have been found more than once, which probably represent independent mutations (Bergoffen et al., 1993a) and founder effects.

Mutations in the open reading frame of Cx32 have not yet been identified in several CMTX kindreds (Bergoffen et al., 1993a; Fairweather et al., 1994; Ionasescu et al., 1994; Bone et al., 1995; Timmerman et al., 1996; Ionasescu et al., 1995). Although linkage to Cx32 was not shown in every kindred, it is plausible that these kindreds have mutations that reduce the expression of Cx32. The possibilities include promoter mutations that reduce the transcription of Cx32 gene, splice site mutations that affect splicing of Cx32 transcripts, or mutations in the 3′ or 5′ untranslated regions that reduce the stability or translation of Cx32 transcripts. Like other connexin genes, *Cx32* has 2 exons separated by

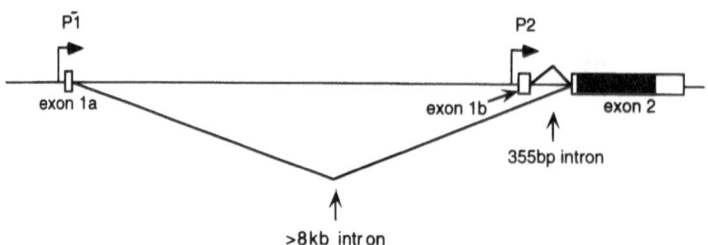

Figure 3. The structure of the *Cx32* gene. Note that exon 2 contains the entire open reading frame. Transcripts are initiated at one of two alternative promoters, P1 and P2.

an intron, and the entire open reading frame is contained within the second exon. Recently, however, an alternative promoter (called P2) has been discovered, located downstream of the previously identified P1 promoter (Neuhaus et al., 1995; Neuhaus et al., 1996; Sohl et al., 1996). Transcripts initiated at the P2 promoter are found in sciatic nerve and brain, but not in other tissues that express Cx32, indicating that Schwann cells and oligodendrocytes probably use this alternative promoter (Fig. 3). Ionasescu et al. (Ionasescu et al., 1996) reported non-coding region mutations in two CMTX families that had moderate neuropathy and no mutations in the open reading frame of *Cx32*. One mutation was just proximal to the start site of transcription, and hence, could affect the transcription of the *Cx32* gene. The other was in the 5' untranslated region, and could affect the transcription, the stability, or the translation of Cx32 mRNA. While the genetic evidence is strong, it remains to be directly shown whether these mutations affect the expression of Cx32 mRNA or protein.

THE EFFECT OF MUTATIONS ON CX32 FUNCTION

Cx32 encodes the gap junction protein Cx32 and belongs to a large family of at least 12 different connexin genes (for reviews, see Bennett et al., 1991; Kumar and Gilula, 1992; Bruzzone et al., 1996). While the structure of Cx32 is known in the most detail (Stauffer and Unwin, 1992; Kumar and Gilula, 1992), all known connexin molecules are predicted to have a similar structure, with four membrane-spanning α-helices connected by two extracellular loops and one intracellular loop (Fig. 2). Six connexin molecules assemble to form a connexon, with their third transmembrane domains lining the central pore. The extracellular loops are thought to mediate protein-protein interactions between connexin molecules (Dahl et al., 1994). Where connexons meet at apposed cell membranes, the opportunity exists for the hemi-channel formed by one connexon to line up with a connexon of the apposed cell membrane, thereby making a channel between the two cells. Ions and small molecules (typically less than 1000 daltons) can pass through the channel; the biophysical characteristics of the channel are largely determined by its constituent connexins. Connexons typically occurring in large aggregates between apposed cell surfaces are called gap junctions.

The open reading frame of the human Cx32 gene is highly homologous to those in the mouse and rat *Cx32* genes. Moreover, all of the connexin genes that have been cloned (in birds, mammals, and frogs) are highly homologous, especially the amino-terminus, intracellular loop, and transmembrane domains, as well as portions of the carboxy-terminus and extracellular loops. Most CMTX mutations are in these conserved regions, so that one might anticipate that these mutations disrupt the function of the Cx32 protein. Conversely, there are no known missense mutations in two of the most divergent regions of the connexin genes, part of the second extracellular loop (residues 110 to 130 of Cx32) and part of the carboxy terminus (residues 240–280 of Cx32). Perhaps missense mutations in these regions are less likely to disrupt the expression or function of Cx32.

The roles of most individual amino acids in the Cx32 protein are unknown, except for six cysteine residues (residues 53, 60, 64, 168, 173, and 179), which are believed to form intramolecular disulfide bonds critical in maintaining the structure of the extracellular loops. These cysteines are absolutely conserved in all connexins, and previous work using site-directed mutagenesis of individual cysteines led to a loss of channel formation in oocytes (Dahl et al., 1992). Two of these cysteines are also mutated in CMTX kindreds (Cys60Phe and Cys179Arg). Besides these mutations, there are 17 missense mutations (Trp3Arg, Trp3Ser, Val63Ile, Try65Cys, Arg75Gln, Arg75TRp, Trp77Ser, Gln80Arg, Pro87Ser, Pre87Ala, Pro172Leu, Pro172Ser, Arg183Cys, Glu186Lys, Lys187Glu, Leu104Val, Glu208Cys) and 1 nonsense mutation (Glu186STOP) affecting 13 other amino acids that are absolutely conserved among all connexins, and many other mutations that affect relatively well conserved amino acids. Mutations that affect conserved residues are probably deleterious.

Other kinds of mutations may have widely different consequences. Deletions and insertions, especially those leading to frameshifts in the open reading frame, might be expected to produce a non-functional protein. The effects of nonsense mutations appear be more predictable, as Rabadan-Diehl et al. (1994) found that the junctional conductance in oocytes fell progressively with the length of truncation - normal with truncations at codons 225, 222, and 219, but progressively smaller with truncations at codons 218, 216, and 214. Their results indicate that truncations at codons 211and 217, but not at codons 220 and 281, will diminish junctional conductance, at least in this assay. How these truncations cause CMTX is thus unexplained; their lack of effect on junctional conductance indicates that they must have other effects.

Is there a relationship between the nature of some mutations and the severity of the peripheral neuropathy in the affected kindred? In the family originally reported by Allan (1939) and subsequently by Rozear et al. (1987), affected patients had a severe phenotype-affected males ultimately required wheelchairs and female carriers were always symptomatic. This kindred has a frameshift mutation beginning at codon 1735 (Bergoffen et al., 1993a). Ionasescu and colleagues (Ionasescu et al., 1995; Ionasescu et al., 1996) have recently reported that families with truncations due to nonsense mutations, large deletions, and frame-shifting mutations tend to have more severe phenotypes than families with missense mutations. This distinction, however, is not absolute, as some missense mutations also cause severe neuropathy. Among missense mutations themselves, no clear-cut relationship has yet emerged between the clinical phenotype and either the location of the mutation or the evolutionary conservation of the affected amino acid residue. For instance, the same amino acid substitution (Trp→Ser) at two different codons, both of which are absolutely conserved across connexins, appears to cause a mild (Trp77Ser) or a moderate to severe (Trp3Ser) phenotype (Ionasescu et al., 1996). Moreover, at the same codon, a theoretically more disruptive substitution of a basic residue (Trp3Arg) appears to cause a milder phenotype than substitution of a more neutral residue (Trp3Ser) (Ionasescu et al., 1996). Finally, the premature truncation at codon 220 is associated with a severe phenotype (Ionasescu et al., 1996), whereas this same mutation forms functional gap junctions in transfected mammalian cells (Rabadan-Diehl et al., 1994; Omori et al., 1996). These discrepancies indicate the difficulties in predicting the effects of missense mutations purely from structural considerations, and underscore the necessity of creating and analyzing appropriate animal models in order to understand the effects of CMTX mutations in peripheral nerve.

CMTX MUTATIONS MAY HAVE LOSS-OF-FUNCTION, DOMINANT-NEGATIVE, AND TOXIC EFFECTS

To determine whether CMTX mutations interfere with the formation of functional gap junctions, Bruzzone et al. (Bruzzone et al., 1994) injected *Xenopus* oocytes with cRNA transcribed either from one of three different CMTX mutations (Arg142Trp, Glu186Lys, or 175 frameshift) or from the normal human Cx32 gene. Oocytes do not normally form gap junctions, but can do so when they are made to express connexins and apposed (Dahl et al., 1987). The gap junctions formed by pairs of oocytes expressing the same connexin are called homotypic, but some connexins (e.g., Cx26) will also form heterotypic gap junctions with Cx32 (Barrio et al., 1991). Bruzzone et al. (1994) observed that no oocytes expressing any of the three CMTX mutants formed functional gap junctions (measured by junctional conductance) with oocytes expressing either Cx26 or wild type Cx32. The mutant Cx32 protein was expressed and properly localized to the membrane by labeling the oocytes with antibodies against Cx32, thus excluding the possibility that mutated Cx32 protein was improperly synthesized in these cells. These data show that, in oocytes, CMTX mutations cause a loss-of-function as these mutant Cx32 proteins fail to make functional gap junctions.

To determine whether *Cx32* mutations could also have a dominant-negative effect, cRNA from individual CMTX mutations was co-injected into oocytes with cRNA encoding Cx26 and apposed to oocytes expressing Cx26 (Bruzzone et al., 1994). Cx26 was chosen as it is frequently co-expressed with Cx32 (Nicholson et al., 1987; Meda et al., 1993) and can form heterotypic channels (Barrio et al., 1991). All of the CMTX mutations significantly reduced the junctional conductance, the Arg142Trp mutation being particularly potent. As controls, cRNA from each of the CMTX mutations was co-injected into oocytes with cRNA encoding Cx40 and apposed to oocytes expressing Cx40, as Cx40 does not form heterotypic channels with Cx32 (Bruzzone et al., 1994). None of the CMTX mutations interfered with junctional conductance of Cx40 homotypic channels. These results indicate that CMTX mutations may have both loss-of-function and dominant-negative effects.

Omori et al. (Omori et al., 1996) have examined the effects of CMTX mutations on the expression and localization of Cx32 in mammalian cells. They permanently transfected HeLa cells to express wild type Cx32, missense mutations (Cys60Phe, Val139Met, Arg215Trp), or a nonsense mutation (Arg220stop). They observed dye-coupling between cells expressing wild type Cx32 and Arg220stop, but not between cells expressing any of the missense mutations. Cells that expressed missense mutations had much less Cx32-immunoreactivity on their cell surface than cells expressing wild type Cx32 or Arg220stop. The finding that the Arg220stop mutation can form functional gap junctions in mammalian cells confirms the observation that this mutation does not disrupt gap junction communication in *Xenopus* oocytes (Rabadan-Diehl et al., 1994). Omori et al. (1996) also looked for dominant-negative interactions by permanently transfecting HeLa cells that already expressed wild type Cx32 protein with these CMTX mutations. All three missense mutations had significant dominant-negative effects, whereas the nonsense mutation and wild-type Cx32 did not decrease coupling.

Our group has also made permanently transfected lines of PC12 cells that express either wild type Cx32 or one of seven CMTX mutations. We find three patterns of expression: (1) Cx32 mRNA is expressed as determined by northern blot analysis, but Cx32 protein is undetectable as determined by western blot analysis (175frameshift); (2) Cx32 protein is expressed, but remains in the cytoplasm and does not reach the cell surface (Fig. 4A; Arg142Trp, Glu186Lys, Glu208Lys); (3) Cx32 protein is expressed and reaches the cell surface and becomes localized between apposed cells to form putative gap junctions (Fig. 4B; wild type Cx32, Arg15Gln, Val63Ile, Arg220stop). These data indicate that in mammalian cells, mutant Cx32 protein may not reach the cell membrane, and raises the possibility that CMTX mutations, like mutations in another intrinsic membrane protein of myelin, proteolipid protein (PLP), may result in the failure of the mutated protein to reach the cell surface in cultured cells (Gow and Lazzarini, 1996).

These results, taken together, show that CMTX mutations can have both loss-of-function and dominant-negative effects in *Xenopus* oocytes and in mammalian cells. The behavior of some CMTX mutations in mammalian cells, however, is different than in oocytes, as the mutant protein does not accumulate and/or reach the cell membrane. The finding that mutant Cx32 protein is abnormally processed raises the possibility that the mutant protein could also have a toxic effect, as has been described for some *PLP* mutations (Nave and Boespflug-Tanguy, 1996). Whether different CMTX mutations can also cause a toxic gain-of-function needs to be resolved by a thorough analysis of the trafficking of mutated protein in myelinating Schwann cells.

The potential for some CMTX mutations to have dominant-negative interactions draws attention to the question of whether myelinating Schwann cells express other connexins. Gap junctions have been found between non-myelinating and denervated Schwann cells (Tetzlaff, 1982; Konishi, 1990), and these gap junctions probably contain Cx46 (Chandross et al., 1996). To determine whether myelinating Schwann cells express other connexins, we have performed RT-PCR with specific primers for the 12 previously cloned

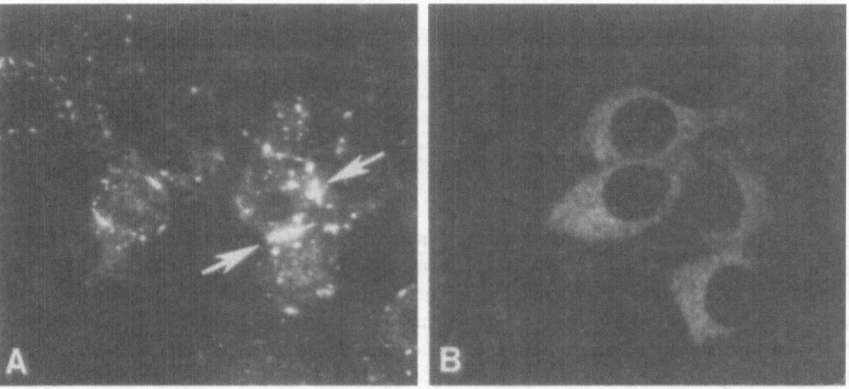

Figure 4. The localization of mutated Cx32. These are 1 μm thick optical sections taken by laser scanning confocal microscopy of PC12 cells that have been permanently transfected with Arg15Trp (A) or Glu208Lys (B) and stained for Cx32. Panel A shows Cx32-immunoreactivity on the cell surface, particularly at apposed cell membranes (arrows; putative gap junctions). Panel B shows intracellular Cx32-immunoreactivity but no cell surface staining.

connexins (Bone et al., 1995). We used RNA from Schwann cells that were treated with forskolin, which increases the levels of Cx32 and other myelin-related transcripts (Scherer et al., 1995), and found expression of Cx32, as well as Cx26, Cx33, Cx37, Cx43, Cx45, and Cx46. Whether any of these other connexins are expressed by myelinating Schwann cells remains to be determined.

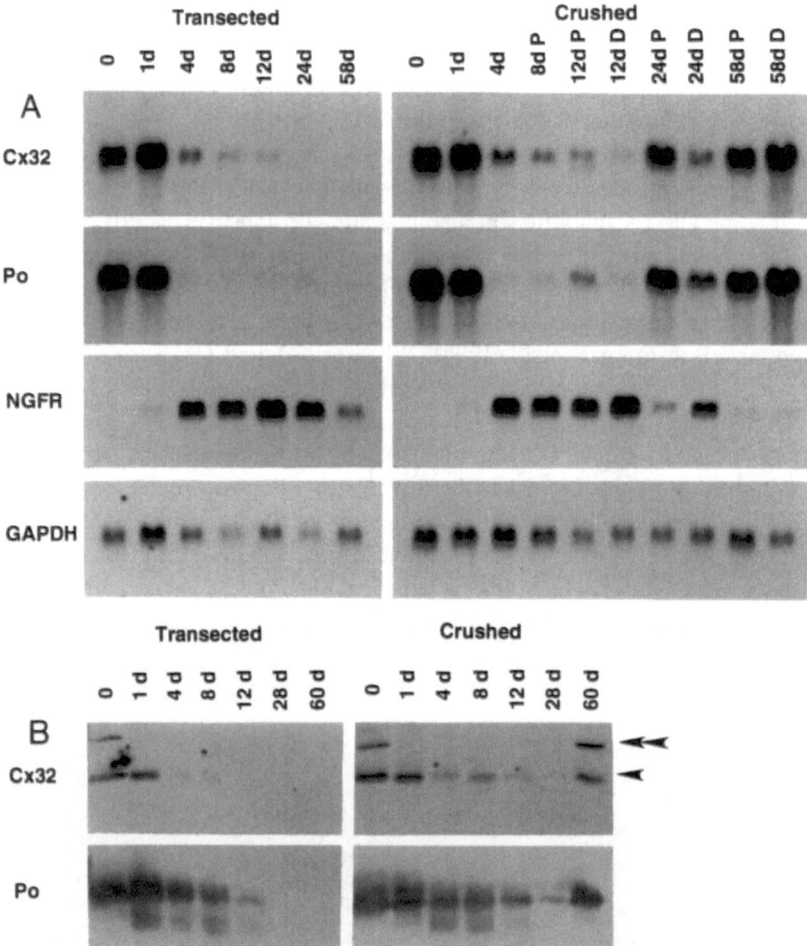

Figure 5. The expression of Cx32 mRNA and protein in normal and lesioned adult rat sciatic nerve. The number of days after injury is indicated; the '0' time point is from unlesioned nerves. Panel A. Each lane contains an equal amount (10 µg) of total RNA isolated from the distal stumps of sciatic nerves that had been transected or crushed. In crushed nerves, the distal nerve-stumps were divided into proximal (P) and distal (D) segments of equal lengths. The blots were successively hybridized with a radiolabeled cDNA probe for Cx32 (A), P_0 (B), NGFR (C), and GAPDH (D), and exposed to film for 14 days (Cx32), 2 hours (P_0), 1 day (NGFR), and 3 days (GAPDH), respectively. Panel B. Each lane contains an equal amount (25 µg) of protein homogenate from the distal nerve-stumps of sciatic nerves that had been transected or crushed. The blots were hybridized together with a rabbit antiserum against rat Cx32, then rehybridized with a rabbit antiserum against rat P_0. The blots were exposed to film for 30 min. (Cx32) or 1 sec. (P_0). The arrow marks the position of the Cx32 monomer, and the double arrowhead marks the position of the Cx32 dimer. Figures reproduced with permission of The Society for Neuroscience (Scherer et al., 1995).

CONNEXIN32 IS A MYELIN-RELATED PROTEIN

The discovery that mutations in *Cx32* cause CMTX led us to examine the expression of Cx32 mRNA and protein in the nervous system. In rat peripheral nerve, the expression of Cx32 mRNA and protein parallels that of other myelin-related genes in development, during Wallerian degeneration, and during axonal regeneration. As shown in Figure 5, following axotomy, the expression of myelin-related mRNAs falls abruptly, but returns if axons regenerate and remyelination occurs. The level of Cx32 protein follows a similar trend following nerve-injury. These findings demonstrate that the expression of Cx32, like many other aspects of Schwann cell biology, is regulated by axon-Schwann cell interactions (Doyle and Colman, 1993; Mirsky and Jessen, 1996; Scherer and Asbury, 1997).

To determine where Cx32 protein is localized, we labeled sections and teased fibers of rat peripheral nerve with anti-Cx32 antibodies (Fig. 6). Cx32-immunoreactivity was found in incisures and paranodes, and to a lesser degree in the inner mesaxon (Bergoffen et al., 1993a; Scherer et al., 1995). The localization of Cx32 adds further evidence to the idea that the Schwann cell myelin sheath is composed of two distinct domains: (1) the compact myelin sheath, which contains P_0, PMP22, and MBP, and (2) incisures and paranodes, which contain Cx32, E-cadherin, myelin-associated glycoprotein (MAG), and oligodendrocyte-myelin glycoprotein (reviewed in Scherer, 1996). The observations that Cx32 is localized to incisures and paranodes and that mutations in *Cx32* cause a demyelinating peripheral neuropathy, provide strong evidence that incisures and paranodes have specialized functions in the myelin sheath.

We also examined Cx32 expression in the CNS. As previously noted (Dermietzel et al., 1989; Micevych and Abelson, 1991; Naus et al., 1990; Belliveau et al., 1991), there was prominent Cx32 mRNA expression in the CNS, and the level of Cx32 mRNA correlated with the degree of myelination, both in development and in rodents with mutations that affect myelination (Scherer et al., 1995). The bulk of the Cx32-immunoreactivity was confined to the membrane of the oligodendrocyte cell bodies and their proximal processes. Cx32-immunoreactivity was not seen in paranodes or incisures, although incisures in the CNS are rare and have not been reported to contain distinctive molecules such as MAG. Thus, there is prominent expression of Cx32 in oligodendrocytes even though CMTX patients do not typically have abnormal findings related to CNS myelin. Perhaps oligodendrocytes express other connexins, so that gap junctional communication is maintained even when *Cx32* is mutated.

GAP JUNCTIONS IN THE MYELIN SHEATH

The finding that a gap junction protein is localized to incisures and paranodes is consistent with previous reports of gap junctions in paranodes (in both the CNS and the PNS) and PNS incisures (Sandri et al., 1982). Whereas gap junctions typically form channels between apposed cells, these would be an example of so-called "reflexive" gap junctions, which are found between apposed layers of the same cell. If these gap junctions formed a radial pathway directly across the myelin sheath, this would provide a significant short-cut for ions and small molecules: for the largest myelin sheaths, this radial pathway is about 1000 times shorter than the potential circumferential pathway within the cytoplasm of the myelin sheath itself (Scherer et al., 1995). If mutations in Cx32 abolish this potentially economical pathway, the failure of ions and small molecules to diffuse directly

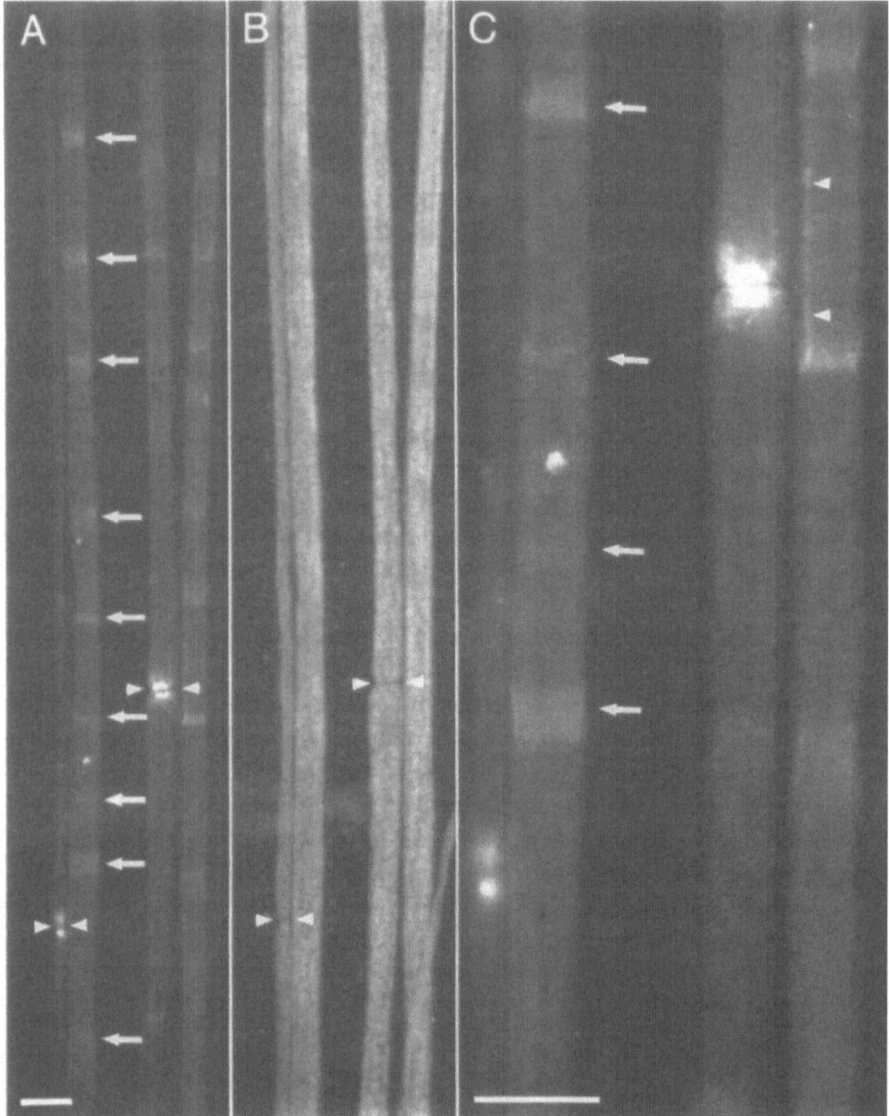

Figure 6. Immunohistochemical localization of Cx32 and P_0 in adult rat sciatic nerve. These are photomicrographs of teased fibers, double-labeled with a monoclonal antibody against rat Cx32 (A,C) and a rabbit polyclonal antibody against rat P_0 (B) and visualized with fluorescein- and rhodamine-coupled secondary antibodies, respectively. Note that Cx32 is predominantly found at the incisures [some of which are indicated by arrows in panel (A)] and paranodes [arrowheads in panels (A) and (B)], whereas the P_0 is found throughout the compact myelin sheath. Panel (C) is an enlargement of the node shown in panel (A), and shows Cx32 staining of the inner mesaxon (arrowheads). Scale bars: 10 μm. Figure reproduced with permission of The Society for Neuroscience (*Scherer et al.*, 1995).

across the myelin sheath could have deleterious effects, and result directly or indirectly in demyelination and axonal loss.

To determine whether there is a radial pathway through the myelin sheath, we have injected living teased fibers from rodent peripheral nerve (Bone et al., 1996; Balice-Gordon, Bone, and Scherer, unpublished observations). Individual myelinating Schwann cells are im-

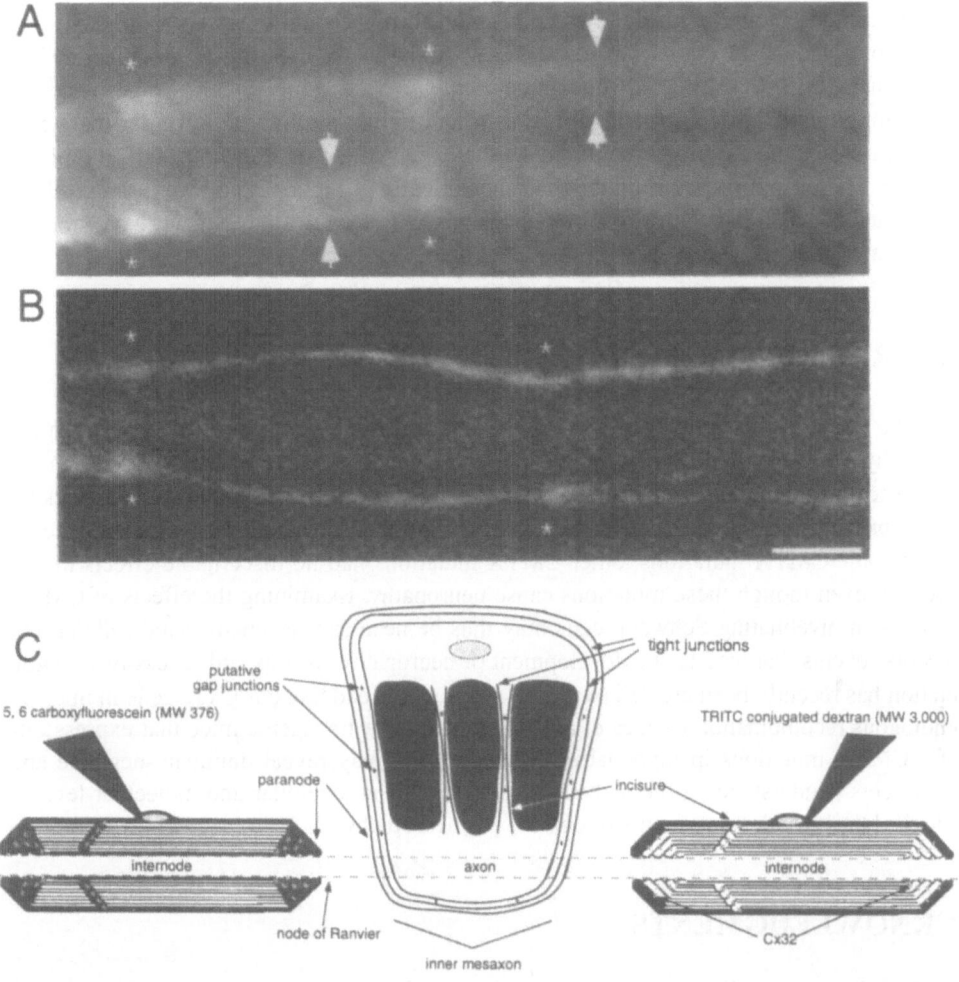

Figure 7. Demonstration of a radial pathway for diffusion of small molecules through the myelin sheath. The positions of incisures are marked by asterisks. Panel A is a photomicrograph of a segment of a myelinating Schwann cell several seconds after injecting 5,6-carboxyfluorescein. Carboxyfluorescein fills the inner and outer collars of cytoplasm (arrows), but not the myelin sheath itself, resulting in a "double line" of fluorescence. Panel B is a confocal micrograph of a myelinating Schwann cell that has been injected with rhodamine-conjugated dextran 10,000 Da. Note the rhodamine signal is confined to the outer rim of Schwann cell cytoplasm surrounding the myelin sheath. Panel C is a schematic view of the different pathways of dye diffusion following perinuclear dye injection. The myelin sheath on the left has been injected with 5,6-carboxyfluorescein; the myelin sheath on the right with rhodamine-conjugated dextran 10,000 kDa. The myelinating Schwann cell in the middle has been "unrolled" to reveal the regions that form compact myelin, incisures, and paranodes. Tight junctions are depicted as two rows of continuous lines; these form a circumferential belt and are also found in incisures. Gap junctions are depicted as ovals; these are found between the rows of tight junctions. Scale bar: 10 μm.

paled in the perinuclear region and injected with fluorescent dyes of different molecular mass; ones that should cross gap junctions (5,6-carboxyfluorescein, 376 Da), and others that should not (rhodamine-conjugated dextran 10,000 Da). Dye movement is recorded with a SIT camera and a computer-assisted imaging system. After injection is completed, the fiber is coverslipped and examined with a laser scanning confocal microscope (Leica TCS).

When 5,6-carboxyfluorescein is injected into the perinuclear region, the dye quickly spreads longitudinally, reaching the node of Ranvier within a few minutes. Before the dye reaches the node, however, it appears to fill both the inner and outer collar of Schwann cell cytoplasm that surrounds the unstained myelin sheath, creating a "double line" of fluorescence (Fig. 7A). Injected rhodamine-conjugated dextran 10,000 kDa also rapidly spreads longitudinally down the myelin sheath, but unlike the lower molecular mass 5,6-carboxyfluorescein, it remains confined to the outer collar of Schwann cell cytoplasm (Fig. 7B). These findings are summarized schematically in Figure 7, and we interpret them as evidence for a radial pathway through the myelin sheath, although we have not yet demonstrated that this pathway is formed by gap junctions.

FUTURE DIRECTIONS

The discovery that mutations in *Cx32* cause CMTX, an inherited demyelinating neuropathy, leads to a reconsideration of the functional importance of gap junctions in the myelin sheath, and the way that different CMTX mutations affect the synthesis and trafficking of Cx32 protein. Whereas cell culture has already proven to be a useful way to analyze the effects of some CMTX mutations, other CMTX mutations had no discernable effects in this paradigm, even though these mutations cause neuropathy. Examining the effects of CMTX mutations in myelinating Schwann cells may thus be necessary to uncover the cellular and molecular events that lead to the development of neuropathy. To this end, a loss-of-function mutation has recently been created by "knocking out" the endogenous Cx32 gene in mice by homologous recombination (Nelles et al., 1996). Similarly, transgenic mice that express different CMTX mutations in myelinating Schwann cells may reveal dominant-negative and toxic effects. Understanding the pathogenesis of CMTX at a cellular and molecular level is the logical basis for developing treatments for this neuropathy.

ACKNOWLEDGMENTS

We thank our collaborator, Dr. David Paul, for his many contributions to the work summarized in this review, Dr. James Garbern for the analysis of homology among connexins, and Dr. Uma Ananth at Athena Diagnostics Inc. (Worcester, MA), as well as Drs. James Garbern, Gerhard Dahl, Emil Janssen, Jeffrey Kant, Catherine Stolle, and Meena Upadhyaya for sharing their unpublished data on CMTX mutations. This work was supported by grants from the Muscular Dystrophy Association, NIH, and the American Academy of Neurology (Murray M. Stokely Award).

REFERENCES

Allan W (1939): Relation of hereditary pattern to clinical severity as illustrated by peroneal atrophy. Arch Int Med
        63:1123–1131.

Barrio LC, Suchyna T, Bargiello T, Xu LX, Roginski RS, Bennett MVL, Nicholson BJ (1991): Gap junctions formed by connexins 26 and 32 alone and in combination are differently affected by applied voltage. Proc Natl Acad Sci USA 88:8410–8414.

Beckett J, Holden JJA, Simpson NE, White BN, MacLeod PM (1986): Localization of X-linked dominant Charcot-Marie-Tooth disease (CMT2) to Xq13. J Neurogenet 3:225–231.

Belliveau DJ, Kidder GM, Vaus CCG (1991): Expression of gap junction genes during postnatal neural development. Dev Genet 12:308–317.

Bennett MVL, Barrio LC, Bargiello TA, Spray DC, Hertzberg E, Saez JC (1991): Gap junctions: new tools, new answers, new questions. Neuron 6:305–320.

Bergoffen J, Scherer SS, Wang S, Oronzi-Scott M, Bone L, Paul DL, Chen K, Lensch MW, Chance P, Fischbeck K (1993a): Connexin mutations in X-linked Charcot-Marie-Tooth disease. Science 262:2039–2042.

Bergoffen J, Trofatter J, Pericak-Vance MA, Haines J, Chance PF, Fischbeck KH (1993b): Linkage localization of X-linked Charcot-Marie-Tooth disease. Am J Hum Genet 52:312–318.

Bone LJ, Dahl N, Lensch MW, Chance PF, Kelly T, Le Guern E, Magi S, Parry G, Shapiro H, Wang S, Fischbeck KH (1995a): New connexin32 mutations associated with X-linked Charcot-Marie-Tooth disease. Neurology 45:1863–1866.

Bone LJ, Scherer SS, Balice-Gordon RJ, Paul DL, Fischbeck KH (1995b): The disease mechanism for X-linked Charcot-Marie-Tooth disease. Am J Hum Genet 57: A236.

Bone LJ, Scherer SS, Balice-Gordon R (1996): The role of the gap junction protein connexin32 in myelinating Schwann cells. Soc Neurosci Abstr 22: 1980.

Bruzzone R, T.W. White, S.S. Scherer, Fischbeck KH, Paul DL (1994): Null mutations of connexin32 in patients with X-linked Charcot-Marie-Tooth disease. Neuron 13:1253–1260.

Bruzzone R, White TW, Paul DL (1996): Connections with connexins: the molecular basis of direct intercellular signaling. Eur J Biochem 238:1–27.

Chance PF (1997): Inherited demyelinating neuropathy: Charcot-Marie-Tooth disease and related disorders. In Rosenberg RN, Prusiner SB, DiMauro S, Barchi RL. "The Molecular and Genetic Basis of Neurological Disease." Butterworth-Heinemann: 807–816.

Chandross KJ, Kessler JA, Spray DC, Simburger E, Kremer M, Bieri P, Dermietzel R (1996): Altered gap junction expression after peripheral nerve injury. Mol Cell Neurosci: 501–518.

Cherryson AK, Yeung L, Dennerson ML, Nicholson GA (1994): Mutational studies in X-linked Charcot-Marie-Tooth disease (CMTX). Am J Hum Genet 55S:1261A.

Cochrane S, Bergoffen J, Fairweather ND, Müller E, Mostaccuiolo ML, Monaco AP, Fischbeck KH, Haites NE (1994): X-linked Charcot-Marie-Tooth disease (CMTX1) - A study of 15 families with 12 highly informative polymorphisms. J Med Genet 31:193–196.

Cowchock RS, Duckett SW, Streletz LJ, Graziani LJ, Jackson LG (1985): X-linked sensory-motor neuropathy type II with deafness and mental retardation: a new disorder. Am J Hum Genet 20:307–315.

Dahl G, Miller T, Paul D, Voellmy R (1987): Expression of functional cell-cell channels from cloned rat liver junction complementary DNA. Science 236:1290–1293.

Dahl G, Nonner W, Werner R (1994): Attempts to define functional domains of gap junction proteins with synthetic peptides. Biophys J 67:1816–1822.

Dahl G, Werner R, Levine E, Rabadan-Diehl C (1992): Mutational analysis of gap junction formation. Biophys J 62:172–182.

de Waegh SM, Lee VM-Y, Brady ST (1992): Local modulation of neurofilament phosphorylation, axonal caliber, and slow axonal transport by myelinating Schwann cells. Cell 68:451–463.

de Weerdt CJ (1978): Charcot-Marie-Tooth disease with sex-linked inheritance, linkage studies and abnormal serum alkaline phosphatase levels. Eur Neurol 17:336–344.

Dermietzel R, Traub O, Hwang TK, Beyer E, Bennett MVL, Spray DC, Willecke K (1989): Differential expression of three gap junction proteins in developing and mature brain tissues. Proc Natl Acad Sci USA 86:10148–10152.

Doyle JP, Colman DR (1993): Glial-neuron interactions and the regulation of myelin formation. Curr Opin Cell Biol 5:779–785.

Dyck PJ, Chance P, Lebo R, Carney JA (1993): Hereditary motor and sensory neuropathies. In Dyck PJ, Thomas PK, Griffin JW, Low PA and Poduslo JF. "Peripheral Neuropathy, 3rd Edn." Philadelphia: W.B. Saunders, pp 1094–1136.

Dyck PJ, Lambert EH (1968a): Lower motor and primary sensory neuron diseases with peroneal muscular atrophy. I. Neurologic, genetic and electrophysiologic findings in hereditary polyneuropathies. Arch Neurol 18:603–618.

Dyck PJ, Lambert EH (1968b): Lower motor and primary sensory neuron diseases with peroneal muscular atrophy. II. Neurologic, genetic and electrophysiologic findings in various neuronal degenerations. Arch Neurol 18:619–625.

Erwin WG (1944): A pedigree of sex-linked recessive peroneal atrophy. J Hered 35:24–26.

Fain PR, Barker DF, Chance PF (1994): Refined genetic mapping of X-linked Charcot-Marie-Tooth neuropathy. Am J Hum Genet 54:229–235.

Fairweather N, Bell C, Cochrane S, Chelly J, Wang S, Mostacciuolo ML, Monaco AP, Haites NE (1994): Mutations in the connexin 32 gene in X-linked dominant Charcot-Marie-Tooth disease (CMT-X1). Hum Mol Genet 3:29–34.

Fischbeck KF, Ritter A, Shi Y (1990): X-linked recessive and X-linked dominant Charcot-Marie-Tooth disease. In Lovelace RE and Shapiro HK. "Charcot-Marie-Tooth diseorders: Pathophysiology, Molecular Genetics, and Therapy." New York: Wiley-Liss, pp 335–341.

Fischbeck KH, ar-Rushdi N, Pericak-Vance M, Rozear M, Roses AD, Fryns JP (1986): X-linked neuropathy: gene localization with DNA probes. Ann Neurol 20:527–532.

Fryns JP, Van den Berghe H (1980): Sex-linked recessive inheritance in Charcot-Marie-Tooth disease with partial clinical manifestation in female carriers. Hum Genet 55:413–415.

Gabreels-Festen AAWM, Joosten EMG, Gabreels FJM, Jennekens FGI, Kempen TWJ (1992): Early morphological features in dominantly inherited demyelinating motor and sensory neuropathy (HMSN Type-1). J Neurol Sci 107:145–154.

Gal A, Mucke J, Theile H, Wieacker PF, Ropers HH, Wienker TF (1985): X-linked dominant Charcot-Marie-Tooth disease: suggestion of linkage with a cloned DNA sequence from the proximal Xq. Hum Genet 70:38–42.

Goonewardena P, Welinhinda J, Anvret M, Gyftodimou J, Haegermark A, Iselius L, Lindsten J, Pettersson U (1988): A linkage study of the locus for X-linked Charcot-Marie-Tooth disease. Clin Genet 33:435–440.

Gow A, Lazzarini RA (1996): A cellular mechanism governing the severity of Pelizaeus-Merbacher disease. Nat Genet 13:422–427.

Hahn AF, Brown WF, Koopman WJ, Feasby TE (1990): X-linked dominant hereditary motor and sensory neuropathy. Brain 113:1511–1525.

Haites N, Fairweather N, Clark C, Kelly KF, Simpson S, Johnston AW (1989): Linkage in a family with X-linked Charcot-Marie-Tooth disease. Clin Genet 35:399–403.

Harding AE, Thomas PK (1980a): The clinical features of hereditary motor and sensory neuropathy types I and II. Brain 103:259–280.

Harding AE, Thomas PK (1980b): Genetic aspects of hereditary motor and sensory neuropathy (types I and II). J Med Genet 176:329–336.

Herringham WP (1889): Muscular atrophy of the peroneal type affecting many members of a family. Brain 11:230–236.

Ionasescu V, Ionasescu R, Searby C (1996a): Correlation between connexin 32 gene mutations and clinical phenotype in X-linked dominant Charcot-Marie-Tooth neuropathy. Am J Med Genet 63:486–491.

Ionasescu VV, Searby C, Ionasescu R, Neuhaus IM, Werner R (1996b): Mutations of noncoding region of the connexin32 gene in X-linked dominant Charcot-Marie-Tooth neuropathy. Neurology 47:541–544.

Ionasescu V, Searby C, Ionasescu R (1994): Point mutations of the connexin32 (GJB1) gene in X-linked dominant Charcot-Marie-Tooth neuropathy. Hum Mol Genet 3:355–358.

Ionasescu VV, Burns TL, Searby C, Ionasescu R (1988): X-linked dominant Charcot-Marie-Tooth neuropathy with 15 cases in a family: genetic linkage study. Muscle Nerve 11:435–440.

Ionasescu VV, Ionasescu R, Searby C (1993): Screening of dominantly inherited Charcot-Marie-Tooth neuropathies. Muscle Nerve 16:1232–1238.

Ionasescu VV, Searby C, Ionasescu R, Meschino W (1995): New point mutations and deletions of the connexin 32 gene in X-linked Charcot-Marie-Tooth neuropathy. Neuromusc Disord 5:297–299.

Ionasescu VV, Trofatter J, Haines JL (1991): Heterogeneity in X-linked recessive Charcot-Marie-Tooth neuropathy. Am J Hum Genet 48:1075–1083.

Ionasescu VV, Trofatter J, Haines JL, Summers AM, Ionasescu R, Searby C (1992): X-linked recessive Charcot-Marie-Tooth neuropathy - clinical and genetic study. Muscle Nerve 15:368–373.

Konishi T (1990): Dye coupling between mouse Schwann cells. Brain Res 508:85–92.

Kumar NM, Gilula N (1992): Molecular biology and genetics of gap junction channels. Semin Cell Biol 3:3–16.

Latour P, Fabreguette A, Ressot C, Blanquet-Grossard F, Antoine JC, Calvas P, et al. (1997): New mutations in the X-linked form of Charcot-Marie-Tooth disease. Eur Neurol 37: 38–42.

Le Guern E, Ravise N, Gugenheim M, Vignal A, Penet C, Bouche P, Weissenbach J, Agid Y, Brice A (1994): Linkage analyses between dominant X-linked Charcot-Marie-Tooth disease, and 15 Xq11-Xq21 microsatellites in a new large family: three new markers are closely linked to the gene. Neuromusc Disord 4:463–469.

Lyon MF (1963): Gene action in the X-chromosome of the mouse. Nature 100:372.

Magyar JP, Martini R, Ruelicke T, Aguzzi A, Adlkofer K, Dembic Z, Zielasek J, Toyka KV. Suter U (1996): Impaired differentiation of Schwann cells in transgenic mice with increased *PMP22* gene dosage. J Neurosci 16:5351–5360.

Meda P, Pepper MS, Traub O, Willecke K, Gros D, Beyer E, Nicholson B, Paul D, Orci L (1993): Differential expression of gap junction connexins in endocrine and exocrine glands. Endocrinol 133:2371–2378.

Micevych PE. Abelson L (1991): Distribution of mRNAs coding for liver and heart gap junction proteins in the rat central nervous system. J Comp Neurol 305:96–118.

Mirsky R. Jessen KR (1996): Schwann cell development, differentiation and myelination. Curr Opin Neurobiol 6:89–96.

Mostacciuolo ML, Muller E. Fardin P. Micaglio GF, Bardoni B, Guioli S, Camerino G. Danieli GA (1991): X-linked Charcot-Marie-Tooth disease: a linkage sudy in a large family by using 12 probes for the pericentromeric region. Hum Genet 87:23–27.

Naus CCG, Belliveau DJ. Bechberger JF (1990): Regional differences in connexin32 and connexin43 messenger RNAs in rat brain. Neurosci Lett 111:297–302.

Nave K-A. Boespflug-Tanguy O (1996): Developmental defects of myelin formation: from X-linked mutations to human dysmyelinating diseases. Neuroscientist 2:33–43.

Nelis E. Simokovic S, Timmerman V, Lofgren A, Backhovens H, De Hoghe P. Martin J. Van Broeckhoven C (1997): Mutation analysis of the connexin32 (Cx32) gene in Charcot-Marie-Tooth neuropathy type 1: identification of five new mutations. Hum Mutat 9: 47–52.

Nelis E. Van Broeckhoven C, co-authors (1996): Estimation of the mutation frequencies in Charcot-Marie-Tooth disease type 1 and hereditary neuropathy with liability to pressure palsies: a European collaborative study. Eur J Human Genet 4:25–33.

Nelles E. Butzler C, Jung D. Temme A. Gabriel H-D, Dahl U, Traub O. Stumpel F. Jungermann K. Zielasek J. Toyka KV, Dermietzel R, Willecke K (1996): Defective propagation of signals generated by sympathetic nerve stimulation in the liver of connexin32-deficient mice. Proc Natl Acad Sci USA 93:

Neuhaus IM. Bone L. Wang S, Ionasescu V, Werner R (1996): The human connexin32 gene is transcribed from two tissue-specific promoters. Biosci Reports 16:239–248.

Neuhaus IM. Dahl G. Werner R (1995): Use of alternative promoters for tissue-specific expression of the gene coding for connexin32. Gene 158:257–262.

Nicholson B. Dermietzel R. Teplow D. Traub O, Willecke K, Revel J-P (1987): Two homologous protein components of hepatic gap junctions. Nature 329:732–734.

Nicholson G. Nash J (1993): Intermediate nerve conduction velocities define X-linked Charcot-Marie-Tooth neuropathy families. Neurology 43:2558–2564.

Omori Y. Mesnil M. Yamasaki H (1996): Connexin 32 mutations from X-linked Charcot-Marie-Tooth disease patients: functional defects and dominant negative effects. Mol Biol Cell 7:907–916.

Orth U. Fairweather N. Wexler MC. Schwinger E. Gal A (1994): X-linked dominant Charcot-Marie-Tooth neuropathy: valine-38-methionine substitution of connexin32. Hum Mol Genet 3:1699–1700.

Oterino A, Monton FI, Cabrera VM, Pinto F, Gonzalez A, Lavilla NR (1996): Arginine-164-tryptophan substitution in connexin32 associated with X linked dominant Charcot-Marie-Tooth disease. J Med Genet 33:413–415.

Phillips LH. Kelly TE. Schnatterly P. Parker D (1985): Hereditary motor-sensory neuropathy (HMSN): possible X-linked dominant inheritance. Neurology 35:498–502.

Rabadan-Diehl C, Dahl G, Werner R (1994): A connexin-32 mutation associated with Charcot-Marie-Tooth disease does not affect channel formation in oocytes. FEBS Lett 351:90–94.

Ressot C, Latour P, Blanquet-Grossard F, Sturtz F, Duthel S, Battin J, Corbillon E, Ollagnon E, Serville F. Vandenberghe A, Dautigny A, Pham-Dinh D (1996): X-linked dominant Charcot-Marie-Tooth neuropathy (CMTX): new mutations in the connexin32 gene. Hum Genet 98:172–175.

Reynolds ML. Woolf CJ (1993): Reciprocal Schwann cell-axon interactions. Curr Opin Neurobiol 3:683–693.

Roa BB, Lupski JR (1994): Molecular genetics of Charcot-Marie-Tooth neuropathy. In Harris H and Hirschhorn L. "Advances in Human Genetics." New York: Plenum Press Div Plenum Publishing Corp, pp 117–152.

Rozear MP, Pericak-Vance MA, Fischbeck K, Stajich JM, Gaskell PC Jr., Krendel DA. Graham DG. Dawson DV, Roses AD (1987): Hereditary motor and sensory neuropathy, X-linked: a half century follow-up. Neurology 37:1460–1465.

Sandri C, Van Buren JM, Akert K (1982): Membrane morphology of the vertebrate nervous system. Prog Brain Res 46:201–265.

Scherer SS (1996): Molecular specializations at nodes and paranodes in peripheral nerve. Microsc Res Tech 34:452–461.

Scherer SS, Asbury AK (1997): Inherited axonal neuropathies and the molecular biology of peripheral neuropathies. In Rosenberg RN, Prusiner SB, DiMauro S and Barchi RL. "The Molecular and Genetic Basis of
Neurological Disease." Butterworth-Heinemann pp 817–843.

Scherer SS, Deschênes SM, Xu Y-t, Grinspan JB, Fischbeck KH, Paul DL (1995): Connexin32 is a myelin-related
protein in the PNS and CNS. J Neurosci 15:8281–8294.

Schiavon F, Fracasso C, Mostacciuolo ML (1996): Novel missense mutation of the connexin 32 (GJB1) gene in X-
linked dominant Charcot-Marie-Tooth neuropathy. Hum Mutat 8:83–84.

Sereda M, Griffiths I, Puhlhofer A, Stewart H, Rossner MJ, Zimmermann F, Magyar JP, Schneider A, Hund E, Meinck HM, Suter U, Nave KA (1996): A transgenic rat model of Charcot-Marie-Tooth disease. Neuron
16:1049–1060.

Skre H (1974): Genetic and clinical aspects of Charcot-Marie-Tooth's disease. Clin Genet 6:98–118.

Sohl G, Gillen C, Bosse F, Gleichmann M, Muller HW, Willecke K (1996): A second alternative transcript of the
gap junction gene connexin32 is expressed in murine Schwann cells and modulated in injured sciatic
nerve. Eur J Cell Biol 69:267–275.

Sorour E, Upadhyaya M (1995): Identification of seven novel mutations in peripheral myelin genes. Am J Hum
Genet 57:A229.

Stauffer KA, Unwin N (1992): Structure of gap junction channels. Sem Cell Biol 3:17–20.

Suter U, Snipes GJ (1995): Biology and genetics of hereditary motor and sensory neuropathies. Annu Rev Neurosci 18:45–75.

Tan CC, Ainsworth PJ, Hahn AF, MacLeod PM (1996): Novel mutations in the connexin 32 gene associated with
X-linked Charcot-Marie tooth disease. Hum Mutat 7:167–171.

Tetzlaff W (1982): Tight junction contact events and temporary gap junctions in the sciatic nerve fibres of the
chicken during Wallerian degeneration and subsequent regeneration. J Neurocytol 11:839–358.

Timmerman V, Dejonghe P, Spoelders P, Simokovic S, Lofgren A, Nelis E, Vance J, Martin JJ, Van Broeckhoven C
(1996): Linkage and mutation analysis of Charcot-Marie-Tooth neuropathy type 2 families with chromosomes 1p35-p36 and Xq13. Neurology 46:1311–1318.

Webster HD, Schroder JM, Asbury AK, Adams RD (1967): The role of Schwann cells in the formation of "onion
bulbs" found in chronic neuropathies. J Neuropathol Exp Neurol 26:276–299.

Woratz G (1964): Neurale Muskelatrophie mit Dominantem X-Chromosomalem Erbgang. Berlin: Akademie-Ver-
lag, 99p.

10

THE ROLE OF MYELIN PO PROTEIN

Marie T. Filbin, Man Har Wong, Kejia Zhang, and Wen Hui Li

Department of Biological Sciences
Hunter College of the City of New York
695 Park Ave.
New York, New York 10021

INTRODUCTION

The compact layers of peripheral nervous system (PNS) myelin are held together by the Po protein, a protein specific to this structure. At over 50% of the total protein, the Po protein is the most abundant protein of PNS myelin (Everly et al., 1973; Kitamura et al., 1976). It is a small molecule of approximately 30 kD, 6% of which is carbohydrate (Everly et al., 1973; Kitamura et al., 1976). Despite its size, in myelin, it is widely believed to be adhesive via both its extracellular sequences, forming the intraperiod line, and its cytoplasmic sequences, forming the major dense line (Kirschner and Ganser, 1980; Braun, 1984; Lemke and Axel, 1985; Filbin and Tennekoon, 1992). The extracellular sequences of Po contain a single immunoglobulin (Ig)-like domain and as all members of the Ig-superfamily of molecules are believed to be involved in recognition/adhesion (Williams and Barclay, 1988), it is not surprising that the extracellular domain engages in homophilic interactions (Filbin et al., 1990; D'Urso et al., 1990; Schneider-Schaulies et al., 1990). On the other hand, the cytoplasmic domain of Po carries an overall basic charge and for this reason is proposed to hold the cytoplasmic leaflets of myelin together by interacting with acidic lipids in the opposing membrane (Braun, 1984; Lemke and Axel, 1985; Filbin and Tennkoon, 1992). The importance of Po in the functioning of the PNS is manifest *in vivo* both in mice engineered by homologous recombination to carry a null mutation in the Po gene, and also in patients suffering from the peripheral neuropathy, Charcot-Marie-Tooth (CMT) 1B disease. In Po-knockout mice, the majority of axons are surrounded by only loose whorls of uncompacted myelin and older mice exhibit tremors and convulsions (Giese et al., 1992). All pedigrees of the disease CMT 1B are heterozygote for a point mutation or a single amino acid deletion in Po and the disease is manifest by weakness in the extremities, with different degrees of severity associated with different pedigrees (Chance and Pleasure, 1993; Chance and Fischbeck, 1994). Until recently, Po protein has only been regarded as an inert, adhesive molecule but a number of reports suggest that this small molecule may have a more dynamic role both in the formation of myelin and in its

Cell Biology and Pathology of Myelin, edited by Juurlink *et al.*
Plenum Press, New York, 1997

maintenance. Here we will examine these novel findings in light of what has been well established for the functioning of Po and what has long been proposed.

THE ADHESIVE INTERACTIONS OF PO PROTEIN

The Extracellular Domain of Po

The Ig-like domain in the extracellular sequences of Po spans Cys21 to Cys98 which form a disulfide bond, characteristic of this family of molecules (Lai et al., 1987; Lemke et al., 1988). As its amino acids can be aligned into 9 β-strands, which come together to form the two β-sheets, the Ig-domain of Po resembles a variable (V)-like Ig-domain (Williams and Barclay, 1988). Within the Ig-domain, at Asn93, is the single N-linked glycosylation site of Po which carries both complex and high-mannose type structures, with considerable variation in the microheterogeneity of the terminal sugars (Everly et al., 1973; Kitamura et al., 1976; Lemke and Axel, 1985; Sakamoto et al., 1987). In addition, in some species, Po has been shown to carry the HNK-1 epitope (Bollensen and Schachner, 1987; Kunemund et al., 1988), a sulfated glucuronic acid shown to be inherently adhesive (Abo and Balch, 1981).

The extracellular domains of Po protein have long been suggested to engage in homophilic adhesion (Kirschner and Ganser, 1980; Braun, 1984; Lemke and Axel, 1985). This was proposed because of its abundance, even before Po was identified as a member of the Ig-superfamily. A number of studies *in vitro* demonstrate directly that indeed Po can behave as a homophilic adhesion molecule (Filbin et al., 1990; D'Urso et al., 1990; Schneider-Schaulies et al., 1990). In our studies, we showed that Chinese hamster ovary (CHO) cells, induced to express Po by transfection of the Po cDNA, were at least two orders of magnitude as adhesive as control transfected CHO cells, not expressing Po (Filbin et al., 1990). The specificity of the Po:Po interaction was demonstrated by the ability of Po antibodies to block adhesion (Filbin and Tennekoon, 1991). At about the same time D'Urso and co-workers (D'Urso et al., 1990), showed that in culture, when two transfected Po-expressing cells met, Po accumulated at the interface of the two cells. In contrast, if a Po-expressing cell was in contact with a non-expressor, Po was evenly distributed over the entire surface of the cell. Similar results from aggregation and immuno-localization studies using Po-expressing cells were reported by two other groups (Schneider-Schaulies et al., 1990; Yazaki et al., 1992).

Once Po had been shown to be a homophilic adhesion molecule, the next step was to identify more precisely its adhesive domains. The CHO cell transfection/aggregation system is ideal for carrying out such studies because it can be manipulated in a number of ways: the Po cDNA can be mutated before transfection; the cDNA can be transfected into a mutant CHO cell line that produces a Po molecule with aberrant post-translational modifications; and, finally, antibodies or peptides can be added to the system to block aggregation. We have applied all these methods in our approach to mapping the functional domains of Po. Initially, the sugar residues were the focus of investigation. The sugar composition of Po was altered in two ways, by preventing their addition completely by mutating Asn93 (Filbin and Tennekoon, 1993) and by changing them from a mix of complex and high-mannose sugars to all high-mannose, by expressing Po in a glycosylation mutant CHO cell line (Filbin and Tennekoon, 1991). After either changing the carbohydrate or their elimination, the adhesive properties of Po were lost. In addition, by mixed aggregation/adhesion assays, with cells expressing unglycosylated-Po and cells expressing fully glycosylated-Po, it was concluded that both Po molecules in the homophilic pair must be glycosylated for adhesion to occur. However, from a molecular model of Po based on the crystal structure of another Ig-domain (the V_H-domain of mouse immunoglobulin M603), the sugar residues are not predicted to be in a position to interact with an opposing Po molecule (Wells et al., 1993). For this reason, we propose that the sugar residues of Po are important indirectly for adhesion, in that they hold the Ig-domain away from the membrane from which it extends, in a position such that other regions of the Ig-domain can interact with an opposing Po molecule (Filbin et al., 1996). Alternatively, or additionally, it is possible that the sugars of Po are required for homophilic *cis* interactions i.e. Po:Po interactions within the same plane of the membrane. Recently we have shown that Po does indeed cluster within the membrane (see below; Wong and Filbin, 1996) and others have suggested that the carbohydrate HNK-1 epitope on Po contributes to this cluster formation (Griffith et al., 1992). It should be noted that expression of HNK-1 on Po is species-dependent (Bollensen and Schachner, 1987; Kunemund et al., 1988) and therefore cannot be ubiquitously responsible for adhesion. However, it is possible that a sulfated glucuronic acid structure similar to, but not identical with, HNK-1 (Chou et al., 1986), which is not recognized by the HNK-1-specific monoclonal antibody, is found on all other species. It is of interest that one of the CMT 1B pedigrees is heterozygote for a mutation at Asn93 (Blanquet-Grossard et al., 1996). This strengthens a role for the sugars in the functioning of Po.

As the interactions of Po with an opposing Po molecule (*trans* interactions), are unlikely to be via the carbohydrate residues, the adhesive domain(s) must be carried by amino acids. The Ig-like domain of Po encompasses the entire extracellular sequence (Lai et al., 1987). Like the majority of Ig-like domains, the domain in Po is stabilized by the formation of a disulfide bond, in this case between Cys21 and Cys98. As a first step in localizing the adhesive domains, the effect of preventing formation of this bond by mutating Cys21, which should prevent conformational integrity, on the homophilic binding of Po was assessed. When expressed in CHO cells, Po mutated at Cys21 reached the cell surface but did not engage in homophilic adhesion. Therefore, as predicted, it was confirmed that the Ig-domain of Po must be conformationally intact for adhesion to take place (Zhang and Filbin, 1994).

To map more precisely the amino acids involved directly in adhesion, a number of antibodies to various peptide sequences in the extracellular domain, and the peptides themselves, were tested for their ability to block homophilic adhesion. Hydrophilic peptide sequences in Po were chosen, based on a hydrophobicity plot, as they were more likely to be exposed at the surface of the molecule and in a position to interact with another Po protein. From these results, one short sequence from Po amino acid # 91–95, SDNGT, appears to be involved in adhesion; antibodies to this sequence and the peptide itself completely block Po:Po-mediated cell adhesion (Zhang et al., 1996). This sequence is of particular interest for a number of reasons: it spans the glycosylation site at Asn93; two amino acids within the sequence, Asp92 and Gly94, are conserved in a large number of V-like Ig domains (Williams and Barclay, 1988); and, this sequence is found in the haemagglutinin protein from the swine flu vaccine which when used, resulted in an outbreak of the demyelinating disease Guillian-Barre syndrome (Schonberger et al., 1979; Safranek et al., 1991). The importance of this sequence in adhesion of Po is supported by the observation that when Asp92 and Gly94 are mutated, adhesion is lost (Zhang et al., 1996). Although SDNGT spans the glycosylation site, Po mutated at both Asp92 and Gly94, was still glycosylated to the same extent as wildtype Po. It is possible that this sequence interacts with the carbohydrates of Po and plays a role in their alignment, optimizing the positioning of other amino acids for interaction. A direct relationship between this sequence and Guillian-Barre syndrome has yet to be described. However, Yazaki and colleagues have also shown that a Po peptide sequence, Po amino acids #91–98, encompassing SDNGT, either glycosylated or not, can inhibit homophilic binding (Yazaki et al., 1992).

The Cytoplasmic Domain of Po

Like its extracellular domain, the cytoplasmic domain of Po has been proposed to have an adhesive role. Also like the extracellular domain, this proposition was made because of the tremendous abundance of Po at this membrane surface relative to other proteins. However, the cytoplasmic domain, because of its strong basic charge, unlike the extracellular domain, is believed to engage in heterophilic interactions with acidic lipids in the opposing membrane (Kirschner and Ganser, 1980; Braun, 1984; Lemke and Axel, 1985; Filbin and Tennkoon, 1992). In support of this suggestion, it has been shown that the isolated cytoplasmic domain of Po induces aggregation of liposomes composed of acidic lipids and that the extent of aggregation is influenced by the phosphorylation state of the Po-cytoplasmic domain (Ding and Brunden, 1994).

The importance of the cytoplasmic domain *in vivo* is exemplified in two strains of mutant mouse, the *shiverer* mouse (Roach et al., 1983) and the Po knockout mouse (Giese et al., 1992). The *shiverer* mutant mouse carries a deletion in myelin basic protein (MBP), which is expressed abundantly in CNS myelin (30–40% of the total protein) but is only a minor constituent of PNS myelin (5–10% of the total protein) (Braun, 1984). In both the CNS and PNS, however, MBP is an extrinsic protein on the cytoplasmic surface where it is believed to maintain the major dense line of myelin. Observations in the *shiverer* mutant mouse confirm this suggestion for the CNS, as little or no CNS myelin is found in these animals, and what little myelin is present contains no major dense line. In contrast, in the PNS, the myelin of *shiverer* animals appears normal, indicating that Po, the only protein present in any abundance, most likely maintains this structure (Kirschner and Ganser, 1980). Similarly, because only loose whorls of membrane, uncompacted at cytoplasmic as well as the extracellular leaflets, are observed surrounding the majority of axons in the Po-knockout mouse, it follows that Po is required for compaction at the major dense line (Giese et al., 1992).

The cytoplasmic domain of Po has been shown to be acylated at Cys153, just at the cytoplasmic-transmembrane domain interface (Agrawal et al., 1983; Bizzozero et al., 1994). Covalent addition of fatty acids at this location may serve to stabilize the interactions of the cytoplasmic domain by holding it in the correct conformation. In addition, Po has been shown to be phosphorylated on a number of serine and threonine residues in the cytoplasmic domain (Brunden and Poduslo, 1987; Suzuki et al., 1990). Phosphorylation and dephosphorylation of Po has been reported to be ongoing in mature myelin and for this reason may affect the adhesive properties of the cytoplasmic domain (Brunden and Poduslo, 1987). Indeed, as described above, this is apparently so, since the aggregation of liposomes by the cytoplasmic domain of Po is strongly influenced by its phosphorylation (Ding and Brunden, 1994). Furthermore, the phosphorylation state of Po may be essential in its putative signal transduction capabilities (see below).

Like other adhesion molecules, such as the cadherins (Nagafuchi and Takeichi, 1988; Nagafuchi et al., 1991) and the integrins (for reviews see, Hynes, 1992; Gumbiner, 1993), it is also possible that the cytoplasmic domain of Po, in addition to being adhesive in its own right, can influence the adhesive function of the extracellular domain. To address this issue, Po proteins truncated in their cytoplasmic domains, missing either 52

(TPo52) or 59 (TPo59) amino acids, were expressed in CHO cells and adhesion of their extracellular sequences assessed. It was found that although each truncated Po protein reached the cell surface and each was glycosylated, neither behaved as a homophilic adhesion molecule (Wong and Filbin, 1994). The question then arises as to how the cytoplasmic domain of a protein can influence the interactions of the extracellular sequences when a lipid bilayer separates the two? It is possible that the conformational integrity of the extracellular domain requires an intact cytoplasmic domain. Alternatively, or additionally, the cytoplasmic domain could influence the interactions of Po within the same plane of the membrane. That is to say, the cytoplasmic domain of Po may be required to induce clustering of the molecule, which in turn is required for strong *trans* adhesion (Fig. 1). It is also possible that clustering is brought about subsequent to an interaction of Po with the cytoskeleton. We have evidence which indeed suggests that the cytoplasmic domain of Po is dynamically involved in the interactions of the extracellular domain. First, by chemical cross-linking and immunoprecipitation, we have shown that Po exists as clusters within the membrane (Wong and Filbin, 1996). Second, approximately 25–30% of intact Po expressed by CHO cells is insoluble in the non-ionic detergent NP40, which is a strong indication of an interaction with the cytoskeleton. In sharp contrast, only 5% of Po missing 52 amino acids from the cytoplasmic domain and none of Po missing 59 amino acids is insoluble in this detergent. Hence, as the cytoplasmic do-

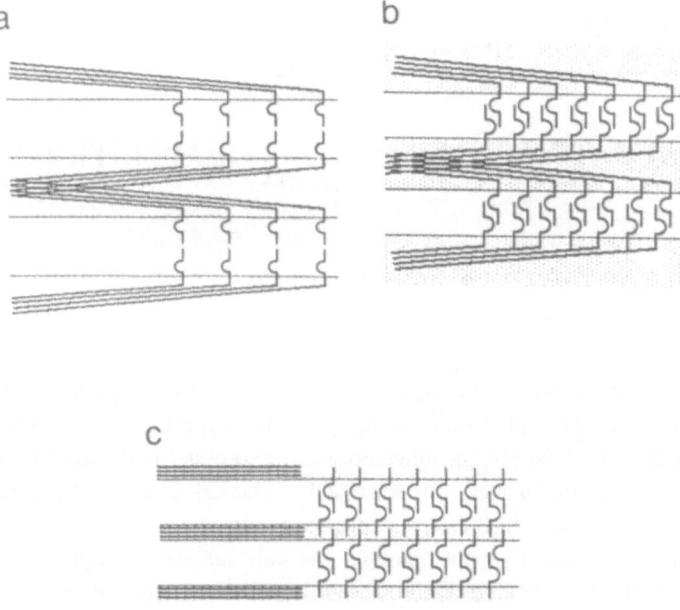

Figure 1. Model of Po's interactions during the compaction of myelin. (a) An initial low affinity Po:Po extracellular domain interaction triggers a change in the interaction of the cytoplasmic domain with the cytoskeleton. (b) The cytoskeleton reorganizes and pulls back towards the nucleus, inducing Po clustering which in turn strengthens the adhesion of the extracellular domains. (c) The cytoskeleton pulls back until the cytoplasmic domains of Po are brought into contact with an opposing membrane. The cytoskeleton disengages. Myelin is compact. Stippled areas represent the cytoplasmic surfaces. Clear areas represent the extracellular surfaces (Taken from Wong and Filbin, 1994).

main of Po is removed, the apparent interaction with the cytoskeleton is lost, coincident with a loss of adhesion of the extracellular domain (Wong and Filbin, 1994). Third, cells expressing full-length Po, when treated with the cytoskeleton disrupting drug, cytochalasin, are no longer adhesive (Wong and Filbin, 1996). As there are no cytoskeleton elements in compact myelin, these putative interactions of Po with the cytoskeleton must occur at the beginning of myelination, when the membranes are first coming together in a compact form, before cytoplasm has been extruded. Taking all of these observations into account, we propose a model for the interactions of Po, both in the dynamics of compaction and in the maintenance of myelin (Fig. 1). In this model, it is suggested that there is an initial low affinity interaction of the extracellular domains of Po, which triggers a rearrangement of the cytoskeleton. The cytoskeleton pulls back and induces clustering of Po within the same membrane and promotes *cis* interactions between either the extracellular and/or the cytoplasmic sequences. The clusters are stabilized by the *cis* interactions and adhesion is strengthened by the clustering. Finally, the cytoskeleton pulls back far enough such that the cytoplasmic membranes are opposed. The cytoplasmic domain of Po disengages from the cytoskeleton and interacts with acidic lipids in the opposing membrane. Myelin is compact and cytoskeleton and cytoplasm have been extruded. It is possible that the cytoskeleton interaction of Po induces clustering before the extracellular domains come into contact with each other. This is suggested by the appearance of Po as large aggregates after chemical cross-linking and immunoprecipitation of Po-expressing cells in a single-cell-suspension, indicating that cluster formation is independent of cell:cell contact (Wong and Filbin, 1996). Regardless of whether clustering occurs before or after extracellular domain interaction, the model dictates that clustering is a prerequisite to strong, stable Po:Po adhesion.

DOMINANT-NEGATIVE INTERACTIONS OF MUTATED PO PROTEINS

Dominant-Negative Effects on Adhesion of Po Truncated in Its Cytoplasmic Domain

Based on the model represented in Fig. 1, it is possible that the presence of mutated, non-adhesive forms of Po affect the functioning of the wildtype molecule. To assess if cytoplasmic domain-truncated Po affects adhesion when expressed with full-length Po, both truncated Po and full-length Po were co-expressed in the same cell and adhesion monitored (Wong and Filbin, 1996). In these co-expressing cells it was ensured that both types of Po, truncated and full-length, were reaching the cell surface in approximately equal amounts before carrying out the aggregation assay. In addition, cell lines were chosen that expressed at least equivalent amounts of full-length Po, which when expressed alone, allows strong adhesion to occur. However, when the adhesion assay was carried out with the cells co-expressing truncated Po (either missing 52 or 59 amino acids), and full-length Po, the cells did not aggregate; they remained as single cells or doublets or triplets (Fig 2 (a)-(c)). Consistent with the microscopic observation of cell-aggregation, when the total particle number was counted at different times (a drop in total particle number with time is indicative of aggregate formation), the total particle number for the co-expressing cells did not change; it was indistinguishable from the control transfected cells, not expressing

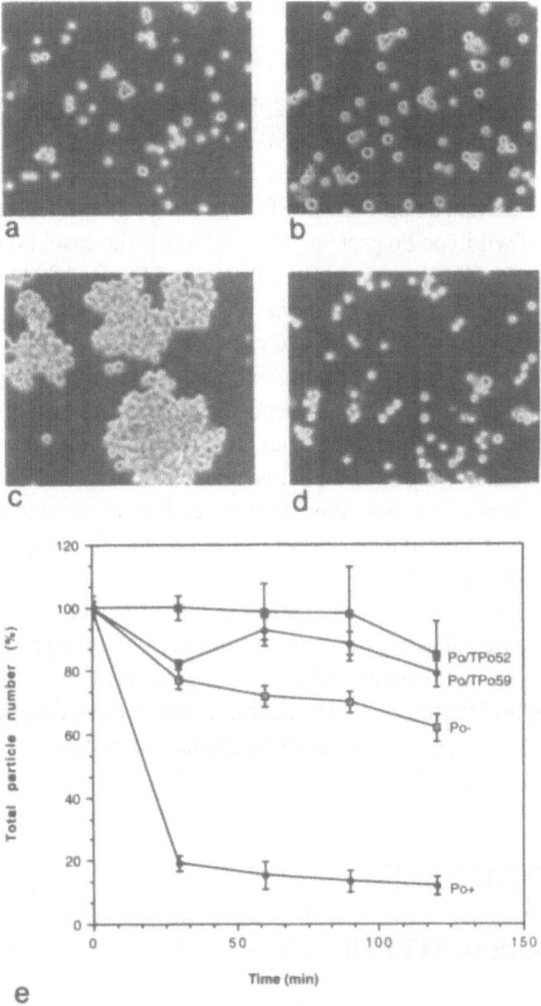

Figure 2. Aggregation properties of Po-expressing cells. Single-cell suspensions of CHO cells expressing (a) full-length Po and TPo52, (b) full-length Po and TPo59, (c) full-length Po only, or (d) control-transfected cells were allowed to aggregate. Samples were withdrawn at intervals and examined under the microscope (a - d) results after 60 min aggregation) and the total particle number was counted in a Coulter counter. The total particle number +/- SE was plotted against time (e) (Taken from Wong and Filbin, 1996).

Po (Fig. 2(d)). These results suggest that the expression of truncated Po with full-length Po prevents adhesion through a dominant-negative mechanism.

If these results are considered in light of the model proposed in Fig. 1, it is possible that the truncated Po is preventing strong adhesion by preventing the clustering of full-length Po. This is likely to be so because after chemical cross-linking and immunoprecipitation, neither full-length Po nor truncated Po expressed by the co-expressors form large aggregates (Wong and Filbin, 1996). This indicates that, unlike the clustering observed when full-length Po is expressed alone, in the co-expressors, Po fails to cluster. How clustering is prevented by truncated Po is not yet known.

Dominant-Negative Effects on Adhesion of Po Mutated in Its Extracellular Domain

For other adhesion molecules (Kitner, 1992; Fujimori and Takeichi, 1993; Balzac et al., 1994; Smilenov et al., 1994), a dominant-negative effect on function has been reported when proteins mutated in the extracellular domain are co-expressed with wildtype protein. In a similar fashion, it is possible that a non-adhesive Po mutated in its extracellular domain, could also affect the adhesion of wildtype Po protein. To test if this is the case, Po mutated at Cys21, which we have shown is not adhesive (see above; Zhang and Filbin, 1994), was co-expressed with wildtype Po. A cell line expressing wildtype Po and already shown to be adhesive was re-transfected with the cDNA for Po mutated at Cys21 carried in a vector with a novel selectable marker. Resistant colonies were screened for expression of both wildtype and Cys21-mutated Po by western analysis in the presence and absence of β-mercaptoethanol; β–mercaptoethanol will distinguish wildtype from Cys21-mutated Po because there is no difference in apparent molecular weight of the latter under reducing and non-reducing conditions. Two individual clones, shown after re-transfection to co-express wildtype and Cys21-mutated Po, lost their adhesive properties. A third clone, although resistant to the second selectable marker, still only expressed wildtype Po and retained its adhesiveness. These results suggest that the introduction of Cys21-mutated Po into a cell line already expressing wildtype Po and shown to be adhesive, prevents the adhesion of wildtype Po (Zhang and Filbin, 1996). It is possible that, in the model in Fig. 1, non-adhesive Cys21-mutated Po joins the *cis* clusters of molecules and "dilutes" out the adhesiveness of the wildtype Po within a cluster. This in turn implies that a critical number of functional Po proteins are required per cluster, for strong adhesion to occur.

PO-ASSOCIATED DISEASES

Charcot-Marie-Tooth (CMT) 1B

The peripheral neuropathy Charcot-Marie-Tooth (CMT) is characterized by demyelination and attempted remyelination. Symptoms can vary tremendously in severity and can have either an early or a late onset (for reviews see, Chance and Pleasure, 1993; Chance and Fischbeck, 1994; Patel and Lupski, 1994). In recent years a number of reports have described a variety of point mutations or single amino acid deletions in different pedigrees of the sub-group CMT 1B (see, Patel and Lupski, 1994). Except for one pedigree with a mutation in the transmembrane region, all of these mutations/deletions are found in the extracellular domain, however, they are not clustered in any one location. Interestingly, all pedigrees of CMT 1B are heterozygotes and as a consequence produce 50% of wildtype Po protein. Initial observations of heterozygote Po-deficient mice suggested that half a gene dose of Po had no effect on myelin formation and maintenance (Giese et al., 1992). However, it was realized that in mice older than 4 months of age, significant demyelination and axonal degeneration occurred (Martini et al., 1995a). Based on these observations of older heterozygote Po-deficient mice, half a gene dose of Po can account for phenotypes of CMT 1B that are mild and have a late onset. For more severe phenotypes with an early onset, gene dose alone is insufficient to explain the symptoms. There are at least three possible mechanisms whereby heterozygote mutations in Po can affect myelin. First, the mutated protein does not reach myelin but is held up in the Schwann cell body. Consequently, only half the normal amount of Po would reach myelin and the disease is likely to be a result of half a gene dose. In addition, accumulation of the mutated Po in the Schwann cell could have a detrimental effect. Second, the mutated Po is not adhesive, reaches myelin, but has no effect on the functioning of the wildtype Po protein. Again, the phenotype would be a consequence of reduced levels of Po gene dose. Third, the mutated Po protein is not adhesive, reaches myelin but has a dominant-negative effect on the functioning of the wildtype protein. Under these conditions the phenotype would be predicted to be more severe and have an early onset.

How individual mutations affect the functioning of Po can be tested directly using the *in vitro* transfection/adhesion assay. The ability of the mutated protein to reach the cell surface and to behave as an adhesion molecule, can be assessed. This we have carried out for two mutations corresponding to two CMT 1B pedigrees, Asn93 (Blanquet-Grossard et al., 1996) and Ser34 (Kulkens et al., 1993). For both we found that the mutated proteins reached the cell surface but were not adhesive (unpublished observations and Filbin and Tennekoon, 1993). Furthermore, by co-expression of wildtype and mutated Po in the same cell, possible dominant-negative effects can be determined. As described above, we have already shown that mutated forms of Po, either truncated in the cytoplasmic domain or mutated at Cys21, can have a dominant-negative effect on the adhesion of wildtype Po (Wong and Filbin, 1996; Zhang and Filbin, 1996). It is surprising that the different point mutations in the patients with CMT 1B are scattered throughout the extracellular domain of Po and are not clustered in any one region. This would suggest that for some, the mutation affects the adhesive binding site directly but for others this may not be the case. Instead, it is likely that some of these mutations disrupt adhesion by affecting the conformation of the molecule. If this is the case, then Po mutated at Cys21 could be the prototype for these conformation-disrupting mutations. Ultimately, the effect of CMT 1B mutations on myelination can be tested *in vivo* by creating transgenic mice heterozygote for each mutation.

Guillian-Barre Syndrome

Guillian-Barre syndrome (GBS) is a demyelinating peripheral neuropathy, usually with axonal sparing. It is believed to be an autoimmune disease, although a definitive autoantigen has yet to be identified (Ropper et al., 1991). It is not known what triggers GBS but it is frequently associated with a post-viral infection. Consistent with this observation, after immunization with a particular strain of swine influenza vaccine, there was an statistically-significant increase in GBS among recipients of the vaccine (Schonberger et al., 1979; Safranek et al., 1991). As noted above, this strain of influenza vaccine carries the SDNGT sequence on its haemagglutinin protein and we have shown that this sequence is important for the adhesive functioning of Po. A direct connection between this sequence and GBS has yet to be demonstrated, however, the demyelinating disease, experimental allergic neuropathy (EAN), has been induced in animals after immunization with Po protein (Milner et al., 1987).

PO PROTEIN AS A DYNAMIC MOLECULE

In the model presented in Fig. 1 we propose that, in addition to holding mature myelin in a stable conformation, Po plays an active role in bringing the membranes together. By interacting with the cytoskeleton, Po helps bring the cytoplasmic leaflets together, which in turn extrudes cytoplasm and cytoskeleton. This implies that Po is more than an inert adhesion molecule.

As well as directly influencing myelin compaction, recently it has been suggested that Po may be able to participate in signal transduction. This suggestion derives from a number of observations. (1) In Po-knockout mice a number of other myelin proteins are mis-regulated;

MBP is under-expressed while myelin associated glycoprotein (MAG) is over-expressed (Giese et al., 1992). This indicates that Po plays a role in regulating the expression of these molecules which can be either direct or indirect. (2) In mice heterozygote for Po expression and also deficient in MBP expression, there is a reduced number of myelin lamellae around large axons (Martini et al., 1995b). The implication of these results is that both these molecules contribute to the determination of the thickness of myelin because mice heterozygous for Po but with the normal levels of MBP, and mice with normal Po but deficient in MBP, each have the usual thickness of myelin (Kirschner and Ganser, 1980). Po could be signaling back to the Schwann cell body when sufficient myelin has been laid down. As Po is dynamically phosphorylated in compact myelin, this could contribute to the signaling process (Brunden and Poduslo, 1987). (3) Expression of Po in a transformed epithelial cell line has been shown to induce reversion to a non-transformed phenotype (Doyle et al., 1995). This reversion is believed to be brought about by the up-regulation of other epithelial-specific adhesion molecules upon expression of Po by these cells and subsequent Po-mediated cell adhesion. It is possible that upon interaction, Po is mediating a signal which, in epithelial cells, results in expression of epithelial-specific proteins. It should be noted, however, that direct demonstration of Po's involvement in signal transduction has not yet been reported.

SUMMARY

The myelin Po protein can be regarded as a "double-adhesive" molecule because of the adhesive properties of both its extracellular and cytoplasmic domains. Furthermore, because of the influence the cytoplasmic domain has on the functioning of the extracellular domain, these domain-specific adhesive interactions are unlikely to be exclusive, independent events. Similarly, the adhesion of Po most likely plays a role in the dynamics of myelin formation by bringing the membranes together and holding them in a stable compact form. Although there is a paucity of data in support of a signal transduction role for Po, the possibility still remains that Po directly dictates the number of myelin lamellae laid down and influences the expression of other myelin proteins. Finally, an association of Po with the peripheral neuropathy CMT 1B is unequivocal but how the individual mutations bring about the disease phenotype remains to be elucidated. A direct connection between Po and GBS is more tenuous but is still possible. For such a small, "sticky" molecule the number of its putative functions appears to be growing daily.

ACKNOWLEDGMENTS

This work was supported by NIH grant NS26242, NSF grant 9319534 and a NIH RCMI core facility grant. MTF is a recipient of a American Heart Association, New York Chapter, Established Investigator Award.

REFERENCES

Abo, T. and Balch, C.M.A. (1981). Differentiation antigen of human NK and K cells identified by a monoclonal antibody (HNK-1). J. Immunol. 127:1024–1029.
Agrawal, H.C., Schmidt, R.E. and Agrawal, D. (1983). *In vivo* incorporation of H³ Palmitic acid into Po protein, the major intrinsic protein of rat sciatic nerve. J.Biol. Chem. 258:6556–6560.

Balzac, F., Retta, S.F., Albini, A., Melchiorri, A., Koteliansky, V.E., Geuna, M., Silengo, L. and Tarone, G. (1994). Expression of beta-1 integrin isoform in CHO cells results in a dominant negative effect on cell adhesion and motility. J. Cell Biol. 127:557–565.

Blanquet-Grossard, F., Pham-Dinh, D., Dautigny, A., Latour, P., Bonnebouche, C., Diraison, P., Chapon, F., Chazot, G. and Vandenbergne, A. (1996). Charcot-Marie-Tooth type 1B neuropathy: a mutation at the single glycosylation site in the major peripheral myelin glycoprotein Po. Human Mutations. In press.

Bollensen, E. and Schachner, M. (1987). The peripheral myelin glycoprotein Po expresses the L2/HNK-1 and L3 carbohydrate structures shared by neural adhesion molecules. Neurosci. Lett. 82:77–82.

Braun, P.E. (1984). Molecular organization of myelin. In Myelin. P. Morell, editor. Plenum Publishing Corp., New York. 97–116.

Bizzozero, O.A., Fridal, K. and Pastuszyn, A. (1994). Identification of the palmitoylation site in rat myelin Po glycoprotein. J. Neurochem. 62:1163–1171

Brunden, K.R. and Poduslo, J.F. (1987). A phorbol ester-sensitive kinase catalyzes the phosphorylation of Po glycoprotein in myelin. J. Neurochem. 46:1863–1872.

Chou, D.K.H., Ilyas, A.A., Evans, J.E., Costello, C., Quarles, R.H. and Jungalwala, F.B. (1986). Structure of sulfated glucuronyl glycolipids in the nervous system reacting with HNK-1 antibody and some IgM paraproteins in neuropathy. J.Biol. Chem. 261: 11717–11725.

Chance, P. and Fischbeck, K.H. (1994). Molecular genetics of Charcot-Marie-Tooth disease and related neuropathies. Hum. Mol. Genet. 3:1503–1507.

Chance, P. and Pleasure, D. (1993). Charcot-Marie-Tooth syndrome. Arch. Neurol. 50:1180–1184.

Ding, Y. and Brunden, K.R. (1994). The cytoplasmic domain of myelin glycoprotein Po interacts with negatively-charged phospholipid bilayers. J. Biol. Chem. 269:10764–10770.

Doyle, J.P., Stempak, J.G., Cowin, P., Colman, D.R. and D'Urso, D. (1995). Protein zero, a nervous system adhesion molecule, triggers epithelial reversion in host carcinoma cells. J. Cell Biol. 131:465–482.

D'Urso, D., Brophy, P.J., Staugaitus, S.M., Gillespie, C.S., Frey, A.B., Stempack, J.G. and Colman, D.R. (1990). Protein zero of peripheral nerve myelin: biosynthesis, membrane insertion and evidence for homotypic interaction. Neuron 2:449–460.

Everly, J. L., Brady, R. O. and Quarles, R. H. (1973). Evidence that the major protein in rat sciatic nerve is a glycoprotein. J. Neurochem. 21:329–334.

Filbin, M.T., and Tennekoon, G.I. (1990). High level of expression of the myelin protein, Po, in Chinese hamster ovary cells. J. Neurochem. 55:500–505.

Filbin, M.T., and Tennekoon, G.I. (1991). The role of complex carbohydrates in adhesion of the myelin protein, Po. Neuron 7:845–855.

Filbin, M.T., and Tennekoon, G.I. (1992). Myelin Po-protein, more than just a structural protein? BioEssays 14:541–547.

Filbin, M.T., and Tennekoon, G.I. (1993). Homophilic adhesion of the myelin Po protein requires glycosylation of both molecules in the homophilic pair. J. Cell Biol. 122:451–459.

Filbin, M.T., Walsh, F.S., Trapp, B.D., Pizzey, J.A. and Tennekoon, G.I. (1990). Role of myelin Po-protein as a homophilic adhesion molecule. Nature (Lond.) 344:871–872.

Filbin, M.T., D'Urso, D., Zhang, K., Wong, M.H., Doyle, J.P. and Colman, D.R. (1996). Protein zero of peripheral nerve myelin: Adhesion properties and functional models. Adv. Mol. Cell Biol. 16:159–192.

Fujimori, T. and Takeichi, M. (1993). Disruption of epithelial cell-cell adhesion by exogenous expression of a mutated nonfunctional N-cadherin. Mol. Biol. Cell 4:37–47.

Giese, K.P., Martini, R., Lemke, G., Soriano, P. and Schachner, M. (1992). Mouse Po gene disruption leads to hypomyelination, abnormal expression of recognition molecules and degeneration of myelin and axons. Cell 71:565–576.

Gumbiner, B. M. (1993). Proteins associated with the cytoplasmic surface of adhesion molecules. Neuron 11:551–564.

Griffith, C.S., Schmitz, B., and Schachner, M. (1992). L2/HNK-1 carbohydrate and protein-protein interactions mediate the homophilic binding of the neural adhesion molecule Po. J. Neurosci. Res. 33:639–648.

Hynes, R.O. (1992). Integrins: Versatility, modulation and signaling in cell adhesion. Cell 69:11–25.

Kirschner, A. and Ganser, A.L. (1980). Compact myelin exists in the absence of basic protein in the shiverer mutant mouse. Nature 283:207–210.

Kitamura, K., Suzuki, M. and Uyemura, K. (1976). Purification and partial characterization of two glycoproteins in bovine peripheral nerve myelin membrane. Biochim. Biophys. Acta. 455:806–809.

Kintner, C. (1992). Regulation of embryonic cell adhesion by cadherins cytoplasmic domain. Cell 69:225–236.

Kulkens, T., Bolhuis, P.A., Wolterman, R. A., Kemp, S., Nijenhuis, S., Valentijn, L. J., Hensels, G. W., Jennekens, F. G. I., Visser, M., Hoogendijk, J. E. and Baas, F. (1993). Deletion of the serine 34 codon from the major peripheral myelin protein Po gene in Charcot-Marie-Tooth disease type 1B. Nature Genetics 5: 35–39.

Kunemund, V., Jungalwala,F.B., Fischer, G., Chou, D.K., Keilhauer, G. and Schachner, M. (1988). The L2/HNK-1 carbohydrate of neural cell adhesion molecules is involved in cell interactions. J. Cell Biol. 106: 213–223.

Lai, C., Brow, M.A., Nave, K.A., Noronha, A.B., Quarles, R.H., Bloom, F.E., Milner, R.J. and Sutcliffe, J.G. (1987). Two forms of 1B236/myelin-associated glycoprotein, a cell adhesion molecule for postnatal neural development, are produced by alternative splicing. Proc. Natl. Acad. Sci. USA 84:4337–4341.

Lemke, G., and Axel, R. (1985). Isolation and sequence of a cDNA encoding the major structural protein of peripheral myelin. Cell 40:501–508.

Lemke, G., Lamar, E. and Patterson, J. (1988). Isolation and analysis of the gene encoding peripheral myelin protein zero. Neuron 1:73–83.

Martini, R., Zielasek, J., Toyka, K.V., Giese, K.P. and Schachner, M. (1995a). Protein zero (PO)-deficient mice show myelin degeneration in peripheral nerves characteristic of inherited human neuropathies. Nature Genetics 11:281–285.

Martini, R., Mohajeri, M.H., Kasper, S., Giese, K.P. and Schachner, M. (1995b). Mice doubly deficient in the genes for Po and myelin basic protein show that both proteins contribute to the formation of the major dense line in peripheral nerve myelin. J. Neurosci. 15:4488–4495.

Milner, P., Lovelidge, C.A., Taylor, W.A. and Hughes, R.A.C. (1987). Po myelin protein produces experimental allergic neuritis in Lewis rats. J. Neurol. Sci. 79:275–285.

Nagafuchi, A. and Takeichi, M. (1988). Cell binding function of E-cadherin-associated protein: similarity to vinculin and posttranscriptional regulation of expression. EMBO J. 7:3679–3684.

Nagafuchi, A., Takeichi, M. and Tsukita, S. (1991). The 102kd cadherin-associated protein: similarity to vinculin and posttranscriptional regulation of expression. Cell 65:849–857.

Patel, P.I. and Lupski, J.R. (1994). Charcot-Marie-Tooth disease: a new paradigm for the mechanism of inherited disease. Trends in Genetics 10:128–133.

Roach, A., Boylan, K., Horvath, S., Prusiner, S.D. and Hood, L.E. (1983). Characterization of a cloned cDNA representing rat myelin basic protein: absence of expression in *shiverer* mutant mice. Cell 34:799–806.

Ropper, A.H., Wijdiks, E.F.M. and Traux, B.T. (1991). Guillian Barre Syndrome. F. A. Davis, Philiadelphia.

Safranek, T.J., Lawrence, D.N., Kurland, L.T., Culver, D.H., Wiederholt, W.C., Hayner, N.S., Osterholm, M.T., O'Brien, P. and Hughes, J.M. (1991). Reassessment of the association between Guillain-Barre syndrome and receipt of swine influenza vaccine in 1976–1977: Results of two-state study. Am. J. Epidemiol. 133:940–952.

Sakamoto, Y., Kitamura, K., Yoshimura, K., Nishijima, T. and Uyemura. K. (1987). Complete amino acid sequence of Po-protein in bovine peripheral nerve myelin. J. Biol. Chem. 262:4208–4214.

Schneider-Schaulies, J., von Brunn, A. and Schachner, M. (1990). Recombinant peripheral myelin protein Po confers both adhesion and neurite outgrowth-promoting properties. J. Neurosci. Res. 27:286–297.

Schonberger, L.B., Bregman, D.J. and Sullivan-Bolyai, J.Z. (1979). Guillain-Barre syndrome following vaccination in the National Influenza Immunization Program, United States, 1976–1977. Am. J. Epidemiol. 110:105–123.

Smilenov, L., Briesewitz, R. and Marcantonio, E.E. (1994). Integrin b1 cytoplasmic domain dominant-negative effects revealed by lysophosphatidic acid treatment. Mol. Biol. Cell 5:1215–1223.

Suzuki, M., Sakamoto, Y., Kitamura, K., Fukunaga, K., Yamamoto, H., Miyamoto, E. and Uyemura, K. (1990). Phosphorylation of Po glycoprotein in peripheral nerve myelin. J. Neurochem. 55:1966–1971.

Wells, C.A., Saavedra, R.A. and Kirschner, D.A. (1993). Myelin Po-glycoprotein: predicted structure and interactions of extracellular domain. J. Neurochem. 61:1987–1995.

Williams, A.F. and Barclay, A.N. (1988). The immunoglobulin superfamily domain for cell surface recognition. Ann. Rev. Immunol. 6:381–405.

Wong, M.H. and Filbin, M.T. (1994). The cytoplasmic domain of the myelin Po protein influences the adhesive interactions of its extracellular domain. J. Cell Biol. 126:1089–1097.

Wong, M.H.and Filbin, MT (1996). Dominant-negative effect on adhesion of the myelin Po protein truncated in its cytoplasmic domain. J. Cell Biol. 134: 1531–1542.

Yazaki, T, Miura, M., Asou, M., Kitamura, K., Toya, S. and Uyemura, K. (1992). Glycopeptide of Po protein inhibits homophilic cell adhesion. FEBS Lett. 310:204–209.

Zhang, K. and Filbin, M.T. (1994). Formation of a disulfide bond in the immunoglobulin domain of the myelin Po protein is essential for its adhesion. J. Neurochem. 63:367–370.

Zhang, K. and Filbin, M.T. (1996). Myelin Po protein mutated at Cys21 has a dominant-negative effect on adhesion of wildtype Po. Submitted

Zhang, K., Merazga, Y., and Filbin, M.T. (1996). Mapping the adhesive domains of the myelin Po protein. J. Neurosci. Res. 45: 525–533.

11

POSTTRANSCRIPTIONAL REGULATION OF MYELIN BASIC PROTEIN GENE EXPRESSION

Anthony T. Campagnoni

Mental Retardation Research Center and Brain Research Institute
UCLA Medical School
760 Westwood Plaza
Los Angeles, California 90095

INTRODUCTION

The myelin basic protein (*mbp*) gene has been the subject of intense study since its structure was first elucidated over a decade ago. Recently, the *mbp* transcription unit has been found to be embedded within a larger unit called the *golli* gene (Campagnoni et al., 1993; Pribyl et al., 1993). This unusual genetic locus, which we call the *golli-mbp* gene, gives rise to two families of alternatively spliced transcripts under the control of different promoters. The mRNAs of both the *golli* gene and *mbp* gene are under independent developmental, cellular and tissue regulation. Interestingly, expression of the *mbp* gene appears to be confined exclusively to the later stages of differentiation of myelin forming cells whereas expression of the *golli* gene is less restricted.

A large body of evidence now indicates that expression of the *mbp* gene is regulated at several levels, which include: (1) alternative promoter utilization, (2) alternative splicing events, (3) mRNA translocation, (4) translational control and (5) posttranslational modifications of the proteins.

The focus of this report will be entirely on the *mbp* gene products and the regulation of the *mbp* portion of the *golli-mbp* gene locus. Furthermore, this report will emphasize posttranscriptional events that may represent appropriate regulatory stages in the expression of the *mbp* gene.

POTENTIAL LEVELS OF POSTTRANSCRIPTIONAL REGULATION OF MBP GENE EXPRESSION

Rates of MBP mRNA Translation

There are several structural features of mRNAs that can affect their rates of translation.

5' CAPs. The presence of a CAP structure on the 5' end of mRNAs has been shown to increase the translational efficiencies of a number of mRNAs (Both et al., 1975; Muthukrishnan et al., 1975). This appears to be true for MBP mRNAs as well. Ueno et al. (1994a) have reported that the CAPPED 14 kDa MBP mRNA is translated 2.25 times more efficiently than the UNCAPPED mRNA in a cell free translation assay. Other studies (Campagnoni et al., 1987) have shown that CAP analogs can inhibit translation of the four major alternatively spliced MBP mRNAs, suggesting that all forms of MBP mRNAs are CAPPED in vivo.

5' Untranslated Region (5' UTR). Structural features of the 5' UTR have been reported to influence the rates of translation of mRNAs. These include the presence of stem-loop structures, AUG sequences, and a number of regulatory sequences (see Jansen et al. [1995] for short review). The scanning model of Kozak (1989) proposes that polypeptide chain initiation begins at the first AUG which sits within the context of a Kozak consensus sequence. In the majority of mammalian mRNAs this AUG is also the first AUG encountered as the 40S ribosomal subunit "scans" the 5' UTR downstream from the CAP. In some instances, the true initiator AUG is not the first AUG downstream from the CAP. This often happens in mRNAs with long 5' UTRs and frequently these "false" initiator AUGs are followed relatively quickly by termination signals. When this occurs the mRNA is frequently less efficiently translated, possibly due to "false starts" at these AUG sequences which slow down translation. This structural feature exists in the MBP mRNA 5' UTR. Campagnoni et al. (1987) reported that MBP mRNAs were less efficiently translated than brain mRNAs, as a whole, and suggested this might be due, in part, to the presence of this structural feature immediately upstream of the initiator AUG in MBP mRNA. This structure is shown in the adjacent diagram with the initiator AUG shown in bold and the upstream AUG shown in italics. Note that the upstream AUG is followed immediately by a termination codon, the last base of which is the first base of the true initiation codon.

> translation begins

GG*AUG*UG**AUG**GC

    init   term

To test this hypothesis, Verdi (1991) mutated several bases in this region of the MBP mRNA and examined the translational efficiencies of the mutated MBP mRNAs vs. non-mutated MBP mRNA. The data given in Table 1, taken from Verdi (1991), show that mutating one or two nucleotides in this region can increase the translational efficiency of the MBP mRNA as much as two fold. These data would support the notion that this structural feature in MBP mRNA reduces its translational efficiency.

The 5' UTR of the MBP mRNAs is relatively short, containing only 48 nucleotides. Ueno et al. (1994a) have examined the efficiencies of translation of MBP mRNAs with systematically truncated 5' UTRs. They found that deletion of 11 nucleotides from the 5' end of the "natural" mRNA decreased the translational efficiency by four fold. The 5' terminal 11 nucleotides appeared to be essential for active translation of the mRNA. Interestingly, just downstream of the 5' terminal sequence, from -29 to -37, a putative element

Table 1. Site-directed changes in the 5'UTR of the 14 kDa
MBP mRNA increases translational efficiency

Sequence	Translation rate (fmoles/min)
...A U G U G *A U* G....	5.4 ± 1.4
...A *G* G U G *A U* G....	8.7 ± 1.2
...A *G* G U *C* A U G....	11.3 ± 2.8

that might be responsible for steroid regulation of MBP mRNA translation has been reported (Verdi and Campagnoni, 1990).

There is substantial evidence that hydrocortisone can increase the expression of MBP protein in vitro. Ved et al. (1989) have shown that hydrocortisone can increase the levels of MBP in developing primary cultures, and Cabacungan et al (1991) have shown that hydrocortisone and triiodothyronine can act cooperatively to regulate MBP levels. Kumar et al. (1989) have shown that hydrocortisone can increase the levels of MBP mRNA without having an effect on transcription. Byravan and Campagnoni (1994) showed that hydrocortisone increased the rate of MBP synthesis in mouse brain primary cultures and that this effect appeared to be developmentally regulated. The action of hydrocortisone and other steroids on MBP mRNA translation may be within the 5' UTR of the MBP mRNA. Verdi and Campagnoni (1990) identified a nine nucleotide sequence in the 5' UTR (from -29 to -37) that was necessary for the cell free stimulation of MBP synthesis by steroids. Thus the 5' UTR of MBP mRNAs harbor several sequence elements that appear to contribute to the intrinsic and extrinsic regulation of their translation.

3' Untranslated Region. We have examined the importance of the 3' UTR for the translation of the MBP mRNAs in cell-free lysates (Ueno et al., 1994a). Removing most of the 3' UTR of four different MBP mRNAs reduced their translational efficiencies by 2–3 fold. In another study (Ueno et al., 1994b) we transfected a chloramphenicol acetyl transferase (CAT) reporter gene with and without the MBP 3'UTR into cultured cells and examined expression of the enzymatic activity of CAT as a measure of the amount of protein synthesized from the transfected construct. CAT activity in the mouse oligodendrocyte cell line, N20.1, transfected with the CAT-MBP3'UTR construct was twice as high as that of the CAT cDNA without the MBP3'UTR. In contrast, CAT activities of the two constructs were the same in transfected NIH 3T3 control cells. Furthermore, the levels of the CAT-MBP3'UTR mRNA were 10 fold higher than CAT mRNA without the MBP3'UTR in transfected N20.1 cells. There were no differences in the levels of the two mRNAs in the NIH 3T3 cells. These results suggest two conclusions: (1) that the MBP 3' UTR confers stability upon mRNAs in the oligodendrocyte N20.1 cell line and (2) that the MBP 3' UTR contains elements that reduce efficient translation of the mRNA, at least in the N20.1 cells. This latter conclusion stems from the observation that in the transfected N20.1 cells there is a 10 fold increase in the CAT-MBP3'UTR mRNA levels, but only a 2 fold increase in CAT activity over controls.

Thus, the 5' and 3' UTRs of MBP mRNAs appear to influence the translation of these mRNAs, and such factors may be involved in the translational regulation of MBP gene expression. The 3' UTRs of many mammalian mRNAs have been reported to bind regulatory proteins that either initiate or prevent degradation of the mRNAs (You et al., 1992; Klausner et al., 1993) as well as to control translation of the mRNA (Kwon and Hecht, 1993; Goodwin et al, 1993). In the case of the MBP mRNA Ainger et al. (1993a) have suggested that an element responsible for the translocation of the MBP mRNA is located within the 3′ UTR. The regulation of MBP mRNA translocation will be discussed in some detail in the next section.

Translocation of MBP mRNAs within Oligodendrocytes

Protein targeting within a cell can occur by several mechanisms, one of which involves the movement of mRNAs to the sites where the protein is to be located followed by synthesis of the protein in situ. There is overwhelming evidence that in the normal, post-myelinating brain MBP mRNAs can be found in myelin sheaths and in the thin processes connecting the sheaths to the oligodendrocyte cell bodies (Colman et al., 1982; Trapp et al., 1987; Verity and Campagnoni, 1988; Landry et al., 1994, also see review by Campagnoni, 1995). The mechanisms governing this translocation process have not yet been elucidated, but evidence has been presented for the existence of MBP mRNA transport particles (Ainger et al., 1993a) as well as a putative targeting signal in the 3′ UTR of MBP mRNAs (Ainger et al., 1993b). We have examined overall regulation of MBP mRNA movement within oligodendrocytes and found that this translocation is under developmental regulation and is also influenced by cell-cell contact and soluble factors, at least in vitro.

Developmental Regulation of MBP mRNA Translocation in Oligodendrocytes. We examined expression of MBP mRNAs by radioactive in situ hybridization in the developing postnatal mouse brain (Verity and Campagnoni, 1988). In general MBP mRNAs were detected prior to the appearance of myelin. We observed MBP mRNA expression as early as 6 hours post partum in the medulla oblongata. In the earliest stages (i.e. P1-P2), MBP mRNA labeling was restricted to oligodendrocyte cell bodies in regions to be myelinated. Shortly thereafter, sometimes within a 24 hour period (the shortest interval examined), there was a striking diffusion of the MBP mRNA grains throughout myelinated areas. While these data were consistent with the notion that MBP mRNA translocation occurred into oligodendrocyte processes and myelin, and that this process was developmentally regulated, the most convincing evidence of this has come from nonradioactive in situ hybridization studies where mRNA can be more precisely localized within the cell. As part of a larger study on translocation of brain mRNAs, Landry et al. (1994) have shown examples of MBP mRNAs confined to oligodendrocyte cell bodies at early stages of postnatal brain development in vivo. At later ages, within the same brain region they observed MBP mRNA localized to the cell bodies, myelin sheaths and the fine processes connecting them. Figure 1 illustrates such a nonradioactive in situ hybridization experiment and shows oligodendrocytes containing MBP mRNA in their cell bodies, processes and myelin sheaths. Amur-Umarjee et al. (1990) have shown that developmental regulation of the translocation of MBP mRNAs can be observed in mixed glial cultures in vitro as a function of time in culture.

Oligodendrocyte contact with (reactive?) astrocytes inhibits MBP mRNA translocation. We have found that MBP mRNA translocation occurs in only approximately 25% of the oligodendrocytes expressing the MBP gene in mixed primary cultures of mouse brains (Amur-Umarjee et al., 1993). However, when enriched oligodendrocytes are prepared from these mixed cultures MBP mRNA translocation occurs in about 95% of the purified oligodendrocytes. When these purified oligodendrocytes were replated on purified primary astrocytes, MBP mRNA translocation was evident in only about 10% of the oligodendrocytes. We have also found that this inhibition of translocation is not due to

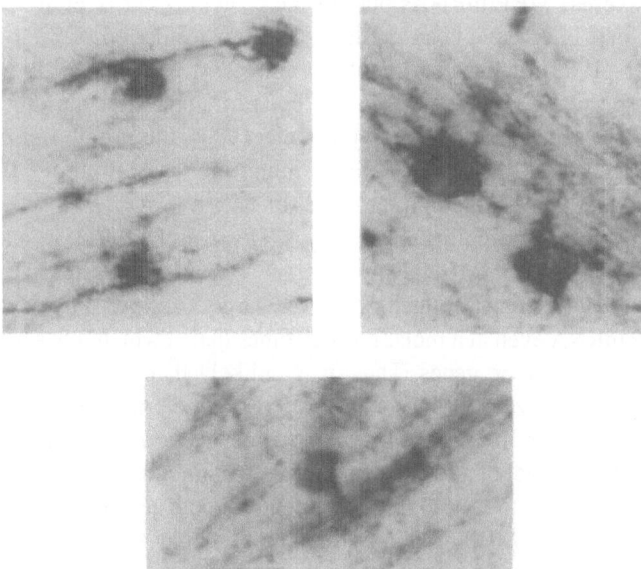

Figure 1. Montage of several mouse brain oligodendrocytes showing the presence of MBP mRNA in cell bodies, processes and in myelin sheaths as revealed by nonradioactive (digoxigenin) in situ hybridization histochemistry.

soluble factors elaborated by astrocytes, but rather through physical contact of the oligodendrocytes with either the live or "fixed" astrocytes.

Neuronal Conditioned Medium and PDGF Can Reverse the Astrocytic Inhibition of MBP mRNA Translocation. The astrocyte findings were interesting because they indicated that although astrocytes could inhibit MBP mRNA translocation within oligodendrocytes in cocultures of the two cell types, they were not as effective in the mixed primary cultures at inhibiting the mRNA movement. We wondered if this could be due to neurons present in the mixed cultures, and we found that soluble factors, secreted by neuronal cultures in vitro could counteract the inhibition of MBP mRNA translocation in oligodendrocytes by astrocytes (Amur-Umarjee et al., 1997). Of several growth factors tested, only platelet derived growth factor (PDGF) was effective in alleviating the astrocytic inhibition of MBP mRNA movement. PDGF appears to mediate its effect via α-receptors on the oligodendrocyte cell surfaces and receptor tyrosine kinases. Thus these studies indicate that cell types other than oligodendrocytes can regulate the movement of MBP mRNA either through cell-cell contact or through secretion of soluble factors, at least in vitro. These potential regulatory mechanisms might have some importance in pathological situations in which prevention of unnecessary myelin membrane elaboration (through interference with the targeting of MBP to the sheath) may be desirable. When conditions are then appropriate remyelination might be encouraged through the action of soluble factors such as PDGF.

Is There Translational Control of MBP Synthesis during Development?

We have inferred that translational regulation of MBP expression may exist during development, in part, because of the phenotype of certain "immortalized" oligodendrocyte cell lines which express MBP mRNAs but not the proteins (Verity et al., 1993; Jensen et al., 1993). These cell lines have been produced by transformation of oligodendrocytes in vivo (Jensen et al., 1993) or in vitro (Verity et al., 1993) with the SV40 large T antigen oncogene. Because the JC virus T antigen has been implicated in the inhibition of MBP mRNA translation in transgenic mice (Trapp et al., 1988), the possibility existed that the SV40 large T antigen might be responsible for inhibiting translation of MBP mRNA and preventing the production of protein in these cell lines.

We investigated this possibility in two ways (Ueno et al., 1995). In one set of experiments we translated the 14 kDa MBP cRNA in rabbit reticulocyte lysates in the presence and absence of purified SV40 large T antigen. The SV40 large T antigen did not inhibit the translation of the MBP mRNA even at a molar ratio 25 times that at which it has been shown to inhibit the transcription of some genes. Thus, it is unlikely that lack of expression of MBP protein in the oligodendrocyte cell lines is due to SV40 large T antigen inhibition of translation. To explore this further, in a second set of experiments, we stably transfected the 14 kDa MBP cDNA under the control of a strong constitutive promoter into COS cells and into N20.1 cells, both of which express the SV40 large T antigen. In both cell lines, it was easy to detect the presence of the 14 kDa MBP protein by immunoblot analysis in the permanently transfected lines. Our data do not support the notion that T antigen inhibits expression of the MBP gene at the translational level. Our results suggest that in the immortalized cell lines of Verity et al. (1993) and Jensen et al. (1993) some other mechanism probably exists for the inhibition of MBP protein expression.

An interesting corollary of the experiments with the permanently transfected N20.1 cells is that this putative translational inhibitory mechanism can be "overridden" by increased levels of MBP mRNA. Normally, little or no MBP protein is produced by the N20.1 cells. When the levels of MBP mRNAs are increased through transfection of a cDNA under the control of a strong, constitutive promoter, then the levels of protein become detectable.

There is other evidence suggesting that the lack of synthesis of MBP protein in the N20.1 cell line is not due to selective inhibition by SV40 large T antigen. For example, when the N20.1 cell line is cultured in vitro with DRG neurons (Newman et al., 1995) or transplanted in shiverer mouse brains (Foster et al., 1995) MBP protein is expressed and the cells elaborate myelin-like membranes. Under appropriate culture conditions, these cells also appear to express MBP protein in vitro (Newman et al., 1995; J. Merrill, personal communication). Whether this is due to the expression of increased levels of MBP mRNAs "overriding" the inhibition mechanism or a "loss" of the inhibition mechanism is not yet clear.

In summary, this brief report outlines a number of potential factors that could contribute to the post transcriptional regulation of MBP gene expression. I suggest that this form of regulation could be important during development and differentiation of the oligodendrocyte and/or in response to certain pathological insults to the oligodendrocyte.

ACKNOWLEDGMENTS

The author wishes to thank his wonderful students and colleagues—Drs. S. Amur-Umarjee, S. Byravan, L. Foster, C. Landry, S-I. Ueno, J. Verdi, and A.N. Verity—without whom this work would never have been possible. Special thanks to Charles Landry for providing the photomicrograph shown in Figure 1 and to Charles, Tom Pribyl and Ernesto Bongarzone for offering suggestions on the manuscript. The author also wishes to acknowledge the support of NIH grants HD 25831, NS 23022, NS 23332, and RG2233A1 from the National MS Society.

REFERENCES

Ainger K, Avossa D, Morgan F, Hill SJ, Barry C, Barbarese E, Carson JH (1993a): Transport and localization of exogenous myelin basic protein mRNA microinjected into oligodendrocytes. J Cell Biol 123:431–41.

Ainger K, Avossa D, Carson JH, Barbarese E (1993b): Transport of microinjected myelin basic protein mRNA in oligodendrocytes. Trans Am Soc Neurochem 24:217.

Amur-Umarjee SG, Hall L, Campagnoni AT (1990): Spatial distribution of messenger RNAs for myelin proteins in primary cultures of mouse brain. Dev Neurosci 12:263- 272.

Amur-Umarjee S, Phan T, Campagnoni AT (1993): Myelin basic protein mRNA translocation in oligodendrocytes is inhibited by astrocytes in vitro. J Neurosci Res 36:99–110.

Amur-Umarjee S, Schonmann V, Campagnoni AT (1997): Neuronal regulation of myelin basic protein mRNA translocation is mediated by platelet-derived growth factor. Dev. Neurosci 19: 143–151.

Both GW, Banerjee AK, Shatkin AJ (1975): Methylation-dependent translation of viral messenger RNAs in vitro. Proc Natl Acad Sci USA 72:1189–1193.

Byravan S, Campagnoni AT (1994): Serum factors and hydrocortisone influence the synthesis of myelin basic proteins in mouse brain primary cultures. Int J Dev Neurosci 12:343–351.

Cabacungan E, Mittal R, Ved HS, Shanker G, Gustow E, Soprano DR, Pieringer RA (1991): Degrees of co-operativity between trithyronine and hydrocortisone in their regulation of myelin basic protein and proteolipid protein during development. Dev Neurosci 13:74–79.

Campagnoni AT, Hunkeler HJ, Moskaitis JE (1987): Translational regulation of myelin basic protein synthesis. J Neurosci Res 17:102–110.

Campagnoni AT, Pribyl TM, Campagnoni CW, Kampf K, Amur-Umarjee S, Landry CF, Handley VW, Newman SL, Garbay B, Kitamura, K (1993): Structure and developmental regulation of Golli-mbp, a 105 Kb gene that encompasses the myelin basic protein gene and is expressed in cells in the oligodendrocyte lineage in the brain. J biol Chem 268:4930–4938.

Campagnoni AT (1995): Molecular biology of myelination. In Ransom B and Kettenmann H (eds): "Neuroglia—a Treatise." London:Oxford University Press, pp 555–570.

Colman DR., Kreibich G, Frey AB, Sabatini DD (1982): Synthesis and incorporation of myelin polypeptides into CNS myelin. J Cell Biol 95:598–608.

Foster LM, Landry C, Phan T, Campagnoni AT (1995): Conditionally Immortalized oligodendrocyte cell lines migrate to different brain regions and elaborate 'myelin-like' membranes after transplantation into neonatal shiverer mouse brains. Dev Neurosci 17:160–170.

Goodwin EB, Okkema PG, Evans TC, Kimble J (1993): Translational regulation of tra-2 by its 3' untranslated region controls sexual identity in C elegans. Cell 75:329–339.

Jansen M, De Moor CH, Sussenbach JS, van den Brande JL (1995): Translational control of gene expression. Pediatr Res 37:681–686.

Jensen NA, Smith GM, Shine HC, Garvey JS, Hood L (1993): Distinct hypomyelinated phenotypes in MBP SV40 large T transgenic mice. J Neurosci Res 34:257–264.

Klausner RD, Rouault TA, Harford JB (1993): Regulating the fate of mRNA: the control of cellular iron metabolism. Cell 72:19–28.

Kozak M (1989): The scanning model for translation: an update J Cell Biol 108:229–241.

Kumar S, Cole R, Chiappeli F, de Vellis J (1989): Differential regulation of oligodendrocyte markers by glucocorticoids: post-transcriptional regulation of proteolipid protein and myelin basic protein and transcriptional regulation of glycerol phosphate dehydrogenase. Proc Natl Acad Sci USA 86:6807–6811.

Kwon YK, Hecht NB (1993): Binding of a phosphoprotein to the 3' untranslated region of the mouse protamine 2 mRNA temporally represses its translation. Mol. Cell. Biol. 13:6547–57.

Landry CF, Watson JB, Kashima T, Campagnoni AT (1994): Cellular influences on RNA sorting in neurons and glia: An in situ hybridization histochemical study. Mol Brain Res 27:1–11.

Muthukrishnan S, Both GW, Furuichi Y, Shatkin AJ (1975): 5'-Terminal 7- methylguanosine in eukaryotic mRNA is required for translation. Nature 255:33–40.

Newman SL, Weikle AA, Neuberger TJ, Bigbee JW (1995): Myelinogenic potential of an immortalized oligodendrocyte cell line. J Neurosci Research 40:680–93.

Pribyl TM, Campagnoni CW, Kampf K, Kashima T, Handley VW, McMahon J Campagnoni AT (1993): The human myelin basic protein gene is included within a 179-kilobase transcription unit: Expression in the immune and central nervous systems. Proc Natl Acad Sci USA 90:10695–10699.

Trapp BD, Moench T, Pulley M, Barbosa E, Tennekoon G, Griffin J (1987): Spatial segregation of mRNA encoding myelin-specific proteins. Proc Natl Acad Sc. USA. 84:7773–7777.

Trapp BD, Small JA, Pulley M, Khoury G and Scangos GA (1988): Dysmyelination in transgenic mice containing JC virus early region. Ann Neurol 23:38–48.

Ueno S, Handley VW, Byravan S, Campagnoni AT (1994a): Structural features of myelin basic protein mRNAs influence their translational efficiencies. J Neurochem 62:1254- 1259.

Ueno S, Kotani Y, Kondoh K, Sano A, Kakimoto Y, Campagnoni AT (1994b): The 3' untranslated region of mouse myelin basic protein gene increases the amount of mRNA in immortalized mouse oligodendrocytes. Biochem Biophys Res Comm 204:1352- 1357.

Ueno S, Kotani Y, Kondoh K, Sano A, Kakimoto Y, Campagnoni AT (1994): The 3' untranslated region of mouse myelin basic protein gene increases the amount of mRNA in immortalized mouse oligodendrocytes. Biochem Biophys Res Comm 204:1352- 1357.

Ved HS, Gustow E, Pieringer RA (1989): Effect of hydrocortisone on myelin basic protein in developing primary brain cultures. Neurosci Lett 99:203–207.

Verdi JM, Campagnoni AT (1990): Translational regulation by steroids: identification of a steroid modulatory element in the 5' untranslated region of the myelin basic protein messenger RNA. J biol Chem 265:20314–20320.

Verdi J M (1991): "Translational regulation of myelin basic protein: Structural features of MBP mRNA that influence its translational efficiency." Ann Arbor:UMI Dissertation Information Service, 121pp.

Verity AN, Bredesen D, Vonderscher C, Handley VW, Campagnoni AT (1993): Expression of myelin protein genes and other myelin components in an oligodendrocytic cell line conditionally immortalized with a temperature-sensitive retrovirus. J Neurochem 60:577–587.

Verity AN, Campagnoni AT (1988): Regional expression of myelin protein genes in the developing mouse brain: In situ hybridization studies. J Neurosci Res 21:238–248.

You Y, Chen CA, Shyu A-B (1992): U-rich sequence-binding proteins (URBPs): interacting with a 20-nucleotide U-rich sequence in the 3' untranslated region of c-fos mRNA may be involved in the first step of c-fos mRNA degradation. Mol Cell Biol 12:2931–2940.

12

MOLECULAR CONTROL OF MYELIN BASIC PROTEIN GENE EXPRESSION

R. Miskimins, R. E. Clark, and A. Zapp

Department of Biochemistry and Molecular Biology
University of South Dakota School of Medicine
Vermillion, South Dakota 57069

1. INTRODUCTION

Expression of the myelin basic protein (MBP) gene has been used as a marker for myelination in the central nervous system (CNS) for many years. This is due to the fact that expression of the gene closely parallels the course of myelination. In mice, MBP RNA can be detected by 7–9 days post-natally, peaks at around 18 days and declines to a steady state thereafter. The gene for MBP was one of the first myelin-specific genes cloned. Since that time, many groups have been working to understand the regulation of MBP gene expression as a means to exploring the underlying mechanisms governing myelination in the CNS. In the interval between the cloning of the gene and now, much has been learned regarding the portions of the MBP promoter region that govern the gene's expression. Additionally, a beginning has been made in discovering signalling pathways that affect MBP gene expression and may therefore play a role in oligodendro- cyte development and myelination.

1.1. Gene Structure

There are several isoforms of MBP in myelin in both the mouse and humans. All arise from differential splicing of a single MBP gene (Fig 1), located at the distal end of chromosome 18. The mouse MBP gene was cloned by Takahashi et al. (1985) and shown to contain seven exons spanning approximately 30 kbp. The human gene also contains seven exons that are distributed over 32–34 kbp (Streicher and Stoffel, 1989). Both the exon sequences and the 5' noncoding sequences of the mouse and human genes have a high degree of homology .

In screening clones for the 14kD MBP cDNA, Kitamura et al. (1990) discovered that RNAs exist that contain a different 5' end from the classical MBP gene cloned by Takahashi et al. (1985). They showed that there exists an additional exon more than 25 kbp upstream of the classical MBP gene exon 1 which was termed exon 0. In addition,

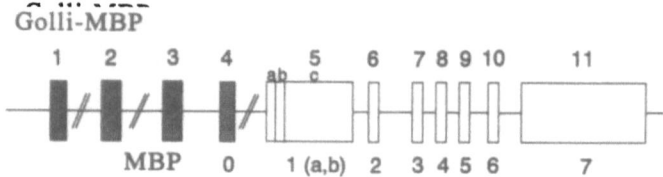

Figure 1. Structure of the MBP gene. The exons (boxes) are numbered in accord with the *Golli-mbp* scheme above and the classical MBP scheme below. Exons belonging exclusively to the *Golli-mbp* gene contain diagonal lines.

transcripts that used the transcriptional start site upstream of exon 0 also contained sequences that are a part of the promoter sequences of the originally described MBP gene. Thus, the originally described exon 1 was further divided into exons 1a and 1b, with a part of the previously untranscribed region of the MBP promoter being assigned as exon 1a and the original exon 1 being called exon 1b.

In addition to the discovery of an additional exon upstream of the originally defined MBP transcriptional start site, Campagnoni's group has described a much larger transcription unit that contains three unique exons spanning a region of 73 kbp upstream of the classic MBP transcription start site that also uses other exons of the traditional MBP gene (Campagnoni et al., 1993) that they have termed Golli-mbp. This gene is expressed much earlier than the MBP gene in the brain and is also present in other tissues in early development of the animal (Pribyl et al., 1993).

It is clear from the complex structure of the MBP and Golli-mbp genes that regulation of the choice of transcriptional start site used at different developmental time points in different tissues is a complicated issue. In rodent myelination, the major transcriptional start site used in producing MBPs is the site originally defined with the cloning of the MBP gene by Takahashi et al. (1985). Thus, most of the information regarding regulation of the expression of the MBP gene has been generated with respect to the regulatory region upstream of this sequence. The information garnered regarding the regulation of the MBP gene through the upstream sequences, with respect to tissue-specific regulation and response of the MBP gene to signalling cues will be discussed.

2. PROMOTER DELETION STUDIES

Deletion analysis of the MBP gene regulatory region (Fig 2) has been done by several groups. Initial studies indicated that the sequences upstream of -300 bp are repressive with respect to their effect on expression of heterologous genes. In one such analysis (Miura et al., 1989) a series of nested deletions of the MBP promoter were subcloned upstream of the β-galactosidase (lacZ) gene. These constructs were transiently transfected into a neuron/glial hybrid cell line. The level of β-galactosidase activity increased as shorter segments of the MBP regulatory region were used as promoters. The activity was highest from a construct containing 253 bp of MBP 5′ flanking sequence. Further deletions resulted in decreasing activity. Further studies have verified and extended these observations. Using a finer set of deletion constructs of the MBP regulatory region linked to the bacterial chloramphenicol acetyltransferase (CAT) gene, it was shown that significant changes in the level of activity occur with removal of sequences in the 5′ flanking region (Zhang et al., 1995). Overall, removal of sequences between -1323 and -333 resulted in in-

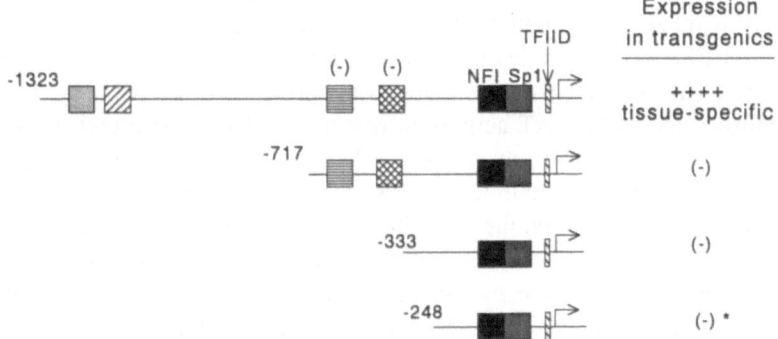

Figure 2. Promoter elements and deletion mutants of the MBP 5′ flanking sequence. The MBP 5′ flanking sequence to -1323 is shown. The transcriptional start site is marked with an arrow. Elements discussed in the text are indicated as boxes. The activity of each of the constructs in transgenic mice is indicated at the right. *Displays low level expression in cultured oligodendrocytes plus PDGF and in transgenic mice.

creasing expression of the CAT gene. Further deletions, beyond -333, again decreased the expression of the CAT gene. The results discussed above have resulted in the division of the regulatory region of the MBP gene into two regions: a positive acting proximal promoter downstream of -333 and a negative acting distal promoter region between -1323 and -333.

2.1. Proximal Promoter Elements

Much work has been done to try and elucidate the sequences within the proximal promoter region of the MBP gene that interact with proteins and contribute to the expression of the gene. Within this region are potential binding sites for TFIID, Sp1 and NF1 transcription factors. In vitro analysis has shown that several of these sites interact with binding proteins that may affect MBP gene expression. Using run-off assays it has been shown that mutations in either the TATA-box element, the Sp1-like sequence or the NF1 site resulted in a dramatic inhibition in transcriptional activity (Tamura et al., 1988). Additional mutational analysis has demonstrated that brain-specific transcription from the MBP proximal promoter in in vitro assays is governed by a brain-specific TFIID-like factor that binds near the TATA-like element (Tamura et al., 1990, 1991).

2.1.1. NF1 Site

Much work has been done to determine the importance of the NF1 site in MBP gene regulation. In addition to its positive role in the mouse MBP gene promoter, the NF1 site has been shown to be involved in transcription of the human MBP gene (Wrabetz et al., 1993) where it is proposed to function in a negative manner. Further work has provided evidence that binding to the NF1 site can be modulated by changes in the levels of cyclic AMP (cAMP). These studies utilized a peripheral neurinoma cell line, D6P2T, in which the endogenous MBP gene can be induced by raising the intracellular cAMP levels (Gandelman et al., 1989; Zhang and Miskimins, 1993). This response to cAMP is a delayed effect, requiring 24–48 hours to increase the expression of the MBP gene. Thus, it is unlikely to be directly mediated through a cAMP response element binding (CREB) protein as are many other such responses. Additionally, the MBP gene contains no consensus binding site for CREB. Using stably transformed D6P2T cells it was shown that increased expression from MBP-CAT constructs required the presence of an NF1 site. Removal of the site abolished increased CAT activity in response to increased cAMP levels (Zhang and Miskimins, 1993). Extracts made from cells treated with isobutyl methylxanthine (IBMX), a phosphodiesterase inhibitor, to raise the intracellular cAMP levels showed a different pattern of binding when the NF1 site is used as a probe in retardation gel analyses (Zhang and Miskimins, 19993). Site specific mutations in the NF1 site both reduces the binding of proteins to this site and decreases the CAT activity obtained in response to elevated cAMP levels (R. Clark and R. Miskimins, unpublished data). Thus it appears that the response of the MBP gene to increased cAMP levels involves a change in the pattern of binding proteins interacting at the NF1 site.

3. DISTAL PROMOTER ELEMENTS

3.1. Elements Involved in the Response to cAMP

Using transgenic animals it has been shown that 1323 base pairs (bp) of the MBP regulatory region is sufficient to drive oligodendrocyte-specific, temporally correct expression of a heterologous gene (Miskimins et al., 1992; discussed below). However, this construct is a relatively poor promoter when used in tissue culture cells of any type. In addition, despite the fact that the expression of the MBP gene can be increased in primary oligodendrocyte cultures when they are treated with cAMP analogs, stably transfected D6P2T cells carrying 1323 bp of the MBP regulatory region linked to the CAT gene cannot be induced with IBMX. Further deletion to -1176 relieves this repression (Anderson and Miskimins, 1994). Thus, there appears to be a sequence upstream of the NF1 site that suppresses the response to increased cAMP. Increasing cAMP levels in D6P2T cells stably transfected with MBPCAT constructs containing from 1176 to 248 bp of MBP 5' flanking sequence induces the expression of the CAT gene. Further deletion removes the NF1 site discussed above and these shorter constructs cannot be induced with cAMP. Conditions must exist in the animal, however, that lead to elevated expression of the CAT constructs containing 1323 bp of the MBP regulatory region that are not mimicked in tissue culture cells strictly by elevating cAMP levels. Indeed, the 1323 bp MBPCAT construct can be highly activated in tissue culture cells by down-regulating protein kinase C (PKC) prior to elevation of cAMP levels (Anderson and Miskimins, 1994). Overcoming the inhibition through down-regulation of PKC requires MBP sequences between -1323 and -1211 while the inhibitory region appears to reside between -1211 and -974. Thus sequences in the distal promoter region play a role in the response of the MBP gene to external signals.

3.2. Negative Regulatory Elements

MBP gene expression is limited to a restricted set of cell types in the post-natal animal: oligodendrocytes in the CNS and Schwann cells in the PNS. Thus, the repression of the gene in other cell types is likely to be under stringent control. Likewise, transcription from the major transcriptional start site is not detectable until about one week after birth, implying repression even in cells destined to produce myelin. Many examples exist of genes that contain negative regulatory elements to "turn off" gene expression or to prevent inappropriate expression. As mentioned above, the distal region of the MBP regulatory region is inhibitory to transcription in tissue culture cells. Until the majority of this region is removed by deletion, activity of reporter constructs is relatively low unless the system is manipulated by changes in signalling pathways as mentioned above. Thus, negative regulatory regions are likely to exist in the MBP 5' flanking sequences.

One such region appears to exist between -333 and -513. Deletion of this 180 bp segment from reporter constructs results in a significant, consistent increase in expression in transfected cells (Zhang et al., 1995). When the region between -720 and -256, containing the proposed negative regulatory element, is internally deleted from within the MBP 5' flanking sequences the activity of transfected MBPCAT constructs rises significantly over constructs containing this region (Zhang et al., 1995). Thus sequences within this region appear to exert a negative effect on the expression of the MBP gene in culture. Gel mobility shift assays have shown that there are specific protein-DNA interactions within the region between -513 and -333. The binding site has been narrowed to a 50 bp segment. When a longer MBPCAT construct, one not containing the protein binding site (extending to -383), was tested for activity in transfected cells it was found to have high level expression, implying a role for the protein binding in the repression of the MBP gene (Zapp and Miskimins, unpublished results).

4. TRANSGENIC MICE STUDIES

4.1. Full Length Constructs

The above studies regarding the influence of signalling pathways and promoter length on the expression of the MBP gene were all done using culture conditions in either primary cultures or immortalized cell lines. While these studies can be useful in pointing out potentially important regions of the 5' flanking sequence they are limited by the fact that none of these culture systems represents the *in vivo* situation accurately. It has thus been important to try to verify the results obtained in these culture systems in the whole animal. Transgenic mice have been used for this type of study.

The earliest use of transgenic mice to express the MBP gene used the entire genomic clone for expression (Readhead et al., 1987). This construct contained 4 kbp of 5' flanking sequence and worked to reverse the phenotype of the shiverer mutation (a mutation in the MBP gene) in mice. It was subsequently shown that only 1323 bp of 5' flanking sequence was necessary to direct high level, tissue-specific expression of a heterologous gene in transgenic mice (Miskimins et al., 1992). This construct was also capable of expression in the sciatic nerve of these mice (Zhang et al., 1995). Thus the important regulatory information for the MBP gene resides within the first 1.3 kbp upstream of the transcriptional start site.

4.2. Deletion Constructs

After the initial success in producing transgenic mice capable of expressing protein(s) directed by the 5' flanking sequence of the MBP gene, no construct shorter than 1.3 kbp has been shown to follow the correct temporal and tissue-specific pattern of expression in transgenic animals. In one report transgenic mice were made using constructs that contained 717 bp, 333 bp or 248 bp of MBP 5' flanking sequence. No expression of the transgene was found in any of the mice tested (Zhang et al., 1995). One report did show a low level of expression in transgenic mice using MBP sequences extending to -256. However, the expression was not limited to the central nervous system (Goujet-Zalc et al., 1993). Thus it is likely that the proximal promoter region is important for basal transcriptional activity, while sequences upstream are required for oligodendrocyte-specific, high level expression of the MBP gene.

REFERENCES

Anderson S and Miskimins R (1994): Involvement of protein kinase C in cAMP regulation of myelin basic protein gene expression. J. Neurosci. Res. 37:604–611.

Campagnoni AT Pribyl TM Campagnoni CW Kampf K Amur-Umarjee S Landry CF Handley VW Newman SL Garbay B and Kitamura K (1993): Structure and developmental regulation of *Golli-mbp*, a 105-kilobase gene that encompasses the myelin basic protein gene and is expressed in cells in the oligodendrocyte lineage in the brain. J. Biol. Chem. 268:4930–4938.

Gandelman K-Y Pfeiffer SE and Carson JH (1989): Cyclic AMP regulation of P0 and myelin basic protein gene expression in semi-differentiated peripheral neurinoma cell line D6P2T. Devel. 106:389–398.

Goujet-Zalc C Babinet CH Monge M Timsit S Cabon F Gansmuller A Miura M Sanchez M Pournin S Mikoshiba K and Zalc B (1993) The proximal region of the MBP gene promoter is sufficient to induce oligodendroglial-specific expression in transgenic mice. Eur. J. Neurosci. 5:624–632.

Kitamura K Newman SL Campagnoni CW Verdi JM Mohandas T Handley VW and Campagnoni AT (1990): Expression of a novel transcript of the myelin basic protein gene. J. Neurochem. 54:2032–2041

Miskimins R Knapp L Dewey MJ and Zhang X (1992): Cell and tissue-specific expression of heterologous gene under control of the myelin basic protein gene promoter in transgenic mice. Dev. Brain Res. 65:217–221.

Miura M Tamura T-a Aoyama A and Mikoshiba K (1989): The promoter elements of the mouse myelin basic protein gene function efficiently in NG108–15 neuronal/glial cells. Gene 75:31–38.

Pribyl TM Campagnoni CW Kampf K Kashima T Handley VW McMahon J and Campagnoni AT (1993): The human myelin basic protein gene is included within a 179 kilobase transcription unit: expression in the immune and central nervous system. Proc. Natl. Acad. Sci. USA 90:10695–10699.

Streicher R and Stoffel W (1989): The organization of the human myelin basic protein gene. Biol. Chem. Hoppe Seyler 370:503–510.

Takahashi N Roach A Teplow DB Pruisner SB and Hood L (1985): Cloning and characterization of the myelin basic protein gene from mouse: one gene can encode both 24kd and 18,5kd MBPs by alternate use of exons. Cell 42:139–148.

Tamura T-a Miura M Ikenaka K and Mikoshiba K (1988): Analysis of transcription control elements of the mouse myelin basic protein gene in HeLa cell extracts: demonstration of a strong NFI-binding motif in the upstream region. Nuc. Acids Res. 16:11441–11459.

Tamura T-a Sumita K Hirose S and Mikoshiba K (1990): Core promoter of the mouse myelin basic protein gene governs brain-specific transcription in vitro. EMBO J 9:3101–3108.

Tamura T-a Sumita K and Mikoshiba K (1991): Sequences involved in brain-specific in vitro transcription from the core promoter of the mouse myelin basic protein gene. Biochim. Biophys. Acta 1129:83–86.

Wrabetz L Shumas S Grinspan J Feltri ML Bozyczko D McMorris FA Pleasure D and Kamholz J (1993): Analysis of the human MBP promoter in primary cultures of oligodendrocytes: positive and negative cis-acting elements in the proximal MBP promoter mediate oligodendrocyte-specific expression of MBP. J. Neurosci. Res. 36:455–471.

Zhang X and Miskimins R (1993): Binding at an NFI site is modulated by cyclic AMP- dependent activation of myelin basic protein gene expression. J. Neurochem. 60:2010- 2017.

Zhang X Anderson S Zapp A and Miskimins R (1995): Identification of regulatory elements of the myelin basic protein gene in cultured cells and transgenic mice. Transgenics 1:505- 514.

13

ROLE OF cAMP DURING THE PROCESS OF DEMYELINATION AND REMYELINATION IN PERIPHERAL NERVE

Randall S. Walikonis and Joseph F. Poduslo

Molecular Neurobiology Laboratory
Departments of Neurology and Biochemistry/Molecular Biology
Mayo Clinic and Mayo Foundation
Rochester, Minnesota 55905

Myelin is a multilamellar structure that surrounds axons to increase axonal conduction velocity and provide trophic support. In the peripheral nervous system, it is formed by the spiral wrapping of the cell membrane of myelinating Schwann cells (SCs) around axons. The SC phenotype is dependent on its association with an axon. Axonal contact induces the expression of myelin-specific proteins, including P_0, myelin basic protein, peripheral myelin protein-22, and myelin associated glycoprotein (Wood and Engel, 1976; Uyemura et al., 1979; Carson et al., 1983; Poduslo, 1984; Gupta et al., 1988; Lemke and Chao, 1988; LeBlanc and Poduslo, 1990). With the onset of myelination, the amount of these myelin proteins and the steady state levels of their mRNAs increases dramatically.

The maintenance of the myelinating phenotype requires continued axon-SC contact. The loss of contact, such as occurs following nerve injury or during isolation in culture, leads to demyelination, characterized by myelin breakdown, decreased expression of myelin genes (Gupta et al., 1988; Trapp et al., 1988; LeBlanc and Poduslo, 1990), and proliferation of SCs (Bradley and Asbury, 1970; Brown and Asbury, 1981; Oaklander et al., 1987). Re-establishment of contact by axonal regrowth or coculture of SCs with neurons results in SC proliferation followed by upregulation of myelin gene expression and remyelination (Brockes et al., 1980; Poduslo, 1984; Gupta et al., 1988; Gupta et al., 1990; LeBlanc et al., 1987; LeBlanc and Poduslo, 1990). Thus axons dramatically modulate SC gene expression, which implies the transmission of a signal from the axon to the SC.

The signaling mechanisms in the SC controlled by the presence of axons are not well understood. While many of the classic signaling systems are present in SCs, most work has focused on the adenylyl cyclase-cyclic AMP pathways (Lemke, 1990; Eccleston, 1992). Numerous studies on the effects of cAMP have been conducted on cultured SCs (Brockes et al., 1980; Sobue and Pleasure, 1984; Sobue et al., 1986; Lemke and Chao, 1988; Jessen et al., 1990; Morgan et al., 1991). These cells are isolated from neonatal rats, purified by immunoselection, and expanded using glial growth factor and forskolin, an ac-

Cell Biology and Pathology of Myelin, edited by Juurlink et al.
Plenum Press, New York, 1997

tivator of adenylyl cyclases (Seamon and Daly, 1986). Following withdrawal of these mitogens, the cells are reexposed to low levels of forskolin. This exposure to cAMP elevating agents causes a partial upregulation of myelin gene expression (Lemke and Chao, 1988; Morgan et al., 1991), expression of some myelin membrane components (Sobue and Pleasure, 1984; Sobue et al., 1986), and flattening of the SCs (Sobue et al., 1986; Morgan et al., 1991). Thus cAMP may mimic some aspects of axonal contact.

The effect of cAMP on myelin gene expression in cultured SCs requires extended incubation in cAMP elevating agents. The upregulation of myelin gene expression in response to cAMP required 36 hours (Lemke and Chou, 1988) to 3 days (Morgan et al., 1991). Furthermore, extended exposure to forskolin leads to SCs that are only minimally responsive to the withdrawal of the drug, indicating that cAMP exposure was in some manner altering SCs (Lemke and Chou, 1988). This extended incubation time indicates that cAMP is probably not directly affecting myelin gene expression. Genes with a cAMP responsive element can be upregulated within minutes of cAMP stimulation (Rickles et al., 1989; Yamamoto et al., 1990). Other genes require protein synthesis for increased expression in response to cAMP (Sasaki et al., 1984). Increased expression of this group requires several minutes to hours to respond to increased cAMP. No cAMP responsive element has been found in any of the myelin genes (Zhang and Miskimins, 1993; Lemke et al., 1988; Roesler et al., 1988). Thus the extended time required for the increased myelin-like characteristics in cultured SCs probably indicates that the exposure to forskolin is causing differentiation of the SCs rather than a directly increasing myelin gene expression. While cis-acting elements are proposed to upregulate myelin genes (Zhang and Miskimins, 1993; Lemke et al., 1988), the effects may not arise from a direct link between cAMP and these elements.

Analysis of the PMP-22 promotor also demonstrates that the response of cultured SCs does not closely mimic *in vivo* events. Suter et al., (1994) analyzed the promotor region of this gene and found two distinct promotors. Promotor 1 is preferentially expressed *in vivo*. This promotor contains a TCAG sequence near the transcription initiation site. This sequence is also located at initiation sites of other myelin genes, including P_0 (Lemke et al., 1988), myelin basic protein (Gow et al., 1992), and the proteolipid protein (Macklin et al., 1987). In culture, however, a different promotor for PMP-22 is preferentially activated in response to forskolin (Suter et al., 1994). Thus the control of myelin gene expression in cultured SCs may not mimic the control mechanism in vivo.

Peripheral nerve injury models have been useful for the study of SC regulation mechanisms, providing an in vivo method for studying myelin gene regulation (Poduslo, 1984; Poduslo, 1985; Poduslo et al., 1985; Trapp et al., 1988; LeBlanc and Poduslo, 1990). In the adult normal peripheral nerve, SCs maintain myelin, with sufficient myelin gene expression to replace myelin protein turn-over. Following crush and transection injury to the nerve, axons degenerate distally to the injury and thus the SCs lose contact with the axolemma (Figure 1). The loss of contact leads to myelin breakdown and downregulation of the expression of myelin genes (Poduslo, 1984; Poduslo, 1985; Poduslo et al., 1985; LeBlanc et al., 1987; LeBlanc and Poduslo, 1990). In the crushed nerve, axons are allowed to regrow into the distal portion of the nerve. High levels of myelin transcripts are again expressed as the axons regenerate and remyelination occurs. Following permanent transection, on the other hand, the transected ends of the nerve are reflected 180° and sutured to adjacent muscle. Demyelination occurs in the distal segment as the axons degenerate, but axonal regrowth into the distal segment is prevented and SCs remain in a nonmyelinating state. These injury paradigms allow the study of changes that occur during the processes of demyelination and remyelination. The transected nerve also provides a

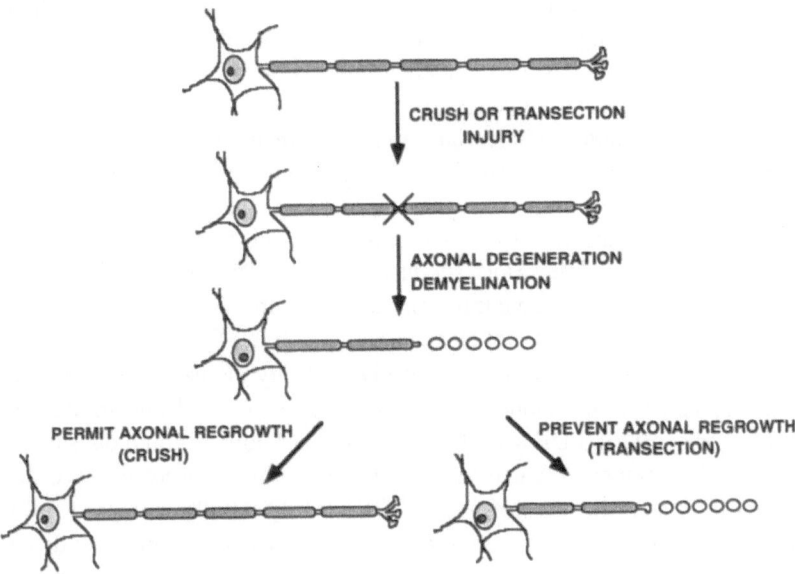

Figure 1. Sciatic nerve injury paradigms. Rat sciatic nerves were injured at the sciatic notch either by a crush or a permanent transection. Axons distal to the injury site degenerate and Schwann cells in this distal segment revert from a myelinating to a nonmyelinating state. Axons regrow following crush injury, with remyelination occurring following axonal contact with Schwann cells. Axonal regrowth is prevented in the permanently transected nerve, thus Schwann cells remain nonmyelinating.

unique model for studying the effect of various agents on myelin gene expression in non-myelinating SCs as approximately 90% of the cellular elements are nonmyelinating SCs (Spencer et al., 1979). This in vivo preparation of SCs avoids the artificial expansion and other manipulation necessary for tissue culture studies. It is thus an ideal model for studying the regulatory mechanisms by which SCs control the expression of myelin genes.

We conducted studies on the relationship between cAMP and myelination using these injury paradigms. Cyclic AMP levels were measured in sciatic endoneurial homogenates following crush and transection injury, during the processes of demyelination following both injuries and remyelination in the crushed nerve. Cyclic AMP levels drop to just 10% of normal following both injuries, concurrent with the decline in myelin gene expression (LeBlanc et al., 1992; Poduslo et al., 1995). Cyclic AMP levels remain low in the nonmyelinating transected nerve. Remyelination occurs in the crushed nerve starting at about 14 days after injury and is complete by 21 days after injury as determined by evaluating myelin-specific transcripts. Cyclic AMP levels only begin to recover in the crushed nerve beginning 21 days following injury, well after remyelination is initiated (Poduslo et al., 1995). Thus remyelination precedes an increase in cAMP, which suggests that cAMP increases do not trigger the process of myelination.

Initial studies on the control of cAMP levels were conducted using agents that stimulate production or inhibit degradation of cAMP. Stimulation of adenylyl cyclases by forskolin significantly elevated cAMP in normal and remyelinating crushed nerves. Surprisingly, forskolin failed to increase cAMP levels in the transected nerve at any time after injury. Cyclic AMP levels could be significantly increased in nonmyelinating nerves only by the inclusion of 3-isobutyl-1-methylxanthine (IBMX), a general phosphodiesterase

(PDE) inhibitor, along with forskolin (Poduslo et al., 1995). This indicates that PDE activity is elevated in nonmyelinating SCs, and that stimulation of adenylyl cyclases alone is insufficient to increase cAMP levels in nonmyelinating SCs.

Other studies provide further evidence that cAMP is not sufficient to elevate myelin gene transcripts in vivo. Sciatic endoneurial explants were incubated in Krebs-Ringers (KR) solution, or KR with forskolin, IBMX, or forskolin and IBMX together, combinations previously shown to elevate cAMP to varying degrees. Incubation of normal, transected, or crushed endoneurium in these agents for either one hour or 48 hours did not increase expression of myelin transcripts (Poduslo et al., 1995). These results contrast with the effect of cAMP on myelin gene transcription in cultured SCs, and further indicate that the response of cultured SCs may not closely mimic in vivo SCs.

The studies using IBMX demonstrated that PDE activity increases following injury. PDE activity was quantified in endoneurial homogenates of normal and injured nerves (Walikonis and Poduslo, 1996). PDE activity increased 3–3.5 fold within 6 days after both injuries. PDE activity remained elevated in the nonmyelinating transected nerve through 35 days following injury. In the crushed nerve, PDE activity rapidly increased following injury, but began to decline as remyelination occurred. Further biochemical studies using PDE isoform-specific inhibitors and stimulators identified the predominant sciatic nerve PDE isoform as the low Km-cAMP specific isoform. This isoform is dramatically upregulated during demyelination. The cGMP-inhibited isoform is also present, but in much lower levels and is not as dramatically upregulated following injury (Walikonis and Poduslo, 1996).

We verified that the PDEs are primarily found in SCs. SCs constitute 85–90% of the endoneurial cells at 35 days following transection (Spencer et al., 1979) and thus likely are the major contributors to the total PDE activity. There are, however, other cellular elements including invading macrophages that could contribute to the total PDE activity, although macrophages account for only 1–3% of total area at 35 days following injury (Spencer et al., 1979). PDEs were localized in sciatic nerve sections by in situ hybridization using a low Km-cAMP specific isoform probe (Walikonis and Poduslo, 1996). SCs were the only cell type labeled by PDE in situ hybridization. Their identity was confirmed by double labeling the same sections using antibodies against S100, a SC specific protein. Only cells that stained heavily for S100 gave a strong signal for PDEs. Thus PDEs in sciatic nerve are primarily located in SCs, and their activity is modulated relative to myelination.

The decline in myelin transcripts occurs concurrently with the upregulation of PDE activity. This increased PDE activity following injury may be a necessary event leading to demyelination, and the manipulation of this PDE activity may alter the course of demyelination. The identification of the low-Km cAMP specific family of PDEs provides a target for pharmacological intervention using isoform specific inhibitors. By inhibiting the activity of these PDE families after nerve injury, demyelination may be delayed or prevented. High levels of cyclic nucleotides may have a role in maintaining myelin structurally, either by inhibiting degradative mechanisms or by promoting myelin assembly. The decline in cAMP levels due to PDE upregulation following injury may be a necessary step in allowing the proliferation of SCs that occurs following injury. Ekström (1995) demonstrated that elevated cAMP levels prevent SC proliferation after injury to the frog sciatic nerve, perhaps through cAMP dependent kinase inhibition of mitogen-activated protein kinases (Graves et al., 1993; Sevetson et al., 1993).

PDEs were thought to be passive participants in the regulation of cyclic nucleotide levels. It has recently become apparent, however, that their activity is specifically and actively regulated. The understanding that individual PDE isozymes are tightly regulated emphasizes their pivotal role in modulating cyclic nucleotide dependent systems. The multiple isozymes with their unique regulatory and catalytic sites are excellent targets for therapeutic intervention in diseases caused by cyclic nucleotide dependent transduction mechanisms. Because many of these isozymes are differentially expressed and regulated in different cell types, inhibition of various PDE isozymes can alter cAMP levels in specific target tissues or cells.

Primary demyelination involves the loss of myelin with the initial sparing of axons. Environmental agents or toxins, such as lead or diphtheria toxin, may induce demyelination. By inhibiting the upregulated PDE activity with isoform specific inhibitors, it may be possible to alter the demyelination process that occurs in response to these agents.

The loss of myelin may also result from an autoimmune attack against SCs, such as occurs in Guillian-Barré syndrome. PDE inhibition has been shown to be effective in preventing demyelination in experimental autoimmune models. Experimental autoimmune encephalomyelitis (EAE), an animal model for multiple sclerosis, is induced by immunization of susceptible animals with CNS tissue. An autoimmune attack against CNS white matter follows, with lesions resulting if autoreactive T cells secrete cytokines. Low Km-cAMP specific PDE inhibition by rolipram blocks the production of cytokines, and thus prevents the demyelinating lesions (Sommer et al., 1995; Genain et al., 1995). Thus PDE inhibition can prevent demyelination during an autoimmune attack by acting directly on the immune cells. PDE inhibition may also prove to be an effective therapy for autoimmune attacks against PNS myelin.

These studies lay the groundwork for further investigation into the role of cAMP in the processes of myelination. The identification of the PDE families involved in hydrolyzing these nucleotides provides a basis for future work on the role of PDEs in demyelination. By manipulating the activity of PDEs following injury, it may be possible to alter the process of demyelination and may ultimately provide a basis for therapeutic approaches to demyelinating neuropathies.

REFERENCES

Bradley WG, Asbury AK (1970) Duration of synthesis phase in neurilemma cells in mouse sciatic nerve during degeneration. Exp Neurol 26:275–282.

Brockes JP, Raff MC, Nishiguchi DJ, Winter J (1980) Studies on cultured rat Schwann cells. III. Assays for peripheral myelin proteins. J Neurocytol 9:67–77.

Brown MJ, Asbury AK (1981) Schwann cell proliferation in the postnatal mouse: timing and topography. Exp Neurol 74:170–186.

Carson JH, Nielson ML, Barbarese E (1983) Developmental regulation of myelin basic protein expression in mouse brain. Dev Biol 96:485–492.

Eccleston PA (1992) Regulation of Schwann Cell Proliferation: Mechanisms Involved in Peripheral Nerve Development. Exp Cell Res 199:1–9.

Ekstrom PAR (1995) Increased cyclic AMP in in vitro regenerating frog sciatic nerves inhibits Schwann cell proliferation but has no effect on axonal outgrowth. J Neurosci Res 42:54–62.

Genain CP, Roberts T, Davis RL, Nguyen MH, Uccelli A, Faulds D, Li Y, Hedgpeth J, Hauser SL (1995) Prevention of autoimmune demyelination in non-human primates by a cAMP-specific phosphodiesterase inhibitor. Proc Natl Acad Sci USA 92:3601–3605.

Gow A, Friedrich VL, Lazzarini RA (1992) Myelin basic protein gene contains separate enhancers for oligodendrocyte and Schwann cell expression. J Cell Biol 119:605–616.

Graves LM, Bornfeldt KE, Raines EW, Potts BC, Macdonald SG, Ross R, Krebs EG (1993) Protein kinase A antagonizes platelet-derived growth factor-induced signaling by mitogen-activated protein kinase in human arterial smooth muscle cells. Proc Natl Acad Sci USA 90:10300–10304.

Gupta SK, Poduslo JF, Mezei C (1988) Temporal changes in P_0 and MBP gene expression after crush-injury of the adult peripheral nerve. Molec Brain Res 4:133–141.

Gupta SK, Poduslo JF, Dunn R, Roder J, Mezei C (1990) Myelin-associated glycoprotein gene expression in the presence and absence of Schwann cell-axonal contact. Dev Neurosci 12:22–33.

Jessen KR, Mirsky R, Morgan L (1990) Role of cyclic AMP and proliferation controls in Schwann cell differentiation. Ann N Y Acad Sci 633:78–89.

LeBlanc AC, Poduslo JF, Mezei C (1987) Gene expression in the presence or absence of myelin assembly. Molec Brain Res 2:57–67.

LeBlanc AC, Windebank AJ, Poduslo JF (1992) P_0 gene expression in Schwann cells is modulated by an increase of cAMP which is dependent on the presence of axons. Mol Brain Res 12:31–38.

LeBlanc AC, Poduslo JF (1990) Regulation of apolipoprotein E gene expression after injury of the rat sciatic nerve. J Neurosci Res 25:162–171.

Lemke G, Lamar E, Patterson J (1988) Isolation and analysis of the gene encoding peripheral myelin protein zero. Neuron 1:73–83.

Lemke G (1990) Glial growth factors. Sem Neurosci 2:437–443.

Lemke G, Chao M (1988) Axons regulate Schwann cell expression of the major myelin and NGF receptor genes. Development 102:499–504.

Macklin WB, Campagnoni CW, Deininger PL, Gardinier MV (1987) Structure and expression of the mouse myelin proteolipid protein gene. J Neurosci Res 18:383–394.

Morgan L, Jessen KR, Mirsky R (1991) The effects of cAMP on differentiation of cultured Schwann cells: progression from an early phenotype (04^+) to a myelin phenotype (P_0^+, GFAP⁻, N-CAM⁻, NGF-receptor⁻) depends on growth inhibition. J Cell Biol 112:457–467.

Oaklander AL, Miller MS, Spencer PS (1987) Rapid anterograde spread of premitotic activity along degenerating cat sciatic nerve. Journal of Neurochemistry 48:111–114.

Poduslo JF (1984) Regulation of myelination: biosynthesis of the major myelin glycoprotein by Schwann cells in the presence and absence of myelin assembly. J Neurochem 42:493–503.

Poduslo JF (1985) Post-translational protein modification: biosynthetic control mechanisms in the glycosylation of the major myelin glycoprotein by Schwann cells. J Neurochem 44:1194–1206.

Poduslo JF, Dyck PJ, Berg CT (1985) Regulation of myelination: Schwann cell transition from a myelin maintaining state to a quiescent state after permanent nerve transection. J Neurochem 44:388–400.

Poduslo JF, Walikonis RS, Domec M, Berg CT, Holtz-Heppelmann CJ (1995) The second messenger, cAMP, is not sufficient for myelin gene induction in the peripheral nervous system. J Neurochem 65:149–159.

Rickles RJ, Darrow AL, Strickland S (1989) Differentiation-responsive elements in the 5' region of the mouse tissue plasminogen activator gene confer two-stage regulation by retinoic acid and cyclic AMP in teratocarcinoma cells. Mol Cell Biol 9:1691–1704.

Roesler WJ, Vandenbark GR, Hanson RW (1988) Cyclic AMP and the induction of eukaryotic gene transcription. J Biol Chem 263:9063–9066.

Sasaki K, Cripe TP, Koch SR, Andreone TL, Petersen DD, Beale EG, Granner DK (1984) Multihormonal regulation of phosphoenolpyruvate carboxykinase gene transcription. J Biol Chem 259:15242–15251.

Seamon KB, Daly JW (1986) Forskolin: its biological and chemical properties. In: Advances in cyclic nucleotide and protein phosphorylation research (Greengard P, Robinson GA eds), pp 1–150. New York: Raven Press.

Sevetson BR, Kong X, Lawrence JC, Jr. (1993) Increasing cAMP attenuates activation of mitogen-activated protein kinase. Proc Natl Acad Sci USA 90:10305–10309.

Sobue G, Shuman S, Pleasure D (1986) Schwann cell responses to cyclic AMP: proliferation, change in shape, and appearance of surface galactocerebroside. Brain Res 362:23–32.

Sobue G, Pleasure D (1984) Schwann cell galactocerebroside induced by derivatives of adenosine 3',5'-monophosphate. Science 224:72–74.

Sommer N, Loschmann PA, Northoff GH, Weller M, Steinbrecher A, Steinbach JP, Lichtenfels R, Meyermann R, Riethmuller A, Fontana A, Dichgans J, Martin R (1995) The antidepressant rolipram suppresses cytokine production and prevents autoimmune encephalomyelitis. Nature Med 1:244–248.

Spencer PS, Weinberg HJ, Krygier-Brevart V, Zabrenetzky V (1979) An in vivo method to prepare normal Schwann cells free of axons and myelin. Brain Res 165:119–126.

Suter U, Snipes GJ, Schoenerscott R, Welcher AA, Pareek S, Lupski JR, Murphy RA, Shooter EM, Patel PI (1994) Regulation of tissue-specific expression of alternative peripheral myelin protein-22 (*PMP22*) gene transcripts by two promoters. J Biol Chem 269:25795–25808.

Trapp BD, Hauer P, Lemke G (1988) Axonal regulation of myelin protein mRNA levels in actively myelinating Schwann cells. J Neurosci 8:3515–3521.

Uyemura K, Horie K, Kitamura K, Suzuki M, Uehara S (1979) Developmental changes of myelin proteins in the chick peripheral nerve. J Neurochem 32:779–788.

Walikonis RS, Poduslo JF (1996) Activity of cyclic AMP phosphodiesterases and adenylyl cyclase in peripheral nerve after crush and permanent transection injuries. Submitted

Wood JG, Engel EL (1976) Peripheral nerve glycoproteins and myelin fine structure during development of rat sciatic nerve. J Neurocytol 5:605–615.

Yamamoto KK, Gonzalez GA, Menzel P, Rivier J, Montminy MR (1990) Characterization of a bipartite activator domain in transcription factor CREB. Cell 60:611–617.

Zhang XP, Miskimins R (1993) Binding at an NFI site is modulated by cyclic AMP-dependent activation of myelin basic protein gene expression. J Neurochem 60:2010–2017.

14

CHARACTERIZATION OF AN AXONAL PROTEIN INVOLVED IN MYELINATION

L. A. Sawant,[1] S. D. Raval,[1] S. J. Quinlivan,[1] A. Ducret,[2] R. Aebersold,[2] and L. H. Rome[1]

[1]Department of Biological Chemistry and
    Mental Retardation Research Center
UCLA School of Medicine
Los Angeles, California 90095
[2]Department of Molecular Biotechnology
University of Washington
Seattle, Washington 98195

INTRODUCTION

The mechanisms controlling myelination at the cellular level are not fully understood. Myelin is formed by an ordered process whereby the myelinating cell, the oligodendroglia in the central nervous system (CNS) and the Schwann cell in the peripheral nervous system (PNS), produces and extends membranous extensions which envelop axonal processes. Studies on peripheral nerve regeneration have indicated that axons are the major regulator of myelin protein gene expression by Schwann cells (Politis et al. 1982). On the other hand, oligodendroglia are capable of myelin membrane lipid and protein synthesis in the absence of neuronal signals in vitro (Poduslo, 1978; Mirsky et al. 1980; Szuchet et al. 1980; Poduslo et al. 1982; Bradel and Prince, 1983; Bressler et al. 1983; Rome et al. 1986), suggesting differences between the PNS and the CNS in the myelination process. Although some of the signals for myelination reside within the oligodendroglia itself or can be replaced by environmental cues (Knapp et al. 1987) present in the cultures, there are several reports of neuronal influences on oligodendroglial proliferation and differentiation both in vivo and in vitro (Hardy and Reynolds 1993).

In order to identify the role of axonal proteins in myelination, our laboratory isolated an anti-axolemma monoclonal antibody, designated G21.3, that inhibits myelination in a functional assay using cerebellar slice cultures (Notterpek and Rome 1994a, Notterpek et al. 1993). The antibody significantly reduced the amount of myelin lipid and protein synthesis in the slice cultures. When examined by immunofluorescence, the G21.3 staining was confined to CNS white matter and proliferative regions, neuronal cultures and sciatic nerve, while isolated oligodendroglia and purified myelin were immunonegative. The antibody selectively identified a doublet at 140 kDa and a single band at ~ 120 kDa in the axolemma preparations (Notterpek and Rome 1994a). Both the proteins were present in tissue homogenates of adult rat spinal cord and brainstem, and were more abundant in sciatic nerve as seen on Western blots. In non-neuronal tissues, the G21.3 antigen is primarily localized to basement membranes. In retinoic acid-induced differentiation of SK-N-SHF neuroblastoma cells, the antigen is up regulated and appears to be associated with the extracellular matrix (ECM). Overall, the expression patterns seen for the G21.3 antigen are consistent with the protein being an ECM component that associates with axolemma in both the CNS and the PNS.

In order to carry out a biochemical characterization of the G21.3 antigen, we attempted to purify the 120 and 140 kDa proteins recognized by the antibody in Western blots. Due to the increased abundance of immunoreactive material seen in sciatic nerve we have concentrated our initial efforts on this tissue. In this study we report on the purification of the 120 and 140 kDa G21.3 immuno-reactive proteins from sciatic nerve and the identification of these proteins as type I collagen chains. The functional significance of these findings is also discussed.

MATERIALS AND METHODS

Tissue

Rats were anesthetized and sacrificed by rapid decapitation. Sciatic nerve, brainstem and spinal cord were dissected. The tissues were frozen in liquid nitrogen and stored at -80°C until use.

Extraction of G21.3 Antigen

Sciatic nerve (1 gm) was dissected from adult rat, cut into pieces and crushed under liquid nitrogen. It was then suspended in 10 ml 0.05 M Tris -HCl pH 8.4 containing 1 mM $MgCl_2$, 1 mM $CaCl_2$, 1 mM DTT, 10 mM EDTA and 1 mM PMSF (Buffer A) and centrifuged at 15,000 rpm for 15 min in a Sorvall SS34 rotor at 4°C. The supernatant (S1), was removed and the pellet (P1) was again crushed under liquid nitrogen, resuspended and centrifuged as described above. The pellet (P2) and supernatant (S2) were separated. P2 was dounce homogenized in buffer A and centrifuged on a clinical centrifuge for 5 min (at setting 5). The pellet (P3) was suspended in 1% SDS, vortexed thoroughly and centrifuged for 5 min in a clinical centrifuge. The pellet (P4) was resuspended in buffer A, vortexed and centrifuged to give pellet (P5) and supernatant (S5) containing the purified G21.3 antigen. Protein content was measured by the BCA protein assay (Pierce., Rockford, IL). Samples were boiled for 5 min in sodium dodecyl sulphate (SDS) sample buffer and proteins were separated on a 5% gel by SDS polyacrylamide gel electrophoresis (PAGE) according to Laemmli (1970). The proteins were transfered to nitrocellulose membrane (Towbin et al. 1979). The membrane was stained with Ponceaus to check for transfer of proteins, blocked in 10% milk in phosphate buffered saline (PBS) for 30 - 60 min. Primary antibodies were added and incubated overnight at 4°C in 5% sheep serum in PBS. G21.3 ascites was used at a 1:250 dilution and monoclonal anti-collagen I at a 1:1000 dilution (Sigma Biosciences, St. Louis, MI). The membrane was then washed with Tris-buffered saline (TBS) and incubated with horse radish peroxidase conjugated sheep anti-mouse Ig (Amersham, IL) at a dilution of 1:2000. After extensive washes, the membrane was developed using ECL Western blotting detection reagents (Amersham, IL) according to the manufacturers instructions.

Frozen Sections and Immunocytochemistry

Frozen sections of ~5–10 μm were cut on a cryostat and dried onto superfrost /plus microscope slides (Fisher Scientific). The sections were fixed in 1% paraformaldehyde in 90% ethanol for 1–2 mins, rinsed with phosphate buffered saline (PBS) for 10 mins with two changes of PBS and blocked for 30 min. in 10% goat serum in PBS. The monoclonal antibody (mAb) G21.3, supernatant or ascites were used at 1:20 or 1:100 dilutions respectively. Monoclonal antibodies to neurofilaments, SMI31 (1:1000 dilution, Sternberger Monoclonals Inc., Maryland), collagen I (1:2000 dilution, Sigma Immunochemicals Inc., MI) and polyclonal antisera against collagen I (1:80 dilution, Chemicon Intl. Inc., CA) were used for immunostaining of rat sciatic nerve, spinal cord and cerebellum. The sections were incubated with primary antibodies overnight at 4°C followed by either FITC-conjugated goat anti-mouse IgG + IgM (Boehringer Mannheim Corporation, IN) or rhodamine conjugated goat anti-rabbit IgG for 1hr at room temperature in the dark.

The tissue sections to be stained with polyclonal and monoclonal antibodies against collagen I were fixed in acetone as the epitopes recognized by these antibodies are sensitive to formaldehyde treatment.

Amino Acid Sequencing and Sequence Comparisons

The purified 120 and 140 kDa G21.3-immunoreactive proteins from rat sciatic nerve were subjected to SDS-PAGE and transferred to a nitrocellulose membrane. In-situ tryptic cleavage was performed essentially as described by Aebersold et al. (1987). Membranes previously blocked with polyvinylpyrrolidone-40 were cut in 1 mm^2 pieces and incubated for 3 hours at 37°C in 25 μl digestion buffer (100 mM NH_4HCO_3 pH 7.8:acetonitrile 90:10 v/v) containing 1 μg of sequencing grade trypsin. The supernatant was collected and the membrane pieces were extracted with another 25 μl digestion buffer. Peptides were purified by reverse-phase high performance liquid chromatography essentially as described by Ducret et al. (1996). 20% of the eluent was deviated to a Finnigan-MAT TSQ 7000 tandem mass spectrometer for further analysis while the remainder was manually collected according to the UV absorbance signal. Automated protein sequencing was performed on an ABI model 477A protein microsequencer using precycled polybrene glass fiber discs following standard protocols.

For identification purposes, the amino acid sequences of selected peptides were compared to the Swiss-Prot database. Sequence alignments were reconstituted by using the standard Smith/Waterman Best Local Similarity algorithm as implemented on the EMBL Blitz server (Biocomputing Research Unit, University of Edinburgh, UK; email: Blitz@EBI.AC.UK).

Collagenase Procedure

Samples were incubated with 1 mg/ml of collagenase D (Sigma Immunochemicals Inc., MI) overnight at 37°C. They were then boiled with SDS-PAGE sample buffer and subjected to electrophoresis and Western blot analysis using G21.3 ascites as described above.

RESULTS

Attempts to solubilize the 120 and 140 kDa G21.3 immunoreactive proteins from sciatic nerve utilizing high salt and/or detergents were unsuccessful. As a consequence of this property, an extraction protocol was developed whereby the contaminating soluble proteins were selectively removed from the apparently insoluble G21.3 antigens. The final purification protocol is illustrated in a flow sheet (Figure 1) and described in Materials and Methods. Selected fractions from the purification protocol run on SDS-PAGE are shown in Figure 2. A Western blot of the most purified fraction (S5, also included in Figure 2) demonstrates that the major proteins are strongly recognized by the G21.3 monoclonal antibody. In addition, higher molecular weight bands with size distributions consistent with dimers and trimers of the 120 and 140 kDa proteins were also detected in coomassie-stained gels and by Western blotting.

The 120 and 140 kDa G21.3-immunoreactive proteins from fraction S5 were transferred to nitrocellulose and subjected to tryptic digestion as described in Materials and Methods. The amino acid sequences of three peptides generated from the 140 kDa band and of six peptides generated from the 120 kDa band were determined and compared to the Swiss-Prot database for identification purposes (Table I). Of the 140 kDa band, two peptides (T10.3 and T10.4) revealed a perfect match with the available partial rat collagen α1 (I) while peptide T9.1 exhibited high homology rates with collagen α2 (IV), α1 (I) and α2 (I) sub-types from several species. Three peptides generated from the 120 kDa band (T11.1/1, T11.1/2 and T13.3) revealed significant homology rates to collagen α2 (I) with peptide T13.3 exhibiting a perfect match to the available partial rat collagen α2 (I) sequence. Interestingly, the amino acid sequence determined for peptide T27.2, while typical for the collagen family, was not sufficiently distinctive to establish a correlation with collagen α1 or α2 (I) subtypes. Finally, peptide T27.1/1 was found to represent an autolytic fragment from trypsin while the identity of peptide T27.1/2 could not be established.

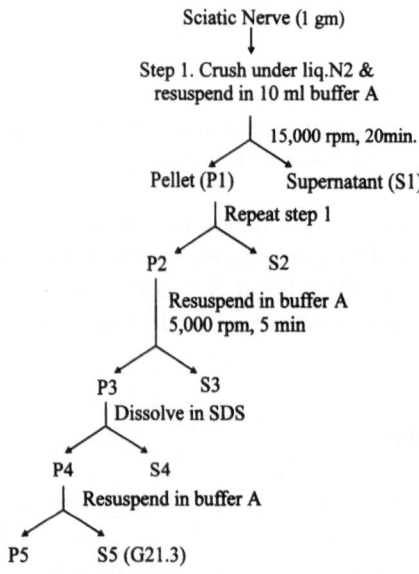

Figure 1. Flow chart of the purification protocol for the G21.3 antigen from adult rat sciatic nerve.

Table I. Microsequence analysis of G21.3 peptides

Protein	Fragment	Peptide sequence	Match
140 kDa	T9.1	GLTGPNGPXG	Collagen α2 (IV), α1 & α2 (I)
140 kDa	T10.3	GEQGPAGSXGFQGL	Collagen α1 (I)
140 kDa	T10.4	GLXGPXGAXGPQGFQ	Collagen α1 (I)
120 kDa	T11.1/1	GAXGAIGAAGPAGA	Collagen α2 (I)
120 kDa	T11.1/2	GPSGPVGAXGATG	Collagen α2 (I)
120 kDa	T13.3	GPXGAVGAXGPQGFQGGPAG E	Collagen α2 (I)
120 kDa	T27.1/1	SIV(H)PSTNSNTLNNA	Trypsin (AA 76-89)
120 kDa	T27.1/2	APYP(L)GQ(S)T	none found
120 kDa	T27.2	TGEIGASGPXGPA(G)E	Collagen α1 or α2 (I)

Sequencing was performed as described in Materials and Methods and peptides were identified using the Swiss-Prot database (Raval-Fernandes et al. *Dev Neursci* In Press)

To further confirm the G21.3 antigen from sciatic nerve as a collagen, we treated the purified proteins (S5) with collagenase overnight and monitored the bands by SDS-PAGE and Western blotting. All of the immunoreactive bands were digested by the collagenase (not shown). Immunochemical staining of sciatic nerve with anti-collagen I and G21.3 antibodies was also carried out (Figure 3). Although the two antibodies showed similar staining of basal lamina (compare Figures 3A and 3B), the G21.3 monoclonal appeared to recognize additional structures in sciatic nerve not stained by the anti-collagen antibody. This additional staining was seen at or near the position of the axonal membrane.

DISCUSSION

As an approach to identify axolemma proteins involved in myelination, we previously pursued an immunological strategy. Rat CNS axolemma was isolated and used to immunize mice for production of monoclonal antibodies. Antibodies specific for axolemma membranes were initially selected and further subselected based on their ability to immunostain white matter regions of cerebellum (Notterpek and Rome, 1994b). Candi-

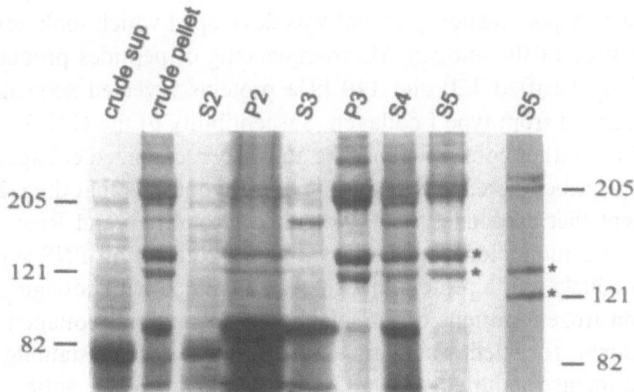

Figure 2. Coomassie stained 5% SDS polyacrylamide gel showing the profile of fractions abtained at each step of purification as described in Materials and Methods. The last lane is a Western blot of fraction S5 probed with G21.3 ascites showing major immunoreactive bands at 120 and 140 kDa.

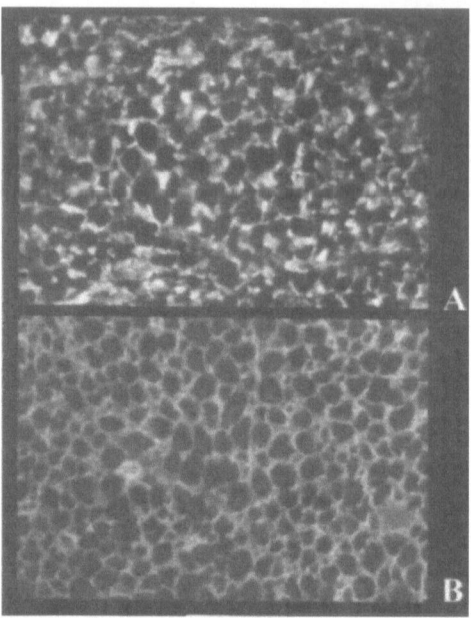

Figure 3. Immunofluorescence staining of adult sciatic nerve using G21.3 ascites (A) and monoclonal antibodies to collagen type I (B). Original magnification was 200X.

date clones were screened functionally by their ability to interfere with myelin membrane deposition in an in vitro cerebellar slice bioassay. This slice culture system permits rapid examination of myelination using both biochemical and morphological procedures (Notterpek et al. 1993). One monoclonal (G21.3) was found to block myelination in the cerebellar slices and the present study was undertaken to further characterize the proteins recognized by this antibody.

Western blots of purified axolemma revealed major G21.3 immunoreactive bands at ~140 kDa and 120 kDa. Similar sized proteins were also stained by the antibody in spinal cord, brain stem, cortex and sciatic nerve (Notterpek and Rome, 1994a). Since sciatic nerve appeared to be the richest source of immunoreactive protein, we began our purification from this tissue. A purification protocol was developed which took advantage of the highly insoluble nature of the antigen. Microsequencing of peptides produced by trypsin digestion of partially purified 120 and 140 kDa proteins revealed sequences consistent with them being derived from type I collagen. Susceptibility of the G21.3 antigens to collagenase degradation further confirmed that the antibody recognized collagen subunits.

The results described here support our previous hypothesis that the G21.3 antigen is an ECM component that associates with axolemma (Notterpek and Rome, 1994b). The finding that the G21.3 monoclonal recognizes a type I collagen in PNS suggests that the antigen is either a collagen or a protein that shares an epitope with collagen. Immunofluorescence studies on frozen sections of rat sciatic nerve using anti-collagen I antisera and the G21.3 monoclonal, revealed a similar, although not identical, staining pattern. Both antibodies showed intense staining of basal lamina, while the G21.3 antibody appeared to react also in the area of the axonal membrane (Figure 3 and Notterpek and Rome, 1994b). This suggests that the G21.3 antigen shares a similar epitope, but is not identical, to type I collagen.

Our previous immunolocalization studies in developing rat cerebellum indicated that the G21.3 antigen was present in the external granular layer and the future Purkinje cell layer in 1-day old animals and by day 10 staining extended to the internal granular layer and the future white matter regions. By postnatal day 18, the highest amount of G21.3 immunoreactive material was in the white matter where it remained high even in the adult. Similar results were obtained in other areas of the developing CNS, where expression of the G21.3 antigen along fiber tracts was found to precede that of myelin markers (Notterpek and Rome, 1994b). This CNS localization of the G21.3 antigen is quite distinct from type I collagen which in the adult rat cerebellum is only seen in blood vessels and in the pial membranes (Raval-Fernades et al., 1997). Few other collagen subtypes have been localized to the CNS except for a recent report describing collagen type IV localization in the embryonic telencephalon of the rat (Eagleson et al., 1996).

By examining Schwann cell development in culture, Bunge and his colleagues determined that myelination is strictly coupled to the ability of Schwann cells to deposit basal lamina on their surface (Eldridge et al., 1987 and 1989). The deposition of basal lamina by Schwann cells depends on the presence of ascorbic acid, a cofactor required for the synthesis of type IV collagen (Eldridge et al., 1988). However, purified type IV collagen does not appear to promote Schwann cell myelination in the absence of other basal lamina constituents (laminin and heparan sulfate proteoglycan) (Eldridge et al., 1989).

Since the G21.3 monoclonal is able to interfere with myelin synthesis in cerebellar slice cultures, the prospect that the antibody specifically recognizes a collagen suggests additional parallels to PNS myelination. We are presently examining the G21.3 antigen in the CNS and the prospect that it may represent a new or previously identified subclass of collagen.

ACKNOWLEDGMENTS

The work referred to in this paper was supported by U.S. Public Health Service Grants HD-06576 and NS34265 and by the National Science Foundation, Science and Technology Center for Molecular Biotechnology.

REFERENCES

Aebersold RH, Leavitt J, Saavedra RA, Hood LE, Kent SB (1987): Internal amino acid sequence analysis of proteins separated by one- or two-dimensional electrophoresis after in-situ protease digestion on nitrocellulose. Proc. Natl. Acad. Sci. USA 84:6970–6974.

Bradel EJ, Prince FP (1983): Cultured neonatal rat oligodendrocytes elaborate myelin membrane in the absence of neurons. J Neurosci Res 9:381–392.

Bressler JP, Cole R, de Vellis J (1983): Neoplastic transformation of newborn rat oligodendrocytes in culture. Cancer Research 43:709–715.

Ducret A, Foyn Bruun C, Bures EJ, Marhaug G, Husby G, Aebersold R (1996): Characterization of human serum amyloid A protein isoforms separated by two-dimensional electrophoresis by liquid chromatography/electrospray ionization tandem mass spectrometry. Electrophoresis 17:866–876.

Eagleson KL, Raymond TF, Levitt P (1996): Complementary distribution of collagen type IV and the epidermal growth factor receptor in the rat embryonic telencephalon.Cerebral Cortex 6:540–549.

Eldridge CF, Bunge MB, Bunge RP (1989): Differentiation of axon-related Schwann cells in vitro: II. Control of myelin formation by basal lamina. J. Neurosci.9:625–638.

Eldridge CF, Bunge RP, Bunge MB (1988): Effects of cis-4-hydroxy-L-proline, and inhibitor of Schwann cell differentiation, on the secretion of collagenous and noncollagenous proteins by Schwann cells. Exp. Cell Res.174:491–501.

Eldridge CF, Bunge MB, Bunge RP, Wood PM (1987): Differentiation of axon-related Schwann cells in vitro. I. Ascorbic acid regulates basal lamina assembly and myelin formation. J. Cell Biol. 105:1023–1034.

Hardy R, Reynolds R (1993): Neuron-oligodendroglial interactions during central nervous system development. J. Neurosci. Res. 36:121–126.

Knapp PE, Bartlett WP, Skoff RP (1987): Cultured oligodendrocytes mimic in vivo phenotypic characteristics: cell shape, expression of myelin-specific antigens, and membrane production. Develop. Biol. 120:356–365.

Laemmli UK (1970): Cleavage of structural proteins during the assembly of the head of bacteriophage T4. Nature 227:681–685.

Mirsky R, Winter J, Abney ER, Pruss RM, Gavrilovic J, Raff MC (1980): Myelin specific proteins and glycolipids in rat Schwann cells and oligodendrocutes in culture. J. Cell Biol. 84:483–494.

Notterpek LM, Rome LH (1994a): Functional evidence for the role of axolemma in CNS myelination. Neuron 13:473–485.

Notterpek LM, Rome LH (1994b): A Protein Involved in CNS Myelination: Localization to the ECM and Induction in Neuroblastoma Cells. Developmental Neuroscience 16:267–278.

Notterpek LM, Bullock PN, Malek-Hedayat S, Fisher R, Rome LH (1993): Myelination in cerebellar slice cultures: development of a system amenable to biochemical analysis. J. Neurosci. Res. 36:621–634.

Poduslo JF (1978): The molecular architecture of myelin: Identification of the external surface membrane components. Adv. Exp. Med. Biol. 100:189–205.

Poduslo SE, Miller K, Wolinsky JS (1982): The production of a membrane by purified oligodendroglia maintained in culture. Experimental Cell Res. 137:203–215.

Politis MJ, Sterberger N, Ederle K, Spencer PS (1982): Studies on the control of myelinogenesis IV. Neuronal induction of Schwann cell myelin-specific protein synthesis during nerve fiber regeneration. J. Neurosci. 2:1252–1266.

Raval-Fernandes S et al. (1997): Develop Neurosci (in Press).

Rome LH, Bullock PN, Chiappelli F, Cardwell M, Adinoff AM, Swanson D (1986): Synthesis of a myelin-like membrane by oligodendrocytes in culture. J. Neurosci. Res. 15:49–65.

Szuchet S, Stefansson K, Wollman RL, Dawson G, Arnason BGW (1980): Maintenance of isolated oligodendrocytes in long-term culture. Brain Res. 200:151–164.

Towbin H, Staehelin T, Gordon J (1979): Electrophoretic transfer of proteins from polyacrylamide gels to nitrocellulose sheets: Procedure and some application. Proc. Natl. Acad. Sci. 76:4350.

15

DO SECRETORY PATHWAY SNARE PROTEINS MEDIATE MYELINOGENESIS?

Dana L. Madison and Steven E. Pfeiffer

Department of Microbiology and Program in Neurological Sciences
University of Connecticut School of Medcine
Farmington, Connecticut 06030-3205

1. INTRODUCTION

Oligodendrocyte development proceeds along a well defined developmental lineage both in vivo and in vitro (Fig. 1). The differentiated oligodendrocyte synthesizes a vast amount of myelin membrane, in the range of 5,000–50,000 mm^2 per cell per day (Pfeiffer et al., 1993). In the neo-natal rat brain, or example, the mg of dry weight or myelin increases 600% from P0 to P30 and 1800% by P180; over this same time period total brain weight increases by only 50–60% (Norton and Cammer, 1984). How do oligodendrocytes accurately target and assemble myelin lipids and proteins into mature myelin membrane?

The transport of lipids and proteins among subcellular compartments, including the plasma membrane, is accomplished by a vesicular trafficking system which vectorially directs membrane and secretory components to specific domains (Rothman, 1994). One basic paradigm in vesicular trafficking is the 'life cycle' of vesicles which includes a) formation, coating and budding of vesicles, b) transport of the vesicles, c) recognition and docking of the vesicles at the proper acceptor membrane, d) fusion and either release or incorporation of the vesicle's contents, and e) recycling of the some of the vesicle's intrinsic components (Rothman, 1994). Intra-Golgi transport is therefore expected to be common among cell types, whereas secretory events will necessarily have differences based on cell type, the molecules secreted and their regulation. One mechanism that mediates the later steps of vesicle recognition and fusion is described by the SNARE hypothesis (SNARE = *SNAP Receptor*; Sollner et al., 1993; Rothman, 1994). According to this model, every transport pathway must contain a cognate pair of proteins that aid in correctly addressing a vesicle to its target membrane. These v-(vesicle) SNAREs and t-(target) SNAREs form the basis for adding other proteins and levels of specificity to the docking and fusion process (Fig. 2). In synaptic vesicles, a v-SNARE (VAMP-2/synaptobrevin 2) and two t-SNAREs (SNAP-25 and syntaxin 1a) form an initial SDS-resistant core docking complex (7S complex) (Hayashi et al., 1994; 1995), upon which the general membrane trafficking proteins NSF and a-SNAP become associated (20S complex). Accessory proteins such as n-sec1, synaptophysin, synaptotagmin, Rab3, and rabphilin-3a are also involved, as presumably are other as yet undescribed proteins. The actual

Oligodendrocyte Development

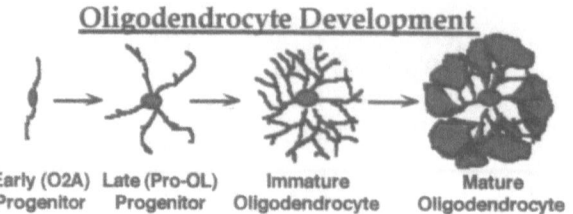

Early (O2A)   Late (Pro-OL)   Immature   Mature
Progenitor   Progenitor   Oligodendrocyte   Oligodendrocyte

Figure 1. A simplified model for oligodendrocyte development. Stages are identified based on the appearance of developmentally regulated lipid and protein markers (reviewed in Pfeiffer et al., 1993).

events that lead to membrane fusion are not yet clear, although numerous in vitro studies have provided insight about these proteins and how they interact (Sollner et al., 1993; Calakos et al., 1994; Pevsner et al., 1994; Chapman et al., 1994; Sogaard et al., 1994; Edelman et al., 1995).

These proteins exist in families and the various homologues may be specific for a particular pathway and/or cell type. SNARE proteins have been best studied in neurotransmitter secretion (Sudhoff, 1995; Bennett, 1995; Calakos and Scheller, 1996), but conservation of the secretory pathway components and mechanisms has also been noted in neurosecretory cells

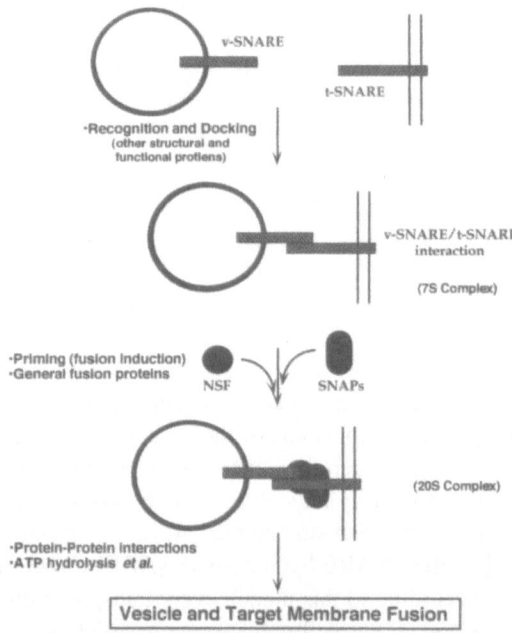

Figure 2. A generalized scheme for vesicular docking and fusion as proposed by the SNARE hypothesis. A vesicle containing a v-SNARE is trafficked to its target membrane containing the cognate t-SNARE. In the first step, recognition, the v- and t-SNAREs come into proximity; small GTP-binding proteins may regulate the fidelity of this step, GTP is hydrolyzed producing a conformational change in a series of protein-protein interactions, perhaps further stabilizing v- and t-SNARE interaction. In the second step, docking, the vesicle and target membrane become physically associated through v- and t-SNARE protein complexes. In the third step the ubiquitous SNAP associates with the previously formed v- and t-SNARE complex, creating enhanced binding of NSF. Following an ATP hydrolytic step, and other unknown events, the vesicle and target membrane fuse (after Bennett, 1995).

such as adrenal chromaffin cells (Holz et al., 1994) and the exocrine pancreas (Sadoul et al., 1995; Braun et al., 1995; Regazzi et al., 1995). The conservation of SNARE proteins among species and cell types (Bennett and Scheller, 1993) suggests that the core proteins may form a general membrane fusion mechanism which is adapted by all cells for different purposes (Bark et al., 1995; Garcia et al., 1995; Calakos and Scheller, 1996; Table 1).

Since myelinogenesis is a special form of membrane biogenesis, we have hypothesized that this process will utilize the principles and mechanisms of the more general vesicular transport machinery. Accordingly, we have examined the expression by oligodendrocytes of certain proteins involved in the transport and fusion of plasma membrane bound vesicles and consider how this may apply to the synthesis of myelin membrane. These include 1) the small GTP-binding proteins, and more specifically, 2) the secretory pathway associated protein Rab3, and 3) the proteins involved in SNARE complex assembly leading to vesicular fusion as described for plasma membrane bound vesicles of the secretory pathway.

2. SMALL GTP-BINDING PROTEINS AND THE RAB PROTEIN CLASS

One protein class important in the regulation of unidirectional membrane transport is the monomeric small GTP-binding proteins (Zerial and Sternmark, 1993; Simons and

Table 1. The SNARE protein families are conserved between mammalian cells and yeast for both major vesicular transport pathways, endoplasmic reticulum (ER) to golgi, and golgi to plasma membrane

		ER-Golgi	post-Golgi
Mammalian	v-SNAREs	?	VAMP, cellubrevin
	t-SNAREs	syntaxin 5	syntaxins 1-4 SNAP-25
	Sec1 family	?	nsec-1,2
	rab family	rab1	rab3a, b, c
		rab2	rab6
			rab8
	General fusion protiens	NSF/α-SNAP	
Yeast	v-SNAREs	Bos1	Snc1
		Sec22	Snc2
		Ykt6/p26	
		Bet1	
	t-SNAREs	Sed5	Sso1
			Sso2
			Sec9
	Sec1 family	Sly1	Sec1
	rab family	Ypt1	Sec4
	general fusion proteins	Sec18/Sec17	

The generalized classes of proteins within the SNARE hypothesis are listed with their specific members in mammalian cells and yeast. Protein homologues for nearly every trafficking step in mammalian cells are found in yeast. Given the conservation within different mammalian cells and among organisms such as mammals, yeast and *Drosophila*, the SNARE hypothesis functions as a generalized framework for events leading to membrane fusion. As pathways leading to fusion steps become more complex, more proteins are being added to this generalized model (Adapted from Bennett, 1995)

Zerial, 1993; Novick and Brennwald, 1993; Nuoffer and Balch, 1994). There are six major groups of GTP-binding proteins, Sar/ARF, *Yptlp/Sec4p*/Rab, Rac/*CDC42*, Rho, dynamin and members of the heterotrimeric G protein family, which play differing roles in vesicular transport. Individual, small GTP-binding proteins are required for specific transport steps (Orci et al., 1993; Balch, 1994). Small GTP-binding proteins contain guanine binding and GTP hydrolytic domains and, by analogy with heterotrimeric GTP-binding proteins and the *ras* proto-oncogene, are believed to act as molecular "switches" mediating key steps of vesicular targeting pathways. It is likely that these proteins function in the assembly and disassembly of protein complexes which are involved in the budding, targeting and fusion of transport vesicles. Post-translational lipid modification at the C-terminus of these proteins function in specific association with the membranes of subcellular compartments (Farnsworth et al., 1991; Yamane et al., 1991; Takai et al., 1993).

The Rab protein family of small GTP-binding proteins is involved in the directed trafficking of vesicles for fusion to target membranes. Rab proteins may be localized at a particular transport step in all cells, *e.g.* Rab6 in intra-Golgi transport (Goud et al., 1990; Antony et al., 1992), or may be restricted to particular cell types which harbor specific transport pathways, such as basolateral restricted transport in polarized epithelial cells by Rab17 (Lutcke et al., 1993). A given step may also contain more than one Rab protein, which may indicate the dynamic nature of organelle flow and responsiveness to multiple functional stimuli.

We have investigated the developmental regulation of membrane associated small GTP-binding proteins in oligodendrocytes by two-dimensional electrophoresis and subsequent binding of radiolabeled GTP and western blot immunodetection (Huber et al., 1994). The results showed a number of developmentally regulated small GTP-binding proteins in oligodendrocytes and purified myelin membrane (Figure 3).

As oligodendrocytes undergo differentiation from Early Progenitors to mature myelin-producing cells, the pattern of expression of small GTP-binding proteins, several of which may be unique to oligodendrocytes and myelin (Fig. 3, designated with letters), becomes substantially more complex. From these analyses we were able to identify the expression of the following proteins during stages of oligodendrocyte development: Rab3, Rap2, RalA and Rho, Rab5a,b,c, Rab7, Rab11, Rab22, Rac, and RhoD; and also tentatively identify Rab1a/1b, Rab4, Rab9, Rab10, and Rab17. The presence of Rab4, 5, 7, 9 and 6 is expected since these proteins are involved in endocytic (4, 5, 7, 9) and intra-Golgi (6) transport, respectively. The number of unidentified proteins, and specifically, unidentified developmentally regulated proteins, is intriguing. Their expression correlates well with changes in other parameters of myelinogenesis, including dramatic changes in morphology, and in the synthesis of myelin-specific lipids and proteins. The expression of several previously unrecognized proteins suggests that oligodendrocytes may utilize cell-type specific GTP-binding proteins in the biogenesis, and maintenance, of the myelin membrane.

3. RAB3 PROTEINS IN OLIGODENDROCYTES

The Rab3 family of small GTP-binding proteins was once thought to be restricted to trafficking pathways in specific cell types involved in regulated secretion within the nervous or neuroendocrine systems (Touchot et al., 1987; Matsui et al., 1988; Zahraoui et al., 1989; Fischer v Mollard et al., 1990; Mizoguchi et al., 1990; Baldini et al., 1992; Thomas-Reetz and DeCamilli, 1994; Li et al., 1996). Rab3a was originally identified as enriched in

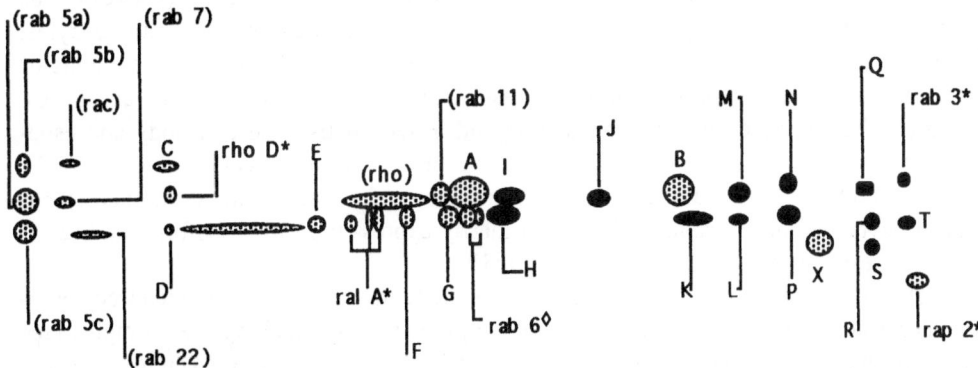

Figure 3. A diagrammatic summary of autoradiographs from two dimensional gel separations probed with a ^{32}P-GTP, of Mature Oligodendrocyte total membrane proteins. The GTP-binding proteins have been labeled with specific names when warranted by identification by co-migration (parenthesis) or immunoblotting (diamond), or both (asterisks). Other signals have been given arbitrary letters pending further identification. The hatched symbols indicate GTP-binding proteins that were detected in all stages of oligodendrocyte development, while the solid symbols indicate GTP-binding proteins that were up-regulated during differentiation, primarily at the Mature Oligodendrocyte stage (Huber et al., 1994).

synaptic vesicles (Fischer v. Mollard *et al.*, 1991), and Rab3 proteins are also present in neuroendocrine and exocrine cells, the adrenal medulla (Darchen et al., 1990) and outside the nervous system in mouse adipose cells (Baldini et al., 1995). Dominant overexpression of Rab3a inhibits secretion from PC12 and bovine chromaffin cells, showing that Rab3a is a negative regulator of secretion prior to the membrane fusion step (Johannes et al., 1994; Holz et al., 1994; Lledo et al., 1994; Weber et al., 1996). Rab3a effector domain peptide injection experiments have produced both stimulatory (Padfield et al., 1992; Richmond and Hayden, 1993; Olszewski, et al., 1994) and inhibitory (Plutner et al., 1990; MacLean et al., 1993) effects in a variety of cell types such as pancreatic b-cells, acinar cells and cultured neurons. Overexpression of wild type Rab3a or Rab3b in PC12 cells produced similar subcellular localizations, and both proteins interacted with the downstream Rab3 effector, Rabphilin-3A; however, despite the fact that Rab3 isoforms are highly homologous proteins, wild type Rab3b markedly stimulated norepinephrine release, while Rab3a modestly inhibited release (Weber et al., 1996). These results are consistent with those obtained in anterior pituitary cells, which express only Rab3b (Lledo et al., 1994). This negative regulation by Rab3a is consistent with decreased synaptic vesicle recruitment following repetitive stimulation in a *rab3a* knockout mouse (Geppert et al., 1994); Therefore, 1) in the case of synaptic vesicles, Rab3a function appears to be proximal to the actual fusion of vesicles mediated by the SNARE complex; and 2) in non-neuronal secretory cells Rab3a plays an important role in secretion, but the molecular mechanisms of its function are unclear.

Both Rab3b and 3c are also thought to be involved in regulated secretory processes. Rab3c co-purifies with Rab3a from synaptic vesicles and may have a similar, possibly redundant, function to Rab3a (Fischer v. Mollard et al., 1994; Li et al., 1994), although it is by far the least expressed in the brain of the three *rab3's* (*a,b,c*; Matsui et al., 1988). Rab3b is predominantly expressed in neurosecretory cells (Lledo et al., 1993, 1994; Weber et al., 1996).

The preliminary identification of a Rab3 protein in oligodendrocytes (above) led us to examine *rab3* isoform expression by PCR analyses in oligodendrocytes and astrocytes. We have found that oligodendrocytes preferentially express *rab3a* and small amounts of *rab3c*, while astrocytes preferentially express *rab3b* (Table 2; Madison et al., 1996). Rab3a can be visualized by immunostaining and is seen within the cell body and especially in the process junctions (Fig. 1). In addition, the Schwann cell line RT4-D6P2T contained both the *rab3a* and *3b* isoforms, but not *rab3c*. These data show that 1) oligodendrocytes, a non-neuronal cell type, primarily express the *rab3a* isoform; and (2) in contrast, astrocytes express predominantly *rab3b*.

The precise role for Rab3a in secretion, let alone its function in oligodendrocytes, is not clear. Rab3's may function in the assembly of the vesicular docking (7S) complex (see Introduction and Fig. 2) since studies in yeast suggest an interaction between low molecular weight GTP-binding proteins and VAMP-2/synaptobrevin-2 and SNAP-25 (Synaptosomal Protein of 25kDa) homologues (Brennwald et al., 1994). However, direct interaction between small GTP-binding proteins and these proteins has been difficult to detect (Scheller, 1995). It is possible that Rab proteins act as proofreading factors for vesicular docking complex assembly, in a manner analogous to increasing fidelity of peptide bond formation by EF-Tu. This may explain the relatively innocuous phenotype of the *rab3a* knockout which displays only decreased vesicle recruitment (Geppert et al., 1994); however, more pronounced phenotypes upon Rab3a disruption in neuroendocrine cells (Holz et al., 1994; Weber et al., 1996) makes Rab3a function all the more difficult to reconcile between cell types. The presence of Rab3a and its upregulation during myelinogenesis is suggestive of a plasma membrane directed vesicular transport pathway.

4. SNARE PROTEINS ARE EXPRESSED IN OLIGODENDROCYTES

Our findings with the secretory pathway associated Rab3a led us to examine other components involved in the trafficking of plasma membrane bound vesicles. These molecules were examined by Northern and Western blots and immunofluorescence. In addition to Rab3a, VAMP-2, *syntaxin-2*, synapsin, synaptophysin, and *n-sec1* were expressed by oligodendrocytes, while *syntaxin-1a*, *rabphilin-3a*, and *synaptotagmin 1* were not detected (Table 3 for summary). The t-SNARE SNAP-25, specifically the SNAP-25a isoform, was cloned from mature oligodendrocyte cDNA, but no protein could be detected by Western blot. Rab3a exhibited the strongest developmental regulation in oligodendrocytes, being

Table 2. A summary of the PCR amplification data for *rab3* isoforms in oligodendrocytes (Oligos), astrocytes (Astros), and the Schwann cell line RT4-D6P2T

	Oligos	Astros	D6P2T
rab3a	Y	L	Y
rab3b	L	Y	N
rab3c	Y	N	Y

All *rab3* isoforms, *rab3a*, *3b* and *3c* were amplified by combinations of degenerate and specific primer sets. Abbreviations: Y= mRNA expressed; N=mRNA not detected; L=low levels of expression (summarized from Madison et al., 1996)

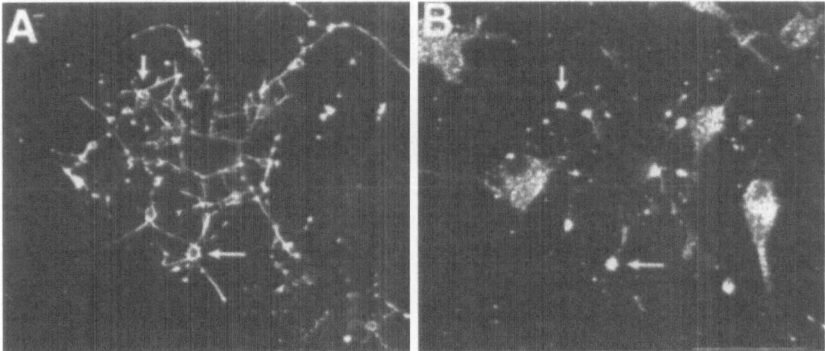

Figure 4. Rab3a is expressed in oligodendrocytes as shown by immunofluorescence microscopy. This cell was double labeled with the oligodendrocyte marker antibody O4 (external membrane staining (A)), and anti-Rab3a (B). Concentrations of Rab3a staining seen in the cells body, in the processes, and within process junctions (arrows). Scale bar is 50 mm

upregulated five-fold during the onset of myelinogenesis. VAMP-2 was upregulated approximately two-fold from Early Progenitors to Mature Oligodendrocytes. The t-SNARE syntaxin isoform detected was *syntaxin 2*, whereas the neuron specific syntaxin *1a* was not detected in either oligodendrocytes or astrocytes. *Synaptotagmin 1* and *rabphilin-3a*, putative regulatory molecules involved in Ca^{2+} gated exocytosis, were also not detected in oligodendrocytes. Since proteins known to be involved in the targeted vesicular delivery of proteins and lipids to the plasma membrane are present in oligodendrocytes, we have proposed a model (below) for the use of this system in myelin membrane delivery.

5. DOES THE BASIC SNARE MECHANISM ADEQUATELY DESCRIBE NASCENT MYELIN MEMBRANE DELIVERY?

Myelination is a process by which large amounts of lipid and protein are synthesized and delivered to a target membrane where an oligodendrocyte process has engaged an axon and actual enwrapping or myelination ensues. The time course and specificity of myelination dictate that a highly organized mechanism for myelin component delivery must exist in oligodendrocytes. This biological problem of membrane delivery argues for a vesicular trafficking system that can accurately target a bulk of vesicles, but does not require many regulatory steps such as stimulus gated secretion. The data presented above details the expression of a number of known synaptic or neuroendocrine secretory vesicle components which are present in oligodendrocytes. These data raise the possibility for multiple, but distinct, functions of these previously known 'neural markers' in both neurons and glia. Interestingly, many of the proteins involved in synaptic vesicle trafficking are now known to be more widely distributed throughout neurons and thought to function in extra-synaptic membrane trafficking (Garcia et al., 1995; Chin et al., 1995).

Myelin synthesis may therefore represent an extension of targeted delivery to the plasma membrane. The SNARE hypothesis predicts a basic set of proteins would be involved in delivery of vesicles to the plasma membrane in all cells; therefore, the goal for each cell is to adapt the basic mechanism to its particular needs by introducing additional

Table 3. Summary of vesicular trafficking components expressed
during oligodendrocyte development

Protein or mRNA	In Oligos?	Devel Reg?
Rab3a[a,b,c,d]	Y	Y
VAMP-2[a,b,d]	Y	S
Syntaxin-2[a]	Y	S
Synaptophysin[a,b,d]	Y	Y
Synapsin la/lb[a,b,d]	Y	Y
nsec-1[a]	Y	N
Syntaxin-1a[a]	N	N
SNAP-25[a,b,d]	?	?
Rabphilin-3a[a,c]	N	N
Synaptotagmin-1[a]	N	N

Each protein/mRNA examined is listed at the left. A "Y" indicates that expression was detected by the method(s) indicated in the superscripts, an "N" indicates no expression was detected, and an "S" indicates some developmental regulation was observed, but was under two-fold. Superscripts are defined as (a) Northern analysis; (b) Western analysis; (c) Immunocytochemistry; (d) PCR/Southern analysis

structural and functional steps and/or proteins. Neurons have achieved a high level of specialization of this system, endocrine cells are slightly less specialized, while oligodendrocytes have adapted the system for the bulk delivery of membrane.

Proper targeting of vesicles to the plasma membrane is likely to involve at least the mechanism and proteins described by the SNARE hypothesis. Although the SNARE mechanism describes both pre- and post-Golgi fusion events on a general basis, there may also be other mechanisms which guide vesicles to their proper domains. In MDCK cells apical transport appeared to proceed via an NSF-independent fusion mechanism, whereas basolateral transport was dependent upon v- and t-SNAREs, a-SNAP and NSF (Ikkonen et al., 1995). There are other examples of membrane fusion that proceed in the absence of NSF which indicates that there are either distant isoforms of these proteins, or that there are unrelated targeting molecules (e.g., GPI-linked proteins) used in these pathways (Wilson, 1995). These data suggest that multiple targeting mechanisms may play a role in directed vesicular trafficking.

The precise roles of certain proteins in myelin biogenesis, such as Rab3a, are still unclear, but the use of VAMP-2 in plasma membrane bound myelin vesicles is consistent with its proposed function as a v-SNARE. Initial experiments suggest that VAMP-2 may play a role in the delivery of membrane in cultured oligodendrocytes. We are currently assessing the usage of VAMP-2 and other SNARE proteins, in order to better understand the processes by which nascent lipids and proteins are rapidly transported during myelination by oligodendrocytes.

REFERENCES

Antony C, Cibert C, Geraud G, Santa-Maria A, Maro B, Maya V, Goud B. (1992): The small GTP-binding protein rab6p is distributed from medial to the trans-Golgi network as determined by a confocal microscopy approach. J Cell Sci 103:785–796.
Balch WE, McCaffery JM, Plutner H, Farquchar MG (1994): Vesicular stomatitis virus glycoprotein is sorted and concentrated during export from the endoplasmic reticulum. Cell 76:841–852.
Baldini G, Hohl T, Lin HY, Lodish HF (1992): Cloning of a rab3 isotype predominantly expressed in adipocytes. Proc Natl Acad Sci USA 89:5049–5052.

Baldini G, Scherer PE, Lodish HF (1995): Non-neuronal expression of Rab3a: Induction during adipogenesis and association with different intracellular membranes that Rab3D. Proc Natl Acad Sci USA 92:4284–4288.

Bark IC, Hahn KM, Ryabinin AE, Wilson MC (1995): Differential expression of SNAP-25 protein isoforms during divergent vesicle fusion events of neural development. Proc Natl Acad Sci USA 92:1510–1514.

Bennett MK, Scheller RH (1993): The molecular machinery for secretion is conserved from yeast to neurons. Proc Natl Acad Sci USA 90:2559–2563.

Bennett MK (1995): SNAREs and the specificity of transport vesicle targeting. Curr Opin in Cell Biol 7:581–586.

Braun JE, Fritz BA, Wong SME, Lowe AE (1994): Identification of a vesicle associated membrane protein (VAMP)-like membrane protein in zymogen granules of the rat exocrine pancreas. J Biol Chem 269:5328–5335.

Brennwald P, Kearns B, Champion K, Keranen S, Bankaitis V, Novick, P. (1994): Sec 9 is a SNAP-25-like component of a yeast SNARE complex that may be the effector of SEC4 function in exocytosis. Cell 79:245–258.

Calakos N, Bennett MK, Peterson KE, Scheller RH (1994): Protein-protein interactions contributing to the specificity of intracellular trafficking. Science 263:1146–1149.

Calakos N, Scheller R (1996): Synaptic vesicle biogenesis, docking, and fusion: A molecular description. Physiol Rev 76:1–29.

Chapman ER, An S, Barton N, Jahn R (1994): SNAP25, a tSNARE which binds to both syntaxin and synaptobrevin via domains that may form coiled-coils. J Biol Chem 269:27427–27432.

Chin LS, Li L, Ferreira A, Kosik KS, Greengard P (1995): Impairment of axonal development and of synaptogenesis in hippocampal neurons of synapsin-I deficient mice. Proc Natl Acad Sci USA 92:9230–9234.

Darchen F, Zahraoui A, Hammel F, Monteils M-P, Tavitian A, Scherman D (1990): Association of the GTP-binding protein Rab3a with bovine adrenal chromaffin granules. Proc Natl Acad Sci USA 87:5692–5696.

Edelmann L, Hanson PI, Chapman ER, Jahn R (1995): Synaptobrevin binding to synaptophysin: a potential mechanism for controlling the exocytic fusion machine. EMBO J 14:224–231.

Farnsworth CC, Kawata M, Yoshida Y, Takai Y, Gelb MH, Glomset JA (1991): Carboxy terminus of the small GTP-binding protein smg-p25A contains two geranylgeranylated residues and a methyl ester. Proc Natl Acad Sci USA 88:6196–6200.

Fischer v. Mollard G, Mignery G, Baumert M, Perin MS, Hanson TJ, Burger PM, Jahn R, Sudhoff T (1990): Rab 3 is a small GTP-binding protein exclusively localized to synaptic vesicles. Proc Natl Acad Sci USA 87:1988–1992.

Fischer v. Mollard G, Sudhoff TC, Jahn R (1991): Rab 3 is a small GTP-binding protein exclusively localized to synapses. Nature 349:79–81.

Fischer v. Mollard G, Stahl B, Khokhlatchev A, Sudhoff T, Jahn R (1994): Rab3c is a synatpic vesicle protein that dissociates from synaptic vesicle after stimulation of exocytosis. J Biol Chem 269:10971–10974.

Garcia EP, McPherson PS, Chilcote TJ, Takei K, DeCamilli P (1995): rbSec1a and b co-localize with syntain1 and SNAP25 throughout the axon, but are not in a stable complex with syntaxin. J Cell Biol 129:105–120.

Geppert M, Bolshakov VY, Slegelbaum SA, Takei K, DeCamilli P, Hammer RE, Sudhoff TC (1994): The role of Rab3A in neurotransmitter release. Nature 369:493–497.

Goud B, Zahroui A, Tavitian A, Saraste, J (1990): Small GTP-binding proteins associated with the Golgi cisternae. Nature 345:553–556.

Hayashi T, McMahon H, Yamasaki S, Binz T, Hata Y, Sudhoff TC, Niemann H (1994): Synaptic vesicle membrane fusion complex: action of clostridial neurotoxins on assembly. EMBO J 13:5051–5061.

Hayashi T, Yamasaki S, Nauenburg S, Binz T, Niemann H (1995): Disassembly of the reconstituted synaptic vesicle membrane fusion complex in vitro. EMBO J 14:2317–2325.

Holz RW, Brondyk WH, Senter RA, Kuizon L, Macara IG (1994): Evidence for the involvement of Rab3A in Ca^{2+}-dependant exocytosis from adrenal chromaffin cells. J Biol Chem 269:10229–10234.

Huber LA, Madison DL, Simons K, Pfeiffer SE (1994): Myelin membrane biogenesis by oligodendrocytes: Developmental regulation of low-molecular weight GTP-binding proteins. FEBS Lett 347:273–278.

Ikonen I, Tagaya M, Ullrich O, Montecucco C, Simons K (1995): Different requirements for NSF, SNAP, and Rab proteins in apical and basolateral transport in MDCK cells. Cell 81:571–580.

Johannes L, Lledo P-M, Roa M, Vincent J-D, Henry J-P, Darchen F (1994): The GTPase Rab3a negatively controls calcium-dependent exocytosis in neuroendocrine cells. EMBO J 13:2029–2037.

Li C, Takei K, Geppert M, Daniell L, Stenius K, Chapman E R, Jahn R, De Camilli P, Sudhoff TC (1994): Synaptic targeting of rabphilin-3A, a synaptic vesicle Ca^{2+}/Phospholipid-binding protein, depends on rab3a/3c. Neuron 13:885–898.

Li J-Y, Jahn R, Hou X-E, Kling-Petersen A, Dahlstrom A. (1996): Distribution of Rab3a in rat nervous system: comparison with other synaptic vesicle proteins and neuropetides. Brain Res 706:103–112.

Lledo P-M, Vernier P, Vincent J-D, Mason WT, Zorec R (1993): Inhibition of rab3b expression attenuates Ca^{2+}-dependant exocytosis in anterior rat pituitary cells. Nature 364:540–544.

Lledo P-M, Johannes L. Vernier P, Zorec R, Darchen F, Vincent J-D, Henry J-P, Mason WT (1994): Rab3 proteins: key players in the control of exocytosis.Trends in Neurosci 17:426–432.

Lutcke A, Jansson S, Parton RG, Chavrier P, Valencia A, Huber LA Lahtonen E, Zerial M (1993): Rab 17, a novel small GTPase, is specific for epithelial cells and is induced during cell polarization. J CellBiol 121:553–564.

MacLean CM, Law GJ, Edwardson JM (1993): Stimulation of exocytic membrane fusion by modified peptides of the rab3 effector domain: re-evaluation of the role of rab3 in regulated exocytosis. Biochem J 294:325–328.

Madison DL, Krueger WK, Kim T, Pfeiffer SE (1996): Differential expression of *rab3* isoforms in oligodendrocytes and astrocytes. J Neurosci Res 45:258–268.

Matsui Y, Kikuchi A, Kondo J, Hishida T, Teranishi Y, Takai Y (1988): Nucleotide and amino acid deduced amino acid sequences of a GTP binding protein family with molecular weights of 25,000 from bovine brain. J Biol Chem 263:11071–11074.

Mizoguchi A, Kim S, Ueda T, Takai Y (1989): Tissue distribution of *smg*-p25A, a *ras*-like GTP-binding protein, studied by use of a specific monoclonal antibody. Biochem Biophys Res Com. 162:1438–1445.

Mizoguchi A, Kim, S, Ueda T, Kikuchi A, Yorifuji H, Hirokawa N, Takai Y (1990): Localization and subcellular distribution of smg-p25A, a *ras* p23-like GTP-binding protein, in rat brain. J Biol Chem 265:11872–11879.

Norton WT, Cammer W (1984): Isolation and characterization of myelin. In Morell P (ed): "Myelin", 2nd Ed. Plenum Press. New York/London. pp 147–195.

Novick P, Brennwald P (1993): Friends and family: the role of the Rab GTPases in vesicular trafficking. Cell 75:597–601.

Nuoffer C, Balch WE (1994): GTPases: Multifunctional molecular switches regulating vesicular traffic. Ann Rev Biochem 63:949–990.

Olszewski S, Deeney JT, Schuppin GT, Williams KP, Corkey BE, Rhodes CT (1994): Rab3a effector domain peptides induce insulin exocytosis via a specific interaction with a cytosolic protein doublet. J Biol Chem 269:27987–27991.

Orci L, Palmer DJ, Amherdt M, Rothman JE (1993): Budding from Golgi membranes requries coatamer complex of non-clatharin coat proteins. Nature 362:648–652.

Padfield PJ, Balch WE, Jamieson JD (1992): A synthetic peptide of the rab3a effector domain stimulates amylase release from permeabilized pancreatic acini. Proc Natl Acad Sci USA 89:1656–1660.

Pevsener J, Hsu SC, Braun JE, Calakos N, Ting AE, Bennett MK, Scheller RH (1994b): Specificity and regulation of a synaptic vesicle docking complex. Neuron 13:353–361.

Pfeiffer SE, Warrington AE, Bansal R (1993): The oligodendrocyte and its many cellular processes. Trends in Cell Biol 3:191–197.

Plutner H, Schwaninger R, Pind S, Balch WE (1990): Synthetic peptides of the rab effector domain inhibit vesicular transport through the secretory pathway. EMBO J 9:2375–2383.

Regazzi R, Wollheim CB, Lang J, Theler J-M, Rossetto O, Montecucco C, Sadoul K, Weller U, Palmer M, Thorens B (1995): VAMP2 and cellubrevin are expressed in pancreatic b-cells and are essential for Ca^{2+} but not for GTPgS induced insulin secretion. EMBO J 14:2723–2730.

Richmond J, Hayden PG (1993): Rab effector domain peptides stimulate the release of neurotransmitter from cultured cell synapses. FEBS Lett 326:124–130.

Rothman JE (1994): Mechanisms of intracellular protein transport. Nature 372:55–63.

Sadoul K, Lang J, Montecucco C, Weller U, Regazzi R, Catsicas S, Wollheim CB, Halban PA (1995): SNAP-25 is expressed in islets of langerhans and is involved in insulin release. J Cell Biol 128:1019–1028.

Scheller RH (1995): Membrane trafficking in the presynaptic nerve terminal. Neuron:893–897.

Simons K, Zerial M (1993): Rab proteins and the road maps for intracellular transport. Neuron 11:789–799.

Sogaard M, Tani K, Ye RR, Geromanos S, Tempst P, Kirchhausen T, Rothman J, Sollner T (1994): A rab protein is required for the assembly of SNARE complexes in the docking of transport vesicles. Cell 78:937–948.

Sollner T, Whiteheart SW, Brunner M, Erdjument-Bromage H, Geromanos S, Tempst P, Rothman JE (1993): SNAP receptors implicated in vesicle targeting and fusion. Nature 362:318–324.

Sudhoff TC (1995): The synaptic vesicle cycle: a cascade of protein-protein interactions. Nature 375:645–653.

Takai Y, Kaibuchi K, Kikuchi A, Kawata T (1992): Small GTP-binding proteins. Int'l Rev Cytol 133:187–230.

Thomas-Reetz AC, DeCamilli P. (1994): A role for synaptic vesicles in non-neuronal cells: clues from pancreatic b-cells and from chromaffin cells. FASEB J 8:209–216.

Touchot N, Chardin P, Tavitian A (1987): Four additional members of the ras gene superfamily isolated by an oligonucleotide strategy: molecular cloning of YPT-related cDNA's from a rat brain library. Proc Natl Acad Sci USA 84:8210–8214.

Weber E, Jilling T, Kirk KL (1996): Distinct functional properties of rab3a and rab3b in PC12 neuroendocrine cells. J Biol Chem 271:6963–6971.

Wilson K (1995): NSF-independent fusion mechanisms. Cell 81:475–477.

Yamane HK, Farnsworth CC, Xie H, Evans T, Howald WN, Gelb MH, Glomset JA (1991): Membrane binding domain of the small GTP-binding protein G25K contains an S-(all-*trans*-geranylgeranyl)-cysteine methyl ester at its carboxy terminus. Proc Natl Acad Sci USA 88:286–290.

Zahraoui A, Touchot N, Chardin P, Tavitian A (1989): The human *rab* genes encode a family of GTP-binding proteins related to yeast YPT-1 and SEC4 products involved in secretion. J Biol Chem 264:12394–12401.

Zerial M, Sternmark H (1993): Rab GTPases in vesicular transport. Curr Opin in Cell Biol 5:613–620.

16

EXPRESSION OF THE PLP GENE AND A PLP-LACZ TRANSGENE IN OLIGODENDROCYTES

Cynthia Duchala and Wendy B. Macklin

Department of Neurosciences
NC30
The Cleveland Clinic Foundation
Cleveland, Ohio 44195

INTRODUCTION

Myelination is essential for normal nervous system development, and a wide variety of environmental and genetic factors lead to dysmyelination, which is a serious neurobiological problem (Raine, 1984). Central nervous system (CNS) myelin is a differentiation of the oligodendrocyte plasma membrane, which is produced during a relatively narrow time period in the early postnatal rodent brain. This differentiation program can be studied extensively in primary culture since cultured oligodendrocyte progenitors can differentiate and express myelin genes, even in the absence of neurons or astrocytes (Dubois-Dalcq et al., 1986; Campagnoni and Macklin, 1988).

The *in vitro* culture of mixed glial cells and enriched oligodendrocytes from brain tissue is a well documented technique (McCarthy and de Vellis, 1980), and primary oligodendrocyte cultures have been extremely important for investigating factors that control and influence glial development and myelination (Raff et al., 1983; Duchala et al., 1995). Oligodendrocyte differentiation both *in vivo* and *in vitro* is characterized by sequential expression of developmental markers that define specific phenotypic stages, and studying myelin gene expression in cultured cells has been particularly useful in defining a number of different stages in the oligodendrocyte lineage (Dubois-Dalcq, 1986; Knapp et al, 1987). Immature progenitor cells in culture have a bipolar appearance and as the cells differentiate, they take on the more characteristic oligodendrocyte morphology of a small round cell body with a well defined network of processes and the extension of a myelin-like membrane. Similar patterns of development and stage-specific antigen expression have been observed in oligodendrocyte cultures derived from several species (Duchala et al., 1995), which has made the cultures very useful for refining and interpreting the vast amount of information currently available on glial cell development and differentiation.

Cell Biology and Pathology of Myelin, edited by Juurlink *et al.*
Plenum Press, New York, 1997

Studies in this laboratory have focused on the myelin proteolipid protein (PLP) and its gene. This hydrophobic protein, along with its alternatively spliced isoform DM20, constitute close to half of the protein in CNS myelin (Eng et al., 1968; Agrawal et al., 1972). We have found that at the very early stages of development, DM20 protein is present prior to PLP (Gardinier and Macklin, 1988), and that DM20 protein is present in the plasma membrane of premyelinating oligodendrocytes, while PLP is present later in compact myelin (Trapp et al., 1997). Other investigators have demonstrated DM20 mRNA in embryos (Ikenaka et al., 1992; Timsit et al., 1992). Overall, these data suggest that DM20 protein may be important in the initial stages of oligodendrocyte differentiation or in immature oligodendrocytes, perhaps in cells that are not yet myelinating, while PLP may be more important in generating or maintaining the structure of compact myelin.

The PLP gene is the site of a number of point mutations that cause dysmyelination and premature death of animals. Mutant animals such as the *jimpy* or *msd* mouse, the *MD* rat or the *shaking* pup are excellent animal models of the human disorder Pelizaeus-Merzbacher disease (Ikenaka et al., 1991), which is caused in some families by point mutations in the PLP protein, and in other families by simple duplications of the normal PLP gene (Hodes et al., 1993; Inoue et al., 1996) . Transgenic models of this form of Pelizaeus-Merzbacher disease have been generated in two laboratories by inserting extra copies of the normal PLP gene into the mouse genome (Readhead et al., 1994; Kagawa et al., 1994). These transgenic mice were originally generated in order to "cure" PLP mutant animals such as *jimpy*, just as it was possible to "cure" the *shiverer* mouse by inserting a normal copy of the myelin basic protein gene (Readhead et al., 1987). Unfortunately, not only did these overexpressor transgenic mice not improve the *jimpy* phenotype when crossed into the *jimpy* background (Schneider et al., 1995), but the transgenic, non-*jimpy* homozygotes suffered from dysmyelination, and died at a young age. One line died at three-four weeks (Kagawa et al., 1994), while the other line died within a few months of birth (Readhead et al., 1994). Thus, in these transgenic models as well as in the human disorder Pelizaeus-Merzbacher disease, overexpression of the normal PLP and/or DM20 protein can be pathological. Synthesizing the normal PLP and/or DM20 protein and the correct amounts of these proteins at the appropriate time in development appears to be an extremely important aspect of normal brain development. Duplications or mutations in the PLP/DM20 gene are far more serious than mutations in other myelin genes. For example, in rodents these mutations are lethal within a few weeks of birth, whereas other dysmyelination mutants survive to adulthood, and some are fertile (Baumann, 1980).

Since mutations in the PLP gene are so serious and the gene appears to be turned on relatively early in development, it has been hypothesized that the mutations are affecting brain development at a site or time prior to that of the other myelin mutations. In order to investigate PLP gene expression both in the early stages of CNS development and in the different PLP mutants, we generated PLP-LacZ transgenic mice, in which the bacterial reporter gene LacZ is expressed using the mouse myelin PLP promoter (Fig. 1) (Wight et al., 1993). In these mice, expression of the transgene appears to be controlled as is the nor-

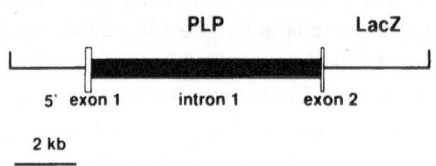

Figure 1. PLP-LacZ transgene.

mal PLP gene, both spatially and temporally. We have used these animals to investigate expression of the PLP gene in the embryo and in cultured oligodendrocytes, in order to understand more fully where and under what control the PLP gene is expressed. The ultimate goal of such studies is to understand how mutations in the PLP gene cause changes in cells expressing this gene at early stages of development.

TRANSGENE EXPRESSION IN MOUSE EMBRYOS

It has been reported that by E10, the PLP gene is expressed in the embryonic mouse nervous system, and that the primary PLP/DM20 transcript in embryos is the DM20 transcript (Ikenaka et al., 1992; Timsit et al., 1992). Techniques to date have not been sensitive enough to detect protein in these cells, so it is presently unclear whether DM20 protein is actually present in these cells. Our studies on E14 PLP-LacZ transgenic animals demonstrated that the transgene was expressed, and that in fact it was expressed at higher levels in the peripheral nervous system than in the central nervous system (Wight et al., 1993). Transgenic protein was detectable, indicating that if DM20 protein is reduced in expression by translational down-regulation, that down-regulation does not extend to the PLP-β-galactosidase transgenic fusion protein. Studies are currently underway to establish in which cells the transgene is expressed, and to understand whether the transgene is regulated normally in PLP mutant animals.

TRANSGENE EXPRESSION IN MOUSE PRIMARY OLIGODENDROCYTES

We have extensively studied PLP gene expression in cultured rat and mouse oligodendrocytes (Duchala et al., 1995). Since the PLP-LacZ transgene is a simple measure of PLP gene transcriptional activation, we began studies on expression of the transgene in cultured oligodendrocytes. β-galactosidase-expressing cells could be identified in mixed glial cultures derived from PLP-LacZ transgenic mice by staining cultures of different developmental ages for β-galactosidase activity with X-gal (Fig. 2). As early as 6 days in culture, β-galactosidase activity was detectable in a small population of undifferentiated bipolar cells. At 12 days in culture the enzyme activity could be identified in cells with small round cell bodies and multiple processes. In mature mixed glial cultures, the enzyme activity was localized to cells with distinct round cell bodies, an elaborate process network and myelin-like membrane. These data indicate that the transgene is expressed *in vitro*, in cells that resemble cultured oligodendrocytes morphologically.

Further studies using immunocytochemical staining investigated the developmental control of the β-galactosidase expression and its cellular and subcellular distribution in cultured cells. In culture, the earliest detectable cells of the oligodendrocyte lineage are the O-2A oligodendrocyte progenitor cells, which express the surface gangliosides recognized by the A_2B_5 antibody (Raff et al., 1984). These cells are bipotential cells that can develop into either oligodendrocytes or type 2 astrocytes, depending on culture conditions (Raff et al., 1983). These cells can be detected by 2 days in culture. Expression of the transgene could be detected immunocytochemically in A_2B_5+ O-2A progenitor cells as early as 5 days in culture, colocalizing with the A_2B_5 antigen in the cell body and short processes. As the cultures matured, A_2B_5 staining disappeared, while the β-galactosidase antibody stained more mature cell types. Rarely was PLP immunoreactivity, a marker for

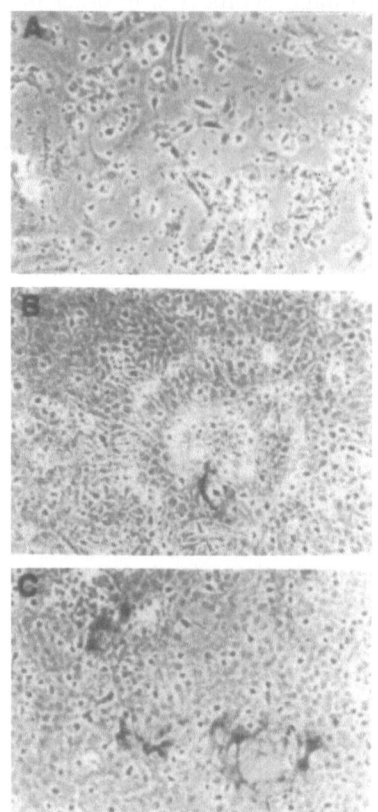

Figure 2. Expression of PLP-LacZ transgene in mixed glial cultures at (A) 6 days in culture; (B) 12 days in culture; (C) 18 days in culture. Mixed glial cultures were fixed with 1.0% glutaraldehyde in PBS, pH 7.3 and stained with X-gal as described by Sanes et al. (1986).

oligodendrocyte maturation, detected at this early culture age, although DM20 protein immunostaining has not been done on these cells. When PLP was first detected in the cell body at 6 days in culture, it colocalized with the β-galactosidase (Fig. 3 A, B). Thus the transgene was turned on at least as early in development as the PLP gene, and it was detectable before PLP protein could be detected.

An intermediate developmental stage of oligodendrocyte differentiation is defined by the presence of the O4 surface antigen (Sommer and Schachner, 1981). The O4+ progenitor represents a transitional stage between the O-2A precursor and the postmitotic differentiated oligodendrocyte, and the O4 antigen continues to be expressed in the mature oligodendrocyte, where it is predominantly localized in the extended myelin-like membrane of the cell (Trotter and Schachner, 1989; Gard and Pfeiffer, 1990; Duchala et al., 1995). In our transgenic cultures, β-galactosidase was detectable in O4+ cells where it was expressed in the cell body and processes (data not shown).

Galactocerebroside (GC) is a surface marker for differentiating postmitotic oligodendrocytes and is first detectable around day 5 in culture (Ranscht et al., 1982; Knapp et al., 1988; Duchala et al., 1995). The GC+ cells are committed to the oligodendrocyte pathway of differentiation, but are still immature. The immature cells express GC in the cell body and developing process network. GC expression persists in the mature oligodendrocyte, localizing in the cell body, processes and the myelin-like membrane. In our studies, β-galactosidase immunoreactivity colocalized with GC staining throughout development (Fig. 3 E,F).

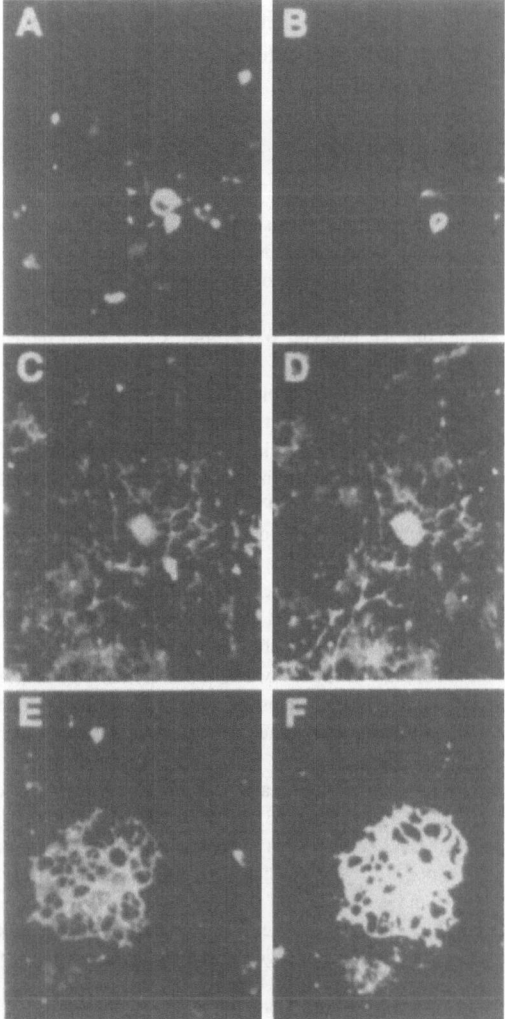

Figure 3. Expression of PLP-LacZ transgene in mixed glial cultures. Immunofluoresence staining for β-galactosidase (A, C, E) ; PLP (B, D); galacto-cerebroside (F) at 6 days in culture (A, B), 18 days in culture (C, D, E, F). The polyclonal β-galactosi-dase antibody (Organon Teknika-Cappel, Durham, NC) was used at a dilution of 1:200; the polyclonal galactocerebroside antibody (Ranscht et al., 1982) was used at a dilution of 1:150 and the polyclonal PLP (M2, Duchala et al., 1995) antibody was used at a dilution of 1:200. The double immunofluore-sence labeling protocol is described in Wight et al. (1993).

PLP is expressed only by mature oligodendrocytes in culture. The protein is observed in the cell body and processes and to a lesser extent in the myelin-like membrane sheet (Duchala et al., 1995). In transgenic mouse cultures, β-galactosidase immunoreactivity colocalizes with PLP (Fig. 3 A-D). One interesting observation is the fact that double immunofluorescence with PLP and β-galactosidase antibodies generates a significantly lower signal in these cultures than is seen for staining of each antibody alone. This may suggest that the antibodies are occupying the same domain and double labeling causes a mutual interference of the fluorescent signal.

The current studies on the PLP gene using a PLP-LacZ transgene as a reporter for PLP gene activity in embryos and cultured oligodendrocytes have established systems in which it is possible to investigate the regulation of PLP transcription both in embryos and in the cultured oligodendrocytes by measuring β-galactosidase activity and protein. This is ideal for studies on the PLP mutant animals, in which PLP and MBP gene transcription are down-regulated. This transgenic model will be useful for investigating many aspects of PLP gene transcriptional control.

REFERENCES

Agrawal HC, Burton RM, Fishman MA, Mitchell RF, Prensky AL (1972): Partial characterization of a new myelin protein component. J Neurochem 19:2083–2089.

Baumann, N (1980): Neurological mutations affecting myelination. Inserm symposium 14, Elsevier, Amsterdam, 565 pp.

Campagnoni AT, Macklin WB (1988): Cellular and molecular aspects of myelin protein gene expression. Mol Neurobiol 2:41–89.

Dubois-Dalcq M, Behar T, Hudson L, Lazzarini RA (1986): Emergence of three myelin proteins in oligodendrocytes cultured without neurons. J Cell Biol 102:384–392.

Duchala CS, Asotra K, Macklin WB (1995): Expression of cell surface markers and myelin proteins in cultured oligodendrocytes from neonatal brain of rat and mouse: a comparative study. Dev Neurosci 17:70–80.

Eng LF, Chao FC, Gerstl B, Pratt D,Tavaststjerna MG (1968): The maturation of human white matter myelin: fractionation of the myelin membrane proteins. Biochem 7:4455–4465.

Gard AL, Pfeiffer SE (1990): Two proliferative stages of the oligodendrocyte lineage (A2B5+O4- and O4+GalC-) under different mitogenic control. Neuron 5:615–625.

Gardinier MV, Macklin WB (1988): Myelin proteolipid protein gene expression in *jimpy* and *jimpy*^{msd} mice. J Neurochem 51:360–369.

Hodes ME, Pratt VM, Dlouhy SR (1993): Genetics of Pelizaeus-Merzbacher disease. Dev Neurosci 15:383–394.

Ikenaka K, Kagawa T, Mikoshiba K (1992): Selective expression of DM-20, an alternatively spliced myelin proteolipid protein gene product, in developing nervous system and in nonglial cells. J Neurochem 58:2248–2253.

Ikenaka K, Ohano H, Tamura T-A, Mikoshiba K (1991):Recent advances in studies on genes for myelin proteins. Dev Growth Differ 33:181–192.

Inoue K, Osaka H, Sugiyama N, Kawanishi C, Onishi H, Nezu A, Kimura K,Yamada Y, Kosaka K (1996): A duplicated PLP gene causing Pelizaeus-Merzbacher disease detected by comparative multiplex PCR. Amer J Hum Gen. 59:32–39.

Kagawa T, Ikenaka K, Inoue Y, Kuriyama S, Tsujii T, Nakao J, Nakajima K, Aruga J, Okano H, Mikoshiba, K (1994): Glial cell degeneration and hypomyelination caused by overexpression of myelin proteolipid protein gene. Neuron 13:427–442.

Knapp PE, Bartlett WP, Skoff RP (1987): Cultured oligodendrocytes mimic *in vivo* phenotypic characteristics: Cell shape, expression of myelin-specific antigens, and membrane production. Dev Biol 120:356–365.

Knapp PE, Skoff RP, Sprinkle TJ (1988): Differential expression of galactocerebroside, myelin basic protein and 2′,3′-cyclic nucleotide 3′phosphohydrolase during development of oligodendrocytes *in vitro*. J Neurosci Res 21:249–259.

McCarthy KD, de Vellis, J (1980): Preparation of separate astroglial and oligodendroglial cell cultures from rat cerebral tissues. J Cell Biol 85:890–902.

Raff MC, Miller RH, Noble M (1983): A glial progenitor cell that develops *in vitro* into an astrocyte or an oligodendrocyte depending on the culture medium. Nature 303:390–396.

Raff MC, Williams BP, Miller RH (1984): The *in vitro* differentiation of a bipotential glial progenitor cell. EMBO J 5:1857–1864.

Raine CS (1984): Morphology of myelin and myelination. In Morell P (ed): "Myelin." New York: Plenum Press, pp 1–50.

Ranscht B, Clapshaw PA, Price J, Noble M, Seifert W (1982): The development of oligodendrocytes and Schwann cells studied with a monoclonal antibody against galactocerebroside. Proc Natl Acad Sci USA 79:2709–2713.

Readhead C, Schneider A, Griffiths I, Nave KA (1994): Premature arrest of myelin formation in transgenic mice with increased proteolipid protein gene dosage. Neuron 12:583–595.

Readhead C, Popko B,Takahashi N, Shine HD, Saavedra RA, Sidman RL, Hood L (1987): Expression of a myelin basic protein gene in transgenic *shiverer* mice: correction of the dysmyelinating phenotype. Cell 48:703–712.

Sanes JR, Rubenstein LR, Nicolas J-F (1986): Use of a recombinant retrovirus to study post-implantation cell lineage in mouse embryos. EMBO J 5:3133–3142.

Schneider AM, Griffiths IR, Readhead C, Nave KA (1995): Dominant-negative action of the jimpy mutation in mice complemented with an autosomal transgene for myelin proteolipid protein. Proc Natl Acad Sci USA 92:4447–4451.

Sommer I, Schachner M (1981): Monoclonal antibodies (O1 to O4) to oligodendrocyte cell surfaces: An immuno-cytological study in the central nervous system. Dev Biol 18:311–327.

Timsit SG, Bally-Cuif L, Colman DR, Zalc B (1992): DM-20 mRNA is expressed during the embryonic development of the nervous system of the mouse. J Neurochem 58:1172–1175.

Trapp BD, Nishiyama A, Cheng D, Macklin WB (1997): Characterization of oligodendrocyte differentiation and death in developing rodent brain. J. Cell Biol, 137: 459–468.

Trotter, J., Schachner, M (1989):Cells positive for the O4 surface antigen isolated by cell sorting are able to differentiate into astrocytes or oligodendrocytes. Dev Brain Res 46:115–122.

Wight PA, Duchala CS, Readhead C and Macklin WB (1993): A myelin proteolipid protein-LacZ fusion protein is developmentally regulated and targeted to the myelin membrane in transgenic mice. J Cell Biol 123:443–454.

17

DEVELOPMENTAL REGULATION OF SCHWANN CELL PRECURSORS AND SCHWANN CELL GENERATION

K. R. Jessen, R. Mirsky, Z. Dong, and A. Brennan

Department of Anatomy and Developmental Biology
University College London
Gower Street
London WC1E 6 BT
England

1. THE NEURAL CREST AND THE DEVELOPMENTAL ORIGIN OF SCHWANN CELLS

Most myelin-forming and non-myelin-forming Schwann cells of peripheral nerves develop from the neural crest (Anderson, 1993; Bronner-Fraser, 1993; Jessen and Mirsky, 1992; Le Douarin et al., 1991; Marusich and Weston, 1991). The other main glial types in the peripheral nervous system (PNS), are also thought to be derived from the crest. This includes the teloglia associated with somatic motor terminals, satellite cells associated with neuronal cell bodies in sensory, sympathetic and parasympathetic ganglia and the enteric glial cells in the intrinsic ganglia of the gut (Georgiou et al., 1994; Gershon et al., 1993; Pannese, 1981). There is evidence that some of the PNS glia found in nerves/ganglia close to the central nervous system (CNS) are not derived from neural crest cells. Thus, some of the Schwann cells in the ventral roots of the spinal cord arise from ventral neural tube and, some satellite and/or Schwann cells of dorsal root ganglia (DRG) are likely to originate from neuroepithelial cells in the spinal cord after migration of the neural crest is complete (Carpenter and Hollyday, 1992; Loring and Erickson, 1987; Lunn et al., 1987; Rickman et al., 1985; Sharma et al., 1995).

The neural crest gives rise to several cell types in addition to PNS glia, including neurons and melanocytes (Le Douarin and Smith, 1988; Weston, 1991). While some crest cells may already be committed to a particular fate, many crest cells are bi- or multi-potent, i.e. they can give rise to more than one differentiated cell type. This has been shown in vivo, by injecting individual crest cells with a tracer that subsequently allows the differentiated progeny to be identified, and in vitro, by clonal culture. Such experiments show that individual crest cells from chick, mouse or rat frequently give rise to clones containing a mixture of differentiated crest derivatives (Frank and Sanes, 1991; Fraser and Bron-ner-Fraser, 1991; Le Douarin et al., 1991; Stemple and Anderson, 1993). Lineage choice could occur stochastically. In this case, extrinsic signals, eg polypeptide growth factors, might act only to amplify or suppress the outcome of a large number of random lineage-entry events by selective regulation of survival and/or proliferation. Alternatively, extrin-sic signals might alter and control such lineage choices directly and therefore channel crest cells towards a particular fate . It is not clear which of these mechanisms operate in the generation of the Schwann cell lineage (Jessen and Mirsky, 1994; Shah et al., 1994).

2. THE FINAL DIFFERENTIATION OF SCHWANN CELLS IS LABILE AND DEPENDS ON EXTRINSIC SIGNALLING

The five PNS glial types mentioned above display distinct phenotypes in the normal, unperturbed nervous system. Yet, the lesson from examining two of them, the Schwann cells, is that among peripheral glia, phenotypic differences are unstable and easily revers-ible. Thus, myelin-forming Schwann cells develop an elaborate structure, the multi-lay-ered myelin sheath, and unique protein expression that enables them to speed up electrical impulse conduction in the axon they envelop. These cells will, however, readily abandon their complex structure and distinctive molecular make-up and revert to a phenotype simi-lar to that of immature Schwann cells prior to myelination, given the right circumstances (Figure 1A). Typically, such regression is triggered by axonal degeneration in vivo or by removing Schwann cells from the influence of axons and placing them in serum-contain-ing media in vitro. Diverse lines of evidence indicate that if such cells are placed in con-tact with appropriate axons they again differentiate and re-form myelin sheaths or, if exposed to axons that are normally unmyelinated in vivo, they adopt the phenotype of ma-ture non-myelin-forming Schwann cells . It is very likely that mature non-myelin-forming Schwann cells react in a similar way, i.e. they regress following axonal degeneration but can re-express the mature phenotype of either non-myelin-forming or myelin-forming cells if subsequently placed in contact with appropriate axons (Jessen et al., 1987). Thus, most workers in the field take the view that myelin-forming and non-myelin-forming Schwann cells are essentially interconvertible, each owing their distinct phenotype to in-structive signals from axons (Bray et al., 1981; Bunge et al., 1990; Jessen and Mirsky, 1992). This plasticity of Schwann cells contrasts with the much stronger phenotypic com-mitment shown by the corresponding glia of the CNS, oligodendrocytes and astrocytes. Understandably, considerable effort has been devoted to examination of signalling mole-cules/growth factors that can regulate the phenotype of these strikingly adaptable cells, and to finding the molecular identity of the axon-associated Schwann cell differentiation signals.

3. DO ALL PNS GLIA SHARE EARLY DEVELOPMENTAL STAGES?

Are the phenotypes of the other main PNS glial types - satellite cells, teloglia and enteric glia - also unstable and dependent on extrinsic signals? Can these cells intercon-vert, can they potentially adopt the Schwann cell phenotypes, and is the opposite true: can Schwann cells be signalled to adopt, for instance, the phenotype of enteric glia? These

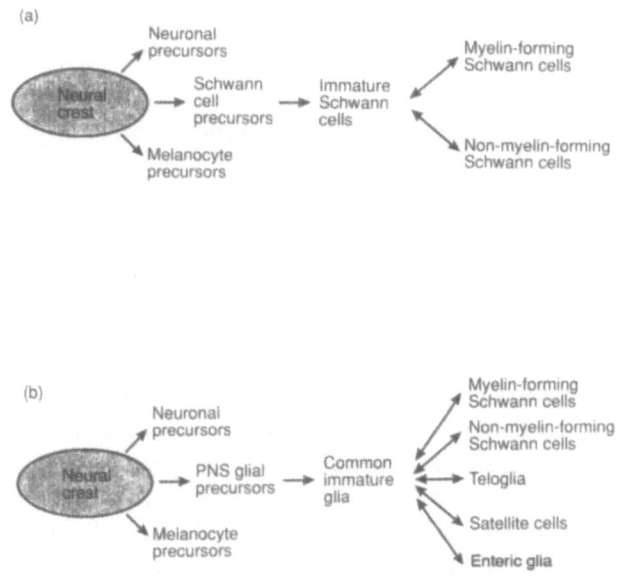

Figure 1. Glial development from the neural crest. (A) Outline of Schwann cell development and de-differentiation. For each lineage, crest cells generate precursors that, in turn, give rise to differentiated crest derivatives. In the rat and mouse, Schwann cell precursors generate cells with the generally recognised phenotype of Schwann cells (immature Schwann cells) before birth. Further progression to form mature Schwann cells - i.e. myelin-forming and non-myelin forming cells - depends on axon-associated signals and is reversible (indicated by double-headed arrows). This scheme would predict that Schwann cells in embryonic nerves just prior to myelination, Schwann cells in the distal stump of transected adult nerves, and Schwann cells removed from animals at any age and grown in neuron-free cultures using media that are free of signalling molecules, should display essentially the same phenotype. (B) Perhaps the model outlined in (A) applies generally to the development of other PNS glia. This model predicts that cell PNS glia derive from a common cell which displays the immature Schwann cell phenotype, and is itself derived from a precursor with the Schwann cell precursor phenotype. This model also predicts, for example, that if enteric glial cells were placed in the path of regenerating axons in the sciatic nerve they would differentiate to form myelin-forming and non-myelin-forming Schwann cells.

questions have not been answered clearly. Many experiments on frogs, chicken and rats suggest, however, that the principles outlined above for rat Schwann cells might hold true for all PNS glia (Cameron-Curry et al., 1993; Georgiou et al., 1994; Jessen and Mirsky, 1983; Woolf et al., 1992). One plausible model of PNS glial development, therefore, is that all types of peripheral glia are derived from a common immature glial cell, with a phenotype similar to that of immature Schwann cells, that they all retain the potential to regress to adopt again the phenotype of this cell, and that such cells can re-differentiate to form any of the PNS glial variants if the appropriate signals are present (Figure 1B). How is such a cell generated from neural crest cells? This problem can be broken down into three related questions. First, what regulates the initial entry of crest cells to the glial lineage (see above); second, what are the developmental stages between lineage entry and the appearance of the putative common immature glial cell; third, how is this lineage progression regulated? As we discuss below, some light has been thrown on the last two questions by recent studies on the Schwann cell precursor.

4. THE EARLY FORMATION OF PERIPHERAL NERVES

Having entered the glial lineage, prospective Schwann cells must find outgrowing neurites, associate with them and become incorporated into embryonic nerves. It is likely that crest cells and axons first meet in significant numbers in the anterior parts of the somites that form along both sides of the neural tube. As neural crest cells migrate from the closing neural tube, many of them are directed through the anterior part of each somite, while the posterior part is avoided. At about this time the first peripheral axons emerge from the ventrolateral part of the neural tube. They also select the anterior part of each somite rather than the posterior part. Thus, it appears that axons and crest cells are channelled to the same place at the same time, although the molecular mechanisms that ensure this temporal and spatial matching of axon outgrowth and crest cell migration are not yet known (Loring and Erickson, 1987; Rickman et al., 1985).

How axons growing through the anterior somite catch cells from the pool of surrounding crest cells is not clear. It is possible that they meet individual crest cells by chance and that the subsequent recruitment of such cells to join with the outgrowing axons is a consequence of greater adhesion between axons and crest cells than between those cells and the surrounding extracellular matrix or other cells. Alternatively, axons might attract crest cells at some distance with the help of secreted molecules.

It is an attractive idea that crest cells are lineage-neutral when they first meet axons, and that the axonal contact provides the first influence that biases them towards glial rather than neuronal development, perhaps by the action of neuron-associated growth factors of the neuregulin family (Shah et al., 1994, see also below). Alternatively, it is possible that many crest cells have already entered the glial lineage before they meet axons and that the crucial encounter in the anterior somite is in fact selective, taking place between axons and the glial-biased sub-population. Such selectivity could be achieved if entry to the glial lineage were accompanied by changes, e.g. in cell surface molecules, that facilitate axonal encounter and attachment. It is not possible to distinguish between these two possibilities at present, but an intriguing observation, first made in the chick (Bhattacharyya et al., 1991) and since confirmed by us in the rat (Lee et al., 1997), might provide a clue. These studies show that the major protein component of Schwann cell myelin, P_0, or its mRNA, is expressed not only in Schwann cell precursors (see below) long before the onset of myelination, but also in a sub-population of migrating neural crest cells located both at the dorsal level of the neural tube and more ventrally. There are no published reports of P_0 expression in cells other than Schwann cells, so it is possible that a low level activation of this gene is an early event in glial crest differentiation. It will be interesting to examine the relationship between P_0 expression and expression of markers of entry to other glial lineages. If P_0 expression emerges as a specific marker of the glial lineage it follows that many crest cells have selected a glial fate even at the early stages of crest migration.

5. SCHWANN CELL PRECURSORS

The identification of Schwann cell precursors in the limb nerves of 14 and 15 day old rat embryos has helped to clarify the question of how crest cells that have joined axons to become part of nascent peripheral nerves progress from this stage to generate cells with the identifiable phenotype of Schwann cells (Dong et al., 1995; Gavrilovic et al., 1995; Jessen et al., 1994).

In the rat, the embryonic period of nerve development in the limbs spans about a week, from embryo day (E) 13/14 to birth at E21. Essentially all the glial cells isolated from these nerves at birth have the well-established phenotype of Schwann cells, some of which are, at this point, in the initial stages of myelination. It is, however, only from about E17 that Schwann cells are present in these nerves in large numbers, and no such cells are present in the nerves that first invade the limbs at E13/14. Instead, essentially all the cells present in E14 nerves show a distinct phenotype that differs from both crest cells and Schwann cells. Since these cells represent a distinct intermediate stage in the genesis of Schwann cells, we refer to them as Schwann cell precursors. The differences between the E14 precursors and Schwann cells include the following: the precursors die abruptly by apoptosis when removed from axonal contact and plated in vitro at moderate/high cell density, whereas Schwann cells survive in similar cultures; the precursors do not express the Schwann cell marker S-100 in the cytoplasm, and they are not induced to synthesize DNA by FGF2 in the presence of forskolin, which is a typical Schwann cell mitogen combination. In vitro, the precursors have a flattened morphology with many cell-cell contacts, while Schwann cells have elongated bi- or tri-polar shape under similar culture conditions. More recently, we hve defined further the differences in survival regulation between precursors and Schwann cells, and found that Schwann cell, but not precursors, support their own survival in vitro by the secretion of autocrine factors. The evidence for this includes the observation that in very sparse cultures (50 cells per coverslip) Schwann cells, like precursors, die by apoptosis. In denser cultures, however, Schwann cells survive, while precursors are not rescued at any cell density unless extrinsic survival factors are added to the medium. Furthermore, medium conditioned by dense Schwann cell cultures rescues Schwann cells in sparse cultures, but does not rescue precursors (C Meier, R. Mirsky, KR Jessen, unpublished).

We have identified a similar precursor cell in embryonic mouse nerves and found that the phenotypic differences between the precursors and immature Schwann cells in most cases parallel those seen in the rat. The mouse PNS develops, however, about 2 days ahead of the rat system so that mouse nerves typically contain precursors at E11/12.

Thus, two distinct members of the Schwann cell lineage figure in the embryonic development of major peripheral nerves: Schwann cell precursors and Schwann cells. The time between E14/15 and E17/18 (mouse E11/12 and E15/16) is particularly important, since during this period Schwann cell precursors progress to generate Schwann cells. This change in glial phenotype extends to a number of different cell properties. Nevertheless, it occurs relatively quickly, so that only during E16 (mouse E14) are significant numbers of both cell types present in the major limb nerves. As mentioned above, E14 rat Schwann cell precursors undergo apoptosis when they are removed from embryonic nerves and placed in neuron-free cultures. DRG neurons secrete protein(s) that prevent this death, and isolated axonal membranes also possess activity that supports precursor survival (Jessen et al., 1994; N Ratner, KR Jessen and R Mirsky, unpublished). We concluded from these experiments that the survival of Schwann cell precursors in vivo was likely to be regulated by signals from axons, a conclusion that was confirmed recently in important in vivo experiments on chick embryos (Ciutat et al., 1996). These signals also drive, or permit, the next step in the development of the lineage, since the precursors mature to generate Schwann cells in vitro in the presence of the neuron- derived proteins. There is now strong evidence that a key component of this signal is neuregulin.

Neuregulins are a family of proteins that are also referred to as neu-differentiation factor (NDF), glial growth factors (GGFs), or heregulins (Ben-Baruch and Yarden, 1994). They are alternatively spliced products of a single gene that has now been cloned from rat fibroblasts (NDF), a human mammary cell line (heregulin), chicken brain (ARIA) and bovine pituitary (GGF). These factors belong to the EGF super family and possess an EGF-like domain that appears to be sufficient for receptor activation. They bind to tyrosine kinase receptors of the EGF receptor family, that includes ErbB-1, -2, -3 and -4. Initially, neuregulin was thought to bind to and phosphorylate ErbB-2, but more recent work has established that direct binding and phosphorylation only takes place with ErbB-3 and -4, although ErbB-2 might act as co-receptor (Sliwkowski et al., 1994; Tzahar et al., 1994). The most prominent site of neuregulin mRNA expression in rat and mouse embryos during the last third of the gestation period is the developing nervous system, where neuregulin expression appears to be largely restricted to neurons and absent from glia (Marchionni et al., 1993; Meyer and Birchmeier, 1994; Orr-Urtreger et al., 1993). Neuregulins are likely to play a very important role in regulation and development in the Schwann cell lineage.

We have found that beta forms of neuregulin, but not alpha forms, support precursor survival in vitro for several days in defined medium in the absence of serum. Furthermore, during this period, the cells in the culture dish alter their phenotype from that of precursors to that of Schwann cells. The time course of this Schwann cell generation in vitro is similar to that with which the glial population of developing nerves switches from the precursor phenotype to the Schwann cell phenotype in vivo. Most importantly, the action of the signal from DRG neurons that supports precursor survival and differentiation is blocked by a soluble hybrid protein containing the extracellular domain of the ErbB-4 neuregulin receptor, a receptor that is not known to bind to any other growth factor (Dong et al., 1995). Neuregulin is also present at the appropriate time and place in rat and mouse embryos: strikingly high expression of neuregulin mRNA is seen over the motor neurons of the ventral horn in the spinal cord and over DRG neurons, the two major sources of axons in embryonic peripheral nerves (Marchionni et al., 1993; Meyer and Birchmeier, 1994; Orr-Urtreger et al., 1993). Lastly, transgenic mice in which the neuregulin gene has been knocked out show greatly reduced numbers of precursor cells in peripheral nerves (Meyer and Birchmeier, 1995). Taken together, these results strongly suggest that neuronally-derived neuregulin is likely to act as a survival/maturation factor for Schwann cell precursors. Recent results using neuron-Schwann cell co-cultures suggest that at least one component of the axon-derived Schwann cell mitogen, first described many years ago, is likely to be a member of the neuregulin beta family, interacting, perhaps indirectly, with the ErbB-2 receptor on Schwann cells (Morrissey et al., 1995).

Fibroblast growth factors (FGFs) also support precursor survival in combination with insulin like growth factor IGF-1 (Gavrilovic et al., 1995). However, FGFs plus IGF-1 only support survival in the short term and most cells die during the second day of exposure to these factors in vitro (Dong et al., 1995).

The lineage commitment of the E14 rat Schwann cell precursor has not yet been tested rigorously. However, studies on nerves from a comparable developmental stage in the chick provide evidence that cells normally destined to generate Schwann cells can be induced to generate melanocytes. The evidence comes from experiments to which explants of embryonic avian nerves were found to give rise to pigmented cells under certain culture conditions (Nichols and Weston, 1977). Melanogenesis in this system is stimulated by the tumor-promoting phorbol ester TPA, which might act via an intracellular FGF2 loop, but is inhibited by transforming growth factor (TGF)-beta (Ciment, 1990; Sherman et al., 1993). This has given rise to the idea that Schwann cell precursors, at least in birds, can be regarded as bipotential melanocyte/Schwann cell progenitors. Alternatively, it remains possible that nerves might contain a small population of non-glial cells with melanogenic potential that can be activated by TPA. So far, melanogenesis has not been observed in cultures of rat Schwann cell precursors (Jessen et al., 1994).

The distinction between Schwann cell precursors and neural crest cells has not yet been drawn in the same detail as that between precursors and Schwann cells. Nevertheless it is already clear that they differ in morphology, cell-cell relationships and location, and also in antigenic phenotype and their requirement for survival factors (Jessen et al., 1994; A Brennan, R Mirsky and KR Jessen unpublished).

REFERENCES

Anderson DJ (1993): Cell and molecular biology of neural crest cell lineage diversification. Curr Opin Neurobiol 3: 8–13.

Ben-Baruch N, Yarden Y (1994): Neu differentiation factors: A family of alternatively spliced neuronal and mesenchymal factors. Proc Soc Exp Biol Med. 206: 221–227.

Bhattacharyya A, Frank E, Ratner N, Brackenbury R (1991): P0 is an early marker of the Schwann cell lineage in chickens. Neuron 7: 831–844.

Bray GM, Rasminsky M, Aguayo AJ (1981): Interactions between axons and the sheath cells. Annu Rev Neurosci. 4: 127–162.

Bronner-Fraser M (1993): Neural crest cell migration in the developing embryo. Trends Cell Biol. 3: 392–397.

Bunge MB, Clark MB, Dean AC, Eldridge CF, Bunge RP (1990) Schwann cell function depends upon axonal signals and basal lamina components. Ann NY Acad Sci 580: 281–287.

Cameron-Curry P, Dulac C, Le Douarin NM (1993): Negative regulation of Schwann cell myelin protein gene expression by the dorsal root ganglionic microenvironment. Eur. J. Neurosci. 5: 594–604.

Carpenter EM, Hollyday M (1992): The location and distribution of neural crest-derived Schwann cells in developing peripheral nerves in the chick forelimb. Dev Biol 150: 144–159.

Ciment G (1990): The melanocyte Schwann cell progenitor: a bipotent intermediate in the neural crest lineage. Comments Dev Neurobiol 1: 207–233.

Ciutat D, Caldero J, Oppenheim RW, Esquerda JE (1996): Schwann cell apoptosis during normal development and after axonal degeneration induced by neurotoxins in the chick embryo. J Neurosci 16: 3979–3990.

Dong Z, Brennan A, Liu N, Yarden Y, Lefkowitz G, Mirsky R, Jessen KR (1995): Neu differentiation factor (glial growth factor, heregulin) acts as a neuro-glia signal and regulates the survival, proliferation and maturation of the rat Schwann cell precursor. Neuron 15: 585–596.

Frank E, Sanes JR (1991): Lineage of neurons and glia in chick dorsal root ganglia: analysis *in vivo* with a recombinant retrovirus. Development 111: 895–908.

Fraser SE, Bronner-Fraser M (1991): Migrating neural crest cells in the trunk of the avian embryo are multipotent. Development 112: 913–920.

Gavrilovic J, Brennan A, Mirsky R, Jessen KR (1995): Fibroblast growth factors and insulin growth factors combine to promote survival of rat Schwann cell precursors without induction of DNA synthesis. Eur J Neurosci 7:77–85.

Georgiou J, Robitaille R, Trimble WS, Charlton MP (1994): Synaptic regulation of glial protein expression in vivo. Neuron 12: 443–455.

Gershon MD, Chalazonitis A, Rothman TP (1993): From neural crest to bowel: development of the enteric nervous system. J Neurobiol. 24: 199–214.

Jessen KR, Mirsky R (1983): Astrocyte-like glia in the peripheral nervous system: an immunohistochemical study of enteric glia. J Neurosci 3: 2206–2218.

Jessen KR, Mirsky R, Morgan L (1987): Axonal signals regulate the differentiation of non-myelin-forming Schwann cells: an immunohistochemical study of galactocerebroside in transected and regenerating nerves. J Neurosci. 7: 3362–3369.

Jessen KR, Mirsky R (1991): Schwann cell precursors and their development. Glia 4: 185–194.

Jessen KR, Mirsky R (1992): Schwann cells: early lineage, regulation of proliferation and control of myelin formation. Curr Opin Neurobiol 2: 575–581.

Jessen KR, Mirsky R (1994): Fate diverted. Curr Biol. 4: 824–827.

Jessen KR, Brennan A, Morgan L, Mirsky R, Kent A, Hashimoto Y, Gavrilovic J (1994): The Schwann cell precursor and its fate: A study of cell death and differentiation during gliogenesis in the rat embryonic nerves. Neuron 12: 509–527.

Le Douarin NM, Smith J (1988): Development of the peripheral nervous system from the neural crest. Ann Rev Cell Biol 4: 375–404.

Le Douarin N, Dulac C, Dupin E, Cameron-Curry P (1991): Glial cell lineages in the neural crest. Glia 4: 175–184.

Lee MJ, Brennan A, Blanchard A, Zoidl G, Doug E, Taganero A, Zoidl C, Dent MAR, Jessen KR, Mirsky R (1997): P_o is constitutively expressed in the rat neural crest and embryomic nerves and is negatively and positively regulated by axons to generate non-myelin-forming and myelin-forming Schwann cells, respectively. Molec Cell Neurosci 8: 336–350.

Loring JF, Erickson CA (1987): Neural crest pathways in the trunk of the chick embryo. Dev Biol 121: 220–236.

Lunn ER, Scourfield J, Keynes RJ, Stern CD (1987): The neural tube origin of ventral root sheath cells in the chick embryo. Development 101: 247–254.

Marchionni MA, Goodearl ADJ, Chen MS, Bermingham-McDonogh O, Kirk C, Hendricks M, Danehy F, Misumi D, Sudhalter J, Kobayashi K, Wroblewski D, Lynch C, Baldassare M, Hiles I, Davis JB, Hsuan JJ, Totty NF, Otsu M, McBurney RN, Waterfield MD, Stroobant P, Gwynne D (1993): Glial growth factors are alternatively spliced erbB2 ligands expressed in the nervous system. Nature 362: 312–318.

Marusich MF, Weston JA (1991): Development of the neural crest. Curr Opin Genes Dev 1: 221–229.

Meyer D, Birchmeier C (1994): Distinct isoforms of neuregulin are expressed in mesen-chymal and neuronal cells during mouse development. Proc Natl Acad Sci USA 91: 1064–1068.

Meyer D, Birchmeier C (1995): Multiple essential functions of neuregulin in development. Nature 378:386–390.

Morrissey TK, Levy A, Nuijens A, Sliwkowski M, Bunge R (1995): Axon-induced mitogenesis of human Schwann cells involves heregulin and P^{185}ErbB2. Proc Natl Acad Sci 92: 1431–1435.

Nichols DH, Weston JA (1977): Melanogenesis in cultures of peripheral nervous tissue. I. The origin and prospective fate of cells giving rise to melanocytes. Dev. Biol. 60: 217–225.

Orr-Urtreger A, Trakhtenbrot L, Ben-Levy R, Wen D, Rechavi G, Lonai P, Yarden Y (1993) : Neural expression and chromosomal mapping of Neu differentiation factor to 8p12-p21. Proc Natl. Acad Sci USA 90: 1867–1871.

Pannese E (1981): The satellite cells of the sensory ganglia. In: Advances in Anatomy, Embryology and Cell Biology, eds. Brodal A, Hild W,. van Limborgh J, Ortmann R, Schiebler TH, Töndury G, Wolff E. Springer-Verlag, Berlin.

Rickman M, Fawcett JW, Keynes RJ (1985): The migration of neural crest cells and the growth of motor axons through the rostral half of the chick somite. J Embryol Exp Morph 90: 437–455.

Shah NM, Marchionni MA, Isaacs I, Stroobant P, Anderson DJ (1994): Glial growth factor restricts mammalian neural crest stem cells to a glial fate. Cell 77: 349–360.

Sharma K, Korade Z, Frank E (1995): Late-migrating neuroepithelial cells from the spinal cord differentiate into sensory ganglion cells and melanocytes. Neuron 14: 143–152.

Sherman L, Stocker KM, Morrison R, Ciment G (1993): Basic fibroblast growth factor (bFGF) acts intracellularly to cause the transdifferentiation of avian neural crest-derived Schwann cell precursors into melanocytes. Development 118: 1313–1326.

Sliwkowski MX, Schaefer G, Aktia RW, Lofgren JA, Fitzpatrick UD, Nuijens A, Fendly BM, Cerione RA, Vandlen RL, Carraway KL (1994): Coexpression of ErbB-2 and ErbB-3 protein reconstitutes a high affinity receptor for heregulin. J Biol Chem 269: 14661–14665.

Stemple DL, Anderson DJ (1993): Lineage diversification of the neural crest: *in vitro* investigations. Dev Biol 159: 12–23.

Tzahar E, Levkowitz G, Karunagaran D, Yi L, Peles E, Lavi S, Chang D, Liu N, Yayon A, Wen D, Yarden Y (1994): ErbB-3 and ErbB-4 function as the respective low and high affinity receptors of all Neu differentiation factor/heregulin isoforms. J Biol Chem 269: 25226–25233.

Weston JA (1991) Sequential segregation and fate of developmentally restricted intermediate cell populations in the neural crest lineage. Curr Top Dev Biol. 25: 133–153.

Woolf CJ, Reynolds ML, Chong MS, Emson P, Irwin N, Benowitz LI (1992): Denervation of the motor endplate results in the rapid expression by terminal Schwann cells of the growth-associated protein GAP-43. J Neurosci 12: 3999–4010.

18

PROTEIN KINASE C REGULATES PROCESS FORMATION BY OLIGODENDROCYTES

V. W. Yong and L. Y. S. Oh

Department of Neurology and Neurosurgery
McGill University
Montreal, Quebec
Canada

INTRODUCTION

In multiple sclerosis (MS) and other demyelinating disorders, the oligodendrocytes (OLs) and the myelin that they produce are the targets of various pathological processes. In many cases, the insults are continuous and recurring but despite this, remyelination can occur, as it does in MS although incomplete in its extent (Raine et al., 1981; Ghatak et al., 1989; Prineas et al., 1989, 1993; Raine and Wu, 1993; Bruck et al., 1994). Stimuli to enhance OL formation, maturation or function may improve the remyelinating capacity of OLs. One such strategy may be to promote surviving OLs, or OLs formed from progenitors, to extend their processes which may then subsequently enwrap axons to form myelin. We have been interested in the promotion of process outgrowth from adult OLs, reasoning that process formation is a prerequisite for any OL to successfully remyelinate axons. Furthermore, understanding the mechanisms by which adult OLs recapitulate process outgrowth lends itself to uncovering mechanisms of process extension during developmental myelination, since the extension of cellular processes from the oligodendrocyte soma is an early and critical event in normal myelin formation. OLs in culture have served as useful models for their *in vivo* counterparts. *In vitro*, OLs extend multiple processes and the plasma membrane of the cultured OL contains all the major proteins and lipids of myelin (Szuchet, 1987; Vartanian et al., 1992). The emergence of myelin proteins and lipids *in vitro*, even in the absence of neurons or astrocytes, follows a time course that corresponds to that *in vivo* (Zeller et al., 1985; Dubois-Dalcq et al., 1986; Zalc et al., 1987). Furthermore, OLs isolated from mutant animals with defective myelination show corresponding abnormalities *in vitro* (Barlett et al., 1988). Many of the experiments to be described have utilized OLs that are cultured from the brains of adult post-myelination animals. These OLs contain no cellular processes initially (myelin is presumably stripped off during the cell isolation procedure) but redevelop their fibers *in vitro* (Fig. 1); this phenomenon has been referred to as palinogenesis (Szuchet, 1987).

Cell Biology and Pathology of Myelin, edited by Juurlink et al.
Plenum Press, New York, 1997

173

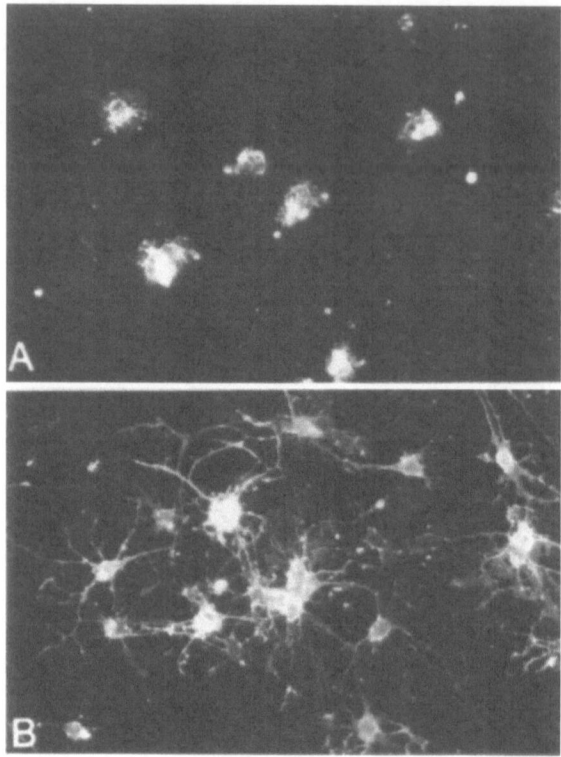

Figure 1. GalC+ adult human OLs *in vitro* normally develop processes very slowly (A). When treated with phorbol esters, bFGF or astrocyte ECM, adult human OLs extend elaborate and long processes very soon thereafter (B).

This chapter describes our findings that the protein kinase C (PKC) family of enzymes is critically involved in process formation by OLs, that astrocytes provide important growth factor signals for oligodendroglial process formation, and that these astroglial-derived factors putatively signal through oligodendroglial PKC.

EVIDENCE THAT PROTEIN KINASE C REGULATES OLIGODENDROGLIAL PROCESS FORMATION

PKC is a family of phospholipid-dependent serine-threonine kinases of which at least 11 isozymes are currently known to exist (reviewed in Dekker and Parker, 1994; Baltuch et al., 1995). These can be broadly classified as the calcium-dependent Group A isoforms (α, β_1, β_2 and γ), the calcium-independent Group B isozymes (δ, ε, η, μ and θ), and the atypical calcium- and diacylglycerol-independent Group C isoforms (z and iota:l is the *mouse* homolog of human iota). The isozymes of PKC are subtly different from one another in their kinetic properties, activator requirements, and mode of activation (Huang, 1987; Nishizuka, 1992; Newton, 1995). It has become increasingly clear that the distribution of these isoforms is cell-type-specific, and that depending upon the cell type, a particular isoform can have different substrate specificities and serve different functions.

A role for PKC in oligodendroglial biology can first be inferred from studies that demonstrate that the corpus callosum, a brain area especially rich in OLs, has very high density of phorbol ester binding sites (now known to be PKC itself) when compared to other brain areas (Shoyab et al., 1976; Girard et al., 1985). Secondly, PKC enzyme levels increase gradually at birth and then peak between 14 and 28 days postnatally in the rodent brain (Shoyab et al., 1976; Hashimoto et al., 1988), a period which correlates with *in vivo* myelination (Norton and Poduslo, 1973; Monge et al., 1986; Bjartmar et al., 1994). Furthermore, Yoshimura et al. (1992) demonstrated that there is high PKC activity in rat CNS myelin and that the PKC levels from postnatal 14 to 42 days parallel the deposition of myelin proteins. Thirdly, several myelin proteins are good substrates for PKC-mediated phosphorylation and these include myelin basic protein (MBP) (Turner et al., 1982; Vartanian et al., 1986), 2',3'-cyclic nucleotide phosphohydrolase (Vartanian et al., 1992) and myelin associated glycoprotein (Kirchoff et al., 1993; Yim et al., 1995). It has also been shown that PKC is involved in the cAMP regulation of MBP gene expression (Anderson and Miskimmons, 1994). Fourthly, when floating OLs in culture (without processes) adhere to a substratum, a PKC-dependent pathway leads to reenactment of myelinogenic metabolism in OLs; adherence, which may be mediated by $\alpha 8 \beta 1$ integrins (Malek-Hedayat and Rome, 1994), also results in the production of myelin lamella by these cells (Vartanian et al., 1986; 1992; Szuchet, 1987). Fifthly, in correspondence with our results (see below), Althaus et al. (1991) have shown that phorbol esters, which are pharmacological activators of PKC, increase the extension of processes of cultured porcine and rodent OLs. At the level of OL precursors, Bhat et al. (1992) provided evidence that the regulation of proliferation of these cells may be through PKC. Finally, localization studies for diacylglycerol kinase, a catabolic enzyme for diacylglycerol, an endogenous ligand for PKC, show that this enzyme is found in OLs *in vivo* in a distribution pattern similar to that for myelin proteins (Goto et al., 1992). Evidence from this laboratory (summarized in Table 1) implicates oligodendroglial PKC in regulating process extension (Yong et al., 1988; 1991; 1994). Firstly, process extension by adult human, rat or bovine OLs is enhanced by phorbol esters that stimulate PKC enzyme activity (Fig. 1): 4β-phorbol-12,13-dibutyrate (PDB) and phorbol-12-myristate-13-acetate (PMA). In contrast, phorbol esters that bind, but do not stimulate, PKC are without effect (4α-PDB or 4α-phorbol-12,13-didecanoate).

The role of PKC in oligodendroglial process formation has been further tested using inhibitors of PKC. In this regard, we have employed inhibitors of PKC ranging from those with low selectivity (heparin, polymixin B and staurosporine) to intermediate selectivity (chelerythrine and H-7) to those that are highly selective for PKC (calphostin C and CGP 41 251). These inhibitors block the basal- or PDB-enhanced fiber outgrowth. The PDB-enhanced oligodendroglial process extension was not blocked by the tyrosine kinase inhibitor genistein (Yong et al., 1994), in agreement with the postulate that the

Table 1. Summary of evidence that PKC regulates process extension by OLs

Phorbol esters that stimulate PKC activity promote process extension
Inhibitors of PKC block basal- or phorbol ester-enhanced process extension
Following treatment of OLs with phorbol esters, PKC translocates from the cytosol to the particulate fraction, in accordance with the mode of PKC activation
PKC enzyme activity in particulate fraction of adult human OLs remains elevated for several days following phorbol ester treatment
The effects of other process-promoting agents, bFGF and astrocyte ECM, are also attenuated by PKC inhibitors

PDB-enhanced oligodendroglial process extension is mediated through intracellular PKC. Measurements of PKC enzyme activity, using a histone phosphorylation assay, have further supported the role of PKC in oligodendroglial process extension. Thus, the treatment of adult human OLs with PDB results in the translocation of PKC from the cytosol to the particulate fraction of cell, in accordance with the mode of PKC activation (Kraft and Anderson, 1983; Chida et al., 1986). PKC enzyme activity in the particulate fraction has also been found to be elevated for prolonged periods (up to 12 days) following PDB treatment, consistent with a temporal relationship between PKC enzyme activity and process extension (Yong et al., 1994).

In studies of the kinetics of PKC stimulation, it has been shown that when cells are treated with phorbol esters, PKC translocates from the cytosol to the plasma membrane as aforementioned. This translocation results in activation of PKC, which however, also becomes sensitive to proteolysis by the calcium-dependent protease, calpain. Such proteolysis leads to degradation of the PKC enzyme; this phenomenon has been referred to as down-regulation of the enzyme (Nishizuka, 1992; Kraft and Anderson, 1983; Chida et al., 1986) and indeed, a single large dose (e.g. 100 to 400 nM) of phorbol ester is commonly used to deplete cells of PKC. In our studies of PKC activity of adult human OLs, a surprising finding is that the treatment with PDB, while resulting in the observed translocation of PKC to the particulate fraction, does not down-regulate enzyme activity, even after 12 days of PDB treatment (Yong et al., 1994). In contrast, down-regulation is observed for astrocytes, glioma cells and fibroblasts. This persistence of oligodendroglial PKC suggests a uniqueness in the manner in which OLs regulate their PKC content. We currently do not know the reason for the inability of adult human OLs to down-regulate their PKC following PDB treatment but this could be related to calpain and phosphatase activities within OLs. The multi-process morphology of adult OLs following their treatment with biologically active phorbol esters is reminiscent of that of myelinating OLs *in vivo* since these must frequently form and maintain in the range of 40 segments of myelin located on different axons (Peters and Vaughn, 1970; Wood and Bunge, 1984). Biochemically, the synthesis of MBP, a prerequisite component for myelinogenesis, is increased by 2-fold in PDB-treated oligodendrocytes when compared to control cells (Yong et al., 1994). We have observed that palinogenesis by control cells (i.e., non-phorbol ester-treated) plated onto a poly-l-lysine substrate is invariably *slowest for adult human* OLs and quickest for the adult rat cells (which itself, is much slower than that for neonatal rodent OLs). However, when treated with PDB or PMA, processes of more than 3 soma diameters in length are observed within 24 hours for all species. In a developmental study, there was a progressive decline with age of basal palinogenesis by OLs cultured from the brains of neonatal rats, or from adult rats of ages 1, 3 and 6 months; growth of processes of the adult cells, however, could be made to approach neonatal levels by treatment with PDB and PMA (Yong et al., 1991), suggesting the potential to invigorate aged OLs. Avossa and Pfeiffer (1993) have reported that progenitor cells of the OL lineage obtained from *newborn rat* brain transiently reverted to a less mature form when incubated with phorbol esters. Asotra and Macklin (1993) have documented that cells from early *neonatal rats* treated with phorbol esters had decreased levels of mRNAs for MBP and proteolipid protein. In our studies, however, all adult human OLs exposed to phorbol esters retained their MBP and MAG immunofluorescence (Oh and Yong, 1996). Furthermore, metabolic labelling with ^{35}S-methionine followed by immunoprecipitation of MBP revealed that the PDB-treated OLs had a 2-fold increase in ^{35}S-methionine incorporation into MBP (Yong et al., 1994). These results indicate that adult human OLs did not lose their maturity upon treatment with PDB; indeed, the morphological alteration was accompanied by biochemical changes that suggest increased myelinogenic potential since increased synthesis of MBP is a prerequisite for myelination *in vivo* (Sternberger et al., 1978; Roach et al., 1983; Shine et al., 1992).

Collectively, our results suggest an important contribution of PKC of OLs to the development of their processes. Studies employing isoform-specific agonists have implicated PKCα as the major determinant of fiber outgrowth by oligodendrocytes (Yong et al., 1994). In correspondence with our results, Schmidt-Schultz and Althaus (1994) concluded, based on calcium and substrate requirements, that PKCα mediates process extension of pig OLs.

ASTROCYTES PROMOTE PROCESS EXTENSION BY OLIGODENROCYTES

As phorbol esters are exogenous compounds, we have sought to identify physiologic growth factors that could promote process extension by OLs. Putatively, these growth factors would engage and activate their receptors on the cell surface, leading to downstream PKC stimulation. To achieve the goal of identifying physiologic growth factors, we addressed whether astrocytes or microglia, which are in close proximity to OLs *in vivo*, could facilitate oligodendroglial process extension *in vitro*, and if so, what growth factors produced by these cell types could be involved.

We have found that adult human OLs in mixed culture with astrocytes and microglia extended processes readily when compared to OLs in purified culture (Oh and Yong, 1996). To address which cell type in the mixed culture promoted process formation by OLs, OLs were seeded onto enriched microglia or astrocyte substrates. Figure 2 shows that astrocytes promoted survival and process outgrowth of OLs; the efficacy of live astrocytes matches that of PDB at 50 ng/ml, a concentration that provides maximal stimulation of oligodendroglial process extension (Yong et al., 1991).

In contrast to astrocytes, microglia were toxic to OLs. Of the surviving OLs in co-culture with microglia, their rate of process extension was also retarded when compared to controls (Oh and Yong, 1996). Rat microglial cells have previously been reported to be toxic to OLs (Merrill and Zimmerman, 1991); in certain cases, however, microglia have been reported to stimulate the myelinogenic program of OLs (Hamilton and Rome, 1994; Loughlin et al., 1994).

Certain characteristics of cells of the OL lineage have previously been shown to be dependent on astrocyte function. Thus, growth factors from astrocytes are reported to cause the proliferation of OL precursors and to modulate their survival and differentiation into OLs (Richardson et al., 1988; McKinnon et al., 1990; Mayer et al., 1994; Barres et al., 1992; 1993; Komoly et al., 1992). The survival of mature oligodendrocytes *in vitro* is also promoted by astrocytes (Gard et al., 1995). Rome and colleagues (Cardwell and Rome, 1988; Malek-Hedayat and Rome, 1994) have demonstrated that OLs adhered well to an astrocyte matrix via integrin-dependent mechanisms. In lesion areas of MS, OLs have been observed to be invested within hypertrophic astrocytes (Ghatak, 1992; Wu and Raine, 1992; Prineas et al., 1990); the role of such glial associations remains controversial and may represent a protective mechanism for OLs by astrocytes (Wu and Raine, 1992) or destruction of OLs by astrocytes (Prineas et al., 1990).

The role of astrocytes in myelin formation by OLs has been the subject of few studies, and some of the results have been conflicting. Bhat and Pfeiffer (1986) reported that soluble extracts from astrocyte cultures increased myelin proteins of OLs. However, astro-

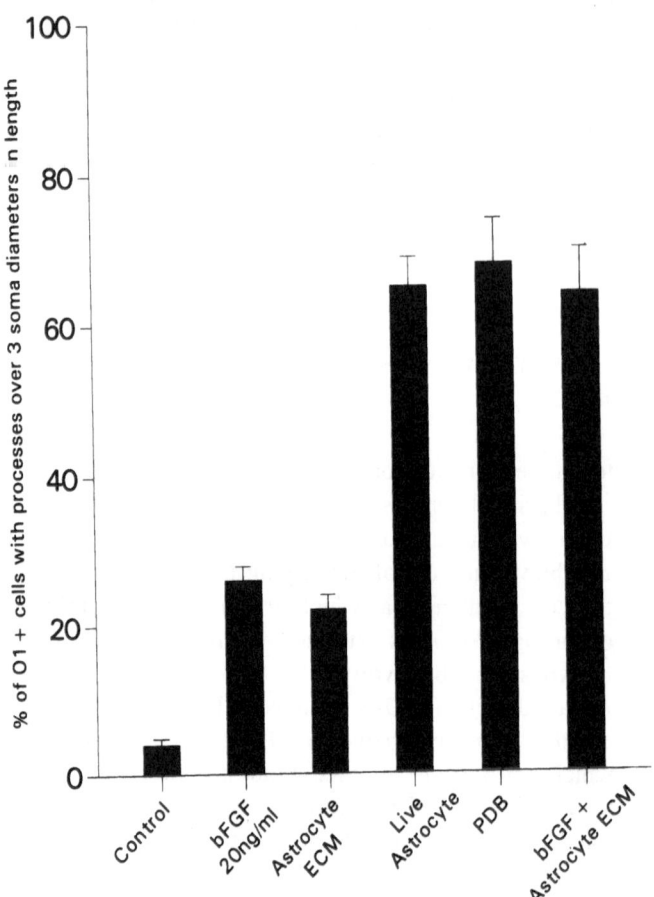

Figure 2. Astrocytes promote process extension by adult human OLs. Adult human OLs were plated onto poly-l-lysine (PL), astrocyte ECM, or live astrocytes. The potency of live astrocytes is equal to that of phorbol esters. By themselves, bFGF on PL, or astrocyte ECM, enhanced process outgrowth; their combination, however, produced process reformation to the extent obtained with live astrocytes.

cytes have been demonstrated to inhibit the myelination of dorsal root ganglion axons by adult rat OLs (Rosen et al., 1989) and to prevent the translocation of MBP mRNA from soma into the processes of OLs (Amur-Umarjee et al., 1993). Nonetheless, in the ethidium bromide model of demyelination, transplants of astrocytes improved remyelination by OLs (Franklin et al., 1993). *In vivo*, astrocytic processes are closely associated with the OL soma and processes (Butt et al., 1995), suggestive of intimate interactions.

While astrocytes have long been known to be excellent substrates for *neurite* extension by many different populations of neurons *in vitro* and *in vivo* (Lindsay, 1979; Silver and Ogawa, 1983; Noble et al., 1984; Fallon et al., 1985; Wujek and Akeson, 1987; Neugebauer et al., 1988; Tomaselli et al., 1988; Kliot et al., 1990; Johnson-Green et al., 1992), a similar role of astrocytes in promoting process formation by OLs has not been reported. In view of this dearth of information, we sought to identify astroglial growth factors that could regulate oligodendroglial process extension. We considered whether

Table 2. Growth factors that do not promote
process extension

Epidermal growth factor
Platelet derived growth factor (AA, AB or BB dimers)
Insulin-like growth factor-1
Insulin-like growth factor-2
Glial derived neurotrophic factor
Ciliary neurotrophic factor
Nerve growth factor
Brain derived neurotrophic factor
Neurotrophin-3
Neurotrophin-4

soluble factors elaborated by astrocytes could enhance oligodendroglial process out-growth. A range of growth factors known to be produced by astrocytes (reviewed in Yong, 1996) was added to purified OLs. Of these, only bFGF and aFGF could significantly en-hance process extension by adult human OLs (Table 2). However, the potency of bFGF was not as marked as that of PDB or live astrocytes (Fig. 2). Furthermore, a range of con-centrations of bFGF, added for varying periods of time, did not result in process extension of the magnitude that was elicited by PDB or astrocytes. Finally, bFGF could not syner-gize with other soluble growth factors, including aFGF, to elicit process extension by adult human OLs (Oh and Yong, 1996).

The process promoting actions of bFGF and live astrocytes are mechanistically linked because when the live astrocyte-OL co-culture was treated with a neutralizing anti-body to bFGF, the facilitatory role of astrocytes could be dose-dependently attenuated by the antibody but not eliminated (Oh and Yong, 1996). This partial attenuation by the neu-tralising antibody, together with the inability of purified bFGF to fully mimic the potency of the live astrocytes, suggests that the astroglial-derived bFGF could be acting in concert with another astrocyte factor; we have considered that this latter factor is a component of the astrocyte extracellular matrix (ECM).

Astrocytes are known to synthesize a number of ECM molecules that include laminin, heparan sulfate proteoglycan, chondroitin sulfate proteoglycan, fibronectin, thrombospondin, tenascin, hyaluronan and vitronectin (Liesi et al., 1983; Selak et al., 1985; Price and Hynes, 1985; Giftochristos and David, 1988; Grierson et al., 1990; Snow et al., 1990; Johnson-Green et al., 1991). Since OLs have integrin receptors (Malek-Hedayat and Rome, 1994; Milner and ffrench-Constant, 1994; Shaw et al., 1996), which would allow them to bind ECM molecules, and given a previous report that rat OLs extended processes on ECM derived from bovine corneal endothelial cells (Ovadia et al., 1984), we plated adult human OLs onto ECM of astrocytes. Process ex-tension was found to be enhanced by astrocyte ECM when compared to cells on poly-L-lysine although the magnitude was not as potent as that of live astrocytes (Fig. 2). However, the co-administration of bFGF and astrocyte ECM promoted process out-growth to the same extent as that of live astrocytes (Fig. 2), underscoring the interplay between bFGF and a component(s) of the astrocyte ECM; the identity of the astrocyte ECM that facilitates process outgrowth is unknown at this time (Oh and Yong, 1996). In reviewing the literature, it is evident that bFGF has many functions on cells of the OL lineage. bFGF is known to be a mitogen for O-2A cells and to block their differ-entiation into OLs (Noble et al., 1988; McKinnon et al., 1990; Pfeiffer et al., 1993; Fressinaud and Vallat, 1994) although this blockade could be overcome by other factors made by astrocytes. In cultures from neonatal rodent brain containing more differentiated GalC+ OLs and their precursors, bFGF was shown to increase the proliferation of both cell types (Saneto and deVellis, 1985; Eccleston and Silberberg, 1985; Deloulme et al., 1992; Vicks and DeVries, 1992). In even more differentiated rodent OL cultures (from neonatal brain and expressing MBP), bFGF administration induced cell proliferation and dedifferentiation of OLs into precursor cells (Grinspan et al., 1993; Fressinaud et al., 1993, 1995). bFGF was recently found to prevent the apoptosis of OLs (Yasuda et al., 1995). Finally, of relevance to this study and in direct contrast, bFGF inhibited process formation of cultured rat (Besnard et al., 1989) and canine (Hoffman and Duncan, 1995) OLs. In our hands, bFGF not only promoted process formation by adult human OLs, as has been reported by Gogate et al. (1994), but bFGF also did not produce proliferation (Yong et al., 1988) of OLs and did not cause MBP+ cells to dedifferentiate (Oh and Yong, 1996). These differences are likely accounted for by species factors. In concordance with this arguement, Gogate et al. (1994) and Yong et al. (1988) reported that bFGF was not a mitogen for adult human white matter derived OLs; Satoh and Kim (1994) reported that bFGF was not a mitogen for OLs from fetal human brains. Finally, we have also observed that astrocytes (Yong et al., 1990; 1992) and microglia (Williams et al., 1992) from human brains are different from their rodent counterparts in their morphology, expression of major histocompatibility antigens and response to mitogenic factors.

In summary, process outgrowth by adult human OLs is promoted by astrocytes through the actions of bFGF and a component of the astrocyte ECM. The findings are relevant to the understanding of myelin formation *in vivo* since FGFs are expressed at several stages of brain development including the postnatal period when myelination occurs (Thomas et al., 1991; Giodano et al., 1992). Furthermore, receptors for FGF have been localized by *in situ* hybridization studies to myelinated process tracts of the rat CNS (Asai et al., 1993). Finally, astrocytes appear chronologically earlier than OLs and the close promixity of both cell types favors interactions that likely include the facilitation of process formation by OLs, an early event in myelin formation.

OLIGODENDROCYTE PROCESS EXTENSION IS PRODUCED BY bFGF AND ASTROCYTE ECM IS BLOCKED BY PKC INHIBITORS

Prior to the finding that astrocyte ECM and bFGF facilitate oligodendroglial process extension, phorbol esters have been the only other promoter of process outgrowth by adult human OLs. If PKC plays a critical role in process outgrowth by OLs, then the efficacy of bFGF and astrocyte ECM in eliciting fiber formation should also involve PKC. To address this question, we have utilized the selective PKC inhibitors, calphostin C and CGP 41251, to determine whether they could attenuate the process promoting action of bFGF and astrocyte ECM. Our findings demonstrate that the outgrowth of processes by adult human OLs produced by bFGF or astrocyte ECM was dose-dependently reduced by calphostin C and CGP 41 251; this inhibition occured at concentrations of drug that did not produce apoptotic or necrotic damage to OLs (Oh et al., in press). These results support the central role of PKC in oligodendroglial process formation, since all promoters of process outgrowth defined to date (phorbol esters, bFGF and astrocyte ECM) appear to transduce signals through PKC of OLs.

FINAL PERSPECTIVES

From the discussion above, several issues remain to be resolved. Firstly, how critical is process extension to the overall scheme of myelination? We take the position that promoting process extension will not necessarily lead to enhanced myelination, since there remains multiple steps between process formation and the end product of myelin; these stages would include axon recognition by oligodendroglial processes, adhesive interactions between oligodendroglial process and axon, spiral wrapping of axons and myelin compaction. Nonetheless, process extension is a critical early step in myelin formation, for without process formation, myelination would not ensue. Secondly, the identity of the phosphorylation substrates of PKC in OLs that leads to process extension is unknown. There exists over 100 PKC substrates in cells (Liu, 1996), and it would be a challenge to decipher the phosphorylated product(s) of PKC that leads to oligodendroglial process formation. Finally, it would be a logical next step to address whether the interference of PKC *in vivo* would lead to the disruption of myelinogenesis, whether during development, normal adulthood or in remyelination.

ACKNOWLEDGMENTS

This work was supported by the Multiple Sclerosis Society of Canada.

REFERENCES

Althaus HH, Schroter J, Spoerri P, Schwartz P, Kloppner S, Rohmann A, Neuhoff V (1991): Protein kinase C stimulation enhances the process formation of adult oligodendrocytes and induces proliferation. J Neurosci Res 29:481–489.

Amur-Umarjee S, Phan T, Campagnoni AT (1993): Myelin basic protein mRNA translocation in oligodendrocytes is inhibited by astrocytes *in vitro*. J Neurosci Res 36:99–110.

Anderson S, Miskimins R (1994): Involvement of protein kinase C in cAMP regulation of myelin basic protein gene expression. J Neurosci Res 37:604–611.

Asai T, Wanaka A, Kato H, Masana Y, Seo M, Tohyama M (1993): Differential expression of two members of FGF receptor gene family, FGFR-1 and FGFR-2 mRNA, in the adult rat central nervous system. Mol Brain Res 17:174–178.

Asotra K, Macklin WB (1993): Protein kinase C activity modulates myelin gene expression in enriched oligodendrocytes. J Neurosci Res 34:571–588.

Avossa D, Pfeiffer SE (1993): Transient reversion of O4$^+$ GalC$^-$ oligodendrocyte progenitor development in response to the phorbol ester TPA. J Neurosci Res 34:113–128. Baltuch GH, Dooley NP, Villemure J-G, Yong VW (1995): Protein kinase C and growth regulation of malignant gliomas. Can J Neurol Sci 22:264–271.

Barlett WP, Knapp PE, Skoff RP (1988): Glial conditioned medium enables jimpy oligodendrocytes to express properties of normal oligodendrocytes: production of myelin antigens and membranes. Glia 1:253–259.

Barres BA, Hart IK, Coles HSR, Burne JF, Voyvodic JT, Richardson WD, Raff MC (1992): Cell death and control of cell survival in the oligodendrocyte lineage. Cell 70:31–46. Barres BA, Schmid R, Sendtner M (1993): Multiple extracellular signals are required for long-term oligodendrocyte survival. Development 118:283.

Besnard F, Perraud F, Sensenbrenner M, et al. (1989): Effects of acidic and basic fibroblastic growth factors on proliferation and maturation of cultured rat oligodendrocytes. Int J Dev Neurosci 7:401–409.

Bhat NR, Hauser KF, Kindy MS (1992): Cell proliferation and protooncogene induction in oligodendroglial progenitors. J Neurosci Res 32:340–349.

Bhat S, Pfeiffer SE (1986): Stimulation of oligodendrocytes by extracts from astrocyte-enriched cultures. J Neurosci Res 15:19–27.

Bjartmar C, Hildebrand C, Loinder K (1994): Morphological heterogeneity of rat oligodendrocytes: Electron microscopic studies on serial sections. Glia 11:235–244. Brück W, Schmied M, Suchanek G, Brück Y, Breitschopf H, Poser S, Piddlesden S, Lassmann H (1994): Oligodendrocytes in the early course of multiple sclerosis. Ann Neurol 35:65–73.

Butt AM, Ibrahim M, Ruge FM, Berry M (1995): Biochemical subtypes of oligodendrocyte in the anterior medullary velum of the rat as revealed by the monoclonal antibody Rip. Glia 14:185–197.

Cardwell MC, Rome LH (1988): Evidence that an RGD-dependent receptor mediates the binding of oligodendrocytes to a novel ligand in a glial-derived matrix. J Cell Biol 107:1541–1549.

Chida K, Kato N, Kuroki T (1986): Down regulation of phorbol diester receptors by proteolytic degradation of protein kinase C in a cultured cell line of fetal rat skin keratinocytes. J Biol Chem 261:13013–13018.

Dekker LV, Parker PJ (1994): Protein kinase C - a question of specificity. Trends Biochem 19:73–77.

Deloulme JC, Janet T, Pettmann B, Laeng P, Knoetgen M-F, Sensenbrenner M, Baudier J (1992): Phosphorylation of the MARCKS protein (P87), a major protein kinase C substrate, is not an obligatory step in the mitogenic signaling pathway of basic fibroblast growth factor in rat oligodendrocytes. J Neurochem 58:567–578.

Dubois-Dalcq M, Behar T, Hudson L, et al. (1986): Emergence of three myelin proteins in oligodendrocytes cultured without neurons. J Cell Biol 102:384–392. Eccleston PA, Silberberg DH (1985): Fibroblast growth factor is a mitogen for oligodendrocytes in vitro. Dev Brain Res 21:315–318.

Fallon JR (1985): Preferential outgrowth of central nervous system neurites on astrocytes and Schwann cells as compared with nonglial cells in vitro. J Cell Biol 100:198–207.

Franklin RJM, Crang AJ, Blakemore WF (1993): The role of astrocytes in the remyelination of glia-free areas of demyelination. Adv. Neurol. 59:125–133.

Fressinaud C, Laeng P, Labourdette G, Durand J, Vallat JM (1993): The proliferation of mature oligodendrocytes in vitro is stimulated by basic fibroblast growth factor and inhibited by oligodendrocyte-type 2 astrocyte precursors. Dev Biol 158:317–329.

Fressinaud C, Vallat JM (1994): Basic fibroblast growth factor improves recovery after chemically induced breakdown of myelin-like membranes in pure oligodendrocyte cultures. J Neurosci Res 38:202–213.

Fressinaud C, Vallat JM, Labourdette G (1995): Basic fibroblast growth factor down-regulates myelin basic protein gene expression and alters myelin compaction of mature oligodendrocytes in vitro. J Neurosci Res 40:285–293.

Gard AL, Burrell MR, Pfeiffer SE, Rudge JS, Williams WC (1995): Astroglial control of oligodendrocyte survival mediated by PDGF and leukemia inhibitory factor-like protein. Development 121:2187–2197.

Ghatak NR (1992): Occurrence of oligodendrocytes within astrocytes in demyelinating lesions. J Neuropathol Exp Neurol 51:40–46.

Ghatak NR, Leshner RT, Price AN, Felton III WL (1989): Remyelination in the human central nervous system. J Neuropathol Exp Neurol 48:507–518.

Giftochristos N, David S (1988): Laminin and heparan sulphate proteoglycan in the lesioned adult mammalian central nervous system and their possible relationship to axonal sprouting. J Neurocytol 17:385–397.

Giordano S, Sherman L, Lyman W, Morrison R (1992): Multiple molecular weight forms of basic fibroblast growth factor are developmentally regulated in the central nervous system. Dev Biol 152:293–303.

Girard PR, Mazzei GJ, Wood JG, Kuo JF (1985): Polyclonal antibodies to phospholipid/Ca^{2+}-dependent protein kinase and immunocytochemical localization of the enzyme in rat brain. Proc Natl Acad Sci USA 82:3030–3034.

Gogate N, Verma L, Zhou M, Milward E, Rusten R, O'Connor M, Kufta C, Kim J, Hudson L, Dubois-Dalcq M (1994): Plasticity in the adult human oligodendrocyte lineage. J Neurosci 14:4571–4587.

Goto K, Watanabe M, Kondo H, Yuasa H, Sakane F, Kanoh H (1992): Gene cloning, sequence, expression and in situ localization of 80 kDa diacylglycerol kinase specific to oligodendrocyte of rat brain. Mol Brain Res 16:75–87.

Grierson JP, Petroski RE, Ling DSF, Geller HM (1990): Astrocyte topography and tenascin/cytotactin expression: correlation with the ability to support neuritic outgrowth.

Grinspan JB, Stern JL, Franceschini B, Pleasure D (1993): Trophic effects of basic fibroblast growth factor (bFGF) on differentiated oligodendroglia: a mechanism for regeneration of the oligodendroglial lineage. J Neurosci Res 36:672–680.

Hamilton SP, Rome LH (1994): Stimulation of in vitro myelin synthesis by microglia. Glia 11:326–335.

Hashimoto T, Ase K, Sawamura S, Kikkawa U, Saito N, Tanaka C, Nishizuka Y (1988): Postnatal development of a brain-specific subspecies of protein kinase C in rat. J Neurosci 8:1678–1683.

Hoffman KL, Duncan ID (1995): Canine oligodendrocytes undergo morphological changes in response to basic fibroblast growth factor (bFGF) in vitro. Glia 14:33–42.

Huang FL, Yoshida Y, Nakabayashi H, Huang KP (1987): Differential distribution of protein kinase C isozymes in the various regions of brain. J Biol Chem 262:15714–15720.

Johnson-Green PC, Dow KE, Riopelle RJ (1991): Characterization of glycosaminoglycans produced by primary astrocytes *in vitro*. Glia 4:314–321.

Johnson-Green PC, Dow KE, Riopelle RJ (1992): Neurite growth modulation associated with astrocyte proteoglycans: influence of activators of inflammation. Glia 5:33–42.

Kirchhoff F, Hofer HW, Schachner M (1993): Myelin-associated glycoprotein is phosphorylated by protein kinase C. J Neurosci Res 36:368–381.

Kliot M, Smith GM, Siegal JD, Silver J (1990): Astrocyte-polymer implants promote regeneration of dorsal root fibers into the adult mammalian spinal cord. Exp Neurol 109:57–69.

Komoly S, Hudson LD, deF Webster H, Bondy CA (1992): Insulin-like growth factor I gene expression is induced in astrocytes during experimental demyelination. Proc Natl Acad Sci USA 89:1894.

Kraft AS, Anderson WB (1983): Phorbol esters increase the amount of Ca^{2+}, phospholipid-dependent protein kinase associated with plasma membrane. Nature 301:621–623.

Liesi P, Dahl D, Vaheri A (1983): Laminin is produced by early rat astrocytes in primary culture. J Cell Biol 96:920–924.

Lindsay RM (1979): Adult rat brain astrocytes support survival of both NGF-dependent and NGF-insensitive neurones. Nature 282:80–82.

Liu JP (1996) Protein kinase C and its substrates. Mol Cell Endocrinol 116:1–29.

Loughlin AJ, Honeggar P, Woodroofe MN, Comte V, Matthieu JM, Cuzner ML (1994): Myelin basic protein content of aggregating rat brain cell cultures treated with cytokines and/or demyelinating antibody: Effects of macrophage enrichment. J Neurosci Res 37:647–653.

Malek-Hedayat S, Rome LH (1994): Expression of beta 1-related integrin by oligodendroglia in primary culture: Evidence for a functional role in myelination. J Cell Biol 124:1039–1046.

Mayer M, Bhakoo K, Noble M (1994): Ciliary neurotrophic factor and leukemia inhibitory factor promote the generation, maturation and survival of oligodendrocytes *in vitro*. Development 120:143–153.

McKinnon RD, Matsui T, Dubois-Dalq M, Aaronson SA (1990): FGF modulates the PDGF-driven pathway of oligodendrocyte development. Neuron 5:603–614.

Merrill JE, Zimmerman RP (1991): Natural and induced cytotoxicity of oligodendrocytes by microglia is inhibitable by TGFb. Glia 4:327–331.

Milner R, ffrench-Constant C (1994): A developmental analysis of oligodendroglial integrins in primary cells: changes in av-associated b subunits during differentiation. Development 120:3497–3506.

Monge M, Kadiisky D, Jacque C, Zalc B (1986): Oligodendroglia expression and deposition of four major myelin constituents in the myelin sheath during development: an *in vitro* study. Dev Neurosci 8:222–235.

Neugebauer KM, Tomaselli KJ, Lilien J, Reichardt LF (1988): N-cadherin, NCAM and integrins promote retinal neurite outgrowth on astrocytes *in vitro*. J Cell Biol 107:1177–1187.

Newton AC (1995): Protein kinase C: structure, function, and regulation. J Biol Chem 270:28495–28498.

Nishizuka Y (1992): Intracellular signaling by hydrolysis of phospholipids and activation of protein kinase C. Science 258:607–614.

Norton WT, Poduslo SE (1973): Myelination in rat brain: Changes in myelin composition during brain maturation. J Neurochem 21:759–773.

Oh LYS, Yong VW (1996): Astrocytes promote process outgrowth by adult human oligodendrocytes *in vitro* through interaction between bFGF and astrocyte extracellular matrix. Glia, 17: 237–253.

Oh LYS, Goodyer CG, Olivier A, Yong VW (1997): The promoting effect of bFGF and astrocyte extracellular matrix on process outgrowth by adult human oligodendrocytes are mediated by protein kinase C. Brain Res, in press.

Ovadia H, Lubetzki-Korn I, Brenner T, Abramsky O, Fridman R, Vlodavsky I (1984): Adult rat oligodendrocytes grown *in vitro* upon an extracellular matrix have the ability to proliferate. Brain Res 322:93–100.

Peters A, Vaughn JE (1970): Morphology and development of the myelin sheath. In Davison AN, Peters A (eds): "Myelination." Springfield, Illinois: Charles C. Thomas, pp 3–79.

Pfeiffer SE, Warrington AE, Bansal R (1993): The oligodendrocyte and its many cellular processes. Trends Cell Biol. 3:191–197.

Price J, Hynes RO (1985): Astrocytes in culture synthesize and secrete a variant form of fibronectin. J Neurosci 5:2205–2211.

Prineas JW, Barnard RO, Kwon EE, Sharer LR, Cho ES (1993): Multiple sclerosis: Remyelination of nascent lesions. Ann Neurol 33:137–151.

Prineas JW, Kwon EE, Goldenberg PZ, Cho E-S, Sharer LR (1990): Interaction of astrocytes and newly formed oligodendrocytes in resolving multiple sclerosis lesions. Lab Invest 63:624–636.

Prineas JW, Kwon EE, Goldenberg PZ, Ilyas AA, Quarles RH, Benjamins JA, Sprinkle TJ (1989): Multiple sclerosis: Oligodendrocyte proliferation and differentiation in fresh lesions. Lab Invest 61:489–503.

Raine CS, Scheinberg L, Waltz JM (1981): Oligodendrocyte survival and proliferation in an active established lesion. Lab Invest 45:534–546.

Raine CS, Wu E (1993): Multiple sclerosis: Remyelination in acute lesions. J Neuropathol Exp Neurol 52:199.

Richardson WD, Pringle N, Mosley MJ, Westermark B, Dubois-Dalcq M (1988): A role for platelet-derived growth factor in normal gliogenesis in the central nervous system. Cell 53:309–319.

Roach A, Boylan K, Horvat S, Prusiner SB, Hood LE (1983): Characterization of cloned cDNA representing rat myelin basic protein: absence of expression in brain of shiverer mutant mouse. Cell 34:799–806.

Rosen CL, Bunge RP, Ard MD, Wood PM (1989): Type 1 astrocytes inhibit myelination by adult rat oligodendrocytes in vitro. J Neurosci 9:3371–3379.

Saneto RP, de Vellis J (1985): Characterization of cultured rat oligodendrocytes proliferating in a serum-free, chemically defined medium. Proc Natl Acad Sci USA 82:3509–3513.

Satoh J, Kim SU (1994): Proliferation and differentiation of fetal human oligodendrocytes in culture. J Neurosci Res 39:260–272.

Schmidt-Schultz T, Althaus HH (1994): Monogalactosyl diglyceride, a marker for myelination, activates oligodendroglial protein kinase C. J Neurochem 62:1578–1585.

Selak I, Foidart JM, Moonen G (1985): Laminin promotes cerebellar granule cells migration in vitro and is synthesized by cultured astrocytes. Dev Neurosci 7:278–285.

Shaw CE, Milner R, Compston AS, ffrench-Constant C (1996): Analysis of integrin expression on oligodendrocytes during axo-glial interaction by using rat-mouse xenografts. J Neurosci 16:1163–1172.

Shine HD, Readhead C, Popko B, Hood L, Sidman RL (1992): Morphometric analysis of normal, mutant, and transgenic CNS: correlation of myelin basic protein expression to myelinogenesis. J Neurochem 58:342–349.

Shoyab M, Warren TC, Todaro GJ (1976): Tissue and species distribution and developmental variation of specific receptors for biologically active phorbol and ingenol esters. Carcinogenesis 2:273–1276.

Silver J, Ogawa MY (1983): Postnatally induced formation of the corpus callosum in acallosal mice in glial-coated cellulose bridges. Science 220:1067–1069.

Snow DM, Lemmon V, Carrino DA, Caplan AI, Silver J (1990): Sulfated proteoglycans in astroglial barriers inhibit neurite outgrowth in vitro. Exp Neurol 109:111–130.

Sternberger NH, Itoyama Y, Kies MW, Webster H de F (1978): Myelin basic protein demonstrated immunocytochemically in oligodendroglia prior to myelin sheath formation. Proc Natl Acad Sci USA 75:2521–2524.

Szuchet S (1987): The plasticity of mature oligodendrocytes: a role for substratum in phenotype expression. In Crescenzi GS (ed): "A Multidisciplinary Approach to Myelin Diseases." New York: Plenum Press, pp 143–159.

Thomas D, Caruelle J-P, Lachapelle F, Barritault D, Boully B (1991): Acidic fibroblast growth factor in normal, injured and jimpy mutant developing mouse brain. Ann NY Acad Sci 648:161–166.

Tomaselli KJ, Neugebauer KM, Bixby JL, Lilien J, Reichardt LF (1988): N-cadherin and integrins: Two receptor systems that mediate neuronal process outgrowth on astrocyte surfaces. Neuron 1:33–43.

Turner RS, Chou CHJ, Kibler RF, et al. (1982): Basic protein in brain myelin is phosphorylated by endogenous phospholipid-sensitive Ca^{2+}-dependent protein kinase. J Neurochem 39:1397.

Vartanian T, Szuchet S, Dawson G (1992): Oligodendrocyte-substratum adhesion activates the synthesis of specific lipid species involved in cell signaling. J Neurosci Res 32:69–78.

Vartanian T, Szuchet S, Dawson G, Campagnoli AT (1986): Oligodendrocyte adhesion activates protein kinase C-mediated phosphorylation of myelin basic protein. Science 234:1395–1398.

Vicks RS, DeVries GH (1992): Mitotic potential of adult rat oligodendrocytes in culture. J Neurosci Res 33:68–74.

Williams K, Bar-Or A, Ulvestad E, Olivier A, Antel JP, Yong VW (1992): Biology of adult human microglia in culture: comparisons with peripheral blood monocytes and astrocytes. J Neuropathol Exp Neurol 51:538–549.

Wood P, Bunge RP (1984): The biology of the oligodendrocyte. Adv Neurochem 5:1–46. Wu E, Raine CS (1992): Multiple sclerosis: Interactions between oligodendrocytes and hypertrophic astrocytes and their occurrence in other non-demyelinating conditions. Lab Invest 67:88–99.

Wujek JR, Akeson RA (1987): Extracellular matrix derived from astrocytes stimulates neuritic outgrowth from PC12 cells in vitro. Brain Res 431:87–97.

Yasuda T, Grinspan J, Stern J, Franceschini B, Bannerman O, Pleasure D (1995): Apoptosis occurs in the oligodendroglial lineage, and is prevented by basic fibroblast growth factor. J Neurosci Res 40:306–317.

Yim SH, Toda K, Goda S, Quarles RH (1995): Comparison of the phosphorylation of myelin-associated glycoprotein in cultured oligodendrocytes and Schwann cells, J Mol Neurosci 6:63–74.

Yong VW (1996): Cytokines, astrogliosis, and neurotrophism following CNS trauma. In: Cytokines and the CNS. RM Ransohoff and EN Benveniste, eds, CRC Press, Boca Raton, FL, pp 309–327.

Yong VW, Cheung JCB, Uhm JH, Kim SU (1991): Age dependent decrease of process formation by cultured oligodendrocytes is augmented by protein kinase C stimulation. J Neurosci Res 29:87–99.

Yong VW, Dooley NP, Noble PG (1994): Protein kinase C in cultured adult human oligodendrocytes: Characteristics, lack of down-regulation and isoform a as a mediator of fiber outgrowth. J Neurosci Res 39:83–96.

Yong VW, Kim SU, Kim MW, Shin DH (1988a): Growth factors for human glial cells in culture. Glia 1:113–123.

Yong VW, Sekiguchi S, Kim MW, Kim SU (1988b): Phorbol ester enhances morphological differentiation of oligodendrocytes in culture. J Neurosci Res 19:187–194. Yong VW, Tejada-Berges T, Goodyer CG, Antel JP, Yong FP (1992): Differential proliferative response of human and mouse astrocytes to gamma-interferon. Glia 6:269–280.

Yong VW, Yong FP, Olivier A, Robitaille Y, Antel JP (1990): Morphologic heterogeneity of human adult astrocytes in culture: Correlation with HLA-DR expression. J Neurosci Res 27:678–688.

Yoshimura T, Kobayashi T, Goto I (1992): Protein kinase C in rat brain myelin. Neurochem Res 17:1021–1027.

Zalc B, Monge M, Jacque C (1987): Oligodendroglial emergence and deposition of four major myelin constituents in the myelin sheath during development: An *in vivo* study. In: A Multidisciplinary Approach to Myelin Diseases. GS Crescenzi, ed, Plenum Press, New York, pp 77–85.

Zeller NK, Behar TN, Dubois-Dalcq ME, Lazzarini RA (1985): The timely expression of myelin basic protein gene in cultured rat brain oligodendrocytes is independent of continuous neuronal influences. J Neurosci 5:2955–2962.

19

MECHANISM OF ADHESION-INDUCED REGENERATION/DIFFERENTIATION OF OLIGODENDROCYTES

Sara Szuchet

Department of Neurology and The Brain Research Institute
The University of Chicago
Chicago, Illinois 60637

INTRODUCTION

Oligodendrocytes (OLGs) synthesize and maintain central nervous system (CNS) myelin, the multilamellar membrane that enwraps axons, thereby facilitating fast nerve conduction. The advent of myelin marked an evolutionary milestone for it allowed the existence of higher organisms; at the same time, an irrevocable seal of myelin dependence was established. Any perturbation of the normal functioning of OLGs that results in dysmyelination will lead to growth arrest and death. Damage to OLGs and/or myelin resulting in demyelination will cause crippling and may result in death. Diseases affecting OLGs and/or myelin range from genetic to autoimmune, from metabolic to traumatic. The molecular mechanisms of some of these diseases have been discovered; for others, e.g., multiple sclerosis, the etiologies are still unknown. Given the dependence on myelin for normal function, the issue of myelin repair, i.e., remyelination, acquires a crucial significance. For a long time the popularly held view was that there is little to no remyelination in the CNS. This notion has now evolved; while CNS remyelination may be scanty and finite, it occurs. This raises the important question as to the origin of the remyelinating cell.

Many *in situ* and *in vitro* models for studying remyelination and the origin of the remyelinating cell have been established (rev. by Dubois-Dalcq, 1995; Ludwin, 1988; Scolding and Lassmann, 1996). No unified answer has emerged. Conceivably there is sufficient plasticity within the nervous system so that the response is dependent on which cell type sustains the brunt of the damage and on the nature of the damage. For example, in those models where OLGs are the primary targets of either a chemical or viral injury, remyelination is associated with the recruitment of healthy immature OLGs. Current evidence indicates that cell division always precedes remyelination. However, there is a finality to the precursor pool. For instance, it was shown with the Cuprizone (bis-cyclooxaldihydrazone) model that the reparative potential declines after each demyelination/remyelination cycle to the point when remyelination can no longer be achieved (Blakemore,

1984; Johnson and Ludwin, 1981). This outcome should not be surprising, considering that Cuprizone represents a prototype model for demyelination arising primarily from OLG degeneration. A new paradigm that is being explored in experimental animals that may eventually find application to humans, is transplantation of precursor cells (Blakemore et al., 1995); (reviewed by, Franklin and Blakemore, 1995) and also cell lines (Franklin et al., 1996).

Wood and Bunge. (1991) have addressed the issue of the capacity (in culture) of mature OLGs — defined as cells that have withdrawn from the mitotic cycle and have differentiated by expressing galactocerebroside — to remyelinate. They concluded that mature OLGs: 1) can be activated to divide; 2) are the major effectors of *in vitro* remyelination; and 3) must undergo cell division in order to remyelinate. However, they remarked that: a) not all mature OLGs are susceptible to induction into the cell cycle, and b) it has yet to be established that such cells play any part in remyelination *in situ*.

I am interested in assessing the potential of mature OLGs — defined as postmyelinating cells — to resynthesize myelin *without* undergoing cell division. To address this point, we needed to establish an experimental model wherein myelin is obliterated but OLGs and axons remain unaffected. As a first approximation to this paradigm, we have developed an *in vitro* model system consisting of pure cultures of OLGs isolated from young but post-myelination brains. This system affords an opportunity to investigate the fate of OLGs deprived of their myelin, in the absence of axons. The approach here is clearly reductionist; the model is a hypothetical CNS consisting of only OLGs. Nonetheless, by learning what these cells *can* or *cannot* do in isolation, we hope to gain insight into the nature of cell-cell and cell-substratum interactions that shape the CNS. Such insight should help to unravel the components critical to these events.

We have found that OLGs recover from the injury and are able to assemble myelin membranes, but to do this they require a signal. *In vitro*, this signal is provided by interaction with an adequate substratum acting, presumably, as a surrogate for an axon (Szuchet, 1987; Szuchet et al., 1986; Yim et al., 1986). We call this process myelin palingenesis, defined as the synthesis of multilamellar membranes by post-myelination OLGs in the absence of neurons and other CNS cells (Arvanitis et al., 1992a; Arvanitis et al., 1992b; Szuchet, 1987; Szuchet et al., 1986). Myelin palingenesis entails OLG regeneration since there is no cell division; this distinguishes it from remyelination, which is customarily used to describe the re-ensheathment of an axon that has lost its myelin, irrespective of whether the myelinating OLG is virgin or has already produced myelin once. Interestingly, these regenerating OLGs are also able to myelinate axons when transplanted into a shiverer mouse as evidenced by the detection of myelin basic protein (MBP)-positive myelin (Ludwin and Szuchet, 1993). Given that the OLGs used in these studies originated from post-myelination brains, this system is as relevant to regeneration as it is to differentiation.

Cell-substratum and/or cell-cell adhesion play critical roles in many aspects of cell function such as response to growth factors, proliferation and differentiation (Boudreau et al., 1995; Mercurio, 1995; Roskelley et al., 1995; Yamada and Miyamoto, 1995). Major advances have been made in the understanding of the molecular mechanisms underlying cell adhesion. Integrins are crucial cell surface receptors that mediate these adhesive interactions; they are part of a signal transduction machinery that carries inside-out and outside-in information. Most cells have a repertoire of integrins that utilize common and cell-type specific signaling pathways (Ruoslahti, 1996). *In situ*, integrins bind to one or more adhesion molecules that are part of an extracellular matrix (ECM). There is a signaling interplay between integrins and the ECM that influences their mutual expression and refines the adhesive specificity (Delcommenne and Streuli, 1995; Juliano and Haskill, 1993; Lin and Bissell, 1993; Meredith *et al.*, 1993; Venstrom and Reichardt, 1993).

We have characterized the events that accompany OLG-substratum interaction biochemically (Szuchet, 1987; Vartanian *et al.*, 1988; Vartanian *et al.*, 1986; Yim *et al.*, 1986), morphologically and immunocytochemically (Arvanitis *et al.*, 1992a; Arvanitis *et al.*, 1992b; Szuchet *et al.*, 1986), and electrophysiologically (Soliven *et al.*, 1988a; Soliven *et al.*, 1988b). Our data are consistent with a model that implicate an integrin-like receptor binding to substratum; the development of cell polarity, followed by the assembly of a cell-associated matrix as the initial and possibly obligatory steps — at least *in vitro* — for the establishment of the myelinogenic phenotype. Herein I present evidence in support of such a model.

EARLY EVENTS FOLLOWING OLG-SUBSTRATUM ADHESION

OLGs exhibit selectivity in their preference for substrata. In previous work, we have purified and partially characterized a horse serum, heparin-binding glycoprotein that effectively functions as an adhesion molecule for OLGs. We have named this novel molecule GRASP (**G**lycine **R**ich **A**dhesion **S**erum **P**rotein) to stress the fact that it is glycine-rich and behaves as an adhesion molecule (Schirmer *et al.*, 1994). We have investigated the biochemical and morphological changes that ensue after 24h of OLG-GRASP adhesion (or OLG adhesion to polylysine in the presence of 20% horse serum). These changes include: 1) activation of protein kinase C (PKC) and protein kinase A (PKA) and the concomitant phosphorylation of MBP and 2',3', cyclic nucleotide phosphodiesterase (CNPase) (Vartanian *et al.*, 1988; Vartanian *et al.*, 1986); 2) synthesis of Plasmalogens with the same class specificity as observed *in situ* (Vartanian *et al.*, 1992); 3) activation of certain receptors and deactivation of others (Vartanian *et al.*, 1986); and 4) synthesis of myelin specific proteins and lipids (Szuchet *et al.*, 1983; Yim *et al.*, 1986). All these events recapitulate, faithfully, steps pertaining to OLG differentiation. Another early sequel to adhesion, that we have uncovered in culture, is the synthesis and vectorial transport to the plasma membrane of heparan sulfate proteoglycans (HSPGs) that bind to surface receptors, thereupon setting up cell polarity (Yim *et al.*, 1993). This has not yet been identified *in situ*. The generation of these distinct biochemical domains on the plasma membrane precedes morphological polarization, i.e., the extension of cellular processes, by 24h to 48h. A class of these HSPGs assemble into a cell-associated matrix (Szuchet *et al.*, 1996). We postulate that this matrix is located on the surface beneath the cells, thus, reinforcing OLG polarity.

Acquisition of Cell Polarity

I hypothesize that the initial phase in the mechanism of adherence-induced differentiation/regeneration of OLGs, is the establishment of cell polarity. This sets the stage for the events that follow. I further hypothesize that the pathways utilized by OLGs resemble those used by epithelial cells in setting up their polarity. Supporting this contention is a recent report stating that the sequence of events leading to cell polarity is common in species as far apart phylogenetically as yeast and epithelial cells (Drubin and Nelson, 1996). The hierarchy of stages involved in establishing cell polarity were listed by Drubin and Nelson (1996) as first, development of surface asymmetry by intrinsic and/or extrinsic spatial cues; second, fixing the cue specific sites of action on the membrane with the aid of receptors and/or cytoskeletal structures; third, assembling a localized signaling network; and fourth, initiating segregated sorting of proteins to maintain and reinforce the specialized domains. This is illustrated in Fig. 1.

What is the indication that OLGs conform to such a model? The following observations support this model, albeit indirectly (Fig. 2). OLGs adhere to GRASP via a surface receptor. Preliminary data indicate that this receptor is an integrin-like molecule and the interaction is cation-dependent (unpublished observations). A signaling cascade ensues that triggers the synthesis, transport and asymmetric binding of HSPGs to the plasmalemma (Yim *et al.*, 1993). Additionally a cell-associated matrix is formed. These steps refine the cell-substratum interaction rendering it cation-independent; they also stabilize the polarized phenotype. A polarized secretory apparatus shaped, presumably, through involvement of the cytoskeleton and formation of targeting patches, directs the secretion of chondroitin sulfate proteoglycans, keratan sulfate proteoglycans and sulfated glycoproteins to the medium (Szuchet and Yim, 1990) while delivering HSPGs beneath the cells. All this takes place prior to structural polarization, i.e., the extension of cellular processes. This molecular surface asymmetry persists and coexists with the morphological polarity (Yim *et al.*, 1993).

The aforementioned observations originated from an *in vitro* system consisting of a single cell type. It is important to ask: What is their physiological relevance? I believe that the findings have physiological significance for the following reasons. First, the morphology that OLGs acquire in these cultures, and their cell to cell alignment with the formation of tight junctions (Massa *et al.*, 1984) resemble the structural organization seen *in situ* (Massa and Mugnaini, 1982; Szuchet, 1995). The importance of providing *in vitro* an environment that will permit cells to maintain a physiologically relevant morphology has been elegantly illustrated for mammary cells; only when this is the case, will mammary cells express the full repertoire of genes characteristic of differentiated cells (Boudreau *et al.*, 1995; Lin and Bissell, 1993). I postulate that the same holds for OLGs. Second, the finding of a differentiation-linked polarization is in line with the knowledge that, both morphologically and biochemically, OLGs are highly polarized. For example, biochemically, this polarization is manifested by the segregation of mRNA to cellular processes, i.e., the sites of myelin assembly (Amur-Umarjee *et al.*, 1990; Colman *et al.*, 1982); it is also manifested by the asymmetric distribution of HSPGs. Third, it is well accepted that for myelination in the peripheral nervous system (PNS) to occur, Schwann cells have to be

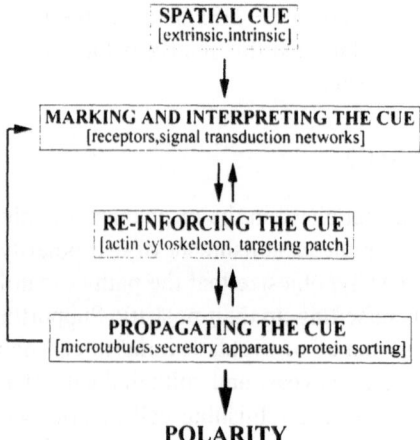

Figure 1. Diagram of events leading to cell polarity. (Based on Fig. 1 from Drubin and Nelson, 1996).

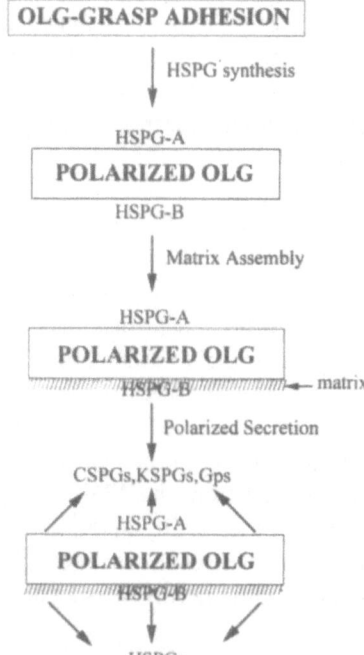

Figure 2. Cartoon depicting sequence of early steps postulated to be involved in the establishment of a myelinogenic phenotype. Abbreviations: OLG, oligodendrocyte; HSPG-A and B, heparan sulfate proteoglycans of class A and B, respectively; CSPGs and KSPGs, chondroitin sulfate and keratan sulfate proteoglycans, respectively; Gps, glycoproteins.

polarized (Bunge *et al.*, 1986; Eldridge *et al.*, 1989; Trapp *et al.*, 1995). These arguments, taken together with our previous demonstration that OLGs in these cultures reenact the ontogenic development of myelin, reinforce my contention that this is a bona fide model system to investigate the mechanism of myelinogenesis.

Identification of OLG HSPG Core Proteins

We have shown that the synthesis of surface HSPGs is modulated by the adhesion of OLGs to GRASP, an event that marks the differentiation of these cells (Yim *et al.*, 1993; Yim *et al.*, 1986). OLG surface proteoglycans (PGs) belong to the heparan sulfate class. Operationally, we have grouped these HSPGs into two classes: 1) those that are transmembrane or belong to a matrix, and require detergent or chaotropic agents for their solubilization [e.g., 4M guanidine.HCl (Gu.HCl)-0.5M NaCl-0.1% Triton X-100], we refer to them as Gu.-extract; and 2) those that are peripherally associated with the plasma membrane and can be displaced by solvents of high ionic strength or by solutions containing 100µg/ml of heparin. The latter class of HSPGs can further be subdivided into two types: type A, removed while the cells remain adhered; and type B, solubilized after OLGs have been detached. No HSPGs were isolated when 1% Triton X-100 at physiological ionic strength was employed. Such behavior would be expected from HSPGs that span the plasma membrane and are linked to the cytoskeleton; this was shown to be the case for a HSPG from Schwann cells (Carey and Todd, 1986).

We have treated highly purified fractions of OLG HSPGs with heparitinase, the enzyme that degrades heparan sulfate, to obtain the size of the respective core proteins (Table 1). Of the two fractions associated peripherally with the plasma membrane (class A and B), we have only identified core proteins for class A (class B identification is in progress). Five core proteins were identified in fraction A with molecular masses ranging from 37 kD to 121 kD. Given the spread in size of the original fraction (80 kD to 250 kD), some of these core proteins must have a high content of glycosaminoglycans (GAGs). It is not presently clear whether these components are products of independent genes or are spliced products of a single gene. According to current notions, this type of HSPGs interacts with surface receptors through their GAG chains (Kelly et al., 1987; Kjellén et al., 1980; Kjellén et al., 1981; Kraemer, 1977; Ruoslahti, 1988). In the specific case of OLGs, these receptors must be confined to the upper surface of the cell (in contact with the medium) because the HSPGs were isolated from adhered cells. Their specific roles have not been established; for this we must first characterize each component at the molecular level. Nevertheless, we may speculate that they function in a manner similar to HSPGs on other cell types. Such roles may include acting as co-receptors for growth factors, being involved in signal transduction, and overall control of the microenvironment of the cell.

A highly purified fraction of the Gu-extract resolves on SDS-PAGE into a 476 kD band and a very broad smear-like band from 110 kD to 220 kD. Seven core proteins were identified in the Gu-extract after heparitinase treatment (Table 1). Three core proteins were derived from the 476K-band; these have M_r of 396K, 263K and 247K. The other four have M_r of 117K, 107K, 98K, and 69K (Szuchet et al., 1996) and arise from the very broad band. The Gu-extract contains also glycoproteins that are not PGs; the fact that they co-purify with the HSPGs is a good indication that they are strongly associated with them.

Assembly of a Cell-Associated Matrix

We wanted to define the nature of the interaction between the material extracted with 4M Gu.HCl-0.5M NaCl-0.1% Triton X-100 (Gu-extract) and the plasma membrane. The solubility properties exhibited by the Gu-extract would be expected from a transmembrane protein that is also linked to the cytoskeleton or from proteins strongly meshed in a cell-associated matrix. To discriminate between these two possibilities, we employed Triton X-114 which, upon a change in temperature, will separate into detergent-rich and water-rich phases, thereby fractionating molecules according to their solubility properties. Transmembrane HSPGs should partition into the detergent phase, whereas HSPGs from a cell-associated matrix should be found in the water layer. When this treatment was applied to the Gu-extract the majority of HSPGs partitioned into the aqueous phase; only a small fraction was detergent soluble (Szuchet et al., 1996). This experiment provided a clear demonstration that upon differentiation (regeneration), OLGs amass a matrix that is cell-bound. Two observations point to a strong interaction between the components of the matrix and the surface of OLGs: 1) the matrix is not extracted with Triton X-100 but pellets with the cells; 2) the need of a chaotropic solvent to dissociate these components from the cell.

Table 1. Identification of oligodendcrocyte heparan sulfate proteoglycans

Fraction	Molecular mass in kD	
	PGs	Core proteins
HSPG-A	80-250	121, 72, 62, 49, 37
HSPG-B	?	250?
Gu.HCl Extract	476	396, 263, 247
	110-220	117, 107, 98, 69

Abbreviations: HSPG, heparan sulfate proteoglycan; PG, proteoglycan

From the findings that freshly isolated OLGs or OLGs maintained in culture under conditions that do not foster adhesion have no matrix (or matrix components are expressed at basal levels); and that after 24h of OLG adhesion to substratum, there is a surge in the synthesis of matrix components that is maintained while the cells are getting poised for myelination, I postulate that OLGs assemble a developmentally regulated matrix with a window of expression that peaks prior to and during myelination. This matrix is kept at a maintenance level, once myelination is complete. Alternatively, one can argue that the severing of the connectivity between OLGs and the myelin sheath during the process of cell isolation is responsible for shutting-off the synthesis of matrix components. While I have no strong arguments to decide between these two models, it is clear that the initiation process requires a signal and this signal is one that also marks differentiation/regeneration. I further postulate that this matrix has polarity; that it is confined to the surface in contact with substratum and that it serves to organize and orient OLG signaling machinery for the commencement of myelination. In support of these conjectures is the report by Kidd et al. (1996) that axons regulate the distribution of Schwann cell microtubules. Future research should permit a distinction between these possibilities.

PARALLEL OR COMPLEMENTARY PATHWAYS?

It is well established that cells have defensive mechanisms whereby they respond to a stressful environment (e.g., excess heat, toxic chemicals, or others) by directing the synthesis and activating protective proteins (Frydman and Hartl, 1996; Hayes and Dice, 1996; Morimoto, 1993; Welch, 1993). Most of these proteins are present constitutively in the cell and have physiological functions beyond those related to the stress response. In the course of defining OLG genes that participate in the induction of a myelinogenic phenotype, we have identified a couple of genes that encode proteins that could be categorized as "stress proteins". A case in point is heavy (H)-chain ferritin that is transcribed both by adhesion and by the action of a cytokine (see below). The issue is whether this is part of a physiological response to the effort of growth and differentiation or it is a reaction to a pathological insult (e.g., the severance from the axon and the destruction of myelin during OLG isolation). I would argue that this is a normal physiological response to the effort of myelination because these proteins are only transcribed after the process of regeneration/differentiation has been initiated, and there is no enhancement of expression in OLGs maintained in the non-adhered state.

Identification of Adhesion-Induced Genes by the Differential Display Technique

Having established that upon substratum-adhesion, OLGs regenerate and reenact the ontogenic development of myelin (Vartanian et al., 1988; Vartanian et al., 1986; Yim et al., 1986), we used the differential display method (Liang and Pardee, 1992, see also Liang et al., 1993; Welsh et al., 1992) to highlight those genes that are activated or deactivated upon adhesion, i.e., the determinants of OLG regeneration/differentiation. One of the genes identified is H-chain ferritin (Fig. 3) (Sanyal et al., 1996). This finding is in agreement with reports of tissue- and cell-specific regulation of the H-chain ferritin gene transcription. Thus in selected cell types, this transcription is iron-independent and is modulated by: a) terminal differentiation (Chou et al., 1986; Liau et al., 1991); b) hormones (Chazenbalk et al., 1990a; Chazenbalk et al., 1990b; Ursini and de Franciscis,

1988); c) cytokines (Miller *et al.*, 1991; Torti *et al.*, 1988) and d) adhesion (Sanyal *et al.*, 1996). To the best of our knowledge, this is the first demonstration of adhesion-induced expression of the H-chain ferritin gene. An ECM-controlled element is operative in the expression of the β-casein gene (Lin and Bissell, 1993).

OLG-substratum adhesion marks the beginning of a myelinogenic metabolism with its concomitant enhanced synthesis of lipids and cholesterol required for the assembly of myelin membranes (Szuchet *et al.*, 1983; Yim *et al.*, 1986). Iron is an indispensable constituent for this onset; it is also an essential component for oxidative metabolism. OLGs have the highest rate of oxidative metabolic activity in the brain (Connor and Menzies, 1996). Fitting these characteristics is the observation that brain iron is localized predominantly in OLGs (Blissman *et al.*, 1996).

The transcription of the H-chain ferritin gene is associated with tissue stress such as might occur during rapid cell growth. The fact that this transcription takes place upon OLG-substratum adhesion can be viewed as another manifestation of the anchorage-induced signal that drives OLGs toward a differentiation program. This entails the synthesis of all the necessary components for the assembly of large quantities of membranes, hence, constituting a period of rapid growth and considerable cell stress.

Role of Tumor Necrosis Factor-α in Gene Expression

Tumor necrosis factor-α (TNF-α) is a polypeptide with a molecular mass of 17 kD that belongs to the cytokine family of factors; it is synthesized by a variety of cells and is expressed in two forms: secreted and transmembrane. By far the action of the secreted protein is better known than that of the transmembrane. Cells have two receptors for TNF-α; they belong to a highly conserved family that includes the NGF receptor (Smith *et al.*, 1994). A tempting speculation is that the pleiotropic action of TNF-α is displayed by a selective use of either one receptor or the other; fitting this viewpoint is the observation that the "death domain" is present in only one of them (Bazzoni and Beutler, 1995). Nonetheless, no solid evidence is available to support such assumption. TNF-α is known for its role in inflammatory processes, in microbial infections and neoplastic diseases. However, there are other features to its function that are more related to cell growth and differentiation than to disease. What the particular action of TNF-α might be may largely depend on the cell type and the specific circumstances (Benveniste and Benos, 1995; Hopkins and Rothwell, 1995; Ihle, 1995; Rothwell and Hopkins, 1995). For example, chronic application of TNF-α produces deleterious effects on 3T3-L1 adipocytes, whereas the same con-

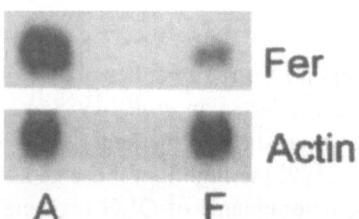

Figure 3. Role of OLG-substratum interaction on the expression of H-chain ferritin. Northern blot analysis. Total RNA isolated from substratum-adhered (A) and non-adhered (F) OLGs were hybridized with OLG cDNA probes from H-chain ferritin and β-actin. Noteworthy is the increase in H-chain ferritin mRNA upon adhesion (cf. lanes A with F in top panel). This adhesion-induced transcription is iron-independent. (Reproduced from Fig. 4, Sanyal et al., 1996, with permission.)

centration, applied acutely, switches-on downstream elements of signal transduction pathways (Guo and Donner, 1996). We made similar observations on cultured OLGs.

TNF-α exerts a broad spectrum of effects on OLGs, (reviewed by (Soliven and Szuchet, 1995). Long term exposure of OLGs to 2.8–28 nM of TNF-α results in process retraction, membrane depolarization and inhibition of MBP and CNPase phosphorylation (Soliven et al., 1991; 1994). These effects can be abrogated by addition of anti-TNF-α antibodies or by prior absorption of TNF-α with the antibodies (Soliven and Szuchet, 1995). We observed neither apoptosis nor cell necrosis. In this context our findings are at odds with those of Selmaj and Raine (1988) who observed OLG death and demyelination in mixed cultures.

TNF-α has been shown to be capable of activating specific genes. The mechanism is complex and a number of transcription factors are involved (reviewed by Collins et al., 1995). One gene that is activated by TNF-α is H-chain ferritin in both adipocytes and human muscle cells; the mechanism has not been elucidated (Miller et al., 1991; Torti et al., 1988). Since we have shown that H-chain ferritin is induced by the adhesion of OLGs to a substratum (Sanyal et al., 1996), it was of interest to examine whether TNF-α had also an effect on the transcription of this gene in OLGs. Two sets of OLG cultures, one adhered to a substratum, the other kept non-adhered (Szuchet and Yim, 1984), were exposed to 10 ng/ml-25 ng/ml of (recombinant human) TNF-α for 18h to 24h. Total RNA was then isolated and probed with a fragment of ferritin cDNA. Interestingly, TNF-α had no effect on the adhered OLGs, but increased 2- to 3-fold the transcription of H-chain ferritin in the non-adhered OLGs. Interleukin 1β was ineffectual. Thus, functionally TNF-α initiated a signal that mimicked adhesion (Sanyal and Szuchet, manuscript in preparation). These results have significant implications for they suggest that common or parallel pathways are operative in the induction of H-chain ferritin. Presumably, the action of TNF-α is effected through its receptor; preliminary experiments indicate that GRASP (Schirmer et al., 1994), binds to an integrin-like receptor. Two alternative mechanisms may be suggested. One, the adhesion- and the TNF-α-induced signals converge to the same pathway. Two, each of them activates different DNA-binding proteins that promote the activation of the H-chain ferritin gene. Research in progress should clarify this issue.

Do Stress-Related Proteins Play a Part in Oligodendrocyte Regeneration?

As part of our search for adhesion-induced genes, we have identified and sequenced a 1600 bp cDNA clone that has high similarity with ovine, human, porcine, and murine α-interferon (INF)- and β-INF-regulated genes (Aebi et al., 1989; Charleston and Stewart, 1993; Horisberger, 1992; Horisberger et al., 1990). Because we are missing sequence at either end of the molecule, it is not certain whether the isolated clone is homologous to the interferon-regulated gene (Szuchet and co-workers; research in progress). However, what seems significant is the fact that the three consensus sequence elements with distinct spacing that defines the GTP-binding domain (Dever et al., 1987) are present in our clone. A GTP-binding domain has been identified in the human interferon-regulated protein (MxA) and shown to have GTPase activity (Horisberger, 1992). The Mx proteins constitute a widely distributed superfamily with representation in yeast, insects and mammals. It has been suggested that they play important roles in protein trafficking and have many characteristic in common with stress proteins (Horisberger, 1992). Should our future research corroborate that adhesion-induced regeneration of OLGs involves the action of stress-related proteins such as ferritin, the 1600bp cDNA and perhaps others (e.g., heat shock proteins), it would be important to corroborate my hypothesis that this is part of the physiological process of myelination.

CONCLUSIONS AND PERSPECTIVES

We have developed a simple and distinct model system where we can turn on or off, at will, the differentiation of OLGs by changing their state of attachment to a selected substratum. This model is as relevant to OLG development as it pertains to OLG regeneration after injury. In this report I have described the early events that characterize the initiation of a myelinogenic phenotype. It is apparent that an orderly sequence of steps ensues after the interaction of an OLG surface receptor (possibly an integrin-like molecule) with GRASP. These steps comprise: 1) synthesis, vectorial transport, and asymmetric association with the plasmalemma of HSPGs that set up a polarized phenotype; 2) a concomitant or sequential assembly of a developmentally regulated cell-associated matrix that has a window of expression peaking during myelination. High molecular weight HSPGs are significant constituents of this matrix; 3) cementing OLG-substratum interaction by rendering it cation-independent; and 4) organization of a polarized secretory machinery that delivers chondroitin sulfate and keratan sulfate proteoglycans plus glycoproteins to the medium while depositing HSPGs beneath the cells. All these events take place before any morphological polarity is visible.

To the best of my knowledge this is the first study of biochemical processes preceding the involvement of myelin specific proteins and/or lipids. It is also the first demonstration that OLGs amass a matrix and the participation of HSPGs in it. These findings have significant implications. First they reaffirm the usefulness of our model system for deciphering the signaling pathways that ultimately lead to myelin formation. Second, they provide a new set of landmarks to search for under both normal and pathological conditions. Third, they alert to the fact that breakdown of myelin and/or failure of remyelination under pathological conditions, e.g., multiple sclerosis, may have its origins either in the absence of a required signal or in a diseased OLG. I believe that this model system of cultured postmyelinating OLGs represents a bona fide tool that can provide invaluable information on the signals that tell OLGs to commence making myelin.

ACKNOWLEDGMENTS

I would like to thank present and past members from my laboratory for their collaboration. I am most grateful to Mr. Thomas Comery for his thorough reading of the manuscript and for invaluable suggestions. I am also thankful to Mr. Paul Polak for his help with the figures. This work was supported by grant # RG-2677A-7/1 from the National Multiple Sclerosis Society. I owe gratitude to Ms. Phillis Mandler and Mr. Gary Elden for their generous gift.

REFERENCES

Aebi, M., Fäh, J., Hurt, N., Samuel, C., Thomis, D., Bazzigher, L., Pavlovic, J., Haller, O., and Staeheli, P. (1989). cDNA structures and regulation of two interferon-induced human Mx proteins. Molecular and Cellular Biology 9:5062–5072.
Amur-Umarjee, S. G., Hall, L., and Campagnoni, A. T. (1990). Spatial distribution of mRNAs for myelin proteins in primary cultures of mouse brain. Developmental Neuroscience 12:263–272.
Arvanitis, D., Dumas, M., and Szuchet, S. (1992a). Myelin palingenesis: 2. Immunocytochemical localization of myelin/oligodendrocyte glycolipids in multilamellar structures. Developmental Neuroscience 14:328–335.

Arvanitis, D., Polak, P. E., and Szuchet, S. (1992b). Myelin palingenesis: 1. Electron microscopical localization of myelin/oligodendrocyte proteins in multilamellar structures by the immunogold method. Developmental Neuroscience 14:313–327.

Bazzoni, F., and Beutler, B. (1995). How do tumor necrosis factor receptors work? Journal of Inflammation 45:221–238.

Benveniste, E., and Benos, D. (1995). TNF-α- and IFN-γ-mediated signal transduction pathways: effects on glial cell gene expression and function. FASEB J. 9:1577–1584.

Blakemore, W., Olby, N., and Franklin, R. (1995). The use of transplanted glial cells to reconstruct glial environments in the CNS. Brain Pathology 5:443–450.

Blakemore, W. F. (1984). The response of oligodendrocytes to chemical injury. Acta Neurologica Scandinavica 70 (suppl 100):33–38.

Blissman, G., Menzies, S., Beard, J., Palmer, C., and Connor, J. (1996). The expression of ferritin subunits and iron in oligodendrocytes in neonatal porcine brains. Developmental Neuroscience 18: 274–281.

Boudreau, N., Myers, C., and Bissell, M. J. (1995). From laminin to lamin: regulation of tissue-specific gene expression by the ECM. Trends In Cell Biology 5:1–4.

Bunge, R. P., Bunge, M. B., and Eldridge, C. F. (1986). Linkage between axonal ensheathment and basal lamina production by Schwann cells. Annual Review of Neuroscience 9:305–328.

Carey, D. J., and Todd, M. S. (1986). A cytoskeleton-associated plasma membrane heparan sulfate proteoglycan in Schwann cells. Journal of Biological Chemistry 261:7518–7525.

Charleston, B., and Stewart, H. (1993). An interferon-induced Mx protein: cDNA sequence and high-level expression in the endometrium of pregnant sheep. Gene 137:327–331.

Chazenbalk, G. D., Wadsworth, H. L., Foti, D., and Rapoport, B. (1990a). Thyrotropin and adenosine 3',5'-monophosphate stimulate the activity of the ferritin-H promoter. Molecular Endocrinology 4:1117–1124.

Chazenbalk, G. D., Wadsworth, H. L., and Rapoport, B. (1990b). Transcriptional regulation of ferritin H messenger RNA levels in FRTL5 rat thyroid cells by thyrotropin. Journal of Biological Chemistry 265:666–667.

Chou, C. C., Gatti, R. A., Fuller, M. L., Concannon, P., Wong, A., Chada, S., Davis, R. C., and Salser, W. A. (1986). Structure and expression of ferritin genes in a human promyelocytic cell line that differentiates in vitro,. Molecular and Cellular Biology 6:566–573.

Collins, T., Read, M. A., Neish, A. S., Whitley, M. Z., Thanos, D., and Maniatis, T. (1995). Transcriptional regulation of endothelial cell adhesion molecules: NF-κB and cytokine-inducible enhancers. The FASEB Journal 9:899–909.

Colman, D. R., Kreibich, G., Frey, A. B., and Sabatini, D. D. (1982). Synthesis and incorporation of myelin polypeptide into CNS myelin. Journal of Cell Biology 95:598–608.

Connor, J., and Menzies, S. (1996). Relationship between iron, oligodendrocytes and myelination. Glia 17:83–93.

Delcommenne, M., and Streuli, C. (1995). Control of integrin expression by extracellular matrix. The Jr of Biological Chemistry 270:26794–26801.

Dever, T., Glynias, M., and Merrick, W. (1987). GTP-binding domain: Three consensus sequence elements with distinct spacing. Proc. Natl. Acad. Sci. USA 84:1814–1818.

Drubin, D., and Nelson, W. J. (1996). Origins of cell polarity. Cell 84:335–344.

Dubois-Dalcq, M. (1995). Regeneration of oligodendrocytes and myelin. Trends in Neurosciences 18:289–291.

Eldridge, C. F., Bunge, M. B., and Bunge, R. P. (1989). Differentiation of axon-related Schwann cells in vitro: II. Control of myelin formation by basal lamina. Journal of Neuroscience 9:625–638.

Franklin, R., Bayley, S., and Blakemore, W. (1996). Transplanted CG4 cells (an oligodendrocyte progenitor cell line) survive, migrate, and contribute to repair of areas of demyelination in x-irradiated and damaged spinal cord but not in normal spinal cord. Experimental Neurology 137:263–276.

Franklin, R. J. M., and Blakemore, W. F. (1995). Glial cell transplantation and plasticity in the O-2A lineage: implications for CNS repair. Trends in Neurosciences 18:151–156.

Frydman, J., and Hartl, F. (1996). Principles of chaperone-assisted protein folding: differences between in vitro and in vivo mechanisms. Science 272:1497–1502.

Guo, D., and Donner, D. (1996). Tumor necrosis factor promotes phosphorylation and binding of insulin receptor substrate 1 to phosphatidylinositol 3-kinase in 3T3-L1 adipocytes. The Journal of Biological Chemistry 271:615–618.

Hayes, S., and Dice, J. (1996). Roles of molecular chaperones in protein degradation. The Journal of Cell Biology 132:255–258.

Hopkins, S. J., and Rothwell, N. J. (1995). Cytokines and the nervous system I: expression and recognition. Trends in Neurosciences 18:83–88.

Horisberger, M. (1992). Interferon-induced human protein MxA is a GTPase which binds transiently to cellular proteins. Journal of Virology 66:4705–4709.

Horisberger, M., McMaster, G., Zeller, H., Wathelet, M., Dellis, J., and Content, J. (1990). Cloning and sequence analyses of cDNAs for interferon- and virus-induced human Mx proteins reveal that they contain putative guanine nucleotide-binding sites: Functional study of the corresponding gene promoter. Journal of Virology 64:1171–1181.

Ihle, J. N. (1995). Cytokine receptor signalling. Nature 377:591–594.

Johnson, E. S., and Ludwin, S. K. (1981). The demonstration of recurrent demyelination and remyelination of axons in the central nervous system. Acta Neuropathologica 53:93–98.

Juliano, R. L., and Haskill, S. (1993). Signal transduction from the extracellular matrix. Journal of Cell Biology 120:577–585.

Kelly, R. B., Carlson, S. S., and Caroni, P. (1987). Extracellular matrix components of the synapse. In "Biology of Proteoglycans" (T. N. Wight and R. P. Mecham, Eds.), pp. 247–265. Academic Press, Inc., Orlando, FL.

Kidd, G., Andrews, S. B., and Trapp, B. (1996). Axons regulate the distribution of Schwann cell microtubules. The Journal of Neuroscience 16:946–954.

Kjellén, L., Oldberg, A., and Höök, M. (1980). Cell-surface heparan sulfate: mechanisms of proteoglycan-cell association. Journal of Biological Chemistry 255:10407–10413.

Kjellén, L., Pettersson, I., and Höök, M. (1981). Cell-surface heparan sulfate: an intercalated membrane proteoglycan. Proceedings of the National Academy of Sciences USA 78:5371–5375.

Kraemer, P. M. (1977). Heparin releases heparan sulfate from the cell surface. Biochemical and Biophysical Research Communications 78:1334–1340.

Liang, P., Averboukh, L., and Pardee, A. B. (1993). Distribution and cloning of eukaryotic mRNAs by means of differential display: refinements and optimization. Nucleic Acids Research 21:3269–3275.

Liang, P., and Pardee, A. B. (1992). Differential display of eukaryotic messenger RNA by means of the polymerase chain reaction. Science 257:967–971.

Liau, G., Chan, L. M., and Feng, P. (1991). Increased ferritin gene expression is both promoted by cAMP and a marker of growth arrest in rabbit vascular smooth muscle cells. Journal of Biological Chemistry 266:18819–18826.

Lin, C. Q., and Bissell, M. J. (1993). Multi-faceted regulation of cell differentiation by extracellular matrix. FASEB Journal 7:737–743.

Ludwin, S. K. (1988). Remyelination in the central nervous system and the peripheral nervous system. Advances in Neurology 47:215–254.

Ludwin, S. K., and Szuchet, S. (1993). Myelination by mature ovine oligodendrocytes in vivo and in vitro: evidence that different steps in the myelination process are independently controlled. Glia 8:219–231.

Massa, P. T., and Mugnaini, E. (1982). Cell junctions and intramembrane particles of astrocytes and oligodendrocytes: a free-fracture study. Neuroscience 7:523–538.

Massa, P. T., Szuchet, S., and Mugnaini, E. (1984). Cell-cell interactions of isolated and cultured oligodendrocytes: formation of linear occluding junctions and expression of peculiar intramembrane particles. Journal of Neuroscience 4:3128–3139.

Mercurio, A. (1995). Laminin receptors: achieving specificity through cooperation. Trends in Cell Biology 5:419–423.

Meredith, J. E., Fazeli, B., and Schwartz, M. A. (1993). The extracellular matrix as a cell survival factor. Molecular Biology of the Cell 4:953–961.

Miller, L. L., Miller, S. C., Torti, S. V., Tsuji, Y., and Torti, F. M. (1991). Iron-independent induction of ferritin H chain by tumor necrosis factor. Proceedings of the National Academy of Sciences USA 88:4946–4950.

Morimoto, R. I. (1993). Cells in stress: transcriptional activation of heat shock genes. Science 259:1409–1410.

Roskelley, C. D., Srebrow, A., and Bissell, M. J. (1995). A hierarchy of ECM-mediated signalling regulates tissue-specific gene expression. Current Opinion in Cell Biology 7:736–747.

Rothwell, N. J., and Hopkins, S. J. (1995). Cytokines and the nervous system II: actions and mechanisms of action. Trends in Neurosciences 18:130–136.

Ruoslahti, E. (1988). Structure and biology of proteoglycans. Annual Review of Cell Biology 4:229–255.

Ruoslahti, E. (1996). Integrin signaling and matrix assembly. Tumor Biology 17:117–124.

Sanyal, B., Polak, P., and Szuchet, S. (1996). Differential expression of the heavy-chain ferritin gene in non-adhered and adhered oligodendrocytes. J Neuroscience Research 46:187–197.

Schirmer, E. C., Farooqui, J., Polak, P. E., and Szuchet, S. (1994). GRASP: A novel heparin-binding serum glycoprotein that mediates oligodendrocyte-substratum adhesion. Journal of Neuroscience Research 39:457–473.

Scolding, N., and Lassmann, H. (1996). Demyelination and remyelination. Trends in Neurosciences 19:1–2.

Selmaj, K. W., and Raine, C. S. (1988). Tumor necrosis factor mediates myelin and oligodendrocyte damage in vitro. Annals of Neurology 23:339–346.

Smith, C., Farrah, T., and Goodwin, R. (1994). The TNF receptor superfamily of cellular and viral proteins: activation, costimulation, and death. Cell 76:959–962.

Soliven, B., and Szuchet, S. (1995). Signal transduction pathways in oligodendrocytes: role of tumor necrosis factor-α. Intl J of Devel Neuroscience 13:351–367.

Soliven, B., Szuchet, S., Arnason, B. G. W., and Nelson, D. J. (1988a). Voltage-gated potassium currents in cultured ovine oligodendrocytes. Journal of Neuroscience 8:2131–2141.

Soliven, B., Szuchet, S., Arnason, B. G. W., and Nelson, D. J. (1988b). Forskolin and phorbol esters modulate the same K^+ conductance in cultured oligodendrocytes. Journal of Membrane Biology 105:177–186.

Soliven, B., Szuchet, S., and Nelson, D. J. (1991). Tumor necrosis factor inhibits K^+ current expression in cultured oligodendrocytes. Journal of Membrane Biology 124:127–137.

Soliven, B., Takeda, M., and Szuchet, S. (1994). Depolarizing agents and TNF-α inhibit MBP and CNPase phosphorylation. Journal of Neuroscience Research 38:91–100.

Szuchet, S. (1987). Myelin palingenesis: the reformation of myelin by mature oligodendrocytes in the absence of neurons. In "Glial-Neuronal Communication in Development and Regeneration" (H. H. Althaus and W. Seifert, Eds.), Vol. 2, pp. 755–777. Springer-Verlag, Berlin.

Szuchet, S. (1995). The morphology and ultrastructure of oligodendrocytes and their functional implications. In "Neuroglia" (H. Kettenmann and B. Ranson, Eds.), pp. 23–43. Oxford, New York.

Szuchet, S., Polak, P., Watanabe, K., and Yamaguchi, Y. (1996). Regeneration/differentiation of oligodendrocytes entails the assembly of a pericellular matrix. In Preparation .

Szuchet, S., Polak, P. E., and Yim, S. H. (1986). Mature oligodendrocytes cultured in the absence of neurons recapitulate the ontogenic development of myelin. Developmental Neuroscience 8:208–221.

Szuchet, S., and Yim, S. H. (1984). Characterization of a subset of oligodendrocytes separated on the basis of selective adherence properties. Journal of Neuroscience Research 11:131–144.

Szuchet, S., and Yim, S. H. (1990). Oligodendrocyte-substratum interaction signals cell polarization and secretion. In "Cellular and Molecular Biology of Myelination" (G. Jeserich, H. H. Althaus, and T. V. Waehneldt, Eds.), Vol. H 43, pp. 231–246. Springer-Verlag, Berlin.

Szuchet, S., Yim, S. H., and Monsma, S. (1983). Lipid metabolism of isolated oligodendrocytes maintained in long-term culture mimics events associated with myelinogenesis. Proceedings of the National Academy of Sciences USA 80:7019–7023.

Torti, S. V., Kwak, E. L., Miller, S. C., Miller, L. L., Ringold, G. M., Myambo, K. B., Young, A. P., and Torti, F. M. (1988). The molecular cloning and characterization of murine ferritin heavy chain, a tumor necrosis factor-inducible gene. Journal of Biological Chemistry 263:12638–12644.

Trapp, B. D., Kidd, G. J., Hauer, P., Mulrenin, E., Haney, C. A., and Andrews, S. B. (1995). Polarization of myelinating schwann cell surface membranes: role of microtubules and the trans-golgi network. Journal of Neuroscience 15:1797–1807.

Ursini, M. V., and de Franciscis, V. (1988). TSH regulation of ferritin H chain messenger RNA levels in the rat thyroids. Biochemical and Biophysical Research Communications 150:287–295.

Vartanian, T., Sprinkle, T. S., Dawson, G., and Szuchet, S. (1988). Oligodendrocyte-substratum adhesion modulates expression of adenylate cyclase linked receptors. Proceedings of the National Academy of Sciences USA 85:939–943.

Vartanian, T., Szuchet, S., and Dawson, G. (1992). Oligodendrocyte-substratum adhesion activates the synthesis of specific lipid species involved in cell signalling. Journal of Neuroscience Research 32:69–78.

Vartanian, T., Szuchet, S., Dawson, G., and Campagnoni, A. T. (1986). Oligodendrocyte adhesion activates protein kinase C-mediated phosphorylation of myelin basic protein. Science 234:1395–1397.

Venstrom, K. A., and Reichardt, L. F. (1993). Extracellular matrix 2: Role of extracellular matrix molecules and their receptors in the nervous system. FASEB Journal 7:996–1003.

Welch, W. J. (1993). How cells respond to stress. In "Scientific American", Vol. 268, pp. 56–64.

Welsh, J., Chada, K., Dalal, S. S., Cheng, R., Ralph, D., and McClelland, M. (1992). Arbitrarily primed PCR fingerprinting of RNA. Nucleic Acids Research 20:4965–4970.

Wood, P. M., and Bunge, R. P. (1991). The origin of remyelinating cells in the adult central nervous system: the role of the mature oligodendrocyte. Glia 4:225–232.

Yamada, K., and Miyamoto, S. (1995). Integrin transmembrane signaling and cytoskeletal control. Current Opinion in Cell Biology 7:681–689.

Yim, S. H., Sherin, J. E., and Szuchet, S. (1993). Oligodendrocyte proteoglycans: modulation by cell-substratum adhesion. Journal of Neuroscience Research 34:401–413.

Yim, S. H., Szuchet, S., and Polak, P. E. (1986). Cultured oligodendrocytes: a role for cell-substratum interaction in phenotypic expression. Journal of Biological Chemistry 261:11808–11815.

20

A PROCEDURE FOR ISOLATING SCHWANN CELLS DEVELOPED FOR ANALYSIS OF THE MOUSE EMBRYONIC LETHAL MUTATION *NF1*

Haesun A. Kim and Nancy Ratner

Department of Cell Biology, Neurobiology, and Anatomy
University of Cincinnati College of Medicine
Cincinnati, Ohio 45267–0521

INTRODUCTION

The study of Schwann cell proliferation and differentiation has been facilitated by the availability of cultured Schwann cells that faithfully mimic Schwann cell *in vivo* maturation, growth, and differentiation. Transgenic mouse models and naturally occurring mouse mutants serve as increasingly important tools for the study of Schwann cell biology. We have developed methods to purify Schwann cells from single embryonic day 12.5 (E12.5) mutant mouse embryos in order to define abnormalities caused by mutations at the type 1 neurofibromatosis (*NF1*) locus. This method can be used to study Schwann cells from any mutant mouse that survives until day 12 of embryonic life.

NF1 null Schwann cells provide a potential mouse model for Schwann cell tumor formation in the common inherited disease, neurofibromatosis type 1 (NF1) (Riccardi, 1991). Adult mice heterozygous for the *NF1* mutation do not show identified peripheral nerve abnormalities (Brannan et al., 1994; Jacks et al., 1994) and homozygous mutants die in utero between embryonic day 11–15, prior to formation of mature peripheral nerve, making it impossible to use peripheral nerves or neonatal DRG as sources for null Schwann cells. Homozygous *NF1* mutant embryos are obtained by mating heterozygous mice: genotypes of each embryo are not known at the time of dissection. Separate cultures, each derived from a single embryo, have to be established for purification of Schwann cells of identified genotype. Further, some types of experiments require highly purified Schwann cells. Our method generates homogeneous (> 99.5% pure) Schwann cells from single embryos.

Cell Biology and Pathology of Myelin, edited by Juurlink *et al.*
Plenum Press, New York, 1997

In vitro studies have used Schwann cells from rat, chicken, mouse, and human peripheral nerves or dorsal root ganglia (DRG) (Askanas et al., 1980; Bhattacharyya et al.,
1993; Cochran and Black, 1985; Kim et al., 1989; Kuhlengel et al., 1990; Levi et al.,
1994; Morrissey et al., 1991; Rutkowski et al., 1992; Wood, 1976; Wood and Bunge,
1975; Yong et al., 1988; Rutkowski et al., 1995). A recurrent problem in preparing
Schwann cell cultures is connective tissue contamination (fibroblastic cells) and overgrowth of Schwann cells by fibroblasts in long-term culture. Two methods (Brockes et al.,
1979; Wood and Bunge, 1975) were developed to obtain viable Schwann cells essentially
free of contaminating fibroblasts from rat peripheral nervous system, and most Schwann
cell purification methods now used are based on either of these two procedures. The original method developed by Wood and Bunge (1975) used dorsal root ganglia as a source of
neurons to expand Schwann cell numbers: our method uses features of this technique. A
dissociation procedure, developed by Brockes (Brockes et al., 1979), used newborn rat
sciatic nerve; fibroblasts were reduced by treating the primary cultures with antimitotic
agents, followed by complement-dependent immunocytolysis of residual fibroblasts by using anti-Thy-1. Purified Schwann cells can be propagated in subsequent culture in the
presence of growth factors. In recent years, availability of recombinant neuregulins, including glial growth factor (Lemke and Brockes, 1984; Levi et al., 1995; Marchionni et
al., 1993; Rutkowski et al., 1995) have enabled significant and selective expansion of
Schwann cell numbers. We tested combinations of these methods for use in embryonic
mouse Schwann cells.

Methods for preparing Schwann cells from normal or neurological mouse mutants
have been reported. Cornbrooks et al (1983) and Cochran and Black (1985) used prenatal
DRG and the explant method to generate enriched Schwann cell preparations in 4 weeks,
but did not report Schwann cell yields. Shine and Sidman (1984) used sciatic nerves from
1 day old mice as a source for Schwann cells but did not attempt purification of Schwann
cells. Krikorian et al (1982) and Manthorpe et al (1980) used differential adhesion to en-
rich for Schwann cells from normal mouse neonatal DRG; maximal purification of 97%
Schwann cells was achieved. Difficulties were reported in removing contaminating fibroblasts from mouse Schwann cell cultures due to limitations in utilizing antimitotics or
complement mediated lysis (Kalderon, 1984; Krikorian et al., 1982; Manthorpe et al.,
1980; Seilheimer and Schachner, 1987). Unlike rat Schwann cells, mouse Schwann cell
proliferation can be stimulated with serum (Komiyama and Suzuki, 1991; Krikorian et al.,
1982; Seilheimer and Schachner, 1987) so that attempts to selectively eliminate fibroblasts using antimitotic drugs also results in Schwann cell death. Kalderon (1984)
achieved 95% purity from new-born mouse sciatic nerve by complement-mediated lysis;
Seilheimer and Schachner (1987) reported 99.5% purity from new-born sciatic nerve by
combining complement mediated lysis using both anti Thy-1.1 and anti MESA-1. None of
these protocols began with early embryonic DRG and at the same time generated nearly
homogeneous Schwann cell cultures.

Here we describe a method to generate highly purified populations of mouse
Schwann cells from single E12.5 mouse embryos by a procedure that includes three steps:
i) culture of Schwann cell precursors derived from embryonic DRG in the presence of
neurons in serum-free medium on tissue culture plastic to selectively stimulate Schwann
cell maturation and proliferation; ii) mechanical separation of the Schwann cell-associated
neuronal network from underlying fibroblasts; iii) enzymatic dissociation of Schwann
cells from neurons and subsequent expansion in secondary culture in rhGGF2-supple-
mented serum-containing media. A schematic illustration of the procedure is shown in
Figure 1.

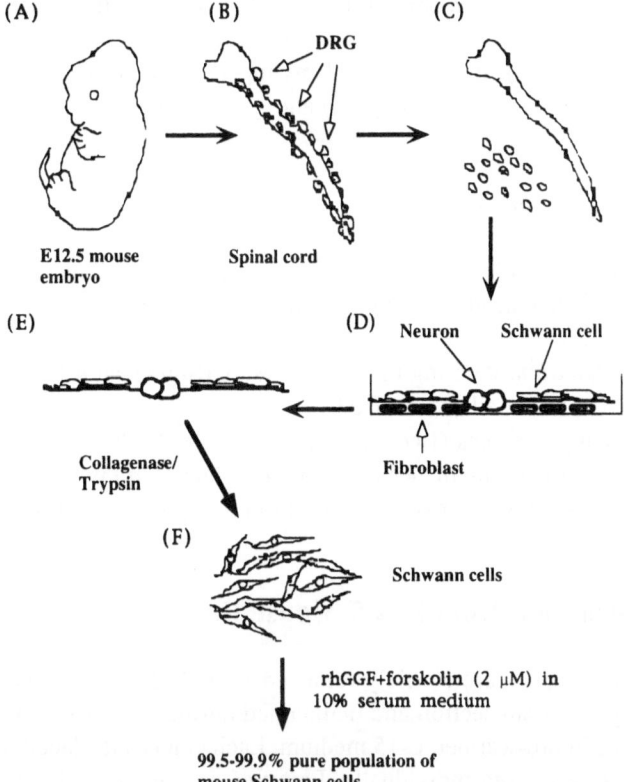

Figure 1. A schematic diagram of a procedure for isolating mouse Schwann cells from E12.5 embryo. (A)-(D): Dorsal root ganglia (DRG) are removed from a E12.5 mouse spinal cord and cells are enzymatically dissociated. The cell suspension, mostly composed of neurons, fibroblasts and Schwann cell precursors, are plated onto un-coated 6-well culture dish and maintained in a serum-free medium supplemented with NGF. (D)-(F) Schwann cells are allowed to expand until the neuritic networks are populated (6–7 days), then subsequently, Schwann cell-neuron networks are mechanically separated from fibroblasts by peeling them off from the dish using a needle. Next, the Schwann cells are enzymatically dissociated from the neurons and plated on a poly-L-lysine coated plate in DMEM+10% FBS containing rhGGF (10 ng/ml) and forskolin (2 μM). After 1 week in culture, > 99.5% of the rhGGF2 cells are positive for both S100 and NGFR.

MATERIALS AND METHODS

Solutions and Media[*]

10% Serum Medium. Dulbecco's modified Eagle's medium (DMEM) with high glucose (GIBCO cat # 11965–050) supplemented with 10 vol.% heat inactivated fetal bovine serum and penicillin/streptomycin (0.1 mg/ml)

C Medium. DMEM with high glucose (GIBCO) supplemented with 10 vol.% human placental serum (provided by local hospital delivery rooms) and nerve growth factor (NGF) (50 ng/ml) (Harlan cat # BT-5017).

---

[*] All media were pre-warmed to 37°C before use.

N2 Medium. 1:1 ratio of DMEM and F-12 supplemented with Na-selenite (5 ng/ml) (Sigma cat # S-526), putrescine (16 µg/ml) (Sigma cat # P-7505), progesterone (125 ng/ml) (Sigma cat # P-0130), transferrin (0.2 mg/ml) (Sigma cat # T-2252), insulin (0.4 µg/ml) (Sigma cat # I-5500), NGF (50 ng/ml) and gentamycin (2.5 µg/ml) (GIBCO cat # 15710–015)).

Leibowitz-15 (L-15) Medium (Gibco)

Trypsin-Collagenase. 0.05% trypsin (GIBCO) and 0.1% collagenase (Worthington Biochemical) in Hank's balanced salt solution (HBSS) (GIBCO).

Poly-L-Lysine Solution. 0.05 mg/ml poly-L-lysine (Sigma cat # P-7890) prepared in 0.15 M sodium borate pH 8.0, filter sterilized.

23_G1 sterile syringe needles (Becton Dickinson); 6-well plastic culture dishes (Falcon); 35 mm and 100 mm culture dishes (Falcon); 4 ml and 15 ml poly styrene tubes (Falcon); Stereoscopic dissecting microscope; Sterile cotton-plugged glass pasteur pipettes (Fischer)

Schwann Cell Isolation from E12.5 Mouse Embryos

Step 1. Mouse embryos are sterilely removed from 12.5 day pregnant female mice under anesthesia by Cesarean section and maintained during the dissection procedure under sterile conditions in pre-warmed L-15 medium. Each embryo is placed in a separate 35 mm culture dish and processed individually. Embryos are killed by decapitation and the head transferred to prelabeled 1.5 ml eppendorf tubes to be used for genotyping. The spinal cord of each embryo is removed and transferred to a new dish, and DRG are dissected from the vertebral column as described for rat DRG at a similar stage (Salzer et al., 1980). We have been unable to collect significant numbers of DRG from mouse spinal cords prior to day 12. The spinal cord is discarded and L-15 medium carefully removed using a glass pasteur pipette. DRG are enzymatically dissociated by adding 20–25 drops (about 1.5 ml) of 0.25% trypsin and incubated at 37°C for 40 min with gentle swirling (40 rpm) in an air incubator. It is important not to over-incubate cells in trypsin. Dissociated cells are transferred to 15 ml tubes containing 10 ml of 10% serum-medium, using a separate glass pasteur pipette for each embryo, and centrifuged for 5 min at 80xg. Pelleted cells are collected by submerging a pasteur pipette under the medium and lifting out the pellet. Cells are transferred to a 4 ml tube containing approximately 0.5 ml C medium. A single cell suspension is generated by triturating the pellet 20–25 times through a glass pipette, being careful to avoid generation of air bubbles. A new glass pipette is used for each embryo. Resuspended cells derived from a single embryo are plated in two wells of an uncoated 6-well plastic culture dish each containing 2 ml of C medium. This is done by dropping same number of cell suspension drops on each well. Culture dishes are then placed at 35°C in a 7.5% CO_2 humidified atmosphere. At this point, cultures are composed of neurons, fibroblasts and Schwann cell precursors. After 24 hours, serum-medium is removed and 2 ml of fresh serum-free N2 medium supplemented with NGF (50 ng/ml) and gentamycin (0.5 µg/ml) added; plates are transferred to 10% CO_2 at 37°C. Cells are observed frequently but the medium is not changed. It is important to use uncoated plastic dishes, as when embryonic cells are plated onto collagen-coated substrates, fibroblasts become bundled together with neurites and Schwann cells, making the subsequent Schwann cell separation difficult.

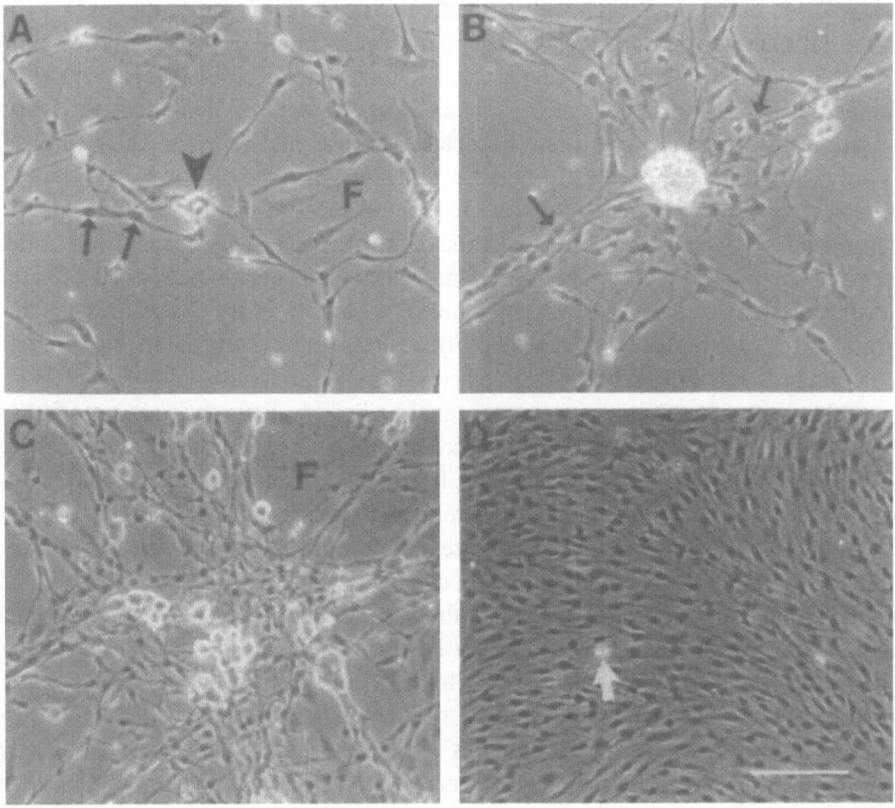

Figure 2. Phase-contrast micrographs of mouse dorsal root ganglion (A-C) and purified Schwann cell (D) cultures prepared from a E12.5 embryo. A: Embryonic mouse DRG cultures 48 hours *in vitro*. Dissociated mouse DRG were plated onto an uncoated 6-well culture dish and maintained in a serum-free medium supplemented with NGF. Most of the NGF supported neurons (arrow head pointing to a neuronal cell body) had extended neurites. Note that the entire neuritic territory appears occupied by tightly associated spindle shaped Schwann cells (arrows). Large flat fibroblasts (F) are also present with no obvious association with neurons. B: Four days *in vitro*. Extensive neuronal networks are being established. Neuron-associated Schwann cells are increased in numbers (arrows). C: Six days *in vitro*. Most of the culture area is populated with extended neurites and virtually the entire neuritic network is occupied with Schwann cells. Fibroblasts (F) had also increased in numbers but with no association with neurons. D: A confluent culture of purified Schwann cells. Neuron-Schwann cell networks were separated from the underlying fibroblasts layer as described in the text. Subsequently, Schwann cells were dissociated from the neurons and expanded in the presence of rhGGF2 and forskolin in 10% serum-medium. The majority of cells have the spindle-shape appearance characteristic of Schwann cells. Arrow points to a mitotically active, dividing Schwann cell. Bar = 140 μm

Figure 2 illustrates the typical appearance of DRG cultures after various times in serum-free defined N2 media. By 24 hours after plating, most cells attach to the plastic (not shown) and by 48 hours (panel A) many neurons extend neurites.

Phase-bright neuronal bodies are observed. Spindle shaped Schwann cells are present in apparent association with the growing neurites. By day 4 (panel B), extensive neuronal networks develop. Schwann cells greatly increase in number and a great proportion of Schwann cells become associated with developing neurites. By 6 days (panel C), virtually the entire neurite network is occupied by Schwann cells. Large flat cells (fibroblasts) are also seen in culture with no obvious association to neurons.

Step 2. On day 6 or 7, each well is marked with the genotype of the embryo from which the cells are derived, and placed under a dissecting microscope in a sterile environment. Under 500X magnification, the neurites and underlying fibroblast layers are visible. Neuron-associated Schwann cells are visible as beaded structures along neurites. Starting from a corner of the plate where few fibroblasts are present, Schwann cell-neuron networks are gently lifted up using a 23_G1 needle, and slowly peeled off toward the middle of the plate. Usually, neuronal networks that are being peeled off can be distinguished from the fibroblast layer left behind at this magnification. If DRG are dissociated well initially, almost complete separation of neuron-Schwann cell networks can be achieved essentially free of fibroblasts. However, if dense populations of fibroblasts are observed in the direction from which neurons are being peeled off, neurons are cut off at that point and the neuronal network remaining on the plate lifted starting from a new point. Sometimes, fibroblasts that are condensed underneath neuronal bodies are difficult to separate from the neurons. In this case, these areas are first excised from the plate using a surgical razor blade, and the separation procedure proceeds. It is crucial to obtain Schwann cell-neuronal networks that are free of contaminating fibroblasts at this point, because removal of fibroblasts from the later Schwann cell cultures is more difficult. Attempts to remove fibroblasts from Schwann cell cultures by complement mediated lysis using several different commercial anti-mouse Thy1.1 antibodies were unsuccessful. Thus, it is important not to over-grow the fibroblasts in the initial DRG cultures; fibroblast overgrowth occurs when DRG cells remained in culture for longer than 7 days prior to purification. After successfully removing the neuron-Schwann cell network from the plates, neurons and Schwann cells are transferred to 15 ml tubes containing 7 ml 10% serum-medium. Cells from a single embryo can be maintained separately, or cells from embryos of the same genotype pooled.

Step 3. Schwann cells, associated with neurons, are centrifuged for 5 min at 80xg and Schwann cells are dissociated from neurons by enzymatically dissociating the resulting pellets in trypsin-collagenase solution in a volume of approximately 0.5 ml per embryo. For example, if cells from 6 embryos (12 wells) of the same genotypes are pooled, 3 ml of the enzyme solution is used to dissociate Schwann cells. If cells from a single embryo are processed separately, a minimum amount of 1.5 ml is used per embryo (corresponding to 2-wells of the DRG culture). Cells are then incubated at 37°C for 30 min with gentle rotation (30–40 rpm) in an air incubator, with tubes laid on their side, not vertically placed. This allows even access of enzymes to the cells. The enzyme reaction is stopped by adding DMEM plus 10% FBS, cells centrifuged, resuspended in 2 ml of 10% serum-medium and counted in a hemacytometer. On average, 0.3×10^6 cells are obtained from one embryo. Cells are plated onto poly-L-lysine-coated 100 mm plates and cultured in 10% serum-medium supplemented with glial growth factor (rhGGF2, 10 ng/ml) (Cambridge Neuroscience) and 2 μM forskolin to suppress contaminating fibroblast growth. After 5–6 days, plates become confluent with approximately 3×10^6 cells/plate (Figure 2 panel D). The majority of cells have the spindle-shape appearance and oval nucleus characteristic of cultured Schwann cells. When cultures are examined for the expression of a Schwann cell antigen, low affinity p75NGF receptor (NGFR) using a rat anti mouse NGFR antibody (obtained from Dr. David Anderson, Caltech), a minimum of 99.5% of the cells are positive for NGFR expression. In three separate experiments, 99.5%, 99.9% and 99.6% of cells were NGFR positive; most of the NGFR positive cells had spindle shaped morphology (Figure 3, panel A). Large flat cells (fibroblasts) are negative for the staining. NGFR expressing cells were also positive for S100 expression (not shown).

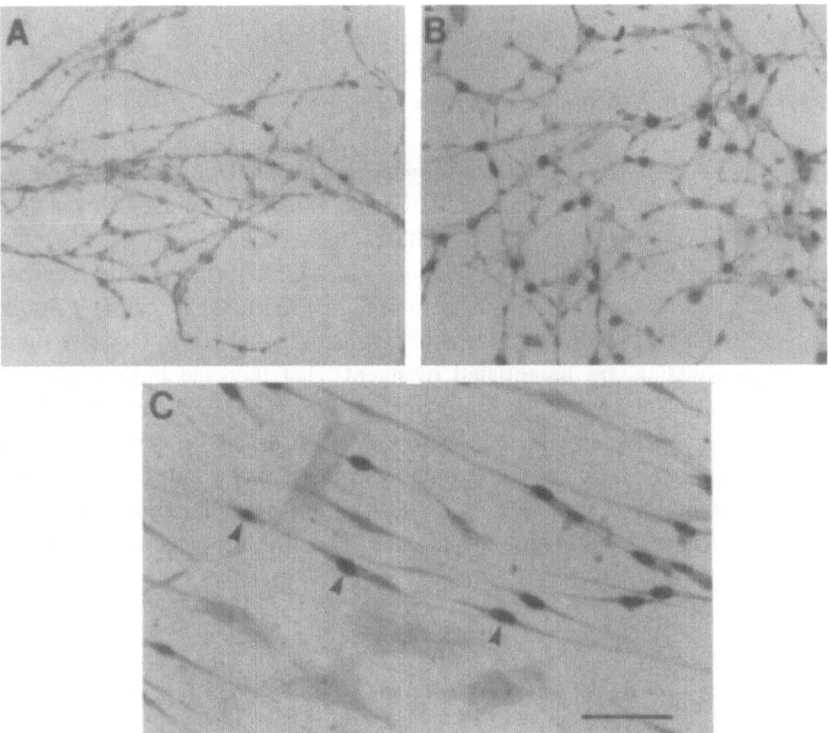

Figure 3. Expression of NGFR by purified mouse Schwann cells. Autoradiograph of Schwann cells stimulated by a soluble growth factor (rhGGF2) or neuronal contacts. A: Purified mouse Schwann cells were fixed in culture and processed for immunostaining with NGFR. DAB stained NGFR positive Schwann cells are shown. B: Schwann cells were stimulated with rhGGF and exposed to 0.5 μCi/ml [³H]-thymidine for 24 hours, fixed, stained with NGFR and processed for autoradiography. Most of the NGFR positive cells (Schwann cells) are overlaid with silver grains upon rhGGF2 stimulation (compare with Schwann cells in panel A (untreated with rhGGF). C: Purified mouse Schwann cells seeded onto cultures of E16 rat DRG neurons. After 24 hours, [³H]-thymidine was added and the cultures were incubated for an additional 24 hours prior to fixation and autoradiography. Cultures were counter-stained with toluidine blue. Neuron-associated Schwann cells with labeled nuclei are shown (arrow heads). Contaminating fibroblasts with no obvious association with neurons are also present (F). Bar = 263 μm.

After a plate becomes confluent with proliferating mouse Schwann cells, cells are trypsinized and replated onto new poly-L-lysine coated plates at a density of 0.5×10^6 cells/100 mm plate in 10% serum containing media. Next day, the medium is changed to 10% serum medium supplemented with rhGGF2 and forskolin (2 μM). If the culture appears contaminated with significant numbers of fibroblasts, serum-free N2 medium can be used during the subsequent Schwann cell expansion. As the passage number increases, each passage being one week in culture, mouse Schwann cells tend to become less responsive to growth factors and begin to die (in contrast to rat Schwann cells). The demise of cells occurs faster in serum-free media than in serum containing media. We recommend use of Schwann cells which have gone through less than 4 passages for analysis.

RESULTS

Characterization of Mouse Schwann Cells

Proliferation in Response to Soluble Mitogens and Axons. Double-label immunocytochemistry assay was used to investigate cell division of isolated mouse Schwann cells in response to rhGGF2 (Figure 3). In control cultures, most NGFR positive cells remained quiescent. Upon rhGGF2 stimulation, the majority of NGFR positive cells incorporated [3H]-thymidine. Quantitation of results is shown in Figure 4; rhGGF2 treatment resulted in proliferation of 50–60% of NGFR positive cells (Figure 3. panel B). In rat Schwann cells, mitogenic effects of GGF are potentiated by agents that increase intracellular cAMP. When purified mouse Schwann cells were treated with rhGGF2 in combination with forskolin, *no* synergy between the two factors was observed. In some experiments, forskolin even slightly inhibited GGF induced mouse Schwann cell proliferation. Forskolin alone had no effect on mouse Schwann cell proliferation, similar to results reported for rat Schwann cells cultured in the absence of serum (Stewart et al., 1991).

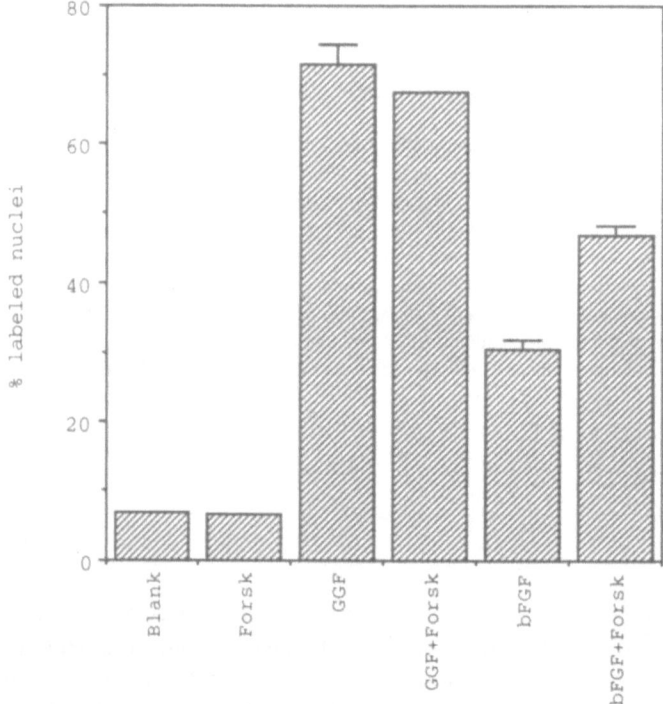

Figure 4. Proliferation of mouse Schwann cells purified from E12.5 embryos in response to forskolin, rhGGF2 or combination of the two. Cells were plated on poly-L-lysine coated 8-chamber glass Lab-Tek slides (25,000 cells/well) in 10% serum-medium. After 48 hours, medium was switched to serum-free N2 and cells were stimulated with either forskolin (5 μM), rhGGF2 (10 ng/ml), bFGF (10 ng/ml) or forskolin in combination with rhGGF2 or bFGF. Cells were pulsed for 24 hours with 0.5 μCi/ml [3H]-thymidine, fixed, immunostained with NGFR and processed for autoradiography. Schwann cell proliferation was determined by the percentage of NGFR positive cells with labeled nuclei.

Basic fibroblast growth factor (bFGF) stimulates rat Schwann cell proliferation only when it is accompanied by agents that increase cAMP in the cells and by itself has no effect on cell division. In contrast, when mouse Schwann cells were treated with bFGF in serum-free medium, 18% of NGFR positive cells were labeled. Thus, bFGF, by itself, is a mitogen for mouse Schwann cells. Addition of forskolin potentiated the mitogenicity of bFGF (labeling index: 42%). It has been shown previously that in serum containing media, bFGF stimulates proliferation of mouse Schwann cells prepared from neonatal mouse DRG (Krikorian et al., 1982). In our analyses, mouse Schwann cells showed slightly increased basal level proliferation in serum containing media (data not shown).

One of the characteristic features of Schwann cells is increased cell proliferation in association with axons. Isolated mouse Schwann cells were added to purified rat DRG neuronal cultures (Wood, 1975; Wood and Bunge, 1976) and incorporation of $[^3H]$-thymidine was examined (Figure 3, panel C). Cells established association with axonal membranes, and nuclear labeling was greatly enhanced in cells with association with axons. Contaminating fibroblasts with flat morphology showed no obvious relationship with axons.

Po Expression and Myelin Formation. Differentiation of rat Schwann cells can be assessed in isolated cultures. Agents that increase intracellular cAMP elevation upregulate expression of myelin proteins (Lemke and Chao, 1988; Morgan et al., 1991; Sobue and Pleasure, 1984). One of these proteins is Po, which has been used as a general marker for differentiating Schwann cells (Morgan et al., 1991). To examine whether mouse Schwann cells also display this phenotype, cells were treated with forskolin, an activator of adenylate cyclase, each day for 5 days, cell extracts prepared and expression of Po analyzed by SDS-PAGE followed by Western blot analysis. As shown in Figure 5, while control cells show no expression of the protein, forskolin treatment induced about a 20 fold increase in Po expression (analyzed by densitometric scan of blots).

Specific association with axons is necessary for the formation of myelin by Schwann cells (Duncan, 1934; Friede, 1972; Matthews, 1968; Voyvodic, 1989). In culture, Schwann cells can be induced to form myelin around axons by co-culturing Schwann cells with neurons (Eldridge et al., 1987). We showed recently that these mouse Schwann cells can also form myelin around axons (Rosenbaum et al., 1997).

Characterization of Phenotypes of Schwann Cells Derived from *NF1* Knock-Out Mice

To study the effect of loss of *NF1* gene product, neurofibromin, in Schwann cells, Schwann cells were isolated from wild type, heterozygous or null mouse embryos at E12.5

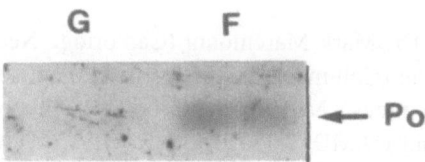

Figure 5. Po expression of mouse Schwann cells purified from E12.5 embryos. Schwann cells were incubated for 4 days in either rhGGF2 (10 ng/ml) or forskolin (5 µM) in serum-free N2 medium and cell lysate prepared. Equal amounts of each lysate was separated on a 10% SDS gel and Po expression analyzed by Western blot using polyclonal antibodies raised against rat Po. Induction of Po expression is seen in forskolin (F) treated mouse Schwann cells, while no Po expression is seen in rhGGF2 (G) treated cells.

using the method described here and phenotypes characterized. The high purity of these cultures enabled us to carry out biochemical experiments to measure levels of Ras-GTP in normal and mutant cells (Kim et al., 1995); GTP-Ras is significantly increased in the absence of *NF1*, substantiating the idea that *NF1* is the major Ras-GAP in Schwann cells (Basu et al., 1992; DeClue et al., 1992). Null Schwann cells also show altered morphology as compared to wild type cells, and decreased cell proliferation in response to axons or rhGGF2 (Kim et al., 1995). Recently, we have used mouse Schwann cells from mutants in studies of perineurium formation (Rosenbaum et al., 1995), of myelination (Rosenbaum et al., 1997) and of cell transformation (Kim et al., 1997). Our results confirm that alterations in mouse Schwann cells resulting from mutations in a single gene can be detected using the purification methods described.

SUMMARY

Several purification methods did not yield satisfactory results for Schwann cells of the mouse. A new method was developed that generates a large quantity of essentially pure mouse Schwann cells. Because fibroblasts are removed at early stage of the isolation procedure, Schwann cells can be maintained in serum containing media, greatly improving survival and proliferation. The method generates 2–3×10^6 Schwann cells/embryo in less than two weeks, providing a powerful new tool for analysis of mouse Schwann cells *in vitro*. Several properties were used to identify isolated cells as Schwann cells. First, the cells are characteristically spindle-shaped in culture. Second, they express immunocytochemical markers, including S100 and p75NGF receptor, characteristic of Schwann cells (Jessen and Mirsky, 1991; Schachner et al., 1981; Stefansson et al., 1982). Third, the cells adhere to rat axons and proliferate in response to axonal contact (Aguayo et al., 1976; McCarthy and Partlow, 1976; Salzer et al., 1980; Wood and Bunge, 1975). The cells also proliferate in response to known Schwann cell mitogens, GGF and bFGF. Regulation of mouse Schwann cell proliferation differs from that in rat (Davis and Stroobant, 1990; Raff et al., 1978; Stewart et al., 1991). Response of mouse Schwann cells to bFGF is independent of cAMP elevation, and increasing cAMP levels in mouse Schwann cells also fails to synergize with GGF mitogenicity. Finally, mouse Schwann cells formed myelin around axons when cultured with rat peripheral neurons and expressed the myelin protein Po after exposure to forskolin. Thus, mouse Schwann cells isolated from E12.5 embryos show only subtle differences from better-studied rat cells.

ACKNOWLEDGMENTS

We are indebted to Dr. Mark Marchionni (Cambridge Neuroscience) for providing rhGGF2, Dr. David Colman (Columbia University) for Po antisera and Dr. David Anderson (Caltech) for rat anti mouse NGFR. This work was supported by grants from RG 2493-82, ROJNS28840 and DAMD17-J3030 to NR. HAK was the recipient of a fellowship from an NCI training grant (CA59268).

REFERENCES

Aguayo AJ, Charron L, Bray GM (1976): Potential of Schwann cells from unmyelinated nerves to produce myelin: A quantitative ultrastructural and autoradiographic study. J. Neurocytol. 5:565–573.

Askanas V, Engel WK, Dalakas MC, Lawrence JV, Carter LS (1980): Human Schwann cells in tissue culture: his- tochemical and ultrastructural studies. Archives of Neurology 37:329–37.

Basu TN, Gutmann DH, Fletcher JA, Glover TW, Collins FS, Downward J (1992): Aberrant regulation of ras pro- teins in malignant tumour cells from type 1 neurofibromatosis patients [see comments]. Nature 356:713–5.

Bhattacharyya A, Brackenbury R, Ratner N (1993): Neuron-Schwann cell signals are conserved across species: purification and characterization of embryonic chicken Schwann cells. Journal of Neuroscience Research 35:1–13.

Brannan CI, Perkins AS, Vogel KS, Ratner N, Nordlund ML, Reid SW, Buchberg AM, Jenkins NA, Parada LF, Copeland NG (1994): Targeted disruption of the neurofibromatosis type-1 gene leads to developmental ab- normalities in heart and various neural crest-derived tissues [published erratum appears in Genes Dev 1994 Nov 15;8(22):2792]. Genes & Development 8:1019–29.

Brockes JP, Fields KL, Raff MC (1979): Studies on cultured rat Schwann cells. I. Establishment of purified popu- lations from cultures of peripheral nerve. Brain Research 165:105–18.

Cochran M, Black MM (1985): PC12 neurite regeneration and long-term maintenance in the absence of exoge- nous nerve growth factor in response to contact with Schwann cells. Brain Research 349:105–16.

Cornbrooks CJ, Mithen F, Cochran JM, Bunge RP (1982): Factors affecting Schwann cell basal lamina formation in cultures of dorsal root ganglia from mice with muscular dystrophy. Brain Research 282:57–67.

Davis JB, Stroobant P (1990): Platelet-derived growth factors and fibroblast growth factors are mitogenic for rat Schwann cells. J. Cell Biol. 110:1353–1360.

DeClue JE, Papageorge AG, Fletcher JA, Diehl SR, Ratner N, Vass WC, Lowy DR (1992): Abnormal regulation of mammalian p21ras contributes to malignant tumor growth in von Recklinghausen (type 1) neurofibroma- tosis. Cell 69:265–73.

Duncan D (1934): A relation between axon diameter and myelination determined by measurement of myelinated spinal root fibers. J. Comp. Neurol. 60:437.

Eldridge CF, Bunge MB, Bunge RP, Wood PM (1987): Differentiation of axon -related Schwann cells in vitro. I. Ascorbic acid regulates basal lamina assembly and myelin formation. J. Cell Biol. 105:1023–34.

Friede RL (1972): Control of myelin formation by axon caliber (with a model of the control mechanism). J. Comp. Neurol. 144:233–252.

Jacks T, Shih TS, Schmitt EM, Bronson RT, Bernards A, Weinberg RA (1994): Tumour predisposition in mice het- erozygous for a targeted mutation in Nf1. Nature Genetics 7:353–61.

Jessen KR, Mirsky R (1991): Schwann cell precursors and their development. Glia 4:9–18.

Kalderon N (1984): Schwann cell proliferation and localized proteolysis: expression of plasminogen-activator ac- tivity predominates in the proliferating cell populations. Proceedings of the National Academy of Sciences of the United States of America 81:7216–20.

Kim HA, Ling B, Ratner N (1997): NF1-deficient mouse Schwann cells are angiogenic, invasive and can be in- duced to hyperproliferate: Reversion of some phenotypes by an inhibitor of farnesyl protein transferase. Mol. Cell Biol. 17:862–872.

Kim HA, Rosenbaum T, Marchionni MA, Ratner N, DeClue JE (1995): Schwann cells from neurofibromin defi- cient mice exhibit activation of p21ras, inhibition of cell proliferation and morphological changes. Onco- gene 11:325–35.

Kim SU, Yong VW, Watabe K, Shin DH (1989): Human fetal Schwann cells in culture: phenotypic expressions and proliferative capability. Journal of Neuroscience Research 22:50–9.

Komiyama A, Suzuki K (1991): Age-related changes in attachment and proliferation of mouse Schwann cells in vitro. Dev. Brain Res. 62:7–16.

Krikorian D, Manthorpe M, Varon S (1982): Purified mouse Schwann cells: mitogenic effects of fetal calf serum and fibroblast growth factor. Developmental Neuroscience 5:77–91.

Kuhlengel KR, Bunge MB, Bunge RP (1990): Implantation of cultured sensory neurons and Schwann cells into le- sioned neonatal rat spinal cord. I. Methods for preparing implants from dissociated cells. Journal of Com- parative Neurology 293:63–73.

Lemke G, Chao M (1988): Axons regulate Schwann cell expression of the major myelin and NGF receptor genes. Development 102:499–504.

Lemke GE, Brockes JP (1984): Identification and purification of glial growth factor. The Journal of Neuroscience 4:75–83.

Levi AD, Bunge RP, Lofgren JA, Meima L, Hefti F, Nikolics K, Sliwkowski MX (1995): The influence of heregulins on human Schwann cell proliferation. Journal of Neuroscience 15:1329–40.

Manthorpe M, Skaper S, Varon S (1980): Purification of mouse Schwann cells using neurite-induced proliferation in serum-free monolayer culture. Brain Research 196:467–82.

Marchionni MA, Goodearl ADJ, Chen MS, Bermingham-McDonogh O, Kirk C, Hendricks M, Danehy F, Misumi D, Sudhalter J, Kobayashi K, Wroblewski D, Lynch C, Baldassare M, Hiles I, Davis JB, Hsuan JJ, Totty

NF, Otsu M, McBurney RN, Waterfield MD, Stroobant P, Gwynne D (1993): Glial growth factors are alternatively spliced erbB2 ligands expressed in the nervous system. Nature 362:312–318.

Matthews MA (1968): An electron microscopic study of the relationship between axon diameter and the initiation of myelin production in the peripheral nervous system. Anat. Rec. 161:337.

McCarthy KD, Partlow LM (1976): Neuronal stimulation of [3H]thymidine incorporation by primary cultures of highly purifed non-neuronal cells. Brain Res 114:415–426.

Morgan L, Jessen RK, Mirsky R (1991): The effects of cAMP on differentiation of cultured Schwann cells: progression from an early phenotype (04$^+$) to a myelin phenotype (P0$^+$, GFAP$^-$, NGF-receptor$^-$) depends on growth inhibition. J. Cell Biol. 112:457–467.

Morrissey TK, Kleitman N, Bunge RP (1991): Isolation and functional characterization of Schwann cells derived from adult peripheral nerve. Journal of Neuroscience 11:2433–42.

Raff MC, Abney E, Brocker JP, Hornby-Smith A (1978): Schwann cell growth factors. Cell 15:813–822.

Riccardi VM (1991): Neurofibromatosis: past, present, and future [editorial; comment]. New England Journal of Medicine 324:1283–5.

Rosenbaum T, Boissy YL, Kombrinck K, Brannan C, Jenkins NA, Copeland NG, Ratner N (1995): Neurofibromin deficient fibroblasts fail to form perineurium *in vitro*. Development 121:3583–3592.

Rosenbaum T, Kim HA, Ling B, Ratner N (1997): Neurofibromin is required for appropriate Po expression and myelination. submitted.

Rutkowski JL, Kirk CJ, Lerner MA, Tennekoon GI (1995): Purification and expansion of human Schwann cells *in vitro*. Nature Medicine 1:80–83.

Rutkowski JL, Tennekoon GI, McGillicuddy JE (1992): Selective culture of mitotically active human Schwann cells from adult sural nerves. Annals of Neurology 31:580–6.

Salzer JL, Bunge RP, Glaser L (1980): Studies of Schwann cell proliferation. III. Evidence for the surface localization of the neurite mitogen. Journal of Cell Biology 84:767–78.

Schachner M, Kim SK, Zehnle R (1981): Developmental expression in central and peripheral nervous system of oligodendrocyte cell surface antigens (O antigens) recognized by monoclonal antibodies. Dev Biol 83:328–338.

Seilheimer B, Schachner M (1987): Regulation of neural cell adhesion molecule expression on cultured mouse Schwann cells by nerve growth factor. EMBO 6:1611–1616.

Shine HD, Sidman RL (1984): Immunoreactive myelin basic proteins are not detected when shiverer mutant Schwann cells and fibroblasts are co-cultured with normal neurons. Journal of Cell Biology 98:1291–5.

Sobue G, Pleasure D (1984): Schwann cell galactocerebroside induced by derivatives of adenosine 3' 5' monophosphate. Science 224:72–74.

Stefansson K, Wollmann R, Jerkovic M (1982): S-100 protein in soft tissue tumours derived from Schwann cells and melanocytes. Am. J. Pathol. 106:261–8.

Stewart JS, Eccleston PA, R. JK, Mirsky R (1991): Interaction between cAMP elevation, identified growth factors, and serum components in regulating Schwann cell growth. J. Neurosci. Res 30:346–352.

Voyvodic JT (1989): Target size regulates calibre and myelination of sympathetic axons. Nature 342:430–433.

Wood PM (1976): Separation of functional Schwann cells and neurons from normal peripheral nerve tissue. Brain Research 115:361–75.

Wood PM, Bunge RP (1975): Evidence that sensory axons are mitogenic for Schwann cells. Nature 256:662–664.

Yong VW, Kim SU, Kim MW, Shin DH (1988): Growth factors for human glial cells in culture. Glia 1:113–23.

21

EXPANSION OF OLIGODENDROCYTE PROGENITORS FOR MYELIN REPAIR

A. Baron-Van Evercooren, V. Avellana-Adalid, S. Vitry, B. Nait-Oumesmar, and F. Lachapelle

U134 INSERM
Laboratoire de Neurobiologie Cellulaire, Moleculaire et Clinique
Hopital de la Salpêtrière
75651 Paris cedex 13
France

INTRODUCTION

The myelin sheath is a multilamellar extension of the cell membrane of the oligodendrocyte in the central nervous system (CNS) and of the Schwann cell in the peripheral nervous system (PNS). Processed mainly during postnatal development, it is considered a stable structure in the adult nervous system. Since its function is to promote and secure electrical conduction, alteration or destruction of the myelin sheath during adult life can cause severe impairement of neurologic function. Recurrent events of demyelination in the CNS occur in multiple sclerosis, leading to the destruction and depletion of the oligodendroglial population. Therefore, the development of strategies to promote myelin repair by replacement of the affected oligodendroglial population could become instrumental in the treatment of demyelinating diseases.

While transplantation has been proposed as a possible means supplying the affected nervous system with newly generated myelin-forming cells, it is also an efficient way to confront cells of the oligodendroglial lineage with the CNS environment and to study their potential for (re)myelination. Whether conceived as a fundamental or therapeutic approach, transplantation requires the ex-vivo expansion of the appropriate donor glial population. To reach this goal, we have developed a reproducible and effective *in vitro* system to expand oligodendrocyte progenitors from newborn rat brain as homotypic aggregates that we have called "oligospheres".

THE OLIGODENDROGLIAL LINEAGE *IN VITRO*

The oligodendroglial lineage has been intensively characterized *in vitro*. A glial progenitor was first isolated from the newborn rat optic nerve (Raff et al., 1983), and then from many regions of the rodent CNS (for review, see Mc Kinnon and Dubois-Dalcq,

Cell Biology and Pathology of Myelin, edited by Juurlink *et al.*
Plenum Press, New York, 1997

1995). Depending on the culture conditions, this cell is capable of generating, oligoden-drocytes or type-2 astrocytes and was thus designated O-2A. The differentiation of O-2A cells into oligodendrocytes or type-2 astrocytes is characterized by the expression of stage specific phenotypes and antigens (Raff,1989; Pfeiffer et al., 1993). The oligodendrocyte progenitor arises from a small unipolar pre-progenitor cell which has a tendency to self-aggregate (Hardy and Reynolds, 1991; Grinspan and Franceschini, 1995). It expresses only stem cell antigens, such as the embryonic form of NCAM (PSA-NCAM) and Nestin (Trotter et al., 1989, Gallo and Armstrong, 1995). The pre-progenitor cell then evolves to-ward the bipolar, motile oligodendrocyte progenitor (Small et al., 1987) acquiring the gan-gliosides GD3 (Goldman 1992) and A2B5 (Eisenbarth et al., 1979; Raff, 1989) but lacking galactolipids and astrogliofilaments. With time and under the appropriate culture conditions, the bipolar progenitor will become a multipolar cell with short processes and express the O4 antigen (Sommer and Schachner, 1981). Losing its capacity to proliferate and to migrate, this cell progressively differentiates into the sheath extending pre-oli-godendrocyte expressing the galactocerebroside O1 (Warrington and Pfeiffer, 1992), and finally into the mature oligodendrocyte expressing structural myelin proteins such as PLP and MBP (Dubois-Dalcq et al., 1986).

In the absence of exogenous growth factors, the isolated O-2A progenitors will timely differentiate over time into mature glia (Raff et al., 1988). However, growth factors can alter the progress of their differentiation and promote their proliferation (Mc Kinnon and Dubois-Dalcq, 1995). In particular, basic fibroblast growth factor (bFGF) prevents O-2A maturation and promotes their proliferation (Mc Kinnon et al., 1990), and platelet de-rived growth factor (PDGF) is a survival and mitogenic factor for these cells and their precursors (Noble et al., 1988, Richardson et al., 1988, Grinspan and Franceschini, 1995). The combined use of bFGF and PDGF induces O-2A progenitors to proliferate indefi-nitely *in vitro* (Bögler et al., 1990). Other factors that promote oligodendrocyte progenitor proliferation are insulin-like growth factor I (IGF-I) (Mc Morris and Dubois-Dalcq, 1988), the neurotrophin NT-3 (Barres et al., 1993), as well as the more recently identified oli-godendroglial mitogen, neuregulin/glial growth factor 2 (GGF2) (Canoll et al., 1996).

While some of these effects on O-2A proliferation and differentiation can be mim-icked by astrocytes, neurons also profoundly influence the biology of the oligodendroglial lineage (reviewed in Barres and Raff, 1994; Hardy and Reynolds, 1993a, Canoll et al., 1996). Some of their mitogenic effect on oligodendroglia appears to be contact-mediated and is related to membrane-bound FGF and to some yet uncharacterized mitogen (Kreider, et al., 1995). Neuronal conditioned medium (CM) from the B103 (Guillian et al., 1991) and the B104 (Hunter and Bottenstein, 1990, 1991) neuroblastoma cell lines, or from cere-bellar granular neurons (Hardy and Reynolds 1993b), are also mitogenic for oligodendro-cyte progenitors. While PDGF seems to be the major mitogen secreted by the B104 cells (Hunter et al., in this chapter), GGF is responsible for the mitogenic activity released by cerebellar neurones (Canoll et al., 1996).

TRANSPLANTATION OF THE OLIGODENDROGLIAL LINEAGE

Transplantation represents another powerful tool to explore the behavior of the oli-godendrocyte lineage in the CNS environment. The use of various models of dysmyelina-tion or demyelination, as well as the development of several cell tracers, has led to the demonstration of the survival, proliferation, migration and differentiation capacity of de-fined glial populations transplanted in the CNS (Gumpel, 1991; Baron-Van Evercooren et al., 1994; Blakemore and Franklin, 1995; Duncan and Milward., 1995). Although all stages of the oligodendroglial lineage were able to myelinate after transplantation, oligodendrocyte progenitors were found to be the cells which migrate and differentiate most efficiently when confronted with the developing CNS (Warrington et al., 1993). Initially, oligodendrocyte progenitors were transplanted as fragments of CNS tissue (Gumpel et al., 1983; Lachapelle et al., 1983–84), but expansion of defined cell suspensions prevailed by supplying large quantities of purified oligodendroglial populations. Oligodendrocyte progenitors were transplanted after expansion on astrocyte feedlayers (Vignais et al., 1993) or after clonal expansion by treatment with FGF and PDGF (Groves et al., 1994). O-2A cell lines were created using the combination of FGF and PDGF (Barnett et al., 1993), the N1-B104 CM (Louis et al., 1992) or by immortalization of the progenitor population with a temperature sensitive SV40 T oncogene (Trotter et al., 1993). From these cell lines, the CG4 cells obtained with the N1-B104 CM, were the only ones capable of extensive differentiation into myelin-forming oligodendrocytes (Tontsch et al., 1994) or astrocytes (Franklin et al., 1995) after transplantation in the CNS.

OLIGOSPHERES AS A SOURCE OF PROLIFERATING OLIGODENDROCYTE PROGENITORS

Expansion of O2A progenitors with characteristics close to their native form is thus a priority for transplantation. We have purified progenitors from the new-born rat brain by sieving, Percoll gradient separation and differential adhesion (Avellana-Adalid et al., 1996 a and b). The resulting population was maintained at high cell density in a mixture of DMEM and B104 CM enriched with N1 supplements and in the absence of adhesive substrate. These culture conditions induce the formation of homotypic aggregates, which increase progressively in number and size, leading to the formation of spheres. By analogy with oligospheres we called these aggregates which generated oligodendrocyte progenitors, "oligospheres". When dissociated, oligospheres produce a monolayer of bipolar cells which remain bipolar and continue to expand in N1-B104 CM. When allowed to adhere to polyornithine-coated substrates, oligospheres give rise to a crown of bipolar progenitors which migrate progressively from the sphere toward the periphery. In several experiments, oligospheres were expanded and maintained over one year in culture.

Immunohistochemical characterization of dissociated oligospheres, indicate that at passage P 2 (2 weeks) oligosphere-derived bipolar cells consisted mainly of GD3+/GFAP-cells (97+1%) and were thus highly enriched in oligodendrocyte progenitors. They remained so during extended periods, since they still contained 89+6% and 86+4% GD3+/GFAP- cells at P12 (4 months) and P40 (1 year), respectively. The presence of rare O4+, Galc+, and Tuj1+ cells indicate that the present culture conditions (N1-B104 CM, high cell density and absence of adhesive substrate) lead to the selection and expansion of a highly enriched O2A population. Combining BrdU incorporation (24 hours) with labeling of GD3, BrdU and Hoechst, showed that after 2 weeks of culture (P2), the majority of GD3+ cells (94+2%) were dividing and that a substantial amount of these cells were still doing so at 4 months (37+6%) and 1 year (35+4%) of culture.

In the above experiment, cells were analyzed after at least one week of culture and mainly after dissociation of the spheres. Since both time in culture and adhesion of the dissociated cells to the polyornithine-coated substrate may induce the cells to differentiate into a more advanced stage of the lineage than cells remaining aggregated in the sphere, we further extended our data by characterizing the cell content of the spheres at various time points of culture. Oligospheres were allowed to adhere to polyornithine-coated dishes for a limited amount of time (2–24 hours) and were immunolabeled with cell stage specific markers at 3 (P0), 7 (P1), and 30 days (P4) after purification. At the earliest time point (P0), spheres were small, and were formed of a core of small aggregated cells occasionally extending one short process and resembling the previously described oligodendrocyte pre-progenitors. These small aggregated cells were surrounded by a few larger bipolar progenitor cells emerging radially from the aggregate. The small cells forming the core of the spheres, expressed essentially the early stem cell antigen PSA-NCAM, but none of the more committed cell antigens such as GD3, GFAP or Tuj1 (Baron-Van Evercooren, unpublished observations). In contrast, the larger bipolar cells that exited the sphere and spread on the substrate, expressed GD3 in addition to PSA-NCAM, but rarely GFAP or Tuj1 antigens. At later time points (P1-P4), when spheres were larger and served as a reservoir of cells, the majority of the cells forming the sphere seemed to express PSA-NCAM while only those cells at the outer surface of the sphere or migrating on the substrate, also expressed GD3 (with rare GFAP or Tuj1+ cells) (Fig. 1). These data suggest that early after purification, aggregates are formed mainly of pre-progenitors (Grinspan et al., 1996) which will evolve over time to the stage of oligodendrocyte progenitor when localized at the surface of the sphere and entering in contact with the soluble growth factors of the B104 CM and the adhesive substrate.

OLIGOSPHERE-DERIVED PROGENITORS REMAIN BIPOTENTIAL AND GENERATE FUNCTIONAL OLIGODENDROCYTES

As indicated in the previous section, the use of B104 CM and the absence of adhering substrate allowed us to maintain the progenitor population in an immature proliferative stage for extended periods in culture. Removal of B104 CM and addition of low concentrations of fetal calf serum (FCS) to the N1 supplemented DMEM medium induced the bipolar progenitor population to differentiate into highly branched GalC+ oligodendrocytes even after one year of culture. When switched to high serum concentrations, the majority of the O2A population differentiated into ramified GFAP+ astrocytes, confirming the great plasticity of the neonatal rat oligodendroglial lineage and indicating that long-term expansion of the O2A population as oligospheres does not alter their bipotentiality.

In previous experiments, immortalization or establishment of cell lines has altered the capacity of the expanded population to differentiate into and function properly as myelin-forming oligodendrocytes. To evaluate this risk, we have transplanted Hoechst-labeled spheres at various time points (P2, P12 and P40) into the developing shiverer mouse brain. Immunohistochemical analysis of the recipient mouse brains indicates that a large proportion of the transplanted Hoechst-labeled cells were associated with MBP-positive myelin-like structures, strongly suggesting that long-term expansion of oligodendrocyte progenitors into oligospheres did not prevent their differentiation into functional oligodendrocytes.

OLIGOSPHERES AS A POTENTIAL STRATEGY TO EXPAND OLIGODENDROCYTE PROGENITORS FROM OTHER SPECIES

The possibility to expand efficiently and reproducibly the rat O2A population as oligospheres opens the perspective of adapting this strategy toward the purification and ex-

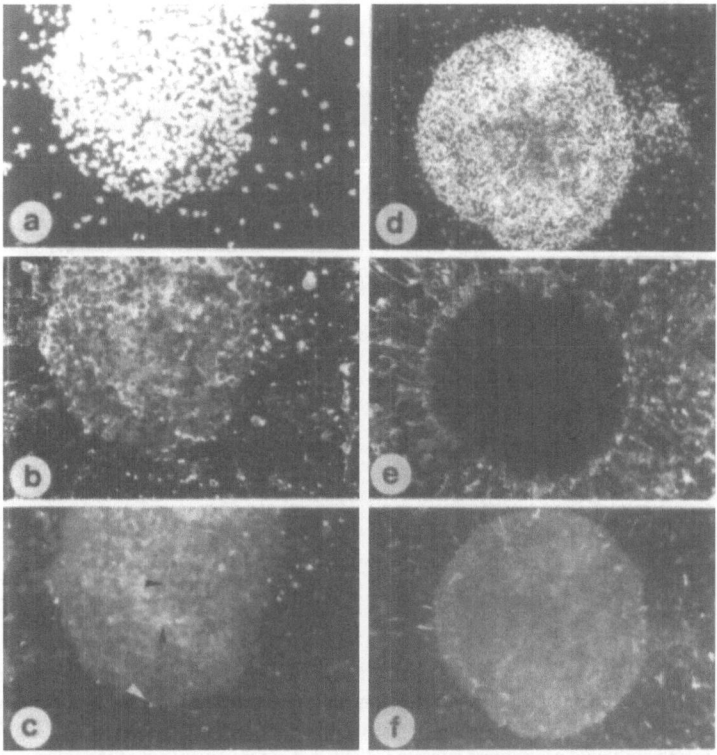

Figure 1. Immunocharacterization of P4 oligospheres from rat (a, b, c) and mouse (d, e, f) newborn brain attached to polyornithine for 24 hours: staining for Hoechst (a, d), for GD3 (b, e) Tuj1 (c) and GFAP (f). Fields in a, b, and c are focused on the sphere surface, while d, e and f are focused on the substrate. Most cells at the surface of the sphere or exiting from it express GD3 while very few of them express Tuj1 (arrows) or GFAP. (X272)

pansion of oligodendrocyte progenitors from other sources. In the same culture conditions, neonate mouse (Vitry, unpublished data) (fig. 2), and adult rat (Avellana-Adalid, unpublished data) progenitors, as well as the rat CG4 cells, were easily induced to aggregate and form oligospheres (Fig. 2). We found that maintenance of CG4 cells as floating oligospheres was, in fact, a very efficient way to amplify the CG4 population which has the tendency to diverge from its original phenotype when maintained as a monolayer for extended periods of culture.

To date, only minor success was obtained with mouse and adult rat aggregates, which declined rapidly in number over time in culture and rarely lasted beyond 3 weeks. Immunocharacterization at P2 showed that as with rat oligospheres, GD3-expressing cells with rare GFAP + cells (fig.1) were essentially localized at the periphery of the mouse spheres, while PSA-NCAM-expressing cells were expressed by the entire sphere population (not shown). Although, GFAP-positive and O4-positive cells were present in minimal amounts early after isolation, these cells increased progressively with time in culture, whether cells were grown in N1-B104, N1–2%FCS or DMEM-20% FCS. Mouse oligosphere-derived O2A thus seem to be less inclined than rat O2A to respond to the B104 CM signaling molecules expected to modulate their differentiation (Vitry, unpublished data). Thus, in addition to expressing a different pattern of antigens (Farranaga et al.,

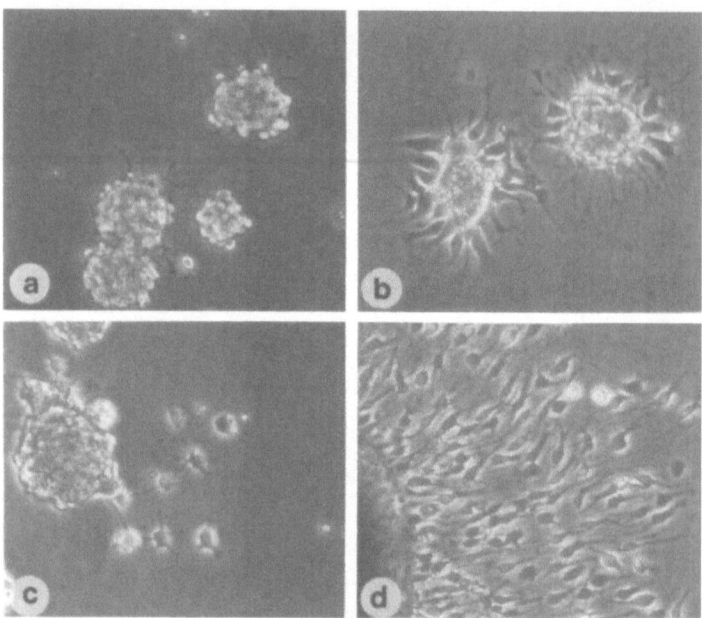

Figure 2. Phase contrast morphology of oligospheres formed of CG4 cells (a, b) or newborn mouse progenitors (c, d) and cultured in N1-B104 medium. In a and c, oligospheres are cultured in the absence of adhesive substrate: a few progenitors are lightly attached to the uncoated plastic; in b and d, oligospheres are adhering for 4 hours (b) and 24 hours (d) to polyornithine: progenitors migrate radially and progressively out of the sphere. (X350)

1996), the mouse oligodendroglial lineage has a plasticity and requirements for growth factors which seem to differ from those of the rat.

Despite these differences, mouse oligosphere-derived progenitors were nonetheless, capable of generating MBP-positive patches when transplanted into the newborn shiverer mouse brain, thus implying that when exposed to the favorable environment of the developing mouse brain, at least some cells of the oligosphere-derived population could differentiate into functional oligodendrocytes (Vitry, unpublished information).

CONCLUSIONS AND PROSPECTS

Oligosphere induction is an effective strategy for expanding neonatal rat oligodendrocyte progenitors without altering their bipotentiality and capacity to function as myelin-forming oligodendrocytes. Although modifications of the established protocol for rat may be required to expand oligodendrocyte progenitor populations from other species (mouse, human or non-human primates), the oligosphere strategy opens the door for studying novel genes or growth factors which control the biology of the oligodendroglial lineage.

Identifying growth factors to expand the glial progenitor population ex-vivo could be instrumental in developing approaches for therapeutic transplantation. While this approach may be hindered, by several limitations (for review see, Franklin and ffrench-Constant, 1995), promoting proliferation of the endogenous pool of oligodendrocyte progenitors in the adult brain may provide a promising source for replacement oligodendrocyte progenitors lost to demyelination. We have obtained evidence in rodents that cells derived from the adult subventricular zone (SVZ) can differentiate into myelin-producing oligodendrocytes. This capacity, which is limited, can be potentiated when donors are treated with bFGF (Lachapelle et al., 1996). Moreover, preliminary data indicate that the SVZ in the adult CNS can be activated by inducing nearby focal demyelination (Nait-Oumesmar et al., 1996). Although it is yet unknown wether these cells contribute to repair of the demyelinated lesion and whether growth factors can promote their involvement in myelin repair, our data suggest that treatment with growth factors either *in vitro* or *in vivo* improves the oligodendrocyte progenitors potential for myelination and could lead to the development of new strategies to repair the demyelinated adult CNS.

ACKNOWLEDGMENTS

The work from our laboratory has been supported by Inserm, the Myelin Project, Arsep and Biomed n°BMH4-CT96–1405. We are grateful to to M.L. Simao and N. Bargey for their contribution to this work.

REFERENCES

Avellana-Adalid V, Nait-Oumesmar B, Bachelin C, Simao ML, Baron-Van Evercooren A (1996a): Expansion of rat oligodendrocyte progenitors into proliferative "oligospheres" that retain differentiation potential. J Neurosci Res 45: 558–570.

Avellana-Adalid,V, Nait-Oumesmar B, Bachelin C, Simao ML, Baron-Van Evercooren A (1996b): Oligodendrocyte progenitors. Isolation and expansion as oligospheres. In Griffiths JB, Doyle A, Newell DG (eds): "Cell & Tissue Culture: Laboratory Procedures." England, John Wiley & Sons Limited. (submitted)

Baron-Van Evercooren A (1994): Future prospects in transplantation. Ann Neurol 36:151–156.

Barnett SC, Franklin RJM, Blakemore WF (1993): *In vitro* and *in vivo* analysis of a rat bipotential O-2A progenitor cell line containing the temperature sensitive mutant gene of the SV40 large T antigen. Eur J Neurosci 5:1247–1260.

Barres BA, Raff, MC (1994): Control of oligodendrocyte number in the developing rat optic nerve. Neuron 12: 935–942.

Barres BA, Schmid R, Sendter M, Raff MC (1993): Multiple extracellular signals are required for long-term oligodendrocyte survival. Development 118: 283–295.

Blakemore WF, Crang A, Franklin R. (1995): Transplantation of glial cells. In Ransom BR. Kettenman H. (eds): "Neuroglial cells." Cambridge, England, Oxford University Press.

Bogler O, Wren D, Barnett SC, Land H, Noble M (1990): Cooperation between two growth factors promotes extended self-renewal and inhibts differentiation of O-2A progenitor cells. Proc Natl Acad Sci USA 87:6368–6372.

Canoll P, Musacchio J, Hardy R, Reynolds R, Marchionni M, Salzer J (1996): GGF/Neuregulin is a neuronal signal that promotes the proliferation and survival and inhibits the differentiation of oligodendrocyte progenitors. Neuron 17:229–243.

Dubois-Dalcq M, Behar T, Hudson LD, Lazzarini R (1986): Energence of three myelin proteins in oligodendrocyte cultured without neurons. J Cell Biol 102: 384–392.

Duncan I, Milward E (1995): Glial cell transplants: experimental therapies of myelin diseases. Brain Pathol 5:301–310.

Eisenberth G, Walsh F, Nirenberg M (1979): Monoclonal antibody to a plasma membrane antigen of neurons. Proc Natl Acad Sci USA 76:4913–4917.

Farranaga M, Sommer I, Griffiths I (1995): O2A progenitors of the mouse optic nerve exhibit a developmental pattern of antigen expression different from the rat. Glia 15:95–104.

Franklin R, ffrench-Constant C (1996): Transplantation and repair in Multiple Sclerosis. In Russel WC (ed): "The Molecular Biology of Multiple Sclerosis." John Wiley and Sons, pp 231 242.

Franklin R, Bayley, S, Milner R, ffrench-Constant C, Blakemore WF (1995): Differentiation of the O-2A progenitor cell line CG-4 into oligodendrocytes and astrocytes following transplantation into glia-deficient areas of CNS white matter. Glia 13:39–44.

Gallo V, Armstrong RC (1995): Developmental and growth factor-induced regulation of nestin in oligodendrocyte lineage cells. J Neurosci 15(1):394–406.

Goldman JE (1992): Regulation of oligodendrocyte differentiation. TINS 10: 359–362.

Grinspan J, Franceschini B (1995): Platelet-derived growth factor is a survival factor for PSA-NCAM+ oligodendrocyte pre-progenitor cells J Neurosci Res 41:540–551.

Groves A, Barnett S, Franklin R, Crang AA, Mayer M, Blakemore WF, Noble M (1993): Repair of demyelinated lesions by transplantation of purified O-2A progenitor cells. Nature 362:453–455.

Guillian D, Johnson B, Krebs JF, Tapscott MJ, Honda S, (1991): A growth factor from neuronal cell lines simulates myelin protein synthesis in mammalian brain. J Neurosci 11:327–336.

Gumpel M, Baumann N, Raoul M, Jacque C (1983): Survival and differenciation of oligodendrocytes from neural tissue transplanted into new-born mouse brain. Neurosci Lett 37:307–312.

Gumpel M (1991): Shiverer and other marker models used in intracerebral transplantations of glial cells. Transplants, implants and grafts. In: Methods in Neurosci., Vol 7, Academic Press, New-York, 378–394.

Hardy R, Reynolds R (1991): Proliferation and differentiation potential of rat forbrain oligodendroglia progenitors both *in vitro* and *in vivo*. Development 111:1061–1070.

Hardy R, Reynolds R (1993a): Neuron-oligodendroglial interactions during central nervous system development. J Neurosci Res 36:121–126.

Hardy R, Reynolds R (1993b): Rat cerebral cortical neurons in primary culture release a mitogen specific for early (GD3+:O4-) oligodendroglial progenitors. J Neurosci Res 34:589–600.

Hunter S.F, Bottenstein JE (1990): Growth factor responses of enriched bipotential glial progenitors. Dev Brain Res 54:235–248.

Hunter SF, Bottenstein JE (1991): O-2A glial progenitors from mature brain respond to CNS neuronal cell-derived growth factors. J Neurosci Res 28:574–582.

Kreider B, Grinspan J, Waterstone M, Bramblett G, Ances B, Williams M, Stern J, Lee V, Pleasure D (1996): Partial purification of a novel mitogen for oligodendroglia. J Neurosci Res 40:44–53.

Lachapelle F, Gumpel M, Baulac M, Jacque C, Duc P, Baumann N (1983): Transplantation of CNS fragments into the brain of shiverer mutant mice: extensive myelination by transplanted oligodendrocytes I: Immunohistochemical studies. Dev Neurosci 6:325–334.

Lachapelle F, Nait-Oumesmar B, Avellana-Adalid V, Bargey N, Seilhan D, Baron-Van Evercooren A (1996): FGF-2, EGF and PDGF A differentially activate *in vivo* the potential of grafted cells derived from the adult subventricular zone to generate myelin-forming oligodendrocytes. J Neurosci 390:14.

Louis JC, Magal E, Muir D, Manthorpe M, Varon S (1992): CG-4, A new bipotential glial cell line from rat brain, is capable of differentiating *in vitro* into either mature oligodendrocytes or type-2 astrocytes. J Neurosci Res 31: 193–204.

McKinnon RD, Matsui T, Dubois-Dalcq M, Aaronson SA (1990): FGF modulates the PDGF-driven pathway of oligodendrocyte development. Neuron 5:603–614.

McKinnon RD, Dubois-Dalcq M (1995): Cytokines and growth factors in the development and regeneration of oligodendrocytes. In Benveniste E, Ransohoff R (eds.): "Cytokines and the CNS; Development, Defense and Disease." Boca-Raton, Florida: CRC.

McMorris FA, Dubois-Dalcq M, (1988): Insulin-like growth factor I promotes cell proliferation and oligodendroglial commitment in rat glial progenitor cells developing *in vitro*. J Neurosci Res 21:199–209.

Nait-Oumesmar B, Lachapelle F, Vignais L, Bachelin C, Rougon G, Baron-Van Evercooren A (1996): Potential of repair of the adult subventricular zone following chemically induced demyelination J Neurosci 390: 13.

Noble M, Murray K, Stroobant P, Waterfield MD, Riddle P (1988): Platelet-derived growth factor promtes division and motility and inhibits premature differentiation of the oligodendrocyte/type-2 astrocyte progenitor cell. Nature 333:560–562.

Pfeiffer S, Warrington A, Bansal R (1993): The oligodendrocyte and its many cellular processes. Trends Cell Biol 3:191–197.

Raff MC (1989): Glial cell diversification in the rat optic nerve. Science 243:1450–1455.

Raff MC, Lillien LE, Richardson WD, Burne JF Noble MD (1988): Platelet-derived growth factor from astrocytes drives the clock that times oligodendrocyte development in culture. Nature 333:562–565.

Raff MC, Miller RH Noble M (1983): A glial progenitor cell that develops *in vitro* into an astrocyte or an oligodendrocyte depending on the culture medium. Nature 303:390–396.

Richardson W, Pringle N, Mosley M, Westermark B, Dubois-Dalcq M (1988): A role for platelet-derived growth factor in normalgliogenesis in the central nervous system Cell 53:309.

Small RK, Riddle P, Noble M (1987): Evidence for migration of oligodendrocyte-type2 astrocyte precursor cells into the developing rat optic nerve. Nature 328:155–157.

Sommer I, Schachner M (1981): Monoclonal antibodies (O1-O4) to oligodendrocyte cell surfaces: an immunological study in the central nervous system. Dev Biol 83:311–327.

Tontsch U, Archer D, Dubois-Dalcq M, Duncan I (1994): Transplantation of an oligodendrocyte cell line leading to extensive myelination. Proc Natl Acad Sci USA 91:11616–11620.

Trotter J, Bitter-Suermann D, Schachner M (1989): Differentiation-regulated loss of the polysialylated embryonic form and expression of the different polypeptides of the neural cell adhesion molecule by cultured oligodendrocytes and myelin. J Neurosci Res 22:369–383.

Trotter J, Crang AJ, Schachner M, Blakemore WF (1993) Lines of glial precursor cells immortalised with a temperature-sensitive oncogene give rise to astrocytes and oligodendrocytes following transplantation into demyelinated lesions in the central nervous system. Glia, 9:25–40.

Vignais L, Nait-Oumesmar B, Mellouk F, Gout O, Labourdette G, Baron-Van Evercooren A, Gumpel M (1993): Migration and differentiation of transplanted oligodendrocyte precursors in the adult demyelinated spinal cord. Int J Dev Neurosci 11 603–611.

Warrington AE, Barbarese E, Pfeiffer SE (1993): Differential myelinogenic capacity of specific developmental stages of the oligodendrocyte lineage upon transplantation into hypomyelinating hosts. J Neurosci Res 34:1–13.

Warrington AE, Barbarese E, Pfeiffer SE (1992): Stage specific (O4+Galc-) isolated oligodendrocyte progenitors produce MBP+myelin *in vivo*. Dev Neurosci 14:93.

GENERATION OF OLIGODENDROCYTE PRECURSORS FROM HUMAN NEURAL CELLS *IN VITRO*

K. Murray,[1] C. Tornatore,[2] M. Hajihosseini,[*1] and M. Dubois-Dalcq[1]

[1]Institut Pasteur
Department of Virologie
Unité NRSN
Paris, France 75724
[2]NINDS
Laboratory of Molecular Medicine
Bethesda, Maryland 20892

INTRODUCTION

We are interested in the molecular mechanisms underlying human oligodendrocyte development as well as regeneration of these myelin forming cells in demyelinating diseases (Dubois-Dalcq 1995). Recent studies have shown that oligodendrocytes and their precursors can be identified in human fetal spinal cord during the 7th week of gestation *in vitro* (Aloisi et al 1992), and on or before 45 days *in vivo* (Hajihosseini et al, 1995 and unpublished observations). Oligodendrocyte precursors first appear in two discrete regions on each side of the ependymal layer of the ventral spinal cord and later emerge in the ventrolateral and dorsal regions (ibidem and Weidenheim, 1993). It is not clear when and where such precursors first appear in the developing human brain and how they are generated. Primary cultures of human fetal brains at 12–15 weeks of gestation (WG) contain a small proportion of cells of the oligodendrocyte lineage (0.1 to 1%) (Satoh and Kim, 1994) reacting with anti-glycolipid antibodies (Pfeiffer et al 1993). Some of these cells actively divide in mixed cultures containing astrocytes and neuronal cells. When anterior and posterior regions of human fetal brain at 5–12 WG were cultured in defined medium with bFGF (basic fibroblast growth factor), the younger brain specimens initially yielded large populations of neuroepithelial cells coexpressing nestin and vimentin (Buc-Caron, 1995). In these culture conditions, such neural precursors maintained this phenotype for

* M.Hajihosseini is presently at the Viral Carcinogenesis Laboratory, ICRF, Lincoln's Inn Field, London, WC2A 3PX.

up to 6 weeks. In contrast, when grown with 10% FCS (fetal calf serum) for 4 weeks, the cultures became more heterogeneous and contained differentiated cells including postmi-- totic neurons, astrocytes and some oligodendrocytes expressing galactocerebroside (GC) (ibidem).

In order to determine how human oligodendrocytes are generated from their precur- sors one would need a system where neural precursors are generated for prolonged periods of time, are synchronized in their growth and can be induced to differentiate into myelin- forming cells *in vitro* and *in vivo*. In the rat, this was accomplished by obtaining purified populations of oligodendrocyte progenitors dividing in the presence of growth factors and differentiating when factors were taken away (Mc Kinnon et al, 1990, Louis et al, 1992). Such homogenous populations of progenitors committed to become oligodendrocytes have not yet been obtained in human because the specificities of ganglioside antibodies -used to purify those cells- are different in rat and human (Kim et al.1986). In addition, the myeli- nation program of the human CNS is much more prolonged and usually not synchronized in a given CNS region. Consequently there are few anatomical sites where enough pro- genitors proliferate in concert so that they can be isolated and purified at one point in time. One alternative would be to isolate early neural precursors (Buc-Caron 1995) which are multipotential and could be directed toward the oligodendrocyte fate. Finally, the re- cent characterization of oligodendrocyte « pre-progenitors » in mixed neonatal rat brain cultures has shown their immunoreactivity with antibodies to the sialylated form of NCAM (PSA-NCAM) and responsiveness to PDGF (platelet derived growth factor), al- lowing enrichment and prolonged survival of these cells (Rougon et al, 1993; Greenspan et al, 1995). N-CAM is a member of the Ig superfamily widely expressed on most CNS cells, and this adhesion molecule is the major carrier of PSA in vertebrates (Rougon 1993a). A specific reactivity distinguishing PSA-NCAM from N-CAM was obtained by raising a monoclonal antibody to meningococcus group B polysaccharides, anti-Men B (Rougon et al, 1986). PSA-NCAM[+] preprogenitors in the rat readily grow in clusters on the underlying astrocyte layer and, if passaged with the astrocytes, resettle on the dividing astrocytes and expand -probably clonally- into new aggregates. Similarly, clusters of round cells and cells with one or two processes were repeatedly observed on an astroglial cell layer derived from a 15 WG fetal brain, giving us the opportunity to characterize and investigate the fate of these putative oligodendrocyte preprogenitors in man.

CHARACTERIZATION OF PRECURSOR CELLS IN BRAIN CULTURES

Brain cultures were derived from a 15 WG fetal cerebral cortex after mechanical dissociation through a 19G needle and establishment of monolayers containing numerous GFAP (glial fibrillary acidic protein) positive astrocytes (Tornatore et al, 1991). Frozen cells were thawed and used in these studies between passages 9 and 13. At passage 12, cy- togenetic analysis was carried out as described (Kato et al, 1992) and revealed that over 90% of the cells nuclei displayed 46 chromosomes when examined in metaphase, demon- strating the genetic stability of these cells.

Mixed brain cultures were maintained in Dulbecco's Modified Eagle Medium with 10% fetal calf serum (DMEM/FCS) of which 50 to 70% was replenished twice weekly as the cells own conditioned medium (CM) appeared to promote their growth. After each passage, clusters of small round cells were growing on the layer of astrocytes expressing GFAP and nestin, an intermediary filament detected early in CNS development (Hockfield

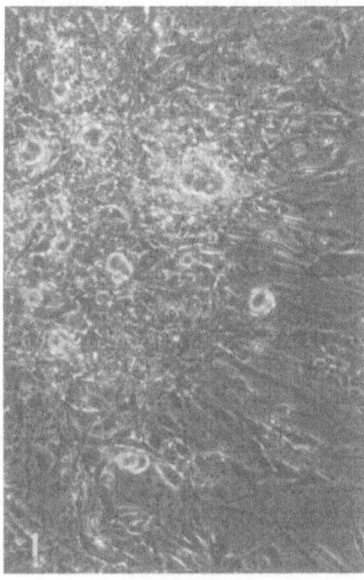

Figure 1. Phase contrast of clusters of precursors on astrocytes in a mixed culture 3 days after passage 12.

and McKay, 1985) (fig. 1).These small cells became bipolar as they migrated out of the clusters (fig. 2).

By double and triple indirect immunofluorescence labelling, these cells showed intense surface labelling for PSA-NCAM, and often expressed nestin but not GFAP (data not shown). Cells associated with PSA-NCAM+ clusters occasionally stained for neurofilament (NF) and,

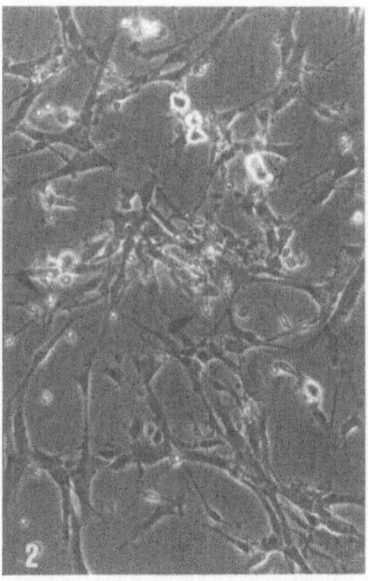

Figure 2. Bipolar cells (at arrows) after 4 days in culture. Photographs taken on a Leica inverted microscope.

more frequently, for the ganglioside GD3 and/or glycolipids characteristic of oligodendrocytes and their precursors. We have used three antibodies to glycolipids to study progression in the oligodendrocyte lineage as studied in the rat: the O4 antibody recognizes sulfatides on proligodendroblasts; Rmab (named anti-GalC by Boehringer) binds sulfatides, seminolipid and galactocerebroside (GC) and identifies a pre-oligodendrocyte stage in the absence of O1 antibody staining; O1 antibody recognizes only GC and binds to postmitotic oligodendrocytes (Bansal et al, 1989; Pfeiffer et al 1993). When immunofluorescence labelling with these antibodies was performed right after passaging the brain cultures, we detected a number of cell bodies intensely labelled with O1 antibody within 8 hours of plating. By 24 hours, such O1+ cells had acquired processes which became more elaborate with time (fig. 3). Rmab and O4 immunoreactivity was detected at 24 hours and its intensity increased thereafter. Rare NF+ cells were also detected from 8 to 72 hours after passing. Taken together, these observations suggested that cells of the oligodendrocyte lineage were generated from PSA-NCAM+/nestin + cells in these mixed brain cultures and that an oligodendrocyte fate may have been favored over neuronal fate in our culture conditions.

To further investigate this question, we used the immunopanning technique to purify PSA-NCAM$^+$ precursors (Barres et al., 1992; Greenspan et al., 1995, Dubois et al., 1995) (fig. 4).

Briefly, 10 cm petri dishes were coated overnight with 50 μg affinity purified goat anti mouse IgM, washed with PBS then incubated with anti-Men B for at least one hour. Cells were removed using 0.125% trypsin and plated at 3–4 million cells per dish for one hour at room temperature. From recently passaged cultures where only the small bipolar cells expressed PSA-NCAM, an average of 7% of the starting population was retrieved in purified cultures, which contained 95% PSA-NCAM$^+$ cells and 5% astrocytes (Pan A) (figs. 5 and 6).

If the cultures were maintained for weeks before passing, we noticed that the number of small bipolar PSA-NCAM + cells decreased with time and that larger flat cells, possibly precursors of astrocytes, acquired weak staining for PSA-NCAM; in that case, immunopanning resulted in 50% immunoselection and the enriched cultures consisted of a mixture of small neural precursors and flat astrocytes (Pan B) (fig. 7). Five immunopanning experiments were performed, 3 Pan A and 2 Pan B. When passage 14 was reached, the mixed brain cultures were no longer able to generate clusters of PSA-NCAM$^+$ precursors.

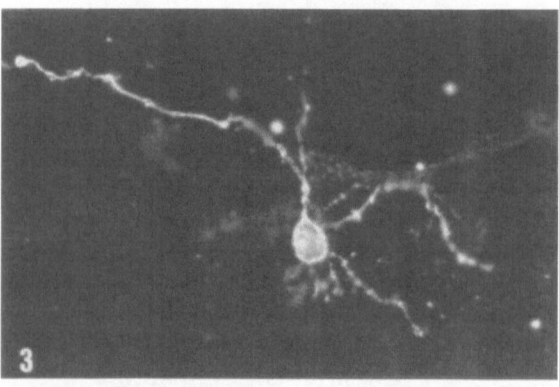

Figure 3. Immunofluorescent labeling of an oligodendrocyte with O1 antibody (supernatant 1:5) 4 days after passage 12.

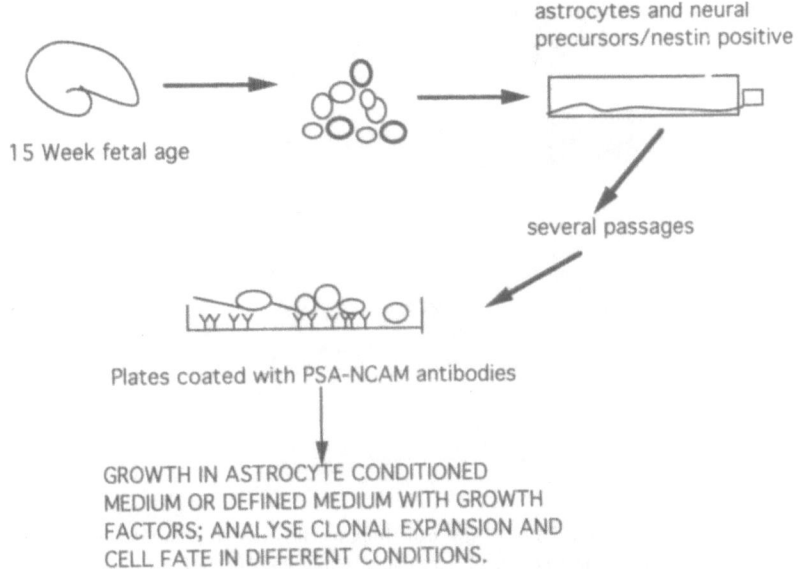

astrocytes and neural
precursors/nestin positive

15 Week fetal age

several passages

Plates coated with PSA-NCAM antibodies

GROWTH IN ASTROCYTE CONDITIONED
MEDIUM OR DEFINED MEDIUM WITH GROWTH
FACTORS; ANALYSE CLONAL EXPANSION AND
CELL FATE IN DIFFERENT CONDITIONS.

Figure 4. Purification of neural precursors from mixed human fetal brain cultures with PSA-NCAM antibody by immunopanning.

GROWTH AND DIFFERENTIATION OF THE PSA-NCAM⁺ PRECURSORS

Cultures enriched in PSA-NCAM$^+$ precursors were maintained in DMEM/FCS supplemented with 50% CM collected from the mixed cultures or from purified human fetal astrocytes (Aloisi et al. 1992). As PDGF and bFGF are mitogenic for rat O-2A progenitor

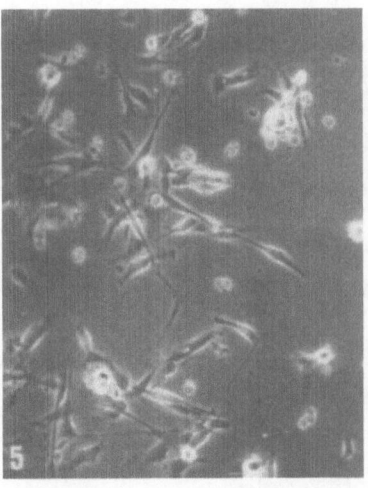

Figure 5. Phase contrast of bipolar and small round cells, 4 days after replating, selected by immunopanning with anti-PSA-NCAM (mouse IgM anti-Men B 1:250) from PanA cultures.

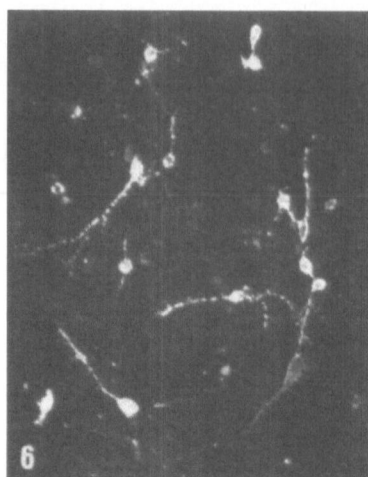

Figure 6. Immunofluorescence labeling of PSA-NCAM⁺ cells from recently passed (PanA) cultures 5 days after replating (anti-Men B 1:250). Photographs taken on a Leica Optovar equipped with epifluorescence.

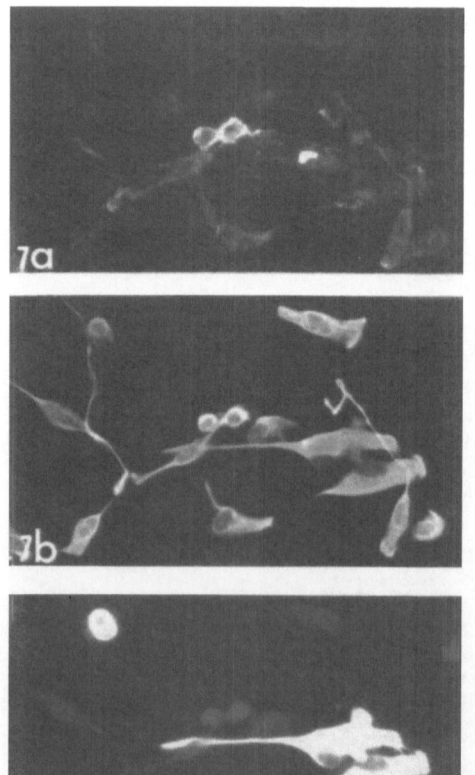

Figure 7. Immunopanning with antibody against PSA-NCAM from mixed brain cells cultured for several weeks (Pan B condition) retrieved small round cells which 2 days after replating, intensely labelled with PSA-NCAM(mouse IgM anti-Men B) (**a**) and Nestin(rabbit IgG) (**b**) but not GFAP(mouse IgG) (**c**). Other larger flat cells weakly PSA-NCAM⁺ and Nestin+/GFAP- may be astrocyte precursors.

cells (Noble et al, 1988, McKinnon et al, 1993), we also investigated whether these two growth factors could stimulate proliferation of PSA-NCAM⁺/nestin+ neural precursors when added to defined medium N1 (Louis et al.,1992) supplemented with biotin 10ng/ml, T3 15nM, hydrocortisone 10nM and insulin 50ng/ml. DMEM/FCS medium with CM was able to sustain steady growth of these cells while more substantial numbers of cells died over time in defined medium supplemented with growth factors. Yet, addition of bFGF promoted growth of clusters of PSA-NCAM⁺ cells as assayed by measurement of DNA synthesis after a BrdU pulse (Mc Kinnon et al 1990). In Pan A cells, BrdU at 1μM was added for 24 hr to PSA-NCAM⁺ cells grown in DMEM/FCS/CM on polylysine or in defined medium with 10 ng/ml bFGF grown on uncoated plastic to promote the formation of clusters. By triple immunofluorescence staining, 9% of PSA-NCAM⁺/Nestin+ cells in DMEM/ FCS/CM had incorporated BrdU between days 3 and 4 in culture. When pulsed from day 11 to 12, the BrdU index was slightly increased to 13% in DMEM/FCS/CM and 11% in the clusters growing in defined medium with bFGF. Although DNA synthesis was not measured in Pan B conditions, we noticed that growth of small PSA-NCAM⁺ cells was more vigorous, perhaps because of the presence of larger numbers of astrocytes. This assumption is supported by the results of an experiment in which PSA-NCAM⁺ enriched cells were detached from the astrocyte layer at passage 12 (by shaking at 150 rpm for 1 hour) and then seeded over a monolayer of pure human fetal astrocytes which were PSA-NCAM negative in DMEM/FCS/CM. (Aloisi et al., 1992). Between one and three days after seeding, the number of bipolar PSA-NCAM⁺ cells had increased by over eight fold, yet, after 13 days, their number had decreased substantially. Although preliminary, these observations suggest that astrocyte derived and/or serum factors as well as bFGF are mitogenic for PSA-NCAM⁺ neural precursors but that these cells also require unidentified survival factors missing in our *in vitro* system.

We then determined the antigenic phenotype of these immunopanned populations early (2 to 5 days) and late (2 to 3 weeks) after plating by using 2 combinations of triple immunofluorescence labelling. We stained for either PSA-NCAM/glycolipids recognized by RmAb/Nestin or PSA-NCAM/nestin/GFAP in order to define which precursors and glial phenotypes developped after immunopurification. In Pan A conditions, at the early time point, small bipolar cells were intensely stained with PSA-NCAM and generally co-expressed nestin (fig. 6). Rare PSA-NCAM-GFAP+ cells represent contaminating astrocytes. At the later time point, many clusters of small round cells had grown in DMEM/FCS/CM and some cells in these clusters were PSA-NCAM⁺/ Rmab+/ Nestin-. The acquisition of glycolipids and loss of nestin expression suggest that the PSA-NCAM⁺ precursors became committed with time to an oligodendrocyte fate while still expressing PSA-NCAM. Other cells underlying the clusters were weakly PSA-NCAM⁺, Nestin+ and sometimes GFAP+ and may consist of astrocytes and their precursors.

In Pan B conditions, at the early time point, large flat cells expressed both PSA-NCAM and Nestin, and a small percentage also stained for GFAP. The comparatively weak PSA-NCAM labelling on this population of flat cells could be due to residual molecules on the membrane blocked by the antibody during the immunopanning procedure, or these cells could represent astrocyte precursors (fig. 7).

Clustered on these cells, some intensely labelled PSA-NCAM⁺ small round and bipolar cells also stained with Rmab but not for Nestin and GFAP, suggesting again their appartenance to the oligodendrocyte lineage (fig. 8a and 8b). The later time point provided the same phenotypic profile but with a decrease in the number of R-mAb positive cells as the flat cells dominated the cultures.

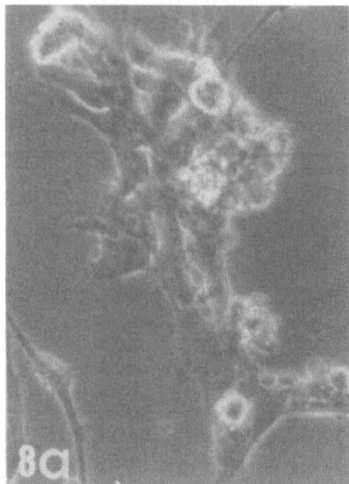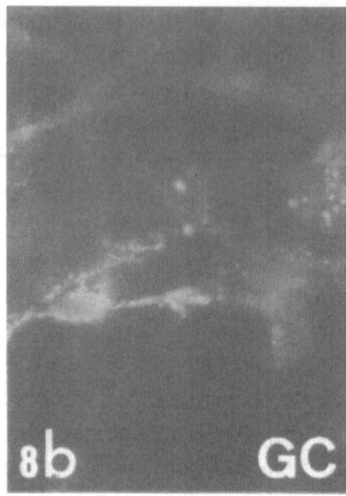

Figure 8. (a) Phase contrast of Pan B condition cells at the early timepoint of 5 days after replating, showing the clusters of small round cells on the flat cells with astrocyte morphology. **(b)** With double immunofluorescence some of the cells stained with PSA-NCAM antibody also express glycolipids as detected by RmAb (shown here).

COMMENTS AND CONCLUSIONS

In these experiments, we have observed high expression of PSA-NCAM on human neural precursors also expressing nestin. Such antigenic phenotype is also characteristic of CNS precursors isolated much earlier in human fetal life at 53–58 days post conception (Murray and Dubois-Dalcq, unpublished observations). This raises the question whether the PSA-NCAM/nestin + precursor cells we describe here are preprogenitors of oligodendrocytes or are multipotential neural precursors (Buc-Caron 1995). Nestin is an intermediate filament protein cloned from the embryonic rat CNS which is expressed in multipotential neural cells isolated from embryonic and adult brain (Lendahl et al, 1990; Gritti et al, 1996). *In vivo*, nestin is down-regulated while neurofilament and GFAP expression increased during differentiation of neurons and astrocytes respectively (Fredericksen and Mc Kay, 1988). Cultured rat astrocytes, however may maintain coexpression of nestin and GFAP (Gallo and Armstrong, 1995). While nestin is still expressed by rat oligodendrocyte progenitors, it is downregulated in differentiated oligodendrocytes (ibidem). Similarly the presence of PSA is generally associated with changes in phenotypes in a particular lineage. The PSA residues on N-CAM are not only thought to play a role in cell plasticity but also in migration such as in neural precursors moving in trains from the anterior ventricular zone to the olfactory bulb (O'Rourke 1996). PSA residues are expressed on early precursors of neurons, astrocytes and oligodendrocytes and later disappears during differentiation of these cells (Rougon, 1993b).

There are advantages and limitations to using *in vitro* systems such as the one described here to study human oligodendrocyte generation. The limitations are the rarity of specimens obtained at 15 WG and the difficulties in precise anatomical identification of the brain fragments from which CNS cells were derived. The latter information may be crucial to the success of future experiments and their reproducibility. This situation could be remediated by fixing a portion of each brain specimen to be cultured and performing histological examination. In addition, the subventricular zone could be dissected when integrity of the brain hemisphere is preserved. The advantage of the present *in vitro* system is that the mixed cultures could go through 13 passages during which precursors could be generated and/or purified for phenotypic characterization. These enriched cultures also provide a means of studying proliferation and differentiation of human neural precursor cells as well as exploring survival factors for these cells. Moreover the passed cells were shown to be genetically stable.

The intensity of antibody binding to PSA on clusters of human precursors was such that it allowed to immunoselect these cells by panning on antibody coated dishes. Such purification of PSA-NCAM/nestin + precursor cells was most efficient early after passage of the mixed CNS cells at a time when weak PSA immunoreactivity on astrocytes was not detected. The latter may result from upregulation of the polysaccharide synthesis and expression with time in culture. As compared to defined medium supplemented with bFGF, growth of PSA-NCAM/nestin + precursors was enhanced in serum containing medium supplemented with conditioned medium from the original mixed cultures or from cultures of purified human fetal astrocytes (Aloisi et al 1992). This suggests that astrocyte-derived factors are essential for growth of these precursors. Our *in vitro* conditions may have favored the precursors glial fate over a neuronal fate which was occasionally observed before purification. Thus multipotential precursors may have evolved into progenitors of oligodendrocytes as suggested by the emergence of cells expressing sets of glycolipids characterisic of oligodendrocyte lineage cells (Pfeiffer et al, 1993). Moreover we have recently observed O1+, GC expressing oligodendrocytes in cultures of PSA-NCAM+/nestin+ precursors isolated from the CNS at 53–58 days post conception (Murray and Dubois-Dalcq unpublished observations). Clonal analysis and/or additional purification experiments on other similar brain specimens are needed to definitely establish the generation of human oligodendrocytes from multi-potential neural precursors.

Although differentiating oligodendrocytes were repeatedly generated from PSA-NCAM immunoselected cultures, their number was rather low and their survival appeared compromised. Cell death is an integral part of CNS development processes and may be accentuated *in vitro* by the lack of essential survival factors or signalling by other cells such as neurons (Barres and Raff 1996). We have therefore begun a series of transplantation experiments in the embryonic rat brain (Brustle et al, unpublished observations). These experiments were inspired by the seminal observation that human fetal CNS fragments grafted in myelin-deficient shiverer mice generated oligodendrocytes myelinating on the mouse schedule (Gumpel et al, 1987). In utero xenografting of mouse neural precursors into the embryonic rat brain (E16-E18) has shown remarkable migratory and differentiating potential of the grafted cells (Brüstle et al.,1995). Since there is immunological tolerance at this time of development, one can also graft human neural cells which can integrate well into the rodent host brain. Indeed, enriched PSA-NCAM/nestin+ human cells (described above) injected into the ventricules of E18 pregnant rats migrated long distances into the developing rat brain and, after 2 to 3 weeks, groups of human oligodendrocyte progenitors and astrocytes had integrated in various regions of the rat CNS including the cerebellum (Brüstle et al., unpublished observations). These preliminary results suggest that the rodent CNS can provide human precursor cells with most signals necessary for their differentiation into oligodendrocytes. Therefore, this *in vivo* model may allow to investigate regional and cellular signals leading to human oligodendrocyte specification early in embryonic life.

ACKNOWLEDGMENTS

We thank Dr. G. Rougon for supplying us with the PSA-NCAM antibodies (anti-Men B) and Dr. F. Aloisi for the human fetal astrocytes.

REFERENCES

Aloisi F, Giampaolo A, Russo G, Peschle C, Levi G (1992): Developmental appearance, antigenic profile, and proliferation of glial cells of the human embryonic spinal cord: an immunocytochemical study using dissociated cultured cells. Glia 5:171–181.

Aloisi F, Borsellino G, Samoggia P, Testa U, Chelucci C, Russo G, Peschle C, Levi G (1992): Astrocyte cultures from human embryonic brain: characterization and modulation of surface molecules by inflammatory cytokines. J Neurosci Res 32:494–506.

Bansal R, Warrington AE, Gard AL, Ranscht B, Pfeiffer SE (1989): Multiple and novel specificities of monoclonal antibodies 01, 04, and R-mAb used in the analysis of oligodendrocyte development. J Neurosci Res 24: 548–557.

Barres B A, Hart I, Coles H, Burne J, Voyvodic J, Richardson W, Raff M (1992): Cell death and control of cell survival in the oligodendrocyte lineage. Cell 70:31–46

Barres B A, Raff M C (1996): Axonal control of oligodendrocyte development. In Jessen KR, Richardson WD (eds): "Molecular and Cellular Neurobiology- Glial Cell Development". London:ßIOS Sci Publ pp 71–81.

Brüstle O, Maskos U, McKay R (1995): Host-guided migration allows targeted introduction of neurons into the embryonic brain. Neuron 15:1–20.

Buc-Caron M-H (1995): Neuroepithelial progenitor cells explanted from human fetal brain proliferate and differentiate *in vitro*. Neurobiology of Disease 2:37–47.

Dubois C, Okandze A, Figarella-Branger D, Rampini C, Rougon G. (1995): A monoclonal antibody against Meningococcus group B polysaccharides used to immunocapture and quantify polysialylated NCAM in tissues and biological fluids. J of Immuno Methods 181:125–135.

Dubois-Dalcq M (1995): "On Regeneration of oligodendrocytes and myelin: Trends in Neurosciences, 18:289–291.

Frederiksen K, McKay RDG (1988): Proliferation and differentiation of rat neuroepithelial precursor cells *in vivo*. J Neurosci 8 : 1144–1151.

Gallo V, Armstrong RC (1995): Developmental and growth factor-induced regulation of nestin in oligodendrocyte lineage cells. J Neurosci 15 : 394–406.

Greenspan J, Franceschini B (1995): Platelet-derived growth factor is a survival factor for PSA-NCAM+ oligodendrocyte pre-progenitor cells. J Neurosci Res 41:540–551.

Gritti A, Parati EA, Cova L, Frolichsthal P, Galli R, Wanke E, Faravelli L, Morassuti DJ, Roisen F, Nickel DD, Vescovi L (1996): Multipotential stem cells from the adult mouse brain proliferate and self-renew in response to basic fibroblast growth factor. J. Neurosci 16(3): 1091–1100.

Gumpel M, LaChapelle F, Gansmuller A, Baulac M, Baron Van Evercooren, Baumann N (1987): Transplantation of human embryonic oligodendrocytes into shiverer brain. Ann. NY Acad Sci.495:71–85

Hajihosseini M, Dubois-Dalcq M (1995): Early expression of MBP and CNP in the fetal human spinal cord;abstract presented at the workshop on "Axonal regrowth in the mammalian spinal cord and peripheral nerve". Deauville, Normandy.

Hockfield S, McKay R (1985): Identification of major cell classes in the developing mammalian nervous system. J Neurosci 5:3310–3328.

Kato H, Nishida J, Honda T, Miyamoto S, Fujinaga K, Wake N (1992): Chromosome alterations contribute to neoplastic progression of transformed rat embryonal fibroblasts. Cancer Genet Cytogenet 58:39–47

Kim SU, Moretto G, Lee V, Yu RK (1986): Neuroimmunology of gangliosides in human neurons and glial cells in culture. J Neurosci Res 15:303–321.

Lendahl U, Zimmermann LB, McKay RDG (1990): CNS stem cells express a new class of intermediate filament protein. Cell 60: 585–595.

Louis JC, Magal E, Muir D, Manthorpe M, Varon S (1992a): CG-4, a new bipotential glial cell line from rat brain, is capable of differentiating *in vitro* into either mature oligodendrocytes or type-2 astrocytes. J Neurosci. Res. 31:193–204.

McKinnon, R., Matsui, T., Dubois-Dalcq, M. and Aaronson, S.: FGF modulates the PDGF- driven pathway of oligodendrocyte development *in vitro*, Neuron. 5:603–614, 1990.

McKinnon RD, Smith C, Behar T, Smith T, Dubois-Dalcq M (1993): Distinct effects of bFGF and PDGF on oligodendrocyte progenitor cells. Glia 7:245–254.

Noble M, Murray K, Stroobant P, Waterfield M, Riddle P (1988): Platelet-derived growth factor promotes division and motility and inhibits premature differentiation of the oligodendrocyte-type-2 astrocyte progenitor cell. Nature 333:560–562.

0'Rourke NA (1996): Neuronal chain gangs: homotypic contacts support migration into the olfactory bulb. Neuron 16:1061–1064.

Pfeiffer SE, Warrington AE, Bansal R (1993): The oligodendrocyte and its many cellular processes. Trends in Cell Biol. 3.6: 191–197.

Ranscht B, Clapshaw PA, Price J, Noble M, Seifert W (1982): Development of oligodendrocytes and Schwann cells studied with a monoclonal antibody against galactocerebroside. Proct Natl Acad Sci USA 79:2709–2713.

Reynolds BA, Tetzlaff W, Weiss S (1992): A multipotent EGF-responsive striatal embryonic progenitor cell produces neurons and astrocytes. J Neurosci 12:4565- 4574.

Rougon G, Dubois C, Buckley N, Magnani J, Zolinger W (1986): A monoclonal antibody against meningococcus group B polysaccharides distinguishes embryonic from adult NCAM. J Cell Biol 103:2429–2437.

Rougon G (1993): Structure, metabolism and cell biology of polysialic acids. Euro J Cell Biol 61:197–207.

Rougon G, Olive S, Figarella-Branger D (1993): Is PSA-NCAM a marker for cell plasticity? Advances in Life Sciences series "Polysialic acid: from microbes to man" p323- 333.J. Roth, U. Rutishauser and F.A. Troy (eds.) Birkhäuser- Verlag Basel/ Switzerland.

Satoh J, Kim S (1994): Proliferation and differentiation of fetal human oligodendrocytes in culture. J Neurosci Res 39:260–272.

Tornatore C, Nath A, Amemiya K, Major E (1991): Persistent human immunodeficiency virus type 1 infection in human fetal glial cells reactivated by T-cell factor(s) or by the cytokines tumor necrosis factor alpha and interleukin-1 beta. J of Virology Nov:6094–6100.

Weidenheim KM, Epstein I, Rashbaum WK, Lyman WD (1993): Neuroanatomical localization of myelin basic protein in the late first and early second trimester human fetal spinal cord and brainstem. J. Neurocytology 22: 507–516.

23

RECENT INSIGHTS INTO THE CELLULAR BIOLOGY OF REMYELINATION

Implications for Multiple Sclerosis

W. F. Blakemore, A. J. Crang, H. S. Keirstead, and R. J. M. Franklin

MRC Cambridge Centre for Brain Repair
Department of Clinical Veterinary Medicine
Madingley Road
Cambridge CB3 OES
United Kingdom

INTRODUCTION

In multiple sclerosis (MS) some areas of demyelination remyelinate while the majority do not. On the basis that it is possible to myelinate myelin-deficient areas by glial cell transplantation in experimental models (Lachapelle et al.1984; Blakemore and Crang, 1988; Blakemore et al.1995; Duncan, 1996), it may be possible to remyelinate areas of persistent demyelination in MS by introducing cells with remyelinating potential. However, if one is to predict the outcome of such a procedure it is necessary to understand the basis for success or failure of endogenous remyelination in MS, since it may be a basic property of this demyelinating disease that some areas of demyelination can be remyelinated while others cannot. If the latter situation pertains in areas of chronic demyelination, introducing cells with remyelinating potential will either have no effect or will only be of limited benefit. In order to understand why remyelination should fail in some lesions and not others, and thereby design a transplant strategy that is likely to be successful, it is necessary to resolve a number of inter-related issues that include: what cell gives rise to remyelinating oligodendrocytes, over what distance can such cells be recruited into an area of demyelination, can cells migrate freely within the adult CNS and within areas of demyelination? This article will address the first of these issues.

Myelin sheaths in areas of remyelination differ from those found in normal tissue in two respects, 1) they are thinner than normal for the axons they surround (Blakemore, 1974) and 2) their internodal length is reduced (Blakemore and Murray, 1981). Since the dimensions of myelin sheaths in normal tissue in adults are a consequence of both areas of initial association and growth it can be predicted that more oligodendrocytes will be required to remyelinate a population of axons than were required to myelinate it. This prediction is supported by the demonstration that the number of oligodendrocytes is increased in areas of remyelination in both experimental models and in multiple sclerosis lesions (Raine et al.1981; Raine et al.1988). We have also found an increase in oligodendrocyte number within remyelinated lesions, using Rip immunohistochemistry. There are a number of possibilities for the source of these extra cells. Firstly, on the basis of observations on acute MS lesions it has been proposed that oligodendrocytes may survive loss of their myelin sheath (Brück et al.1994; Ozawa et al.1994) and thus could either directly remyelinate the demyelinated axons, or divide with or without some degree of dedifferentiation to generate new oligodendrocytes (Wood and Bunge, 1986; 1991; Wood and Mora, 1993). Secondly, oligodendrocytes from outside the area of demyelination may be induced to divide (Ludwin and Bakker, 1988; Ludwin, 1984). An alternative proposition considers that an oligodendrocyte that has made myelin sheaths cannot divide and that remyelination requires a new population of oligodendrocytes generated by division of progenitor cells. Such cells are present in the adult nervous system (ffrench-Constant and Raff, 1986; Scolding et al.1995) and following demyelination may either survive within the area of demyelination or migrate into the area of demyelination from surrounding normal tissue. Integral to the idea of a recruited cell is the concept that the remyelinating cell has an immature phenotype capable of migration; however, the cell which undergoes mitosis to generate this cell is not unequivocally resolved since many studies implicate both precursors and differentiated oligodendrocytes as potential candidates (Prayoonwiwat and Rodriguez, 1993).

In this paper we present a series of observations indicating that: 1) mitosis is a requirement for remyelination, 2) oligodendrocytes which survive an episode of demyelination are not stimulated to divide by the presence of demyelinated axons and may not have the capacity to reform myelin sheaths.

MITOSIS IS A PREREQUISITE FOR REMYELINATION

There are many studies which indicate that the cells which remyelinate areas of demyelination are generated by mitosis (Ludwin, 1979; Prayoonwiwat and Rodriguez, 1993; Rodriguez et al.1991; Carroll and Jennings, 1994). Since it is possible to cause the death of dividing cells by x-irradiation we have conducted a series of experiments to resolve whether mature oligodendrocytes divide in response to demyelination.

When ethidium bromide is injected into white matter which has been subjected to 40 Grays of x-irradiation an area of demyelination is produced which shows no evidence of remyelination (Blakemore and Crang, 1988; Franklin et al.1993). This failure to remyelinate is due to a failure to generate remyelinating cells rather than a change in the demyelinated axons which renders them incapable of remyelination since these axons are easily remyelinated by transplanted myelinating cells (Blakemore and Crang, 1988). To demonstrate that this transplant-mediated remyelination is dependent on mitosis we undertook an experiment in which the area of spinal cord into which the transplants were made was subjected to 40 Grays of x-irradiation the day *after* transplantation, a procedure that removes the mitotic potential from both the spinal cord and the implanted cells. No remyelination was observed in these lesions and the transplanted animals could only be distinquished from uninjected animals by the presence of small numbers of astrocytes within the areas of demyelination. This result indicates that mitosis is a prerequisite for remyelination and confirms the results of a previous study in which we demonstrated removal of remyelinating potential of cultures by 40 Grays of x-irradiation (Franklin et al. 1993).

Since the aim of the current studies was to examine the remyelinating potential of mature oligodendrocytes, we examined whether it was possible to remove mitotically active members of the lineage by x-irradiation to leave post mitotic oligodendrocytes. When cultures of mixed glial cells were subjected to 40 Grays of x-irradiation the number of cells labelled with the O4 antibody was rapidly reduced to a constant low number (Fig. 1). The radiation resistent O4 positive cells had extensive processes and sheets of membrane which also labeled with O1 and O10 monoclonal antibodies, all characteristics of mature oligodendrocytes (Fig. 2). The number of these cells remained unchanged even when such cultures were passaged and grown for a further 8 days *in vitro*.

When such cultures were transplanted into x-irradiated ethidium bromide lesions astrocytes and a small number of cells with oligodendrocyte morphology were found within the area of demyelination (Fig 3). Associated with these oligodendrocytes were a small number of axons surrounded by a thin myelin covering (Fig 4). Although some of the axons appeared to have complete myelin sheaths, most were abnormal in that the myelin was not in the form of a membranous spiral around the axon. Instead, it appeared as if a membranous sheet was apposed to the axons, as there were two or more sets of oligodendrocyte loops present rather than the single cytoplasmic tongue observed in normal myelin sheaths. A similar situation was observed when preparations of oligodendrocytes prepared from white matter of adult rats were implanted into ethidium bromide lesions and x-irradiated post transplantation.

OLIGODENDROCYTES WHICH SURVIVE AN EPISODE OF DEMYELINATION ARE NOT STIMULATED TO DIVIDE BY THE PRESENCE OF DEMYELINATED AXONS

When lysolecithin (Blakemore and Patterson, 1978) or a combination of anti-galactocerebroside antibodies plus serum complement proteins were used to produce an area of

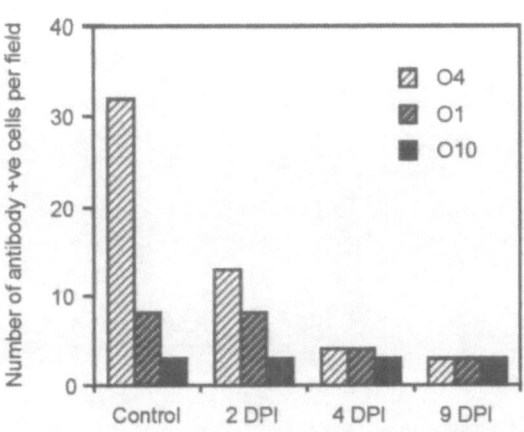

Figure 1. Oligodendrocyte-lineage-enriched cultures were isolated from mixed glial cell cultures prepared from two-day-old PVG rat cerebral hemispheres by shaking-off and harvesting the top-dwelling cells after 7 days *in vitro* (DIV). The effect of 40 Grays of X-irradiation at 9 DIV on cells of the oligodendrocyte lineage was examined by immunostaining cultures with O4, O1 and O10 mAbs at various times following X-irradiation. Compared to control cultures irradiated cultures exhibit a rapid reduction in the number of oligodendrocyte lineage cells by 2 days post irradiation (DPI) to reach an essentially stable level by 4 DPI.

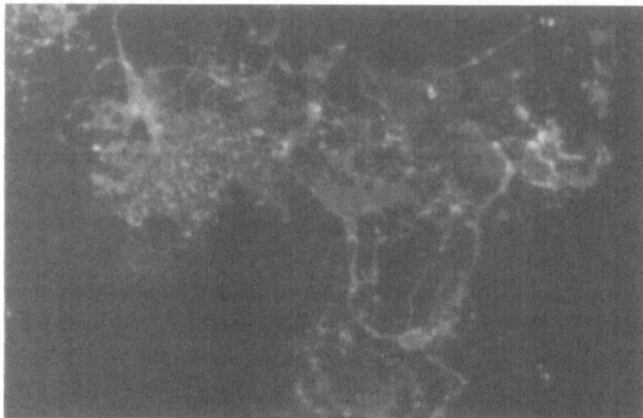

Figure 2. Mixed glial cell cultures prepared from cerebral hemispheres of two-day-old rats were exposed to 40 Grays X-irradiation at 14 DIV and were passaged and replated onto equivalent culture areas at 18 DIV. Surviving oligodendrocytes were assessed by immunostaining cultures after 25 DIV. Irradiation-resistant O1+ve cells have extensive processes and sheets of membrane, characteristic of mature oligodendrocytes. x150.

demyelination in spinal white matter which has been subjected to 40 Grays of x-irradiation there was no evidence of remyelination in animals killed up to 4 weeks after injection. A small number of cells with oligodendrocyte morphology could be found in the region of demyelination closest to normal tissue (Fig 5), where one would expect the dose of demyelinating agent would be lower than in the centre of the lesion. (This is in contrast to the situation that occurs following injection of ethidium bromide which can only cause demyelination by killing oligodendrocytes.)

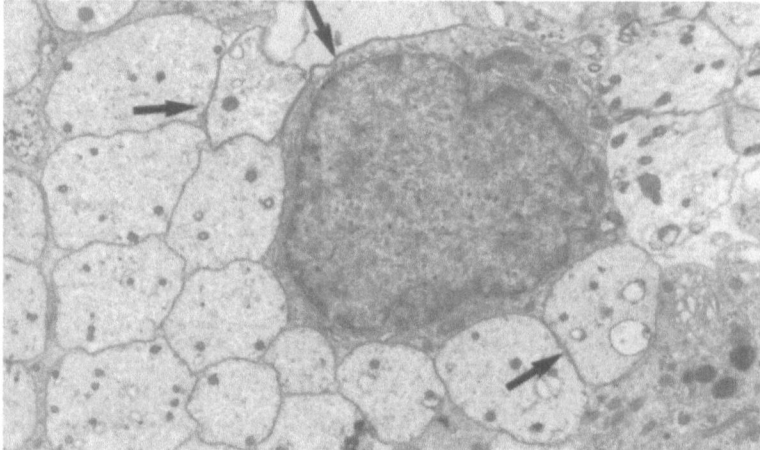

Figure 3. A cell with the ultrastructural appearance of an oligodendrocyte 28 days after transplantation of a mixed glial cell culture exposed to 40 Grays of x-irradiation into an ethidium bromide lesion made in the x-irradiated spinal cord. Some myelin membranes can be seen adjacent to the cell and between the demyelinated axons (arrows). x9500.

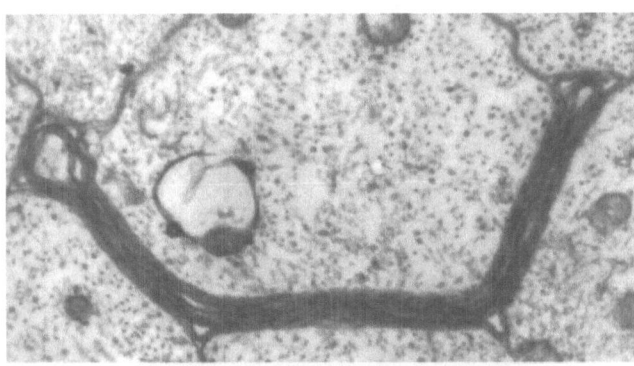

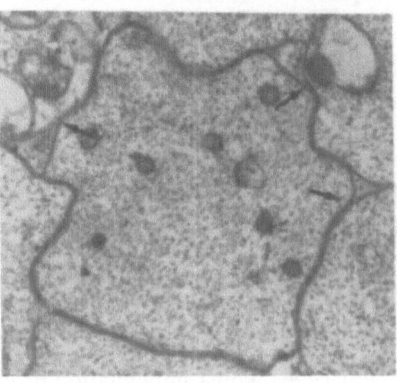

Figure 4. (Top) In the vicinity of cells such as that illustrated in figure 3, an occasional myelin-like structure can be found and (Bottom) compacted oligodendrocyte processes associate with the demyelinated axons (arrows indicate points of cytoplasmic fusion). x 19000.

Associated with these surviving cells were cell processes that associated with de-myelinated axons in a manner characteristic of oligodendrocyte association with axons during myelination and remyelination (Fig 6). In some rare instances sheets of myelin membrane could be detected. However, in no instances were myelin spirals observed around the demyelinated axons.

On examination of immunostained sections small numbers of Rip+ve cells were de-tected in the periphery of the area of demyelination providing confirmation of our ultras-tructural identification of these cells as oligodendrocytes (Fig 7). When lesions were examined 24 and 72hrs after injection of anti-galactocerebroside antibodies and comple-ment, oligodendrocytes were absent from the centre of the lesion but could be detected in areas of demyelination close to normal tissue (Fig 8). These observations demonstrate that even though some oligodendrocytes survived the loss of their myelin sheaths, such cells were not stimulated to divide by the presence of demyelinated axons since they persisted in tissue exposed to a dose of x-irradiation that results in death of dividing cells.

Further evidence for the inability of oligodendrocytes that have made myelin sheaths to remyelinate demyelinated axons comes from contrasting the results of implanting em-bryonic and adult human brain glia into rodents. The implantation of fragments of embry-onic human brain into newborn shiverer mice results in formation of myelin sheaths

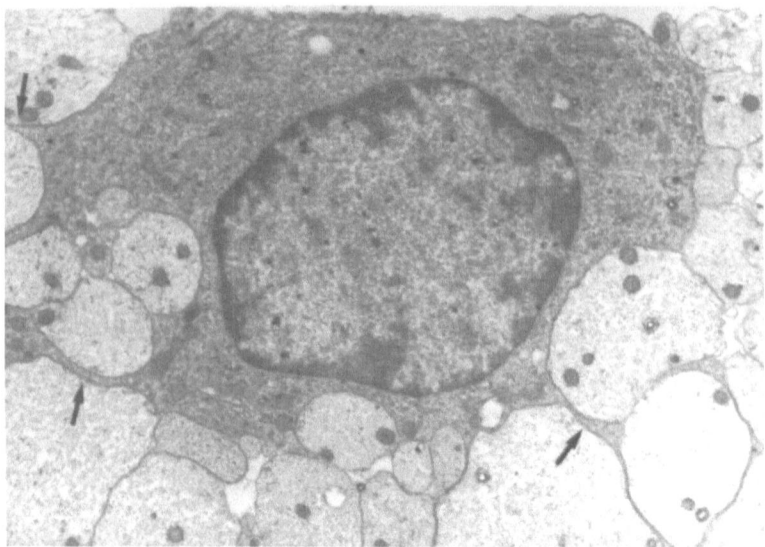

Figure 5. 3 weeks after injecting anti-galactocerebroside antibodies and complement into spinal tissue which has been subjected to 40 Grays of x-irradiation a small number of oligodendrocytes can be found within the area of demyelination. Processes from these cells separate demyelinated axons (arrows). x 12000.

(Gumpel et al.1987; Seilhean et al.1996) whilst implantation of oligodendrocytes derived from the human adult brain into areas of demyelination in immunosuppressed rats results in survival of cells but no myelin sheath formation (Targett et al.1996). Of particular relevance to the present discussion is the similarity of the relationship those non-irradiated cells adopted with the demyelinated axons to that seen in the x-irradiated anti-galactocere-

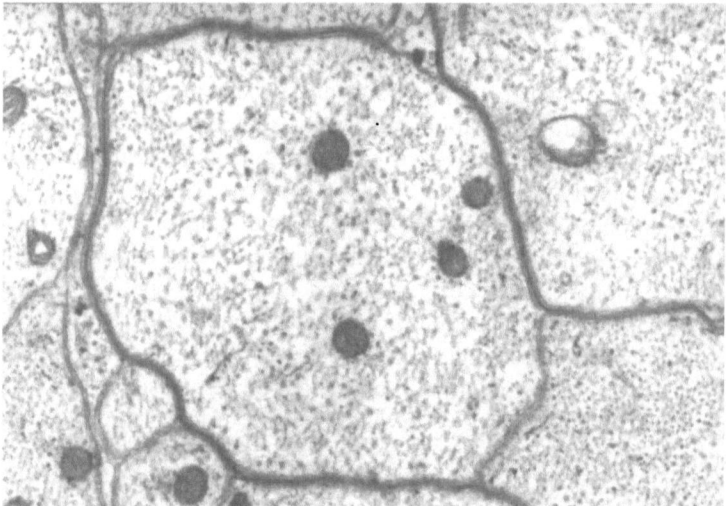

Figure 6. In association with the cells found in the x-irradiated lesions oligodendrocyte processes, which often show compaction, associate with demyelinated axons but no myelin sheaths are formed. The myelin membrane associated with this axon does not enclose the axon. x 30,000.

 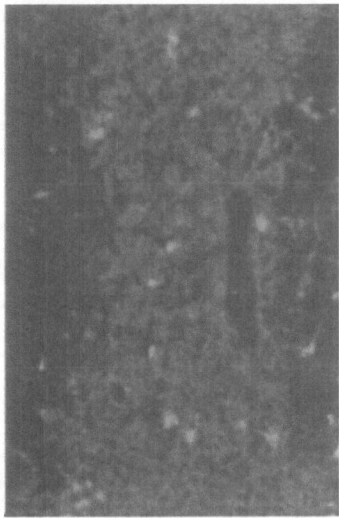

Figure 7. Rip immunoreactive oligodendrocytes within a demyelinated region of the dorsal column 3 weeks after irradiation and intraspinal injection of anti-galactocerebroside antibodies plus complement. A small number of irradiation resistant oligodendrocytes can be seen in the region of demyelination closest to normal tissue. The number and distribution of Rip+ve cells confirms our ultrastructural identification of these cells as oligodendrocytes.

Figure 8. Rip immunoreactive oligodendrocytes within a demyelinated region of the dorsal column 48 hours after intraspinal injection of anti-galactocerebroside antibodies plus complement. The number and distribution of Rip+ve cells at early time points indicates that some oligodendrocytes survive within regions of demyelination, despite loss of their myelin sheaths.

broside lesions and following the implantation of x-irradiated oligodendrocytes. In all three situations the cells have processes which engage axons, have an ability to make sheets of myelin membrane but fail to form myelin sheaths.

Taken together the observations recorded above support the conclusion that mitosis is a prerequisite for remyelination (Wood and Bunge, 1991; Ludwin, 1979; Prayoonwiwat and Rodriguez, 1993; Rodriguez et al.1991; Carroll and Jennings, 1994) and indicate that oligodendrocytes which have supported myelin sheaths or reached a certain degree of differentiation in tissue culture do not divide in response to the presence of demyelinated axons. Although these non-dividing cells are able to associate with axons and elaborate myelin membranes, they do not have the ability to form myelin sheaths. Because of the power of the irradiation paradigm to suppress cell division it remains a possibility that there exists a sub-population of oligodendrocytes that will divide in response to demyelination and give rise to cells capable of myelin sheath formation. However, the behaviour of the adult derived oligodendrocyte transplants do not provide compelling evidence for the existence of such a cell. Thus in the context of understanding cellular mechanisms of remyelination and thereby developing logical transplantation strategies, the series of studies described in this paper lead to the conclusion that 1) the overwhelming majority (if not all) remyelinating oligodendrocytes are generated from progenitor cells, 2) transplantation of preparations that only contain mature oligodendrocytes that have achieved a myelinating phenotype is unlikely to result in remyelination, and 3) there is little point in attempting to promote remyelination by enhancing the survival of oligodendrocytes that have lost their myelin sheaths.

ACKNOWLEDGMENTS

This work was supported by the Multiple Sclerosis Society of Great Britain and Northern Ireland. HSK holds a postgraduate award from the Canadian NSERC. RJMF holds a Wellcome Trust Career Development Fellowship.

REFERENCES

Blakemore WF (1974) Pattern of remyelination in the CNS. Nature 249: 577–578

Blakemore WF, Crang AJ (1988) Extensive oligodendrocyte remyelination following injection of cultured central nervous system cells into demyelinating lesions in adult central nervous system. Dev Neurosci 10: 1–10

Blakemore WF, Crang AJ, Franklin RJM (1995) Transplantation of glial cells. In: Ransom BR, Kettenmann H (eds) "Neuroglia". Oxford University Press, New York, pp 869–882

Blakemore WF, Murray JA (1981) Quantitative examination of internodal length of remyelinated nerve fibres in the central nervous system. J Neurol Sci 49: 273–284

Blakemore WF, Patterson RC (1978) Suppression of remyelination in the CNS by x- irradiation. Acta Neuropathol 42: 105–113

Brück W, Schmied M, Suchanek G, Brück Y, Breitschopf H, Poser S, Piddlesden S, Lassmann H (1994) Oligodendrocytes in the early course of multiple sclerosis. Annal Neurol 35: 65–73

Carroll WM, Jennings AR (1994) Early recruitment of oligodendrocyte precursors in CNS demyelination. Brain 117: 563–578

Duncan ID (1996) Glial cell transplantation and remyelination of the central nervous system. Neuropath Appl Neurobiol 22: 87–100

ffrench-Constant C, Raff MC (1986) Proliferating bipotential glial progenitor cells in adult optic nerve. Nature 319: 499–502

Franklin RJM, Crang AJ, Blakemore WF (1993) The reconstruction of an astrocytic environment in glia-deficient areas of white matter. J Neurocytol 22: 382–396

Gumpel M, Lachapelle F, Gansmuller A, Baulac M, Baron-Van Evercooren A, Baumann N (1987) Transplantation of human embryonic oligodendrocytes into shiverer brain. Annal N Y Acad Sci 495: 71–85

Lachapelle F, Gumpel M, Baulac M, Jacque C, Duc P, Baumann N (1984) Transplantation of CNS fragments into the brain of shiverer mutant mice: Extensive myelination by implanted oligodendrocytes. Dev Neurosci 6: 325–334

Ludwin SK (1979) An autoradiographic study of cellular proliferation in remyelination of the central nervous system. Amer J Pathol 65: 683–696

Ludwin SK (1984) Proliferation of mature oligodendrocytes after trauma to the central nervous system. Nature 308: 274–275

Ludwin SK, Bakker DA (1988) Can oligodendrocytes attached to myelin proliferate. J Neurosci 8: 1239–1244

Ozawa K, Suchanek G, Breitschopf H, Brück W, Budka H, Jellinger K, Lassmann H (1994) Patterns of oligodendroglia pathology in multiple sclerosis. Brain 117: 1311–1322

Prayoonwiwat N, Rodriguez M (1993) The potential for oligodendrocyte proliferation during demyelinating disease. J Neuropath exp Neurol 52: 55–63

Raine CS, Scheinberg L, Waltz JM (1981) Multiple sclerosis: Oligodendrocyte survival and proliferation in an active established lesion. Lab Invest 45: 534–546

Raine CS, Moore GRW, Hintzen R, Traugott U (1988) Induction of oligodendrocyte proliferation and remyelination after chronic demyelination. Relevance for Multiple Sclerosis. Lab Invest 59: 467–476

Rodriguez M, Pierce ML, Thiemann RL (1991) Immunoglobulins stimulate central nervous system remyelination: electron microscopic and morphometric analysis of proliferating cells. Lab Invest 64: 358–370

Scolding NJ, Rayner PJ, Sussman J, Shaw C, Compston DAS (1995) A proliferative adult human oligodendrocyte progenitor. NeuroReport 6: 441–445

Seilhean D, Gansmüller A, Baron-Van Evercooren A, Gumpel M, Lachapelle F (1996) Myelination by transplanted human and mouse central nervous system tissue after long-term cryopreservation. Acta Neuropathol 91: 82–88

Targett MP, Sussman J, O'Leary MT, Compston DAS, Blakemore WF (1996) Failure to achieve remyelination of demyelinated rat axons following transplantation of glial cells obtained from the adult human brain. Neuropath Appl Neurobiol 22: 199- 206

Wood PM, Bunge RP (1986) Evidence that axons are mitogenic for oligodendrocytes isolated from adult animals. Nature 320: 756–758

Wood PM, Bunge RP (1991) The origin of remyelinating cells in the adult central nervous system - The role of the mature oligodendrocyte. Glia 4: 225–232

Wood PM, Mora J (1993) Source of remyelinating oligodendrocytes. Adv Neurol 59: 113–123

24

CHARACTERISING THE ADULT HUMAN OLIGODENDROCYTE PROGENITOR

Neil Scolding,* Chris Shaw, and Alastair Compston

University of Cambridge Neurology Unit
Addenbrooke's Hospital
Hills Rd Cambridge CB2 2QQ and
MRC Cambridge Centre for Brain Repair
University Forvie Site
Robinson Way, Cambridge, CB2 2SR
United Kingdom

INTRODUCTION

The demonstration that the normal adult human CNS harbours a proliferative oligodendrocyte progenitor may have significant implications for remyelination in multiple sclerosis. Here, we describe the experimental background to this cell, and discuss our findings regarding two properties important in the context of repair: firstly (and most obviously) its capacity for laying down new myelin, and secondly, its proliferative response.

In multiple sclerosis, repeated episodes of inflammatory and immunological damage targeted upon myelin and oligodendrocytes ultimately results in the formation of multiple scar-like plaques within the CNS (Scolding *et al.* 1994). Within each lesion, myelin is lost and oligodendrocytes are absent; a variable astrocytic overgrowth completes the picture. Demyelination disrupts saltatory conduction along previously myelinated axons traversing the lesion, paralysing normal function within these pathways and causing permanent neurological disability.

Until very recently, most neurological clinicians and experimental neurobiologists in the field shared the view that the prospects for developing therapeutic strategies for promoting myelin repair in multiple sclerosis were rather remote. However, advances in three key related but separate areas have substantially altered this position.

Firstly, significant and important progress has been made in the treatment of multiple sclerosis (Jacobs *et al.* 1996). β-interferon, now in clinical use, was the first agent demonstrated clearly to affect the course of multiple sclerosis (Duquette *et al.* 1993). Its

---

* Current address: Department of Neurology, Institute of Psychiatry, De Crespigny Park, London.

effect is modest, with a decrease in the frequency of relapses of approximately one third, but appears consistent. Other agents ending their period of evaluation, such as copolymer-1, appear to have an effect very similar in magnitude (Johnson *et al.* 1995); yet others of promise remain under investigation (Moreau *et al.* 1994). Few neurologists now doubt that, within the medium term future, agents will be available which, alone or in combination, will allow the inflammatory and autoimmune processes responsible for myelin damage to be substantially suppressed, creating the immunologically quiescent environment crucial for the success of any reparative treatment.

Secondly, it is now clear that, given an immunologically stable environment, the de- or dysmyelinated rodent CNS can be very effectively remyelinated by the transplantation of neonatal glial progenitors (Duncan, 1996). Prior *in vitro* expansion of these cells, by exposure to defined growth factors, does not appear to impair their reparative capacity (Groves *et al.* 1993). A number of studies in fact suggest that progenitors are significantly better at remyelination than fully differentiated cells (Warrington *et al.* 1993), and that the capacity for proliferation is an important prerequisite for successful repair [reviewed in Chapter by, Bill Blakemore et al.].

The final area of important progress is in our understanding of the pathology of multiple sclerosis. A number of studies over the past few years have independently confirmed the presence of thinly remyelinated axons in a minority of acute lesions in multiple sclerosis (Raine & Wu, 1993; Prineas *et al.* 1993). This indicates that the limitations imposed upon repair within the demyelinated human CNS are not absolute, and emphasises the feasibility of pursuing lasting remyelination as a therapeutic goal. Significantly, however, the cell ultimately responsible for this partial remyelination remains to be identified. One suggestion is that oligodendrocytes unaffected by the inflammatory process might re-extend processes and remyelinate denuded axons. Others describe oligodendrocyte loss in acute lesions; it has been suggested in consequence that a putative oligodendrocyte progenitor migrates into lesions from unaffected areas, differentiates, and then elaborates new myelin. In either case, the observation of spontaneous repair clearly indicates that the adult human CNS not only provides an environment not wholly hostile to myelin repair, but also contains a cells with significant potential for remyelination.

We therefore sought to establish whether the adult human CNS, in common with the adult rodent brain (FfrenchConstant & Raff, 1986), does indeed contain a proliferative oligodendrocyte progenitor. Since the rodent progenitor has largely been defined by its *in vitro* properties, we studied cultured glial cells derived from adult human temporal lobe, removed surgically to treat drug-resistant epilepsy. Cultured cells, prepared using the method of Armstrong et al (Armstrong *et al.* 1992) were characterised after 48–72 hours *in vitro*. At this stage, 95–99% glial cells were round or multi-processed O4-positive cells; most were GalC-positive, but we also found 3% were O4-positive/GalC-negative pre-oligodendrocytes as previously described (Figure1 A,B); others also have described this adult human pre-oligodendrocyte (Prabhakar *et al.* 1995).

In addition, however, we found a very small proportion (1.3 + 0.7%) of oligodendrocyte-lineage cells which comprised uni-, bi- or tripolar cells positive for A2B5 (Figure1 C) but negative for GalC and GFAP (Scolding *et al.* 1995). These cells were also negative for antibody markers of microglia and neurones, and therefore closely resembled the rat O-2A progenitor in both morphology and immunophenotype (Raff *et al.* 1983; Ffrench-Constant & Raff, 1986). We found these putative progenitors to exhibit the capacity *in vitro* for differentiation into A2B5-positive/GFAP-positive astrocytes (the phenotype of the so-named type-2 astrocyte) or oligodendrocytes (initially A2B5 and GalC co-positive), indicating a third feature shared with the rat oligodendrocyte progenitor, a bipotential differentiation capacity *in vitro*.

However, unlike the rodent progenitor, the adult human cell failed to proliferate in response to basic fibroblast growth factor (bFGF), platelet derived growth factor (PDGF) and neurotrophin-3 (NT-3), the three principal rat progenitor mitogens, alone or in combination. Interestingly, this observation finds parallels in studies of embryonic human oligodendrocyte progenitors, two groups independently having shown no specific proliferative response to either PDGF or bFGF (Satoh & Kim, 1994; Aloisi *et al.* 1992). In common with rodent oligodendrocyte progenitor cells, however, human progenitors were found to proliferate when growing on a monolayer of (adult human) astrocytes (Scolding *et al.* 1995) (Figure 1 D,E, small arrow). This top-dwelling layer also included a population of proliferating GalC-ve/O4+ve pre-oligodendrocytes (not shown).

These *in vitro* findings of a cell which therefore convincingly exhibited the characteristics of an adult human oligodendrocyte progenitor could not necessarily be extrapolated to the human brain *in vivo*. We therefore prepared tissue prints of fresh temporal lobe and parietal cortex white matter using previously described techniques (Armstrong *et al.* 1992), and found process-bearing A2B5-positive cells (Figure1 F,G). Approximately two thirds of these cells were co-positive for GFAP, the phenotype of the type-2 astrocyte, and the remainder were GFAP-negative, providing strong support for the existence *in vivo* of the progenitor characterised *in vitro*.

MYELIN SYNTHESIS BY CELLS OF THE ADULT HUMAN OLIGODENDROCYTE LINEAGE

As described in the Chapter by (Blakemore et al.), we studied the myelinating potential of adult human oligodendrocytes by transplanting mixed populations of oligodendrocyte lineage cells into gliotoxic demyelinated lesions in the irradiated rodent spinal cord (Targett *et al.* 1996). Histopathological analysis of the transplanted lesions revealed obvious survival of human oligodendrocytes, with positive staining for the myelin proteins MBP and MAG, and clear areas of axo-glial contact. Ensheathment of demyelinated axons by oligodendrocyte processes occurred, but large areas of compact myelin formation were not seen.

We did not formally establish whether differentiated adult oligodendrocytes or progenitors effected these initial stages of the process of myelination, but the presence of only very low numbers of progenitors in the mixed transplanted population implies strongly that terminally differentiated oligodendrocytes were responsible. Our findings contrasted with those of Seilhean et al, who found myelin synthesis by fetal human glia transplanted into a dysmyelinating rodent (Seilhean *et al.* 1996). The proliferative properties of the transplanted cells was not established, but their fetal origin might suggest a higher proportion of progenitors, of greater mitotic activity than adult CNS-derived cells.

We attempted to develop an *in vitro* system to observe in more detail the interactions between axons and human glia, and to address some of the questions raised by the failure of significant elaboration of compact myelin sheaths in our transplantation experiments. Comparable mixed populations of adult human glia - comprising 90% oligodendrocytes, 5% pre-oligodendrocytes, 4% astrocytes and 1% progenitors - were seeded onto cultures of embryonic rat dorsal root ganglia (DRG) established according to conventional protocols. Rat oligodendrocytes or their precursors successfully myelinate axons extended by these neurones (Wood & Bunge, 1986).

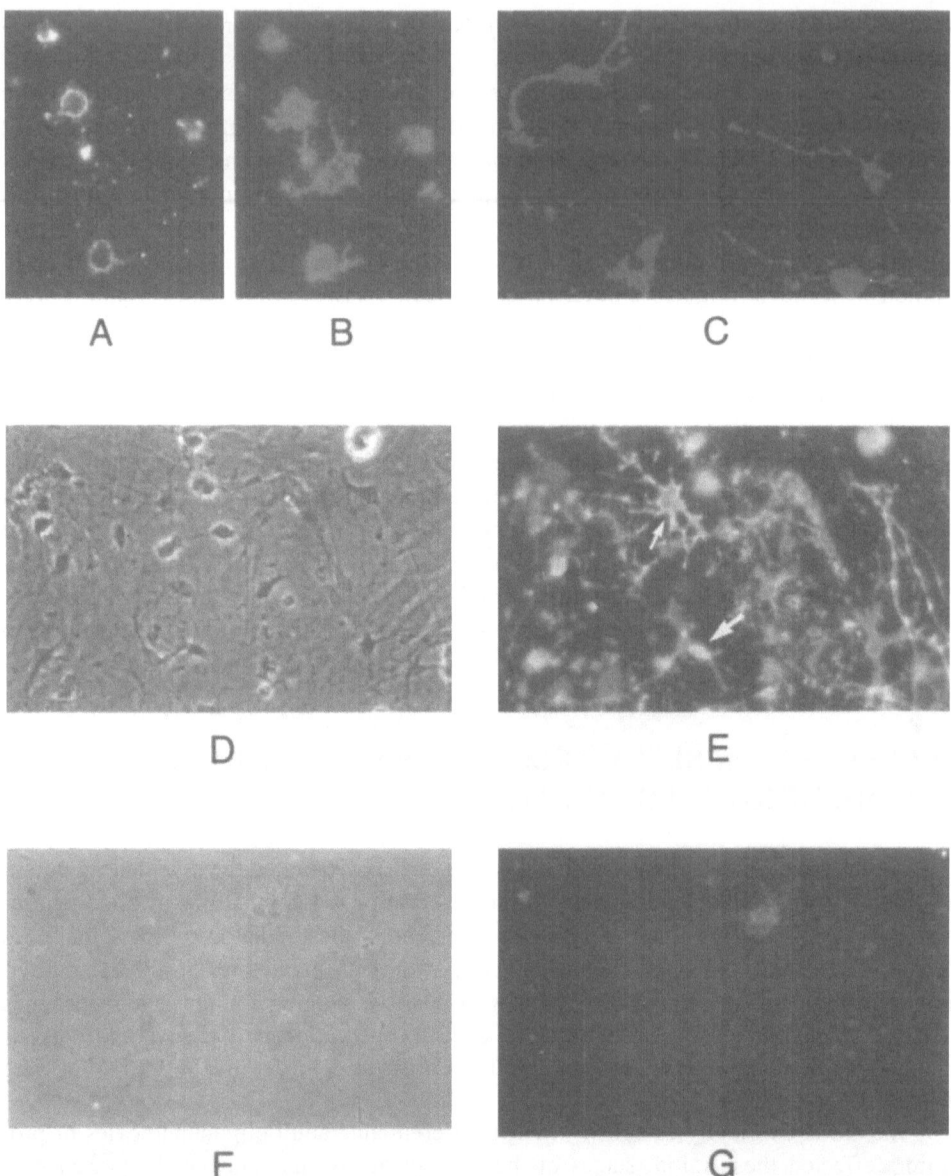

Figure 1. A human glial cell with similarities to the rat O-2A progenitor. **A,B.** In the first few days of culture, most of the cells replated from the initial non-adherent population were round and GalC+ve/O4+ve oligodendrocytes or GalC-ve/O4+ve pre-oligodendrocytes as previously described. Here GalC is tagged with fluorescein (**A**), O4 with rhodamine (**B**); one preoligodendrocyte is shown. In addition, bi-and tripolar cells were present which were A_2B_5 positive (**C**; rhodamine optics, x400) and GalC and GFAP negative (not shown). **D,E.** After 25-50 days in vitro, a monolayer of astrocytes had appeared in cultures of initially adherent cells, upon which a sparse layer of cells with progenitor phenotype was seen (**D**); after 48 hours' incubation with $10^{-4}M$ BrDU, numerous cells were positive for A_2B_5 and had incorporated BrDU (**E**; large arrow; double exposure: A_2B_5 - rhodamine optics; BrDU - fluorescein optics, x400). Surface staining with anti-GalC antibody (applied before cell permeabilisation and also demonstrated with a fluorescein conjugate) reveals a mixed population of process-bearing BrDU-negative/ A_2B_5-negative/GalC positive oligodendrocytes, A_2B_5-positive/BrDU-positive/GalC-negative progenitors (which were also GFAP-negative; not shown), and newly-derived oligodendrocytes (small arrow; A_2-B_5-positive/BrDU-negative/GalC-positive). **F,G.** Fresh tissue print preparations of adult human temporal lobe white matter confirmed the presence of processed, A_2B_5-positive cells (phase contrast and rhodamine optics). Approximately 75% of these were GFAP-positive, the remainder GFAP-negative (not shown).

04-positive/GalC-ve oligodendroglial cells extended processes towards axons; contact and adherence between oligodendrocyte processes and axon surfaces was observed, and an extensive network of processes aligned with axon bundles was established. After four weeks *in vitro*, extensive ensheathment of axons by myelin membranes positive for myelin basic protein and galactocerebroside was found. After six weeks, loose wraps of myelin around axons were identifiable by transmission electron microscopy (TEM); myelin compaction around axons was not seen.

We attempted to develop a novel human co-culture system by using neurones derived from the hNT_2 human choriocarcinoma cell line, rather than embryonic rodent DRGs, as a source of axons for adult human oligodendrocytes. Differentiated hNT_2 cells extend both axonal and dendritic processes, and have the antigenic phenotype - expressing neurofilament and other neuronal markers - and electrophysiological properties of CNS neurones (Pleasure & Lee, 1993). After 4–6 weeks in co-culture, numerous areas of axoglial contact and ensheathment were seen by light microscopy, but TEM examination was hampered by difficulty in identifying hNT axonal profiles. Human oligodendrocyte-neurone interactions broadly similar to those in the rodent-human co-cultures and, indeed, to the human-rodent transplantation experiments were, however, apparent on TEM. The formation by oligodendrocytes of loose wraps of myelin around neurite-like profiles was seen, but there was no evidence of compaction of myelin sheaths around axons.

If, as mentioned above (and see also Chapter by, Blakemore et al.), proliferation of a myelinating precursor is the key to successful compact myelin sheath synthesis, then more vigorous myelin synthesis in a human co-culture system might have been predicted, based on the extrapolation of the observation in rodent co-culture systems that axons - and neuronal conditioned medium - stimulate a brisk proliferative response in (rodent) oligodendrocyte progenitors. We therefore sought evidence of proliferation within human oligodendrocyte lineage cells, but found none, either with rodent DRG neurones, or with hNT cells.

THE PROLIFERATIVE RESPONSE OF THE ADULT HUMAN OLIGODENDROCYTE PROGENITOR

These investigations of the capacity for myelin production *in vitro* and *in vivo* are therefore clearly consistent with, and help to consolidate the contention that proliferation is the key to a successful remyelinating cell. We attempted to define further the proliferative properties of the adult human oligodendrocyte progenitor.

The initial observations showed that growth on a monolayer of astrocytes provided a stimulus to progenitor division (Scolding *et al.* 1995). Although a number of studies have shown that astrocytes are capable of synthesising and secreting a wide range of growth factors, including the potent neonatal rat oligodendrocyte progenitor mitogens bFGF, PDGF and NT-3, these findings almost invariably relate to rat astrocytes, usually embryonic or neonatal, and often following exposure to various exogenous stimuli (Eddelston & Mucke, 1993). Little is known of growth factor production by unstimulated adult human astrocytes.

Using an immunocytochemical approach, we found that GFAP-positive cultured human astrocytes (and also microglia) synthesised demonstrable PDGF-AA protein. However, neither astrocytes nor contaminating microglia in these cultures synthesised detectable quantities of bFGF or NT-3. When live cells were stained using the same anti-growth factor antibodies, neither PDGF-AA, NT-3, nor bFGF could be demonstrated on the surface of human astrocytes.

Using a complementary approach to seek growth factor receptors on the surface of A2B5-positive adult human oligodendrocyte progenitors, we found no evidence of either NT-3 or bFGF receptors on these cells, but do have preliminary results indicating that surface PDGF-α receptor is present. We have not yet sought receptor message using *in situ* techniques - clearly the presence of mRNA does not necessarily indicate translated product.

These observations support a role for PDGF in human oligodendrocyte progenitor proliferation; the very substantial body of evidence concerning the activity of PDGF as a mitogen for neonatal and adult rodent oligodendrocyte progenitors also helps to support this suggestion (Pfeiffer *et al.* 1993). However, neither soluble recombinant human PDGF, nor human astrocyte-conditioned medium stimulated adult human oligodendrocyte progenitor proliferation, suggesting that contact with the astrocyte surface is a necessary component of the proliferative signal. We were unable to demonstrate detectable PDGF on the astrocyte surface, and it may therefore be that a combination of PDGF produced by astrocytes acting with as-yet unidentified astrocyte cell surface components, may represent the stimulus for adult human oligodendrocyte progenitor division.

Other, simpler, explanations can also be offered, however. It is commonly accepted that the proliferative response of neonatal rodent progenitor populations *in vitro* is at least partially dependent on cell density: sparsely plated cells exhibit a less pronounced response to growth factor stimuli than densely populated cultures, while single cells are still less responsive. As yet undefined homotypical contact or autocrine factors active only an a micro-environmental level are thought to be responsible for this effect. The adult human progenitor is present in mixed oligodendrocyte lineage cultures only at very low densities, and usually as single cells not in contact with other progenitors (or, indeed, with glia of any other persuasion). This alone might explain the apparent lack of proliferative effect of combinations of bFGF, PDGF, and NT-3, a possibility allowing one to avoid invoking a rather striking species difference between rodent and human cells in their mitotic response to defined growth factors.

CONCLUSION

Observations in other laboratories on the histopathological changes in tissue from patients with multiple sclerosis or other demyelinating conditions clearly indicate that the adult human CNS contains a population of cells of the oligodendrocyte lineage which have a substantial remyelinating capacity. Experimental findings in rat studies suggest that the adult mammalian CNS does harbour a proliferative oligodendrocyte progenitor and, at least in regard to neonatal cells, the progenitor is a much more efficient remyelinating cell than the fully differentiated oligodendrocyte.

In a series of experiments, we have attempted to bridge the gap between observations based on human pathology on the one hand, and on experimental animals on the other. We have studied oligodendrocyte lineage cells from adult human white matter, and our findings have led to two principle conclusions. First, that the adult human CNS also contains a proliferative oligodendrocyte progenitor, and that this cell has marked similarities to its rodent counterpart. Astrocyte-derived PDGF may, at least partially, act as a proliferative stimulus. Second, mixed populations of adult human oligodendrocyte lineage cells, consisting in the main of differentiated oligodendrocytes, are capable both *in vitro* and *in vivo* of synthesising myelin membranes which may be assembled into loosely compacted oligolamellar sheath-like structures. However, the available evidence indicates that they are not able to elaborate in any significant quantity properly compacted multi-lamellated myelin sheaths around axons.

Future work further defining the proliferative response and the myelinating potential of the adult human oligodendrocyte lineage will be aided respectively by the finding that unstimulated human astrocytes directly trigger adult human oligodendrocyte progenitor mitosis, and by the development of an allogeneic co-culture system allowing us more closely to study and to manipulate glial-neuronal interactions by adult human oligodendrocytes and their progenitors cultured with axon-forming human neurones.

ACKNOWLEDGMENTS

NJS is an MRC Clinician Scientist Fellow; CS is supported by The Wellcome Trust. We also acknowledge the support of the Multiple Sclerosis Society of Great Britain and Northern Ireland. This work was made possible through the help and co-operation of the neurological and neurosurgical epilepsy teams, and particularly Professor F. Scaravilli and staff in the Department of Pathology, at the National Hospital for Neurology and Neurosurgery, London, and the neurosurgical theatre and research staff, particularly Lynn Maskell, at Addenbrooke's Hospital, Cambridge. We thank Genentech for the provision of neurotrophin-3 and Amgen Inc. for the provision of IGF-1, bFGF, PDGF and CNTF. We are particularly grateful to Paula Rayner for her expert technical assistance.

REFERENCES

Aloisi F, Giampaolo A, Russo G, Peschle C, Levi G. (1992): Developmental appearance, antigenic profile, and proliferation of glial-cells of the human embryonic spinal-cord - an immunocytochemical study using dissociated cultured-cells. Glia 5:171–181.

Armstrong RC, Dorn HH, Kufta CV, Friedman E, Dubois Dalcq ME. (1992): Pre- oligodendrocytes from adult human CNS. Journal of Neuroscience 12:1538–1547.

Duncan ID. (1996): Glial cell transplantation and remyelination of the central nervous system. Neuropath App Neurobiol 22:87–100.

Duquette P, Girard M, Despault L, Dubois R, Knobler RL, Lublin FD, Kelley L, Francis GS, Lapierre Y, Antel J, Freedman M, Hum S, Greenstein JI, Mishra B, Muldoon J, Whitaker JN, Evans BK, Layton B, Sibley WA, Laguna J, Krikawa J, Paty DW, Oger JJ, Kastrukoff LF, Moore GRW, Hashimoto SA Morrison W, Nelson J, Goodin DS, Massa SM, Gutteridge E, Arnason BGW, Noronha A, Reder AT, Martia R, Ebers GC, Rice GPA, Lesaux J, Johnson KP, Panitch HS, Bever CT, Conway K, Wallenberg JC, Bedell L, Vandennoort S, Weinshenker B, Weiss W, Reingold S, Pachner A, Taylor W. (1993): Interferon beta-1b is effective in relapsing-remitting multiple-sclerosis - clinical-results of a multicenter, randomized, double-blind, placebo-controlled trial. Neurology 43:655–661.

Eddelston M, Mucke L. (1993): Molecular profile of reactive astrocytes - Implications for their Role in neurologic disease. Neuroscience 54:15–36.

Ffrenchconstant C, Raff MC. (1986): Proliferating bipotential glial progenitor cells in adult rat optic nerve. Nature 319:499–502.

Groves AK, Barnett SC, Franklin RJM, Crang AJ, Mayer M, Blakemore WF, Noble M. (1993): Repair of demyelinated lesions by transplantation of purified O-2A progenitor cells. Nature 362:453–455.

Jacobs LD, Cookfair DL, Rudick RA, Herndon RM, Richert JR, Salazar XX, Fischer JS, Goodkin DE, Granger CV, Simon JH, Alam JJ, Bartoszak XX, Bourdette DN, Braiman J, Brownscheidle CM, Coats ME, Cohan SL, Dougherty DS, Kinkel RP, et al. (1996): Intramuscular interferon beta-1a for disease progression in relapsing multiple sclerosis. Annals of Neurology 39:285–294.

Johnson KP, Brooks BR, Cohen JA, Ford CC, Goldstein J, Lisak RP, Myers LW, Panitch HS, Rose JW, Schiffer RB, Vollmer T, Weiner LP, Wolinsky JS, Bird SJ, Kolson DL, Gonzalezscarano F, Brennan D, Mandler RN, Rosenberg GA, et al. (1995): Copolymer 1 reduces relapse rate and improves disability in relapsing- remitting multiple sclerosis: Results of a phase III multicenter, double- blind, placebo- controlled trial. Neurology 45:1268–1276.

Moreau T, Thorpe J, Miller D, Moseley I, Hale G, Waldmann H, Clayton D, Wing M, Scolding N, Compston A. (1994): Preliminary evidence from magnetic resonance imaging for reduction in disease activity after lymphocyte depletion in multiple sclerosis. Lancet 344:298–301.

Pfeiffer SE, Warrington AE, Bansal R. (1993): The oligodendrocyte and its many cellular processes. Trends in Cell Biology 3:191–197.

Pleasure SJ, Lee VMY. (1993): Ntera 2 cells: A human cell line which displays characteristics expected of a human committed neuronal progenitor cell. Journal of Neuroscience Research 35:585–602.

Prabhakar S, Dsouza S, Antel JP, Mclaurin JA, Schipper HM, Wang E. (1995): Phenotypic and cell-cycle properties of human oligodendrocytes in- vitro. Brain Research 672:159–169.

Prineas JW, Barnard RO, Kwon EE, Sharer LR, Cho ES. (1993): Multiple sclerosis: Remyelination of nascent lesions. Annals of Neurology 33:137–151.

Raff MC, Miller RH, Noble M. (1983): A glial progenitor cell that develops *in vitro* into an astrocyte or an oligodendrocyte depending on culture medium. Nature 303:390- 396.

Raine CS, Wu E. (1993): Multiple sclerosis: Remyelination in acute lesions. Journal of Neuropathology and Experimental Neurology 52:199–204.

Satoh J, Kim SU. (1994): Proliferation and differentiation of fetal human oligodendrocytes in culture. Journal of Neuroscience Research 39:260–272.

Scolding NJ, Zajicek JP, Wood N, Compston DAS. (1994): The pathogenesis of demyelinating disease. Progress in Neurobiology 43:143–173.

Scolding NJ, Rayner PJ, Sussman J, Shaw C, Compston DAS. (1995): A proliferative adult human oligodendrocyte progenitor. Neuroreport 6:441–445.

Seilhean D, Gansmuller A, Baronvan Evercooren A, Gumpel M, Lachapelle F. (1996): Myelination by transplanted human and mouse central nervous system tissue after long-term cryopreservation. Acta Neuropathologica 91:82–88.

Targett MP, Sussman J, Scolding N, O'Leary MT, Compston DAS, Blakemore WF. (1996): Failure to achieve remyelination of demyelinated rat axons following transplantation of glial cells obtained from the adult human brain. Neuropath App Neurobiol 22:199–206.

Warrington AE, Barbarese E, Pfeiffer SE. (1993): Differential myelinogenic capacity of specific developmental stages of the oligodendrocyte lineage upon transplantation into hypomyelinating hosts. Journal of Neuroscience Research 34:1–13.

Wood PM, Bunge RP. (1986): Myelination of cultured dorsal root ganglion neurons by oligodendrocytes obtained from adult rats. Journal of the Neurological Sciences 74:153–169.

25

THE REPAIR OF CENTRAL NERVOUS SYSTEM MYELIN

Cell Biology and Pharmacologic Action of Remyelination-Promoting Autoantibodies

Samuel F. Hunter, Kunihiko Asakura, David J. Miller, and Moses Rodriguez

Departments of Neurology and Immunology
Mayo Clinic and Foundation
Rochester, Minnesota 55905

MYELIN REPAIR IS POOR FOLLOWING INFLAMMATORY DEMYELINATION OF THE CENTRAL NERVOUS SYSTEM

In this symposium, we will review our rationale and experience with pharmacologic strategies for the promotion of myelin repair in inflammatory demyelinating disease. The most common clinical form of these illnesses in humans is multiple sclerosis (MS). MS and most other forms of inflammatory demyelination consist of an idiopathic, inflammatory syndrome which affects predominantly the densely myelinated white matter of the central nervous system (CNS). The inflammatory response is associated with damage of oligodendrocytes (ODC) and myelin, resulting in loss of these tissue elements (reviewed in Hunter & Rodriguez, 1995). Similar diseases exist in animals of both idiopathic and viral origin. Theiler's murine encephalomyelitis virus (TMEV), a naturally occurring, murine, picornaviral pathogen, produces inflammatory demyelinating lesions in spinal cord and brainstem with progressive neurologic disability in susceptible mouse strains.

Disability in both MS and TMEV disease results from multifactorial axonal dysfunction due to inflammation, demyelination, and a less prominent axonal loss. Although inflammation and the underlying immunopathy can be treated by a number of pharmacologic strategies, restoration of tissue structure and function is required following any immune insult to the CNS. This process has been called "remyelination," and spontaneous CNS myelin repair has long been recognized histologically in humans. Although CNS myelin repair can occur despite ongoing inflammatory processes (Rodriguez et al., 1993), it is often limited, especially as disease progresses. An extensive analysis of MS lesions has suggested that a single, small inflammatory event is remyelinated well, but recurrent inflammation results in failure of remyelination (Prineas et al., 1993). Similarly, SJL/J mice (a highly susceptible strain) with TMEV-induced demyelination spontaneously repair only about 5–8% of the injured areas (Miller et al., 1994). Therefore, facilitation of myelin repair has become a major therapeutic goal, complementing other strategies for the management of the recurrent and chronic leukoencephalitis of MS.

A number of potential obstacles to spontaneous remyelination have been proposed. These include inadequate numbers of ODC or progenitors, chronic ODC or myelin injury ("dying-back oligodendrogliopathy"), deficiency of molecular signals or substrates for myelination, or suppression of repair by immune or glial signals. Our laboratory has approached solutions to the remyelination problem from two directions: trophic factor-mediated modulation of ODC behavior in vitro, and stimulation of myelin repair in TMEV-infected mice by autoantibodies.

TROPHIC FACTOR-MEDIATED REGULATION OF THE OLIGODENDROCYTE LINEAGE

Insights from the Neuronal Cell Line-Model of Neuron-ODC Interaction

In vitro studies of ODC development have demonstrated important actions of trophic signals in ODC development. Similar mechanisms could underlie myelin repair and restoration of ODC and myelin following inflammatory demyelination. Early observations indicated that both astroglial and neuronal lineage cells could regulate gliogenesis of neonatal rat brain glia in vitro. Noble and colleagues (1988) demonstrated platelet-derived growth factor (PDGF) to have a limited mitogenic action on immature ODC lineage, explaining the earlier observations that astroglia could also prolong proliferation of progenitors in vitro. A more profound effect on neonatal ODC production was produced by soluble factors (conditioned medium, CM) from the B104 rat CNS neuronal cell line, resulting in large increases in ODC in vitro (Bottenstein et al., 1988). This system was conceived originally as a model of neuroglial interaction mediated by soluble factors from a differentiated neuronal cell line. The oligodendropoetic action of B104 CM was demonstrated to be due to proliferation of the immature ODC progenitor (Hunter & Bottenstein, 1989). As shown in Figure 1, when the neonatal ODC progenitor was exposed for a prolonged period to B104 CM, it became slowly differentiating, multipolar ODC progenitor. Unlike the neonatal progenitor, this cell remained proliferation-competent even after several days in the absence of growth factors (Hunter & Bottenstein, 1990). B104 CM also was mitogenic for glial progenitors isolated from mature rat brain (Hunter & Bottenstein, 1991). This confirmed that these neuronal cell line-derived soluble factors could not only commit neonatal glial progenitors to the mature brain glial progenitor phenotype, but remained active in regulating ODC production beyond the neonatal period. Attempts to neutralize the action of B104 factors with anti-PDGF-AB antibody suggested that PDGF alone was not responsible for the action of B104 CM (Hunter & Bottenstein, 1990, 1991). In contrast to the action of purified PDGF, ODC progenitors could be continually expanded with B104 CM, permitting production of a clonal cell line (Louis et al., 1992). The response of O-2A progenitors to B104 CM was similar to that

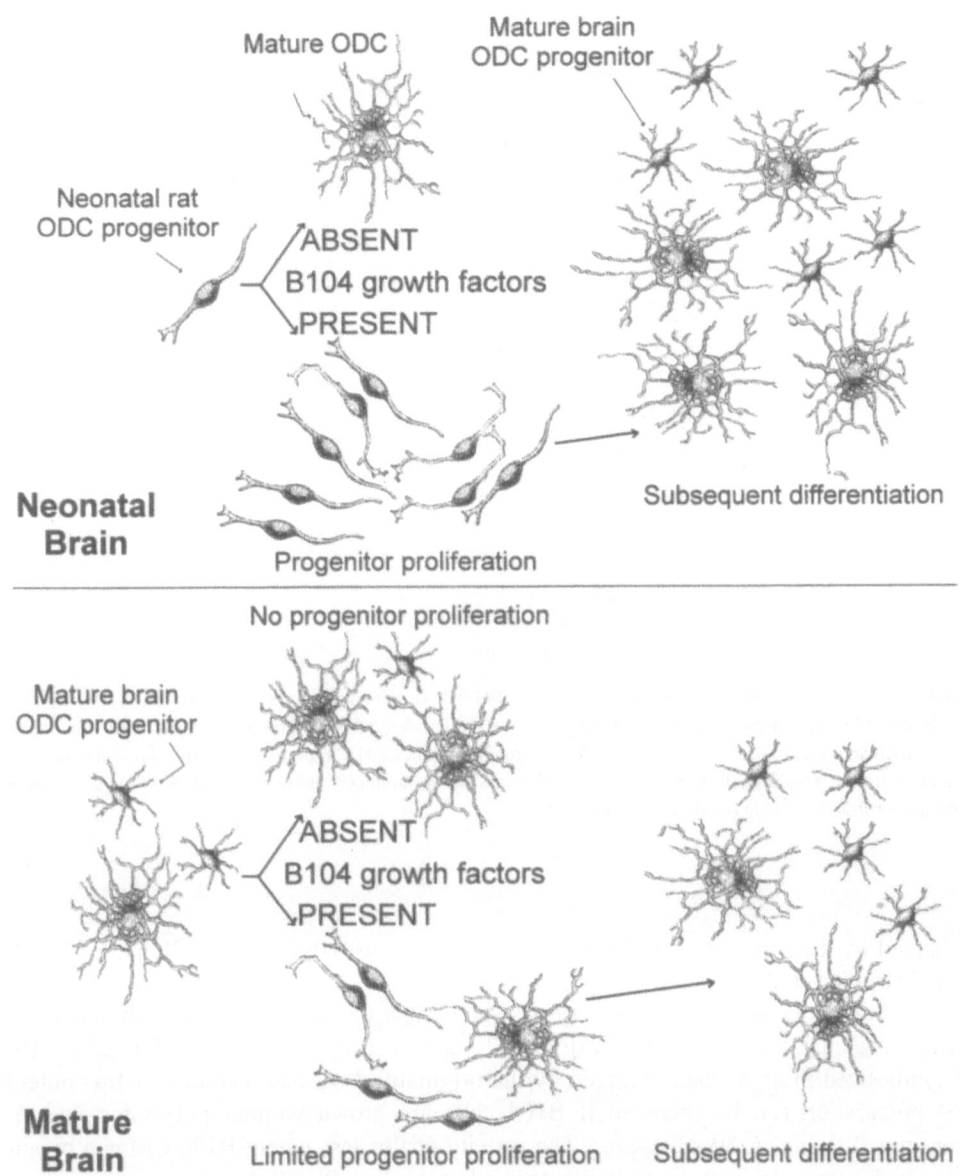

Figure 1. Action of B104 CM on glial development. (**upper**) B104 CM has a profound oligodendropoetic action on neonatal rat brain glia. This action is due to a block in differentiation and sustained proliferation of the stem cell-like ODC progenitor. As cell density increases, differentiation is initiated into both ODC and multipolar ODC progenitors. The latter cells are mitogen-responsive and slowly differentiating. (**lower**) ODC progenitors prepared with B104 CM and those derived from mature rat brains remain responsive to B104 CM after a period in the absence of growth factors; differentiation into GC-positive ODC occurs slowly. Repeat stimulation with B104 CM evokes a modest proliferative response of progenitors and subsequent increases in ODC.

256 S. F. Hunter *et al.*

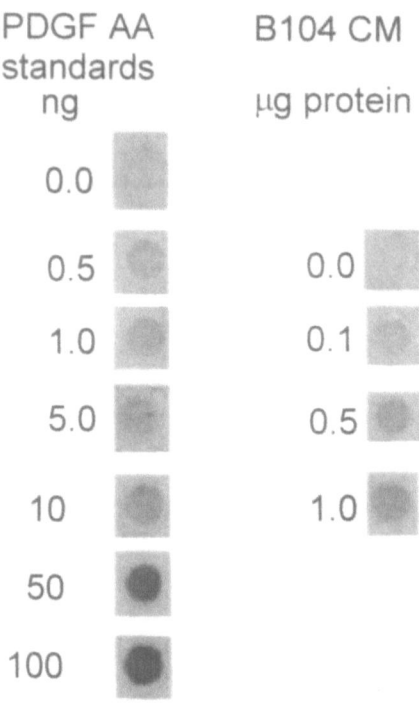

Figure 2. Immunoreactive PDGF-AA is present in B104 CM. An immunoblot with purified PDGF-AA standards and B104 CM protein demonstrates that about 1% of the protein derived from B104 CM (Bottenstein et al., 1988) is immunoreactive as PDGF-AA, which is about 180 ng /ml in this B104 CM preparation with a protein concentration of 13 μg/ml. This finding is confirmed by SDS-immunoblot, immunostaining for PDGF-AA, and detection of message for PDGF A-chain (Asakura et al, submitted).

produced by the synergistic action of basic fibroblast growth factor (bFGF, FGF-2) with PDGF (Bögler et al., 1990).

Recent work from our laboratory has shown that the B104 line synthesizes and secretes large quantities of PDGF-AA (Figure 2), up to 0.2 μg/ml in B104 CM, or about 1% of synthesized protein when prepared by the originally described methods. A less potent CM preparation can be prepared if B104 cells are grown without polylysine and fibronectin (PDGF-AA 30–40 ng/ml). The activity of this less potent B104 CM can be antagonized with sufficient neutralizing antibodies against PDGF-AA. Using the reverse transcription-polymerase chain reaction, the message for PDGF A-chain (but not B-chain) can be demonstrated; however, B104 cells do not express message for basic fibroblast growth factor (bFGF) or neurotrophin-3 (Asakura et al., 1997). Since PDGF concentrations in B104 CM greatly exceed those required for progenitor response (usually 1–10 ng/ml), another factor which maintains the progenitor phenotype or PDGF response is likely present. PDGF α-receptor expression by neonatal rat progenitors can be sustained by B104 CM for at least one month, and expression falls after removal of B104 CM. Thus an unknown factor is present which acts similarly to bFGF, although the addition of bFGF augments the response to the less potent B104 CM (Asakura et al., 1997). Further investigation is needed to determine what other trophic factors are responsible for the action of B104 CM on glial progenitors from both neonatal and mature brains.

Adult Human Oligodendroglial Lineage Cells May Respond Differently to Trophic Signals

Although mature rat brain ODC preparations have numerous proliferation-competent glia, larger, long-lived animals probably have far fewer proliferative ODC lineage cells. Unfortunately, assessment of developmental lineage markers (galactocerebroside [GC], gangliosides, etc.) in cells derived from older animals is more difficult. Problems arise in discriminating cells arrested in maturation from those which lose differentiation markers as a consequence of disturbing their structure or environment, in other words "dedifferentiation." Whereas many "GC-negative" ODC lineage cells are present in cultures from human brain, these cells do not respond with proliferation to PDGF (Armstrong et al., 1992; Prabhakar et al., 1995). The developmental identity of these "less differentiated" ODC remains inconclusive, as ODC markers may be environmentally influenced. For example, bFGF dramatically decreases expression of differentiated ODC markers in a non-dividing population of ODC (Fressinaud et al., 1995). Recently, Scolding and colleagues (1995) described a small minority of cells in human ODC cultures (1%) which possess similar morphology and immunohistochemical markers to the rat glial progenitor. Proliferation of these cells could be demonstrated only on a feeder layer of astroglia, and purified trophic factors had no effect. Other possibly significant trophic actions on mature brain ODC lineage cells include promotion of ODC differentiation (by IGF-1) and ODC survival (by neurotrophins such as NT-3). While trophic factors have important actions on developing ODC and antigenic expression in vitro, their role in the setting of myelin repair is less clear.

Since current data indicate that adult human brains yield few proliferation-competent cells in vitro, future studies need to focus on the trophic stimuli required for progenitor proliferation, factors controlling survival of mature brain ODC, and the events initiating myelin repair. In vitro cultures of adult human ODC lineage cells represent a valuable model for examining the biological behavior of ODC surviving within the inflamed, recently demyelinated white matter of the CNS. Culture conditions approximate the CNS environment of inflammatory demyelination: myelin sheath loss, disrupted axonal contact, and an excess of activated microglia.

In our laboratory, we have studied adult human brain ODC and find rare (much less than 1% of HNK-1$^+$) proliferative, bipolar cells in cultures treated with B104 CM. However, the vast majority of ODC lineage cells are heterogenous in their expression of GalC, which can be augmented by culture in the presence of B104 CM. Adult human ODC lineage cells in vitro are also heterogenous for expression of HNK-1 antigen (presumably myelin-associated glycoprotein). ODC highly expressing HNK-1 generally have larger nuclei and processes. B104 CM treatment not only increases GC expression, but it decreases nuclear size, suggesting a maturation-promoting action (Hunter & Rodriguez, 1996). This trophic factor stimulus also decreases ODC process number and promotes a more linear, rather than radial, process organization. Others have reported that NGF (Althaus et al., 1992) or protein kinase C stimulation (Yong et al., 1988) promote ODC process extension. Trophic factors might act on persisting cells in demyelinated areas to promote differentiation of less mature ODC and encourage ODC to extend processes or move into nearby regions of naked axons. Although a mitogenic action on an immature cell of the ODC lineage can be demonstrated for B104 CM, it is difficult to assess these cells further due to infrequency of proliferation in this preparation.

THE PHARMACOLOGIC PROMOTION OF MYELIN REPAIR IN AN ANIMAL MODEL

Myelin Repair Can Be Greatly Increased in TMEV-Infected Mice with Inflammatory Demyelination

In order to study actual myelin repair, an animal model is required. TMEV infection in mice represents an ideal system for studying the events of inflammatory myelin destruction and subsequent repair. Mice of susceptible strains are intracerebrally infected with TMEV at age 4–6 weeks and fail to clear the viral infection. Over months, a slowly progressive neurologic disability develops with widespread inflammatory infiltrates and demyelination in the spinal cord white matter; brainstem is also involved but to a lesser degree. Animals survive for more than one year, developing worsening signs of spinal cord dysfunction. In our laboratory, histological parameters of myelin destruction and repair are quantitatively assessed. Demyelination (Figure 3A) and remyelination (Figure 3B; areas of thin myelin sheaths) are assessed at ten spinal cord levels using digital planimetry of semi-thin, myelin-stained, spinal cord sections to determine the total area of demyelination, remyelination, and percentage of repair in demyelinated areas.

Our work has demonstrated that myelin repair can be greatly augmented by immunotherapy in chronically TMEV-infected mice. Remyelination can be increased sixfold with immunoglobulin (Ig) from mice hyperimmunized with spinal cord homogenate (Rodriguez & Lennon, 1990). Ig against kidney homogenate is far less effective. Anti-myelin basic protein Ig stimulates a two- to fourfold increase in myelin repair (Rodriguez et al., 1996). In remyelinating lesions, radiothymidine-labelled, immature glial cells (presumably ODC progenitors) are found by both light and electron microscopy (Rodriguez, 1991). In

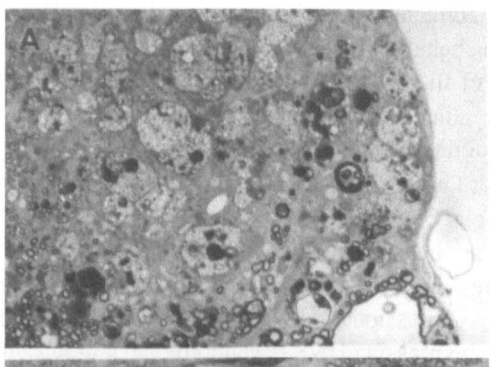
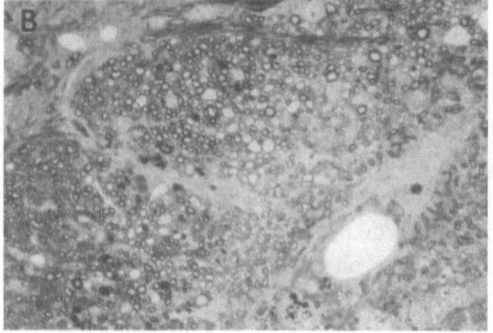

Figure 3. Histological evidence of myelin repair following treatment with remyelination-promoting autoantibody. (A) Characteristic demyelinated lesion in the spinal cord of a mouse with chronic TMEV-induced demyelination. In this case the animal has been treated for five weeks with a control monoclonal antibody (IgMκ), and this lesion remains filled with demyelinated axons, debris-filled macrophages, and minimal evidence of myelin repair. Normal, uninvolved myelin is seen at the lesion edge. (B) Spinal cord lesion in TMEV-infected mouse which received a total of 100 µg of SCH 94.03 over five weeks. Extensive CNS-type myelin repair (thin, pale myelin) has occurred with a great decrease in the presence of macrophages. Some areas of demyelination and inflammation persist in the perivascular area.

situ hybridization for proteolipid protein/DM20 message demonstrates a threefold increase in response to Ig treatment (Rodriguez et al., 1994). The Ig treatment probably does not act directly on the viral infection or in limiting the size of demyelinating lesions, since demyelination scores, viral antigen, and viral titers do not correlate with the amount of myelin repair (Patick et al., 1991). Nevertheless, agents which impair the cellular immune response can also produce a similar remyelination phenomenon with immature glial cell proliferation at the edges of remyelinating lesions (Rodriguez & Lindsley, 1992). Mice with a defect of the cytotoxic T cell response (β2-microglobulin -/-) have remarkably effective myelin repair, spontaneously repairing 40–80% of demyelinated areas, with increasing percentages with time (Miller et al., 1995).

In order to clarify the action of the spinal cord autoantibody, monoclonal antibodies were prepared using spinal cord homogenate as an immunogen, and subsequently screened by ELISA and immunoblot assays. Positive clones were tested for their activity in promoting CNS myelin repair. To date, two effective monoclonal antibodies have been produced in this manner. We have termed these "remyelination-promoting autoantibodies" (RPA); other monoclonal RPAs have been identified by screening hybridomas obtained from elsewhere. Many inactive monoclonal antibodies have also been identified. Monoclonal RPA (thus far all of the IgMκ isotype) generated in our laboratory belong to a family known as "natural autoantibodies," characterized by a germline-encoded immunoglobulin variable domain and reactivity with multiple autoantigens (spectrin and hemoglobin) and chemical haptens. These antibodies are a major constituent of plasma proteins in mammals, and do not seem to be pathogenic (Avrameas & Ternynck, 1993). The best characterized antibody developed in our laboratory, SCH 94.03, promotes myelin repair fourfold better than control antibodies (MOPC 104E, ABPC 22, TB5–1), is highly effective in cumulative doses of 100–1000 μg (2–20 mg/kg), and does not react with any viral determinants. In addition, animals are neurologically improved by Ig treatment, as the increase in myelin repair caused by SCH94.03 correlates with less severe clinical manifestations of the disease (Miller et al., 1994). In addition to spectrin, tubulin, and actin, SCH 94.03 recognizes a diverse set of antigens from both identifiable and novel cDNA sequences derived by screening a neonatal rat brain cDNA (bacteriophage) expression library (Asakura et al., 1996b). A second antibody developed in the laboratory, SCH 79.08, not only is a natural autoantibody but also reacts with myelin basic protein on immunoblots (Asakura and Rodriguez, in preparation).

Pharmacokinetic Considerations of Agents Used to Promote Remyelination

Important pharmacologic considerations for candidate agents to enhance remyelination include penetration into the CNS, a practical half-life of elimination, and the binding to intended cellular targets. Fortunately, inflammatory diseases of the CNS parenchyma produce a local disturbance of the blood brain barrier, permitting circulating proteins (even the relatively large IgM molecules) access to the area where myelin breakdown has occurred. Antibodies usually have a long half-life of days or weeks. To address this issue we grew the SCH 94.03 hybridoma in the presence of ^{35}S-methionine and assessed its pharmacokinetics and penetration into the CNS. We found that radiolabeled SCH 94.03 penetrates into the area of demyelinating lesions, binds to oligodendrocytes and myelin within the lesion, has a peak serum concentration of 3–10 μg/ml, and disappears from serum with half-life of 4.5 days (Hunter et al, in press). When one considers strategies to promote myelin repair, RPA has the advantages of being proven effective in an animal, entering into the CNS, and having a half-life far greater than trophic factors. Furthermore, since similar Ig are present in human Ig preparations, it is possible to test the benefit of these antibodies in multiple sclerosis. Currently, double-blinded, placebo-controlled, clinical trials are underway at our institution to study the action of intravenous Ig on visual and motor function in MS patients (Noseworthy et al., 1994).

Potential Pharmacologic Actions of Remyelination-Promoting Antibodies

Penetration of antibody into the CNS does not explain how these RPA act. One possible action is directly at the ODC. Immunohistochemical studies are complicated because monoclonal RPA react with a wide variety of tissues via intracellular antigens, often appearing as a cytoskeletal staining pattern, for example in astrocytes (Figure 4A). In the CNS, immunofluorescence of frozen sections label principally astroglia and ependyma and to a lesser extent macrophages and ODC intracellularly (Miller et al., 1996b). Cultures of neonatal mouse ODC maintained for three weeks contain mature-appearing ODC which have a punctate, cell-surface staining for SCH94.03 antigen. Furthermore, SCH 94.03 does not react with lipids from ODC and staining patterns differ greatly from the usual markers of ODC lineage (Figure 4B; Asakura et al., 1996a). Approximately 50% of ODC expressing 94.03 antigen in vitro also express myelin basic protein, a marker of more mature ODC. We have stained unfixed, live brain slices with monoclonal RPA and

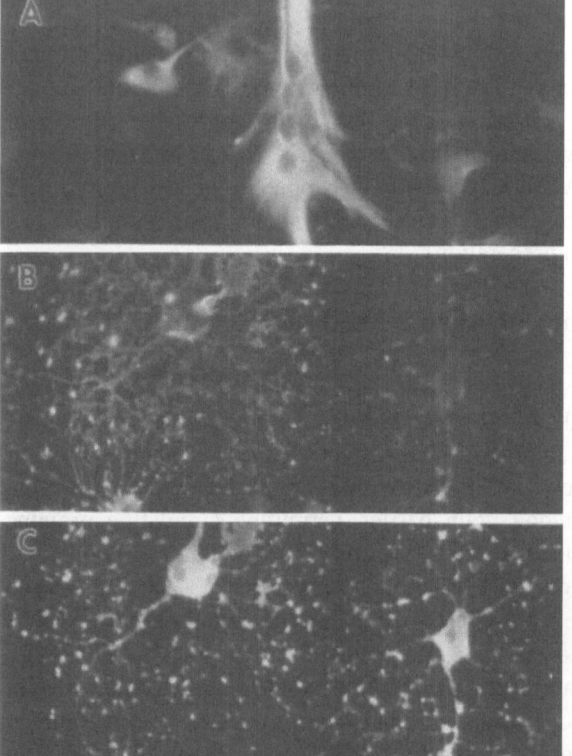

Figure 4. A monoclonal remyelination-promoting autoantibody reacts with different cellular epitopes. (A) SCH 94.03 stains the cytoskeletal elements of many cells, including neonatal rat astroglia following fixation by air drying and permeabilization with Triton X-100. (B) SCH 94.03 reacts with a subpopulation of differentiated ODC which have been selected to eliminate less mature ODC lineage cells. This cell membrane-associated staining pattern is obtained only by staining live cells. (C) Same cells labeled for myelin basic protein following fixation and permeabilization. Only a subpopulation of SCH 94.03+ ODC have reactivity for MBP (about 50%). ODC progenitors do not have surface reactivity with SCH 94.03 (not shown).

showed labeling of both ODC and neurons (Hunter et al, in preparation). Thus, the ODC cell surface is one important potential target for the direct action of RPA. Although traditional concepts of antibody action specify a role in neutralizing and opsonizing infectious agents, other roles have been found for these molecules.

Antibodies have been documented to have receptor-stimulating or blocking activity in many diseases. Direct action of some Ig on ODC have been previously reviewed (Benjamins & Dyer, 1990). White matter antisera without complement can stimulate myelin lipid synthesis in vitro (Lehrer et al., 1979). Antibodies to spinal cord homogenate stimulate MBP expression and proliferation of neonatal rat oligodendroglia in vitro (Rodriguez et al., 1987). Antibodies against ODC membrane lipids can modulate membrane and process organization and produce a Ca^{2+} influx (Dyer & Benjamins, 1988, 1991; Dyer et al., 1991). ODC progenitor maturation is blocked by antibodies to glycolipids of more mature, differentiated ODC (Bansal & Pfeiffer, 1989). Monoclonal antibody O4, reacting with ODC progenitors and ODC, hastens differentiation, possibly by stimulating aggregation (Bansal et al., 1988). Antibodies against the reovirus serotype 3 receptor on ODC can also induce markers of differentiation on immature ODC lineage cells (Cohen et al., 1990), possibly by activation of adenylate cyclase.

RPA could also assist in the clearance of injured or dead ODC in regions of demyelination, providing a more suitable substrate for repair by glial progenitor cells or persisting ODC. Infected ODC in the region might be cleared more efficiently by the immune response. Opsonization of cellular fragments could achieve more rapid phagocytosis and ultimately eliminate a sustained or intense immune activation in the parenchyma. Finally, the presence of persisting ODC in the lesion could impair proliferation of glial progenitor cells, which are associated with myelin repair (see above). Previously, in vitro studies have demonstrated that progenitor proliferation is decreased following ODC differentiation in vitro or in the presence of mature brain ODC (Hunter & Bottenstein, 1989, 1991). Further investigations are ongoing to determine whether there is a direct action of RPA on ODC in demyelinating lesions.

Potential mechanisms are not limited to purely an effect on the ODC. Administration of exogenous Ig modulates the immune response in a number of human and animal immune diseases, and an immunomodulatory action has been proposed for natural autoantibodies (Avrameas & Ternynck, 1993). SCH 94.03 has effects on both CNS and systemic immune parameters. RPA treatment of TMEV-infected mice decreases infiltration of T-lymphocytes into the CNS by two- to threefold, and RPA impair the humoral immune response to T cell-dependent antigens (Miller et al., 1996a). Unexpectedly, RPA treatment increases the number of cells expressing viral antigens, although infectious virus titers and areas of demyelination remain unchanged. Systemic actions of the antibody might occur through an anti-idiotypic antibody network or by interaction with specific immune cells, as SCH 94.03 binds the dendritic cells of the thymus which are crucial for antigen presentation and T cell development. Potential RPA actions both systemically and localized to the CNS are summarized in Figure 5.

Further clarification of RPA action could improve therapeutic strategies for treating the chronic and recurrent inflammatory injury to the CNS occurring in diseases like MS, as well as for other immune-mediated neurologic diseases. Restoration of white matter organization and subsequent improvement in axonal function can be effectively studied using the TMEV mouse model of MS. The study of trophic factor actions in vitro on mature brain ODC complements the in vivo model and may clarify remyelination-promoting actions at the cellular level.

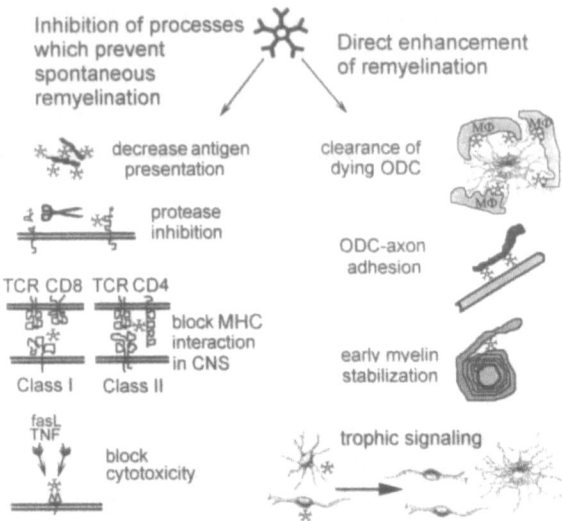

Figure 5. Potential remyelination-promoting actions of SCH 94.03 and 79.08. Hypothetical actions of remyelination-promoting autoantibodies which may occur by action within the CNS parenchyma include those which act indirectly to modulate the immune response and permit myelin spontaneous myelin repair, or those which act on the axon, ODC, or myelin to provide a direct stimulation of myelin repair.

ACKNOWLEDGMENTS

    The research work discussed in this paper has been supported by the National Institutes of Health (NS24180); S.F.H. is supported by a Clinician Investigator Award by Mayo Foundation. We are grateful to Mr. and Mrs. Eugene Appelbaum for their support of K.A. as a Neuroscience Fellow.

REFERENCES

Althaus HH Kloppner S Schmidt-Schultz T Schwartz P (1992): Nerve growth factor induces proliferation and enhances fiber regeneration in oligodendrocytes isolated from adult pig brain. Neurosci. Lett. 135: 219–223.

Armstrong RC Dorn HH Kufta CV Friedman E Dubois-Dalcq ME (1992): Pre-oligodendrocytes from adult human CNS. J. Neurosci. 12: 1538–1547.

Asakura K Miller DJ Murray K Bansal R Pfeiffer SE Rodriguez M (1996a): Monoclonal autoantibody SCH94.03, which promotes central nervous system remyelination, recognizes and antigen on the surface of oligodendrocytes. J. Neurosci. Res. 43: 273–281.

Asakura K Pogulis R Pease LR Rodriguez M (1996b): A monoclonal autoantibody which promotes central nervous system remyelination is highly polyreactive to multiple known and novel antigens. J. Neuroimmunol. 65: 11–19.

Asakura K Hunter SF Rodriguez M (1997): The effects of transforming growth factor-beta and platelet-dervied growth factor on oligodendrocyte precursors: insights gained from a neuronal cell line J. Neurochem. in press.

Avrameas S and Ternynck T (1993): The natural autoantibodies system: between hypotheses and facts. Mol. Immunol. 30: 1133–1142.

Bansal R and Pfeiffer SE (1989): Reversible inhibition of oligodendrocyte progenitor differentiation by a monoclonal antibody against surface galactolipids. PNAS USA 86: 6181–6185.

Bansal R Gard AL Pfeiffer SE (1988): Stimulation of oligodendrocyte differentiation in culture by growth in the presence of a monoclonal antibody to sulfated glycolipid. J. Neurosci. Res. 21: 260–267.

Benjamins JA and Dyer CA (1990): Glycolipids and transmembrane signaling in oligodendroglia. Ann. NY Acad. Sci. 605: 90–100.

Bögler O Wren D Barnett SC Land H Noble M (1990): Cooperation between two growth factors promotes extended self-renewal and inhibits differentiation of oligodendrocyte-type-2 astrocyte (O-2A) progenitor cells. PNAS USA 87: 6368–6372.

Bottenstein JE Hunter SF Seidel M (1988): CNS neuronal cell line-derived factors regulate gliogenesis in neonatal rat brain cultures. J. Neurosci. Res. 20: 291–303.

Cohen JA Williams WV Weiner DB Geller HM Greene MI (1990): Ligand binding to the cell surface receptor for reovirus type 3 stimulates galactocerebroside expression by developing oligodendrocytes. PNAS USA 87: 4922–4926.

Dyer CA and Benjamins JA (1988): Antibody to galactocerebroside alters organization of oligodendroglial membrane sheets in culture. Journal of Neuroscience 8: 4307–4318.

Dyer CA and Benjamins JA (1991): Galactocerebroside and sulfatide independently mediate Ca2+ responses in oligodendrocytes. J. Neurosci. Res. 30: 699–711.

Dyer CA Hickey WF Geisert EE Jr (1991): Myelin/oligodendrocyte-specific protein: a novel surface membrane protein that associates with microtubules. J. Neurosci. Res. 28: 607–613.

Fressinaud C Vallat JM Labourdette G (1995): Basic fibroblast growth factor down-regulates myelin basic protein gene expression and alters myelin compaction of mature oligodendrocytes in vitro. J. Neurosci. Res. 40: 285–293.

Hunter SF and Bottenstein JE (1989): Bipotential glial progenitors are targets of neuronal cell line-derived growth factors. Dev. Brain Res. 49: 33–49.

Hunter SF and Bottenstein JE (1990): Growth factor responses of enriched bipotential glial progenitors. Dev. Brain Res. 54: 235–248.

Hunter SF and Bottenstein JE (1991): O-2A glial progenitors from mature brain respond to CNS neuronal cell line-derived growth factors. J. Neurosci. Res. 28: 574–582.

Hunter SF and Rodriguez M (1995): Multiple sclerosis: a unique immunopathological syndrome of the central nervous system. Springer Sem. Immunopath. 17: 89–105.

Hunter SF and Rodriguez M (1996): In vitro myelin antigen expression and response to trophic factors indicate at least two phenotypes of oligodendroglia in mature human brain. Neurology (supp) 46:A294 (abstract).

Hunter SF, Miller DJ, Rodriguez M (1997): Monoclonal remyelination-promoting auto antibody SCH94.03: pharmacokinetics and in vivo targets within demyelinated spinal cord in a mouse model of multiple sclerosis. J. Neurosci. Res. In press.

Lehrer GM Maker HS Silides DJ Weiss C Bornstein MB (1979): Stimulation of myelin lipid synthesis in vitro by white matter antiserum in absence of complement. Brain Res. 172: 557–560.

Louis JC Magal E Muir D Manthorpe M Varon S (1992): CG-4, a new bipotential glial cell line from rat brain, is capable of differentiating in vitro into either mature oligodendrocytes or type-2 astrocytes. J. Neurosci. Res. 31: 193–204.

Miller DJ Sanborn KS Katzmann JA Rodriguez M (1994): Monoclonal autoantibodies promote central nervous system repair in an animal model of multiple sclerosis. J.Neurosci. 14: 6230–6238.

Miller DJ Rivera-Quinones C Njenga MK Leibowitz J Rodriguez M (1995): Spontaneous CNS remyelination in beta-2-microglobulin-deficient mice following virus-induced demyelination. J. Neurosci. 15: 8345–8352.

Miller DJ Njenga MK Murray PD Leibowitz J Rodriguez M (1996a): A monoclonal natural autoantibody that promotes remyelination suppresses central nervous system inflammation and increases virus expression after Theiler's virus-induced demyelination. Int. Immunol. 8: 131–141.

Miller DJ Njenga MK Parisi JE Rodriguez M (1996b): Multi-organ reactivity of a monoclonal natural autoantibody that promotes remyelination in a mouse model of multiple sclerosis. J. Histochem. Cytochem. 42: (in press).

Noble M Murray K Stroobant P Waterfield MD Riddle P (1988): Platelet-derived growth factor promotes division and motility and inhibits premature differentiation of the oligodendrocyte/type-2 astrocyte progenitor cell. Nature 333: 560–562.

Noseworthy JH O'Brien PC van Engelen BGM Rodriguez M (1994): Intravenous immunoglobulin therapy in multiple sclerosis: progress from remyelination in the Theiler's virus model to a randomised, double-blind, placebo-controlled clinical trial. J. Neurol. Neurosurg. Psych. 57 Suppl: 11–14.

Patick AK Thiemann RL O'Brien PC Rodriguez M (1991): Persistence of Theiler's virus infection following promotion of central nervous system remyelination. J. Neuropathol. Exp. Neurol. 50: 523–537.

Prabhakar S D'Souza S Antel JP McLaurin J Schipper HM Wang E (1995): Phenotypic and cell cycle properties of human oligodendrocytes in vitro. Brain Res. 672: 159–169.

Prineas JW Barnard RO Revesz T Kwon EE Sharer L Cho ES (1993): Multiple sclerosis. Pathology of recurrent lesions. Brain 116: 681–693.

Rodriguez M (1991): Immunoglobulins stimulate central nervous system remyelination: electron microscopic and morphometric analysis of proliferating cells. Lab. Invest. 64: 358–370.

Rodriguez M and Lennon VA (1990): Immunoglobulins promote remyelination in the central nervous system. Ann. Neurol. 27: 12–17.

Rodriguez M and Lindsley MD (1992): Immunosuppression promotes CNS remyelination in chronic virus-induced demyelinating disease. Neurology. 42: 348–357.

Rodriguez M Lennon VA Benveniste EN Merrill JE (1987): Remyelination by oligodendrocytes stimulated by antiserum to spinal cord. J. Neuropathol. Exp. Neurol. 46: 84–95.

Rodriguez M Scheithauer BW Forbes G Kelly PJ (1993): Oligodendrocyte injury is an early event in lesions of multiple sclerosis. Mayo Clin. Proc. 68: 627–636.

Rodriguez M Prayoonwiwat N Howe C Sanborn K (1994): Proteolipid protein gene expression in demyelination and remyelination of the central nervous system: a model for multiple sclerosis. J. Neuropathol. Exp. Neurol. 53: 136–143.

Rodriguez M Miller DJ Lennon VA (1996): Immunoglobulins reactive with myelin basic protein promote CNS remyelination. Neurology. 46: 538–545.

Scolding NJ Rayner PJ Sussman J Shaw C Compston DA (1995): A proliferative adult human oligodendrocyte progenitor. Neuroreport 6: 441–445.

Yong VW Sekiguchi S Kim MW Kim SU (1988): Phorbol ester enhances morphological differentiation of oligodendrocytes in culture. J. Neurosci. Res. 19: 187–194.

26

INHIBITION OF HUMAN AND RAT GLIAL CELL FUNCTION BY ANTI-INFLAMMATORY CYTOKINES, ANTIOXIDANTS, AND ELEVATORS OF cAMP

Barbara St. Pierre,[1] Joyce L. Wong,[1] and Jean E. Merrill[2]

[1]Department of Neurology
UCLA School of Medicine
Los Angeles, California 90024
[2]Department of Immunology
Berlex Biosciences
Richmond, California 94804

INTRODUCTION

There is accumulating evidence for the involvement of tumor necrosis factor alpha (TNFα) and nitric oxide (NO), produced as a consequence of activation of inducible nitric oxide synthase (iNOS), in Multiple Sclerosis (MS). The same evidence has been found in the animal model of MS, experimental allergic encephalomyelitis (EAE). EAE brains and spinal cords have elevated levels of iNOS and NO correlating with the severity and stage of the disease. Aminoguanidine, an inhibitor of iNOS, ameliorates EAE. Inducible NOS mRNA and protein have been detected in MS brain and footprints of NO seen in serum and spinal fluid of MS patients (reviewed in Parkinson et al., in press). TNFα is also elevated in MS patients' central nervous system (CNS) as well as in EAE, where the interference with the TNFα receptor or TNFα itself antagonizes the disease in the mouse model. The functional removal of macrophages or the pretreatment of EAE animals with Interleukin 4 (IL4), Interleukin 10 (IL10), or Interleukin 13 (IL13), which downregulate class II major histocompatability molecules, IL1 and TNFα, inhibit clinical and histological EAE (de Waal Melefyt et al., 1991; Khoury et al., 1992; reviewed in Merrill and Benveniste, in press).

The mononuclear phagocyte system includes monocytes, tissue macrophages, and CNS microglia. Activated monocytes, macrophages, and microglia secrete various cytokines, including Interleukin 1 (IL1), TNFα, and NO. Microglia are cytotoxic mediators of oligodendrocyte cell death *in vitro*. Microglial cytotoxicity and NO production appear to be mediated partially and indirectly by TNFα. That is, TNFα does not induce NO by itself, nor can soluble TNFα directly kill oligodendrocytes; TNFα associated with the cell surface of microglia may play a role in the cytotoxicity (Zajicek et al., 1992; Chao et al., 1993; Merrill et al., 1993). TNFα secretion, NO production, and/or cytotoxicity by these cells can be regulated by several types of agents, including antiinflammatory cytokines, antioxidants, and compounds which raise intracellular cAMP. Since a complete study on the effects of these agents on rodent and human glial cell functions as they pertain to events we believe occur in MS has not been performed, we assessed the effects of these three classes of antiinflammatory drugs on TNFα production, NO induction, and cytotoxicity of oligodendrocytes *in vitro*.

MATERIALS AND METHODS

Primary Cultures of Glial Cells

Rat. Primary rat brain cultures were established from 2 day old rat cerebra and the cells purified and identified according to previously published methods of this laboratory (Merrill et al., 1993). Such cultures contained 90–95% microglia as identified by nonspecific esterase staining and DiI-Ac-LDL (acetylated low density lipoprotein conjugated to 1,1'- dioctadecyl-3,3,3'3-tetramethyl-indocarbocyanine perchlorate) and were at least 95% viable as detected by trypan blue dye exclusion. Oligodendrocytes were harvested at day 10; they were 90–95% galactocerebroside (GalC) positive as determined by anti-GalC staining as previously described (Merrill et al., 1993).

Human. Primary human mixed glial cultures were established for separate experiments from a total of 9 different second trimester fetal brains (Koka, et al., 1995). These cultures consisted of about 40% parenchymal microglia and 60% astrocytes as identified by the standard markers used in the rat cultures.

Reagents

Pro- and Antiinflammatory Recombinant Cytokines. The murine and human cytokines were all titrated, and unless otherwise indicated, were used at optimal doses for stimulation or inhibition. Murine Interferon beta (IFNβ) (Research Manufacturing Group of Berlex Biosciences, Alameda, CA @ 25 ng/ml); murine IL13 (R+D Systems, Minneapolis, MN @ 50 ng/ml); murine Interferon gamma (IFNγ) (Boehringer Mannheim, Indianapolis, IN @ 10U/ml); murine IL10 (gift of DNAX, Palo Alto,CA @ 25ng/ml); murine IL4 (gift of Immunex, Seattle, WA @ 25 ng/ml); Lipopolysaccharide (LPS) (Sigma Chemical corp., St. Louis, MO @ 10μg/ml); human IFNγ (R+D Systems @ 10 ng/ml); human IL1β (R+D Systems @10 ng/ml); human IL13 (Intergen, Milford MA @ 10^{-7}–10^{-11}M); human IL4 (R+D Systems @ 10^{-7}–10^{-11}M); human IFNβ (Research manufacturing Group of Berlex Biosciences @ 10^{-7}–10^{-11}M).

Antioxidants. The following were purchased from Sigma Chemical Corp.: Niacinamide (nicotinamide) and nicotinic acid (both used @ 5×10^{-3} - 5×10^{-8}M); N-acetyl-L-cysteine (NAC @ 5×10^{-3}- 5×10^{-8}M).

Elevators of cAMP. Pentoxifylline (PTX, Trental®, gift of Hoechst-Roussel, Sommerville, NJ @ 0.1–5 mg/ml); isobutyl methyl xanthine (IBMX) (Sigma Chemical Corp. @ 10^{-2}–10^{-9}M); Iloprost (Schering AG, Berlin @ 10^{-5}–10^{-11}M).

Functional Assays for Rat Microglia and Human Mixed Glia

Cytotoxicity. Rat microglia effectors, at 10^5 cell/well in quadruplicate wells , were allowed to attach for approximately 3 hrs, at which time (in some experiments) some inhibitors were added for an additional 24 hr incubation (-24 hr time point). During this period, enriched oligodendrocyte targets were radiolabeled as previously described in detail (Merrill et al., l993). Labeled oligodendrocytes were added at 10^4 cells/well in the 96-well plate containing treated or untreated microglial effectors giving an effector:target (E:T) cell ratio of 10:1. IFNγ and LPS were added to some wells to activate killing. In some experiments, inhibitors were added at this time (0 time point). After 24 hrs, total and spontaneous label release were measured and percent specific cytotoxicity was calculated as previously described (Merrill et al., l993).

NO Production. Five x 10^5 rat microglia or human mixed glia per well were added to 24-well plates and allowed to adhere for 2 hrs as previously described (Merrill et al., l993). Cells were then left untreated or pretreated with antiinflammatory cytokines or cAMP elevators (-24 hr time point). At time 0, glia were left unstimulated as controls or were stimulated with IL1β and IFNγ (human cultures) or IFNγ and LPS (rat microglial cultures). At time 0, in some wells not previously treated, antiinflammatory cytokines, antioxidants, or cAMP elevators were added as stated in Figure legends. Supernatants for quantitation of NO were collected at 48 hours from rat microglial cultures or day 5 from human glial cultures. Total NO (NO_x^-) was determined by reduction of nitrate and nitrite to NO_x^- in boiling vanadium and read using chemiluminescence as previously described (Ignarro et al., l981; Merrill et al., l993).

TNFα Production. Supernatants for detection of TNFα were harvested 24 hours after stimulation in both rat and human cultures, and TNFα was quantitated using the L929 cytotoxicity assay as previously described (Merrill et al., l993). With this assay, minimal detection was 0.25 pM TNFα so that any value lower than this was considered 0 pM.

Statistical Analysis

Analysis of variance procedures and post-hoc Tukey-Kramer (Honestly Significant Difference) analyses using the software Abacus Concepts, SuperANOVA (Abacus Concepts, Inc., Berkeley, CA, 1989) were performed on a Macintosh Quadra 650 computer. Level of significance was set at $\alpha = 0.05$.

RESULTS

Effects of Antiinflammatory Cytokines on Glial Cell Functions

Rat Microglia. The stimulated cultures which were not treated with inhibitors produced 1.5±0.5 ng/ml TNFα, 48± 10 μM NO_x^-, and 85± 10% cytotoxicity (data not shown).

Figure 1 shows that all four antiinflammatory cytokines, when used at optimal concentrations, were effective at inhibiting 80–100% of the TNFα produced. IL13 inhibited cytotoxicity and NO by 60% and this was not augmented by preincubation with the cytokine. In contrast, IFNβ, IL10, and IL4 had little or no significant effect on NO production or cytotoxicity by rat microglia. Again, this was true even if the cells were preincubated with cytokines. These data suggest that the induction of NO and microglial cell-mediated oligodendrocyte cytotoxicity do not require TNFα, and that the mechanism of NO and TNFα induction are different. The correlation of inhibition (or lack thereof) of NO production and cytotoxicity suggests a link between these two microglial cell-mediated events.

Human Mixed Glia. Human mixed glia produced 11± 2 ng/ml TNFα and 51± μM NO$_x^-$ when there were no inhibitors present. Figure 2 demonstrates that, unlike rodent microglia, human mixed glial cell NO production is significantly inhibitable by IFNβ and IL4. IFNβ inhibited NO at lower concentrations than were required for inhibition by IL4 and IL13, while inhibiting less well at higher concentrations. As in rodent cultures, TNFα was more easily and completely inhibited by cytokines than was NO. The inhibition of TNFα was achieved at lower concentrations of cytokines than were required for NO inhibition. TNFα was equally sensitive to all 3 antiinflammatory cytokines. These findings support the observations that TNFα does not directly induce NO and that some antiinflammatory cytokines may work differently in their regulation of rodent versus human glial cells. IL13 is a good inhibitor of NO and TNFα in both species. IFNβ may work in a complex, biphasic manner in its regulation of human glia.

Effects of Antioxidants on Rat Microglial Cell Functions

Both nicotinamide and NAC significantly inhibited TNFα and oligodendrocyte cytotoxicity produced by rat microglia (Figure 3). TNFα was inhibited 80–90% while cytotoxicity was inhibited 60–70% at pharmacologically relevant doses of either compound. At the lowest concentration tested, NAC was a slightly better inhibitor than nicotinamide. Nicotinic acid had no inhibitory effect at the concentrations where nicotinamide was ef-

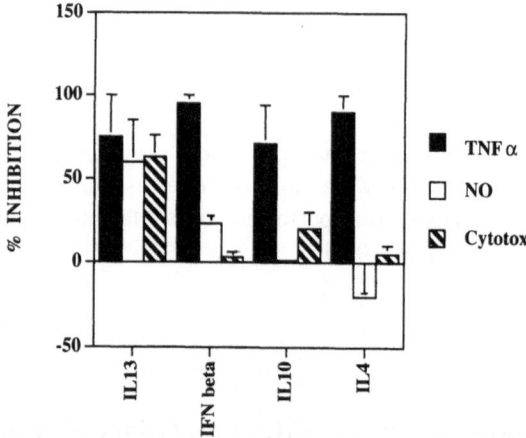

Figure 1. Effect of cytokines on rat microglial cell function. Antiinflammatory cytokines were added at time 0 of stimulation with IFNγ and LPS. Bars represent means ± sd of three each of IL13 and IFNβ experiments and four each of IL10 and IL4 experiments.

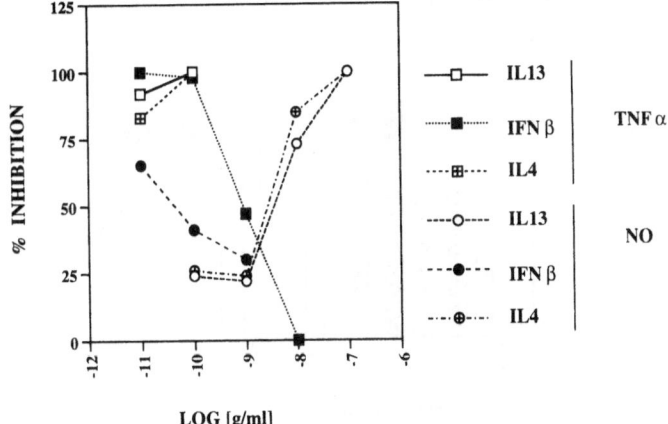

Figure 2. Effect of cytokines on human glial cell function. Antiinflammatory cytokines were added at time 0 of stimulation with IFNγ and IL1β. This is a representative experiment of three experiments.

fective. At low concentrations, neither nicotinamide nor NAC inhibited NO; at mM concentrations, nicotinamide was a weak inhibitor of NO.

Effects of cAMP Elevators on Glial Cell Functions

Rat Microglia. A titration of PTX on rat microglial cell cultures demonstrated that TNFα production and oligodendrocyte cytotoxicity could be completely inhibited at 0.5 mg/ml, a concentration at which NO was only inhibited by 35% (Figure 4). NO was not completely inhibited until PTX was at a concentration of 5mg/ml. Pretreatment of cultures with PTX did not augment this inhibition. These data, once again, suggest that NO induction is independent of TNFα and that microglial cell cytotoxicity of oligodendrocytes is partly independent of NO production.

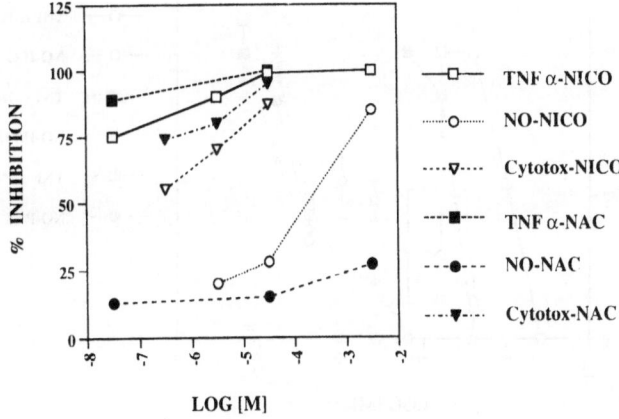

Figure 3. Effects of nicotinamide and NAC on rat microglial cell function. Antioxidants werer added at time 0 of stimulation performed as in Figure 1. This is a representative experiment of two experiments.

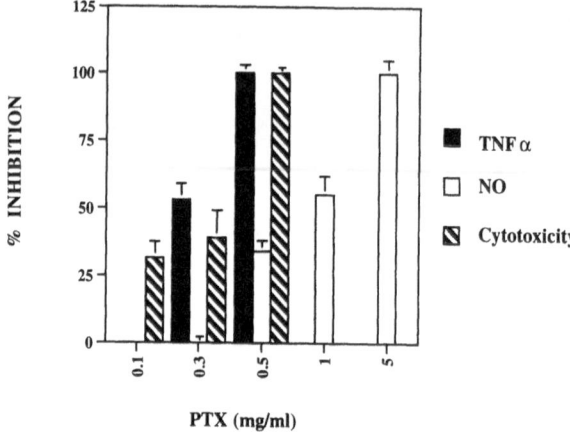

Figure 4. Effect of PTX on rat microglial cell function. PTX was added at time0 of stimulation as in Figure 1. Bars represent means ± SD of three experiments.

In a comparative study using PTX , another phosphodiesterase inhibitor IBMX, and the prostacyclin analogue Iloprost, we found that PTX was the weakest inhibitor of TNFα and NO (Figure 5). Iloprost was a more potent inhibitor than either IBMX or PTX. With all three inhibitors, NO production was not as easily inhibited as TNFα production. TNFα could be completely blocked by IBMX at 10^{-7}M and by Iloprost at 10^{-9}M. There was a 5 log difference in the IC_{50} inhibition of TNFα and NO by Iloprost and a 6 log difference in the IC_{50} inhibition of TNFα and NO by IBMX (Figure 5). Thus, in rat microglia, TNFα is more sensitive to the elevation of cAMP than is NO, and an increase in cAMP through exposure to a prostanoid was more effective than the inhibition of degradation of cAMP.

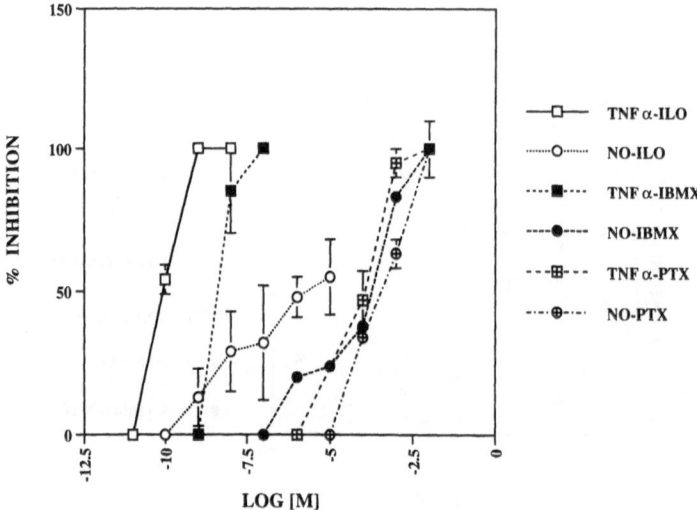

Figure 5. Inhibition of TNFα and NO production by rat microglia. Drugs were addded at time 0 of stimulation as in Figure 1. Symbols represent means ± SD (where indicated) of three experiments.

Human Glia. Human glial cells were more sensitive to IBMX and Iloprost than rodent cells. TNFα was more easily inhibited than NO as with rodent cells. Compared to the effect on rat microglia, Iloprost was even more potent than IBMX in the inhibition of human glial cell TNFα and NO. For TNFα inhibition, there was a 7 log difference between the two different agents in human mixed glial cell cultures (Figure 6).

DISCUSSION

IL4 inhibited human glial cell TNFα at ng/ml and NO production at μg/ml concentrations. While inhibiting rat TNFα at ng/ml concentrations, IL4 had no effect on rat NO production. The induction of the iNOS gene in rodents and humans is different, suggesting that the regulation of the gene may also be species-specific. When mouse macrophages are pretreated with IL4, NO production and cytotoxicity of microbes or cells are suppressed (Bogdan et al., 1994; Cenci et al., 1993; Chao et al., 1993;Liew et al., 1991; Oswald et al., 1992a). However, if IL4 is added at the time of stimulation, NO induction may not be inhibited (Bogdan et al., 1994). The effect of IL4 on TNFα release from induced macrophages or microglia may be stimulatory, weakly inhibitory, or suppressive (Bogdan et al., 1991a,1994; Chao et al., 1993; Oswald et al., 1992b). IL10 blocks the release of TNFα from stimulated macrophages (Bogdan et al., 1991b; Fiorentino et al., 1991; Oswald et al., 1992b), and suppresses macrophage killing of pathogens and tumor cells(Cenci et al., 1993; Gazzinelli et al., 1992; Oswald et al., 1992a,b; Nabioullin et al., 1994). As with IL4, IL10 has complex effects on NO induction. IL10 inhibits NO production by stimulated macrophages (Bogdan et al., 1991b; Cenci et al., 1993; Cunha et al., 1992; Gazzinelli et al., 1992; Oswald et al., 1992a). Nevertheless IL10 can synergize with IFNγ and TNFα to increase NO production in these same cells (Betz Corradin et al., 1993). In our studies, rat microglia are in an activated state *in vivo* at birth; they therefore may have been refractory to inhibition by IL4 and IL10 in the *in vitro* studies performed here. IL10 has been shown to prevent the induction or reduce the severity of the demyelinating inflammatory condition EAE (Rott et al., 1993; 1994). In addition, high concentrations of IL10 mRNA have been detected in the CNS of animals recovering from EAE (Kennedy et al., 1992). IL13

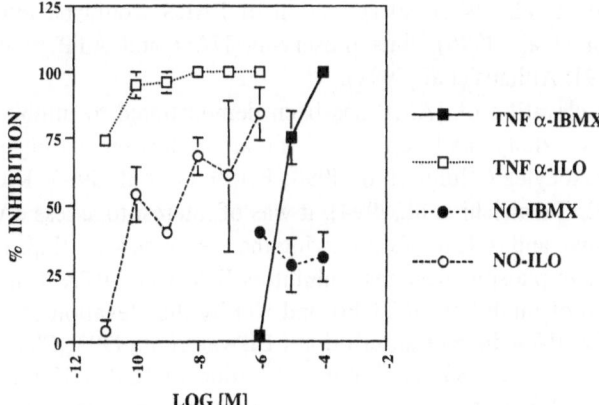

Figure 6. Inhibition of TNFα and NO production by human glia. Drugs were added at time 0 of stimulation as in Figure 2. Symbols represent means ± SD (where indicated) of three experiments.

suppresses NO in human mesangial cells and rodent macrophages and microglia (Cash et al., 1994; Doyle et al., 1994; Saura et al., 1996) and inhibits EAE (Cash et al., 1994). Here we confirm the ability of IL13 to inhibit NO in rodent glia and demonstrate that human glial NO is inhibited as well. The effect of IL13 on TNFα and cytotoxicity may be unrelated: IL13 downregulation of IL1 may account for the depression in TNF production while cytotoxicity is inhibited partially because of the NO inhibition (Cash et al., 1994; Doyle et al., 1994). By itself, IFNβ does not induce NO in mouse macrophages (Fast et al., 1993; MacMicking et al., 1995: Zhang et al., 1994; Zhou et al., 1995). IFNβ inhibits IFNγ +/- TNFα- or virus-induced NO (Deguchi et al. 1995; Kreil et al., 1995) but amplifies IFNγ plus LPS-induced NO in mouse macrophages (Fujihara et al., 1994; Kreil et al., 1995; Zhang et al., 1994; Zhou et al., 1995) . As presented here, IFNγ/LPS induced rat microglial cell TNFα was stringly inhibited by IFNβ but NO was only weakly inhibited. This suggests that there is a big difference between mice and rats. Furthermore, the significant inhibition of human glial cells' NO (especially at the lowest concentrations) and TNFα point out the complex mechanisms of actions and species-specific differences in IFNβ regulatory pathways.

Antioxidants will also inhibit mediators of inflammation. Nicotinamide, an oxygen free radical scavenger (Andersen et al., 1995) and poly ADP ribose synthase (PARS) inhibitor (Zingarelli et al., 1996) has been reported to inhibit macrophage mediated cytotoxicity (Kröncke et al., 1991) and nitric oxide production through transcriptional (Hauschildt et al., 1991; Pellat-Deceunynck et al., 1994) and posttranscriptional effects (Andersen et al., 1995; Cetkovic-Cvrlje et al., 1993). N-acteyl cysteine (NAC) , a precursor of glutathione, is another oxygen radical scavenger (Althaus et al., 1994) and inhibits TNFα production by blocking oxidation of the IκB-NFκB complex and thereby preventing NFκB from translocating to the nucleus (Dröge et al., 1994; Galter et al., 1994; Hayashi et al., 1993; Schreck et al., 1992). Mayer and Noble (1994) demonstrated that NAC protected oligodendrocytes from programmed cell death. In this study, both antioxidants inhibited cytotoxicity and TNFα but not NO production. This suggests that neither NFkB nor significant ADP ribosylation are required for NO induction in rat microglia, unlike what has been described for normal macrophages and macrophage cell lines in the mouse (Hauschildt et al., 1991; PellatCeceunynch et al., 1994). On the other hand, TNFα gene expression and microglial mediated cytotoxicity probably require oxygen radicals, which should then be scavenged by NAC or nicotinamide (Schreck et al., 1992; Hayashi et al., 1993; Dröge et al., 1994). Additionally, NO mediates DNA strand breaks in oligodendrocytes (Parkinson et al., 1996), so activation of PARS would be inhibited by nicotinamide (Zingarelli et al., 1996), thus preserving NAD and ATP pools in these cells (Kröncke et al., 1991; Althaus et al., 1994).

Because the elevation of cAMP has been demonstrated to inhibit both TNFα and NO induction, cytotoxicity, and superoxide anion production variously in human and mouse glia and leukocytes (Bulut et al., 1993; Feinstein et al., 1993; Kozaki et al., 1993; Marotta et al., 1992; Takahashi et al., 1994), it was of interest to assess cAMP elevation in the function of human and rodent glia following the treatment with Iloprost, a prostacyclin analogue, and the two phosphodiesterase inhibitors IBMX and PTX. It has been proposed that the mechanism of inhibition of TNFα and NO by the elevation of cAMP is through the inhibition of NFκB-mediated transcription (Biswas et al., 1993; Chao et al., 1992; Han et al., 1990; Strieter et al., 1988) or through elevation of IL10 (Kambashi et al., 1995; Strassmann et al., 1994). Both Rolipram, another phosphodiesterase inhibitor, and PTX have been shown to inhibit EAE in rodents and primates (Genain et al., 1995; Rott et al., 1993; Sommer et al., 1995). In the experiments performed here, we found that the elevation of cAMP by Iloprost was much more effective than the inhibition of cAMP degradation by IBMX or PTX. Why TNFα production was more sensitive to elevation of cAMP than iNOS induction was in both rat and human glial cultures is not known. Several reports have suggested that chronic elevation of high levels of cAMP is required for NO inhibition (Marotta et al., 1992; Feinstein et al., 1993; Bulut et al., 1993); we may have not achieved such levels in our cultures with these agents.

In the assessments described here, with all three classes of inhibitors, we wish to re-emphasize that when TNFα was completely inhibited, NO and cytotoxicity were only partially or not at all inhibited. This supports our contentions 1) that microglial cell killing of oligodendrocytes is not mediated by soluble TNFα; 2) that TNFα does not induce NO; and 3) that microglial cell-mediated killing is partly NO-dependent (Merrill et al., 1993). Nevertheless, all three microglial cell-mediated events may be involved at some point or other in lesion formation in MS, and therefore interference with these three processes might be of benefit in treating MS.

ACKNOWLEDGMENTS

The authors would like to thank Dr. Sean Murphy, University of Iowa College of Medicine and Dr. Harald Dinter, Berlex Biosciences for helpful discussions as well as the laboratory of Dr. Louis Ignarro, UCLA School of Medicine for NO measurements. This work was supported by the Conrad Hilton Foundation grant L890523 and an NIH RO-1 NS30768 (JEM), and an NIH NIMH postdoctoral fellowship 5T32MH17140-11 and an Oncology Nursing Foundation/Bristol Myers research grant (BAS)

REFERENCES

Althaus JS Oien TT Fici GJ Scherch HM Sethy VH and vonVoigtlander PF (1994): Structure activity relationships of peroxynitrite scavengers: an approach to nitric oxide neurotoxicitry. Res. Comm. Chem. Pathol. Pharmacol. 83: 243–252.

Andersen HU Larsen PM Fey SJ Karlsen AE Mandrup-Poulsen T and Nerup J (1995): Two dimensional gel electrophoresis of rat islet proteins. Interleukin1β-induced changes in protein expression are reduced by L-arginine depletion and nicotinamide. Diabetes 44:400–407.

Betz Corradin S Fasel N Buchmüller-Rouiller Y Ransijn A Smith J and Mauël J (1993): Induction of macrophage nitric oxide production by interferon-γ and tumor necrosis factor -α is enhanced by interleukin-10. Eur. J. Immunol. 23: 2045–2048.

Biswas DK Dezube BJ Ahlers CM and Pardee AB (1993): Pentoxifylline inhibits HIV-1 LTR-driven gene expression by blocking NF-κB action. J. AIDS 6: 778–786

Bogdan C Stenger S Röllinghoff M and Solbach W (1991a): Cytokine interactions in experimental cutaneous leishmaniasis. Interleukin 4 synergizes with interferon-γ to activate murine macrophages for killing of Leishmania major amastigotes. Eur. J. Immunol. 21: 327–333.

Bogdan C Vodovotz T and Nathan C (1991b): Macrophage deactivation by interleukin 10. J. Exp. Med. 174:1549–1555.

Bogdan C Vodovotz T Paik J Xie Q and Nathan C (1994): Mechanism of suppression of nitric oxide synthase expression by interleukin-4 in primary mouse macrophages. J. Leukoc. Biol. 55: 227–233.

Bulut V Severn A and Liew FY (1993): Nitric oxide production by murine macrophages is inhibited by prolonged elevation of cyclic AMP. Biochem. Biophys. Res. Comm. 195: 1134–1138.

Cash E Minty A Ferrara P Caput D Fradelizi D and Rott I (1994): Macrophage-inactivating IL-13 suppresses experimental autoimmune encephalomyelitis in rats. J. Immunol. 153: 4258–4267.

Cenci E Romani L Mencacci A Spaccapelo R Schiaffela E Puccetti P and Bistoni F (1993): Interleukin-4 and interleukin-10 inhibit nitric oxide-dependent macrophage killing of Candida albicans. Eur. J. Immunol. 23: 1034–1038.

Cetkovic-Cvrlje M Sandler S and Eizirik DL (l993): Nicotinamide and dexamethasone inhibit interleukin-1 in-
    duced nitric oxide production by RINm5F cells without decreasing messenger ribonucleic acid expression
    for nitric oxide synthase. Endocrinol. 133: 1739–1743.
Chao CC Hu S Close K Choi CS Molitor TW Novick WJ and Peterson PK (1992): Cytokine release from micro-
    glia: Differential inhibition by pentoxifylline and dexamethasone. J. Infect. Dis. 166: 847–853.
Chao CC Molitor TW and Hu S (1993): Neuroprotective role of IL-4 against activated microglia. J. Immunol. 151:
    1473–1481.
Cunha FQ Moncada S and Liew FY (1992): Interleukin-10 (IL-10) inhibits the induction of nitric oxide synthase
    by interferon-γ in murine macrophages. Biochem. Biophys. Res. Comm. 182: 1155–1159.
Deguchi M Inaba K and Muramatsu S (l995): counteracting effect of interferon-α and-β on the interferon-γ-in-
    duced production of nitric oxide which is suppressive for antibody response. Immunol. Letts. 45: 157–162.
de Waal Malefyt R Abrams J Bennett B Figdor CG and de Vries JE (1991): Interleukin 10 (IL-10) inhibits cytok-
    ine synthesis by human monocytes: An autoregulatory role of IL-10 produced by monocytes. J. Exp. Med.
    174; 1209–1220.
Doyle AG Herbein G Montaner LJ Minty AJ Caput D Ferrara P and Gordon s (l994): Interleukin-13 alters the acti-
    vation state of murine macrophages *in vitro*: comparison with interleukin-4 and interferon-γ. Eur. J. Immu-
    nol. 24:1441–1445.
Dröge W Schulze-Osthoff K Mihm S Galter D Schenk H Eck H-P Roth S and Gmünder H (l994): Functions of
    glutathione and glutathione disulfide in immunology and immunopathology. FASEB J. 8: 1131–1138.
Fast DJ Lynch RC Leu RW (l993): Interferon-γ, but not interferon-αβ, synergizes with tumor necrosis factor-α
    and lipid A in the induction of nitric oxide production by murine L929 cells. J. Interferon Res. 13:
    271–277.
Feinstein, DL Galea E and Reis DJ (1993): Norepinephrine suppresses inducible nitric oxide synthase activity in
    rat astroglial cultures. J. Neurochem. 60: 1945–1948.
Fiorentino DF Zlotnik A Mosmann TR Howard M and O'Garra A (1991): IL-10 inhibits cytokine production by
    activated macrophages. J. Immunol. 147: 3815–3822.
Fujihara M Ito N Pace JL Watanabe Y Russell SW and Suzuki T (l994): Role of endogenous interferon-β in
    lipopolysaccharide-triggered activation of the inducible nitric oxide synthase gene in a mouse macrophage
    cell line, J774. J. Biol. Chem. 269: 12773–12778.
Galter D Mihm S and Dröge W (1994): Distinct effects of glutathione disulphide on the nuclear transcription fac-
    tors kB and the activator protein-1. Eur. J. Biochem. 864: 639–648.
Gazzinelli, R. T., Oswald, I. P., James, S. L., and Sher, A. (1992). IL10 inhibits parasite killing and nitrogen oxide
    production by IFN-γ-activated macrophages. J. Immunol., 148, 1792–1796.
Genain CP Roberts T Davis RL Nguyen M-H Uccelli A Faulds D Li Y Hedgpeth J and Hauser SL (l995): Preven-
    tion of autoimmune demyelination in non-human primates by a cAMP-specific phosphodiesterase inhibi-
    tor. Proc. Natl. Acad. Sci. 92: 3601–3605.
Han J Thompson P and Beutler B (1990): Dexamethasone and pentoxifylline inhibit endotoxin-induced
    cachectin/tumor necrosis factor synthesis at separate points in the signaling pathway. J. Exp. Med. 172:
    391–394.
Hauschildt S Scheipers P and Bessler WG (1991): Inhibitors of poly (ADP-ribose) polymerase suppress
    lipopolysacchardie-induced nitrite formation in macrophages. Biochem. Biophys. Res. Comm.
    179:865–871.
Hayashi T ueno Y and Okamoto T (l993): Oxidoreductive regulation of nuclear factor κB. Involvement of a cellu-
    lar reducing catalyst thioredoxin. J. Biol. Chem 268: 11380–11388.
Ignarro LJ Lipton H Edwards JC Baricos WH Hyman AL Kadowitz PJ and Gruetter GA (1981): Mechanism of
    vascular smooth muscle relaxation by organic nitrates, nitrites, nitroprusside, and nitric oxide: Evidence for
    the involvement of S-nitrosothiols as active intermediates. J. Pharmacol. Exp. Therapeutics 218: 739–743.
Kambashi T Jacob CO Zhou D Mazurek N Fong M and Strassmann G (l995): cyclic nucleotide phosphodiesterase
    type IV participates in the regulation of IL-10 and in the subsequent inhibition of TNF-α and IL-6 release
    by endotoxin-stimulated macrophages. J. Immunol. 155: 4909–4916.
Kennedy MK Torrance DS Picha KS and Mohler KM (1992): Analysis of cytokine mRNA expression in the cen-
    tral nervous system of mice with experimental autoimmune encephalomyelitis reveals that IL-10 mRNA
    expression correlates with recovery. J. Immunol. 149: 2496–2505.
Khoury SJ Hancock WW and Weiner HL (1992): Oral tolerance to myelin basic protein and natural recovery from
    experimental autoimmune encephalomyelitis are associated with downregulation of inflammatory cytoki-
    nes and differential upregulation of transforming growth factor β, interleukin 4, and prostaglandin E ex-
    pression in the brain. J. Exp. Med. 176: 1355–1364.

Koka P He K Zack JA Kitchen S Peacock W Fried I Tran T Yashar SS and Merrill JE (1995): Human immunodeficiency virus 1 envelope proteins induce interleukin 1, tumor necrosis factor α, and nitric oxide in glial cultures derived from fetal, neonatal, and adult human brain. J. Exp. Med. 182: 941–952.

Kozaki H Egawa H Bermudes L Feduska NJ So S and Esquivel CO (1993): Pentoxifylline inhibits production of
superoxide anion and tumor necrosis factot by kupffer cells in rat liver preservation. Transplant. Proc. 25:
3025–3026.

Kreil TR and Eibl MM (1995): Viral infection of macrophages profoundly alters requirements for induction of nitric oxide snthesis. Virol. 212: 174–178.

Kröncke KD Funda J Berschick B Kolb H and Kolb-Bachofen V (1991): Macrophage cytotoxicity towards isolated rat islet cells: neither lysis nor its protection by nicotinamide are beta cell specific. Diabetol. 34:
232–238.

Liew FY Li Y Severn A Millot S Schmidt J Salter M and Moncada S (1991): A possible novel pathway of regulation by murine T helper type-2 (T_h2) cells of a T_h1 cell activity via the modulation of the induction of nitric
oxide synthase on macrophages. Eur. J. Immunol. 21: 2489–2494.

MacMicking JD Nathan C Hom G Chartrain N Fletcher DS Trumbauer M Stevens K Xie Q-W Sokol K Hutchin-
son N Chen H and Mudgett JS (1995): Altered responses to bacterial infection and endotoxic shock in mice
lacking inducible nitric oxide synthase. Cell 81: 641–650.

Marotta P Sautebin L and Di Rosa M (1992): Modulation of the induction of nitric oxide synthase by eicosanoids
in the murine macrophage cell line J774. Brit. J. Pharmacol. 107: 640–641.

Mayer M and Noble M (1994): N-acetyl-L-cysteine is a pluripotent protector against cell death and enhancer of
trophic factor-mediated cell survival in vitro. Proc. Natl. Acad. Sci. 91: 7496–7500.

Merrill JE and Benveniste EN (1996): Cytokines in inflammatory brain lesions: helpful and harmful. TINS
19:331–338.

Merrill, JE Ignarro LJ Sherman MP Melinek J and Lane TE (1993): Microglial cell cytotoxicity of oligodendrocytes is mediated through nitric oxide. J. Immunol.,151,;2132–2141.

Nabioullin R Sone S Mizuno K Yano S Nishioka Y Haku T and Ogura T (1994): Interleukin-10 is a potent inhibitor of tumor cytotoxicity by human monocytes and alveolar macrophages. J. Leukoc. Biol. 55: 437–442.

Oswald IP Gazzinelli RT Sher A and James SL (1992a): IL10 synergizes with IL-4 and transforming growth fac-
tor-β to inhibit macrophage cytotoxic activity. J. Immunol. 148: 3578–3582.

Oswald IP Wynn TA Sher A and James SL (1992b): Interleukin 10 inhibits macrophage microbicidal activity by
blocking the endogenous production of tumor necrosis factor α required as a costimulatory factor for interferon γ- induced activation. Proc. Natl. Acad. Sci. USA 89: 8676–8680.

Parkinson JF Mitrovic B and Merrill JE (1996): The role of nitric oxide in multiple sclerosis. J. Molec. Med in
press.

Pellat-Deceunynck C Wietzerbin J and Drapier J-C (1994): Nicotinamide inhibits nitric oxide synthase mRNA induction in activated macrophages. Biochem. J. 297: 53–58.

Rott O Cash E and Fleischer B (1993): Phosphodiesterase inhibitor pentoxifylline, a selective suppressor of T
helper type 1- but not type 2-associated lymphokine production, prevents induction of experimental
autoimmune encephalomyelitis in Lewis rats. Eur. J. Immunol. 23: 1745–1751.

Rott O Fleischer B and Cash E (1994): Interleukin-10 prevents experimental allergic encephalomyelitis in rats.
Eur. J. Immunol. 24: 1434–1440.

Saura M Martinez-Dalmau R Minty A Perez-Sala D and Lamas S (1996): Interleukin-13 inhibits inducible nitric
oxide synthase expression in humnan mesangial cells. Biochem. J. 313: 641–646.

SchreckR Meier B Männel DN Dröge W and Baeuerle PA (1992): Dithiocarbamates as potent inhibitors of nuclear
factor κB activation in intact cells. J. Exp. Med. 175: 1181–1194.

Sommer N Löschmann P-A Northoff GH Weller M Steinbrecher A Steinbach JP Lichtenfels R Meyermann R
Riethmüller A Fontana A Dichgans J and Martin R (1995): The antidepressant rolipram suppresses cytokine
production and prevents autoimmune encephalomyelitis. Nature Med. 1: 244–248.

Strassmann G Patil-Koota V Finkelman F Fong M and Kambayashi (1994): Evidence for the involvement of interleukin 10 in the differential deactivation of murine peritoneal macrophages by prostaglandin E_2 J. exp.
Med. 180: 2365–2370.

Strieter RM Remick DG Ward PA Spengler RN Lynch III JP Larrick J and Kunkel SL (1988): Cellular and molecular regulation of tumor necrosis factor-alpha production by pentoxifylline. Biochem. Biophys. Res.
Comm. 155: 1230–1236.

Takahashi GW Montgomery RB Stahl WL Crittenden CA Valentine MA Thorning DR Andrews III DF and Lilly
MB (1994): Pentoxifylline inhibits tumor necrosis factor-α-mediated cytotoxicity and cytostasis in L929
murine fibrosarcoma cells. Int. J. Immunopharmac. 16: 723–736.

Zajicek JP Wing M Scolding NJ and Compston DAS (1992): Interactions between oligodendrocytes and microglia. Brain 115: 1611–1631.

Zhang X Alley EW Russell SW and Morrison DC (1994): Necessity and sufficiency of beta interferon for nitric oxide production in mouse peritoneal macrophages. Infect. Immunity 62: 33–40.

Zhou A Chen Z Rummage JA Jiang H Kolosov M Kolosova I Stewart CA and Leu RW (1995): Exogenous interferon-γ induces endogenous synthesis of interferon-α and-β by murine macrophages for induction of nitric oxide synthase. J. Interferon Res. 15:897–904.

Zingarelli B O'Connor M Wong H Salzman AL and Szabó C (1996): Peroxynitrite-mediated DNA strand breakage activates poly-adenosine diphosphate ribosyl synthetase and causes cellular energy depletion in macrophages stimulated wwith bacterial lipopolysaccharide. J. immunol. 156: 350–358.

27

OLIGODENDROCYTES AND AXONAL REGROWTH

A Double-Edged Sword

M. Schwartz,[*] S. Eitan, D. L. Hirschberg, O. Eizenberg, and P. Beserman

Department of Neurobiology
The Weizmann Institute of Science
76100 Rehovot
Israel

INTRODUCTION

Oligodendrocytes are postmitotic cells that develop from rapidly dividing precursor cells (Temple and Raff, 1986). Their survival depends on exogenous signaling molecules, such as growth factors, cytokines and neurotrophic factors (Arakawa et al., 1990; Barres et al., 1992, 1993; Buchman and Davies, 1993; Louis et al., 1993; McKinnon et al., 1990; Noble et al., 1988; Richardson et al., 1988).

Oligodendrocytes appear to play an important role in limiting postmaturational growth of axons by expressing growth inhibitors (Caroni and Schwab, 1988). While this function seems to be of benefit to intact axons, it imposes a severe disability on mature injured axons, as it interferes with regrowth. In fact, in the absence of a mechanism for overcoming, bypassing or eliminating these inhibitors after injury, the CNS axons fail to regenerate. It therefore seems that on one hand oligodendrocytes are indispensable for the physiological activity of mature axons, and their inactivation would therefore result in pathologies such as multiple sclerosis; on the other hand, in the event of axonal injury their elimination would have favorable consequences.

In this article we describe our results regarding the cell cycle regulation of oligodendrocytes under normal and pathological conditions, and discuss the possible implications for regrowth following trauma.

---

[*] To whom correspondence should be addressed.

Cell Biology and Pathology of Myelin, edited by Juurlink *et al.*
Plenum Press, New York, 1997

REGENERATION IS ACCOMPANIED BY ELIMINATION OF OLIGODENDROCYTES

Nerves of the mammalian central nervous system (CNS), unlike their counterparts in lower vertebrates, have a low regeneration capability (reviewed by Schwartz et al., 1989). The ability of nerves to regenerate has been correlated with the ability of sections of these nerves to support neuronal attachment and axonal growth in vitro. Thus, sections of rat sciatic nerve and fish optic nerve, both of which are capable of regeneration, support neuronal attachment and axonal growth of neuroblastoma cells or embryonic neurons, whereas sections of the nonregenerative rat optic nerve do not (Carbonetto et al., 1987; Savio and Schwab, 1989). This by itself might lead one to conclude that the fish CNS and the mammalian peripheral nervous system (PNS) differ constitutively from the mammalian CNS with respect to inhibitors. Studies carried out by Bedi et al. (1992) and in our laboratory (Sivron et al., 1994) point to the likelihood that this might be true in part, and that there is an additional mechanism that may be responsible for the acquisition of regeneration-supportive properties, and is critically dependent on the nature of the post-injury response. Thus for example, fish optic nerve sections, which support the growth of embryonic neurons, did not support neuritic outgrowth from adult neurons, i.e., regenerating retina (Sivron et al., 1994). Once injured, however, both the fish optic nerve (Sivron et al., 1994) and the rat sciatic nerve (Bedi et al., 1992) became permissive to growth of adult neurons. The injury-induced increase in growth-permissiveness of the fish optic nerve could be the result of elimination of inhibitory molecules and/or an increase in growth-supportive molecules.

AN INTERLEUKIN-2 (IL-2)-LIKE FACTOR FOUND IN SPONTANEOUSLY REGENERATING CNS IS CYTOTOXIC TO OLIGODENDROCYTES

The post-injury changes in permissiveness to neuronal growth observed in the fish optic nerve suggest that the environment of the nerve undergoes cellular and molecular changes as a result of the injury. Early studies indicated that, following injury, there is a change in the repertoire of soluble substances originating from the regenerating fish optic nerve (Harel et al., 1989, 1990, Schwartz et al., 1985). We postulated that among these soluble substances are some that are actively involved in making the nerve permissive to growth; more specifically, that they regulate the glial response to injury in such a way that inhibitory elements are eliminated and supportive elements are acquired or enhanced. Application of a preparation containing such substances to injured rabbit optic nerves in vivo resulted in regenerative growth, which significantly exceeded the growth occurring spontaneously in untreated injured nerves (Lavie et al., 1990; Schwartz et al., 1985). On the basis of this finding we proposed that the preparation might contain glia-modulating factors, and we examined this possibility in vitro in cultures of dissociated adult rat optic nerve. The most pronounced effect of the preparation was on the number of mature oligodendrocytes (Cohen et al., 1990). Application of the preparation to cultures of mixed glial cells verified that its effect was specific for oligodendrocytes. Further studies revealed that the active molecule within the preparation is an IL-2-like compound, as its activity was neutralized by antibodies directed against IL-2 (Eitan et al., 1992). However, it was found to be twice the molecular weight of IL-2 (Eitan et al., 1992). The same preparation was also found to contain an enzyme which is capable of dimerizing IL-2 and is present in increased amounts in the regenerating fish optic nerve (Eitan and Schwartz, 1993). The resulting dimer was cytotoxic to oligodendrocytes. The enzyme was identified as a nerve-derived transglutaminase (TGase) and was purified to homogeneity (Eitan and Schwartz, 1993). Dimeric IL-2, produced in the presence of the purified enzyme and Ca2+, was found to be cytotoxic to oligodendrocytes. Taken together, the above findings suggest that oligodendrocytes may be prevented from inhibiting post-injury axonal growth of the fish optic nerve as a result of their elimination by the IL-2-like compound. It follows that the presence in the injured nerve of a factor cytotoxic to oligodendrocytes may be crucial for regeneration. It is possible that mammals lack the mechanism for eliminating oligodendrocytes, or at least that the mechanism does not operate at the right time. The elimination of oligodendrocytes by dimeric IL-2 was found to be apoptotic in nature (Eizenberg et al., 1995). Interestingly, the protein p53 was found to be involved in differentiation and death of both oligodendrocytes and neurons (Eizenberg et al., 1996).

RELATIONSHIP BETWEEN REGENERATION, INFLAMMATION, AND ELIMINATION OF OLIGODENDROCYTES

The above findings, coupled with the fact that oligodendrocyte elimination in the fish was prevented when the nerves were excised immediately after injury and kept in vitro until use (Sivron et al., 1992), led us to propose that a physiological event such as the post-injury inflammatory response might be a prerequisite for operation of the mechanism for elimination of oligodendrocytes and myelin. Studies in our laboratory (Hirschberg et al., 1994) and elsewhere (Perry et al., 1987) revealed a relationship between inflammation and regeneration. They also suggested the possibility that the mammalian CNS possesses an intrinsic mechanism that limits inflammation, either because of a lack of activating mechanisms or via the presence of inhibitory mechanisms.

Our recent studies revealed the presence in the mammalian CNS of an active mechanism that suppresses macrophage recruitment (Hirschberg and Schwartz, 1995) and activation (Lazarov Spiegler et al., 1996). The link between the inefficient recruitment and inadequate activation of macrophages and the failure of regeneration was further substantiated by our observation that local transplantation of macrophages, preincubated with segments of regenerating peripheral (sciatic) nerve, into transected optic nerve (nonregenerative CNS), leads to regrowth of the transected nerve. The induced growth is very likely related to the increased macrophage phagocytic activity that we observed (Lazarov-Spiegler et al., 1996). This would suggest that the delay in myelin clearance from the injury site after CNS injury is possibly overcome by the transplantation of macrophages preexposed to segments of regenerating sciatic nerve. Such macrophages are possibly a source of factors cytotoxic to oligodendrocytes and provide efficient removal of the dead cells and myelin debris (Lazarov-Spiegler and Schwartz, 1996).

Active macrophages have plural activities in healing processes, in general. One of these activities is phagocytosis, needed for removal of inhibitory elements including myelin debris, which would otherwise block the process of regrowth (Bandtlow et al., 1990; Cadelli et al., 1992; Schnell and Schwab, 1990). Activated macrophages may also provide apolipoproteins, the lipid-associated proteins involved in lipid metabolism and membrane reconstruction (Harel et al., 1989; Muller et al., 1985; Stoll and Muller, 1986), as well as the growth factors and cytokines needed to make the neuronal environment permissive to and supportive of regeneration (Benveniste, 1992; David et al., 1990; Eitan and Schwartz,

1993; Eitan et al., 1992, 1994; Faber-Elman et al., 1995; Heumann et al., 1987; Lotan et al., 1994; Merrill and Jonakait, 1995; Schwartz et al., 1989). It is conceivable that the inhibitor of macrophage migration that we recently described in the optic nerve (Hirschberg and Schwartz, 1995) and the observed inhibition of macrophage phagocytosis (Lazarov-Spiegler et al., 1996) are mediated by the same CNS-resident factor. Moreover, because the phagocytic activity is a marker for the state of macrophage activation, other activities of the macrophages are likely to be inhibited by the CNS as well. This macrophage-inhibiting activity might be the biochemical basis of CNS immunosuppression, i.e., CNS immune privilege (Streilein, 1993). Abundance of microglia in an immunosuppressed state in the CNS (Thomas, 1992) is another aspect of the immune privilege that might be explained by the CNS-resident inhibitory activity.

Immunosuppression of the CNS and its status as an immune-privileged system (Streilein, 1993) might be viewed as an evolutionary development that arose from the need to avoid or minimize remodeling of the complex mammalian brain (Hirschberg and Schwartz, 1995). The loss of ability of the mammalian CNS to regenerate its injured axons is presumably a direct consequence of this immunosuppression, mediated by the macrophage inhibiting activity, and may be viewed as a side effect of the attempt to protect the brain from the immune system. If so, transplantation of suitably preactivated macrophages might compensate for this physiological immunosuppression.

As a potential treatment, transplantation of macrophages has further important advantages in that the cells can be derived autologously and noninvasively. An effective multipotent therapy for CNS injuries might therefore involve optimizing this procedure in order to achieve the maximal number of growing axons, possibly in conjunction with treatment designed to improve nerve survival and/or delay nerve degeneration. In addition to achieving the practical objective of regrowth, the studies described here substantiated a novel view of CNS regeneration and CNS immune privilege, and may well lead to a novel view of the role of macrophages in the healing processes of the CNS.

CONCLUSIONS

It is well established that in the normally functioning CNS the oligodendrocytes play an important role. We have shown that treatment of mature oligodendrocytes with IL-2 in its dimeric form leads to their apoptotic death, and that pS3 is involved in this process.

Several experimental models indicate that pS3 can be regulated by differential subcellular localization. This notion is strongly supported by recent observations confirming that pS3 acts as a cell-cycle regulatory protein, and as such may require a well-controlled nuclear translocation mechanism.

Oligodendrocytes express growth-associated inhibitors and therefore act as major obstacles to axonal regrowth following injury. Accumulating evidence suggests that elimination or neutralization (Schnell and Schwab, 1990; Sivron and Schwartz, 1994) of these inhibitors represents a mechanism that enables regrowth to occur. Interestingly, elimination of oligodendrocytes was found to be associated with the post-injury inflammatory response. Ironically, what seems to be devastating for healthy nerves is essential for injured ones; for example, brain demyelination leading to development of an autoimmune disease such as multiple sclerosis is essential in cases of trauma. In this report we discussed the growth-related mechanisms of oligodendrocytes in a system that regenerates spontaneously, in order to gain insight into the mechanism underlying the physiological and pathological death of oligodendrocytes, i.e., the way to protect these cells in pathological cases, such as multiple sclerosis, and how to achieve their transient elimination, thus paving the way to regeneration.

ACKNOWLEDGMENTS

M. Schwartz is the incumbent of the Maurice and Ilse Katz Professorial Chair in Neuroimmunology.

REFERENCES

Arakawa Y, Sendtner M and Thoenen H (1990): Survival effect of ciliary neurotrophic factor (CNTF) on chick embryonic motoneurons in culture: comparison with other neurotrophic factors and cytokines. J. Neurosci. 10:3507–3515.

Bandtlow C, Zachleder T and Schwab ME (1990): Oligodendrocytes arrest neurite growth by contact inhibition. J. Neurosci. 10:3837–3848.

Barres BA, Hart IK, Coles HS, Burne JF, Voyvodic JT, Richardson WD and Raff MC (1992): Cell death and control of cell survival in the oligodendrocyte lineage. Cell 70:31~6.

Barres BA, Schmid R, Sendtner M and Raff MC (1993): Multiple extracellular signals are required for long-term oligodendrocyte survival. Development 118:283–295.

Bedi KS, Winter J, Berry M and Cohen J (1992): Adult rat dorsal root ganglion neurons extend neurites on predegenerated but not on normal peripheral nerves in vitro. Eur. J. Neurosci. 4: 193–200.

Benveniste EN (1992): Inflammatory cytokines within the central nervous system: sources, function, and mechanism of action. Am. J. Physiol. 263:Cl-C16.

Buchman VL and Davies AM (1993): Different neurotrophins are expressed and act in a developmental sequence to promote the survival of embryonic sensory neurons. Development 118:989–1001. 10

Cadelli DS, Bandtlow CE and Schwab M (1992): Oligodendrocytes and myelin-associated inhibitors of neurite outgrowth: their involvement in the lack of CNS regeneration. Exp. Neurol. 115: 189–192.

Carbonetto S, Evans D and Cochard P (1987): Nerve fiber growth in culture on tissue substrates from central and peripheral nervous system. J. Neurosci. 7:610–620.

Caroni P and Schwab ME (1988): Antibody against myelin-associated inhibitor of neurite growth neutralizes nonpermissive substrate properties of CNS white matter. Neuron 1:85–96.

Cohen A, Sivron T, Duvdevani R and Schwartz M (1990): Oligodendrocyte cytotoxic factor associated with fish optic nerve regeneration: implications for mammalian CNS regeneration. Brain Res. 537:24–32.

David S, Bouchard C, Tsatas O and Giftochristos N (1990): Macrophages can modify the nonpermissive nature of the adult mammalian central nervous system. Neuron 5:463~69.

Eitan S and Schwartz M (1993): A novel nerve-regeneration-associated transglutaminase that converts interleukin-2 to a factor cytotoxic to oligodendrocytes. Science 261: 106–108.

Eitan S, Zisling R, Cohen A, Belkin M, Hirschberg DL, Lotan M and Schwartz M (1992): Identification of an interleukin-2-like substance as a factor cytotoxic to oligodendrocytes and associated with central nervous system regeneration. Proc. Natl. Acad. Sci. USA 89:5442–5446.

Eitan S, Solomon A, Lavie V, Yoles E, Hirschberg DL, Belkin M and Schwartz M (1994): Recovery of visual response of injured adult rat optic nerves treated with transglutaminase. Science 264: 1764–1768.

Eizenberg O, Faber-Elman A, Gottlieb E, Oren M, Rotter V and Schwartz M (1995): Direct involvement of p53 in programmed cell death of oligodendrocytes. EMBO J. 14: 1136–1144.

Eizenberg O, Faber-Elman A, Gottlieb E, Oren M, Rotter V and Schwartz M (1996): pS3 plays a regulatory role in differentiation and apoptosis of central nervous system-associated cells. Mol. Cell. Biol. 16 (in press).

Faber-Elman A, Miskin R and Schwartz M (1995): Components of the plasminogen activator system in astrocytes are modulated by tumor necrosis factor-a and interleukin- 1 β through similar signal transduction pathways. J. Neurochem. 65: 1524–1535.

Harel A, Fainaru M, Schafer Z, Hernandez M and Schwartz M (1989): Optic nerve regeneration in adult fish and apolipoprotein A-1. J. Neurochem. 52:1218–1228.

Harel A, Fainaru M, Rubinstein M, Tal N and Schwartz M (1990): Fish apolipoprotein-A-1 has heparin-binding activity: implication for nerve regeneration. J. Neurochem. 55: 1237–1243.

Heumann R, Lindholm D, Bandtlow C, Meyer M, Radeke MJ, Miko TP, Shooter E and Thoenen H (1987): Differ-
    ential regulation of mRNA encoding nerve growth factor and its receptor in sciatic nerve during develop-
    ment, degeneration, and regeneration: role of macrophages. Proc. Natl. Acad. Sci. USA 84:8735–8739.
Hirschberg DL and Schwartz M (1995): Macrophage recruitment to acute injury in the CNS is inhibited by a resi-
    dent factor: a basis for an immune-brain barrier. J. Neuroimmunol. 61 :89–96.
Hirschberg DL, Yoles E, Belkin M and Schwartz M (1994): Inflammation after axonal injury has conflicting con-
    sequences for recovery of function: rescue of spared axons is impaired but regeneration is supported. J.
    Neuroimmunol. 50:9–16.
Lavie V, Murray M, Solomon A, Ben-Bassat S, Rumelt S, Belkin M and Schwartz M (1990): Growth of injured
    CNS axons within their degenerating optic nerve. J. Comp. Neurol. 298:293–314.
Lazarov-Spiegler O, Solomon AS, Ben Zeev-Brann A, Hirschberg DL, Lavie V and Schwartz M (1996): Trans-
    plantation of activated macrophages overcomes central nervous system regrowth failure. FASEB J. 10 (in
    press).
Lotan M, Solomon A, Ben-Bassat S and Schwartz M (1994): Cytokines modulate the inflammatory response and
    change permissiveness to neuronal adhesion in injured mammalian central nervous system. Exp. Neurol.
    126:284–290.
Louis J-C, Magal E, Takayama S and Varon S (1993): CNTF protection of oligodendrocytes against natural and tu-
    mor necrosis factor-induced death. Science 259:689–692.
McKinnon RD, Matsui T, Dubois-Dalcq M and Aaronson SA (1990): FGF modulates the PDGF driven pathway of
    oligodendrocyte development. Neuron 5:603–614.
Merrill JE and Jonakait GM (1995): Interactions of the nervous and immune systems in development, normal
    brain, homeostasis, and disease. FASEB J 9:611–618.
Muller HW, Gebicke-Harter PJ, Hangen DH and Shooter EM (1985): A specific 37000-dalton protein that accu-
    mulates in regenerating but not in nonregenerating mammalian nerves. Science 228:499–501.
Noble M, Murray K, Strobant P, Waterfield MD and Riddle P (1988): Platelet-derived growth factor promotes di-
    vision and motility and inhibits premature differentiation of the oligodendrocyte/type-2 astrocyte progeni-
    tor cell. Nature 333:560–562.
Perry VH, Brown MC and Gordon S (1987): The macrophage response to central and peripheral nerve injury. J.
    Exp. Med. 165:1218–1223.
Richardson WD, Pringle N, Mosley MJ, Watermark B and Dubois-Dalcq M (1988): A role for platelet-derived
    growth factor in normal gliogenesis in the central nervous system. Cell, 53:309–319.
Savio T and Schwab ME (1989): Rat CNS white matter, but not grey matter, is nonpermissive for neuronal cell ad-
    hesion and fiber outgrowth. J. Neurosci. 9: 1126–1129.
Schnell L and Schwab ME (1990): Axonal regeneration in the rat spinal cord produced by an antibody against
    myelin-associated neurite growth inhibitors. Nature 343:269–272.
Schwartz M, Belkin M, Harel A, Solomon A, Lavie V, Hadani M, Rachailovich I and Stein-Izsak C (1985): Regen-
    erating fish optic nerves and a regeneration-like response in injured optic nerves of adult rabbits. Science
    228:601–603.
Schwartz M, Cohen A, Stein-Izsak C and Belkin M (1989): Dichotomy of the glial cell response to axonal injury
    and regeneration. FASEB J. 3:2371–2378.
Sivron T and Schwartz M (1994): The enigma of myelin-associated growth inhibitors in spontaneously regenerat-
    ing nervous systems. Trends Neurosci. 17:277–281.
Sivron T, Jeserich G, Nona S and Schwartz M (1992): Characteristics of glial cells in culture: possible implication
    as to their lineage. Glia 6:52–66.
Sivron T, Schwab ME and Schwartz M (1994): Presence of growth inhibitors in fish optic nerve myelin: post-in-
    jury changes. J. Comp. Neurol. 341:1–10.
Stoll G and Muller H (1986): Macrophages in the peripheral nervous system and astroglia in the central nervous
    system of rat commonly express apolipoprotein-E during development but differ in their response to injury.
    Neurosci. Lett. 72:233–238.
Streilein JW (1993): Immune privilege as the result of local tissue barriers and immunosuppressive microenviron-
    ments. Curr. Opin. Immunol. 5:428–432. 15
Temple S and Raff MC (1986): Clonal analysis of oligodendrocyte development in culture: evidence for a devel-
    opmental clock that counts cell divisions. Cell 44:773–779.
Thomas WE (1992): Brain macrophages: evaluation of microglia and their functions. Brain Res. Rev. 17:61–74.

28

FUNCTIONAL REPAIR OF MYELINATED FIBERS IN THE SPINAL CORD BY TRANSPLANTATION OF GLIAL CELLS

S. G. Waxman[1] and J. D. Kocsis[2]

[1]Department of Neurology
Yale University School of Medicine
P.O. Box 208018
New Haven, Connecticut 06520-8018
[2]PVA/EPVA Center for Neuroscience
VA Medical Center
West Haven, Connecticut 06516

INTRODUCTION

Axons fulfill the essential role of conducting action potentials from the neuron's receptive pole, i.e., the dendrites and cell body, to synaptic terminals which impinge on other neurons. In myelinated axons the myelin provides a high-resistance, low-capacitance shield, and conduction occurs in a saltatory manner at a conduction velocity which is linearly related to fiber diameter. In contrast, in non-myelinated fibers, the action potential propagates in a continuous manner, and conduction velocity is proportional to (diameter)$^{1/2}$. The two conduction velocity-diameter relationships intersect so that, above a critical diameter of 0.2 μm, myelinated axons conduct more rapidly than their non-myelinated counterparts of the same size (Waxman and Bennett, 1972). Myelination thus results, in practice, in an increase in conduction velocity. In axons that have sustained damage to their myelin sheaths (i.e., *demyelinated* axons), and in axons that have not acquired normal myelin sheaths (i.e., *dysmyelinated* axons), conduction velocities are reduced compared to those of normally myelinated axons of the same diameter. In addition, the refractory period for transmission is increased in demyelinated axons (Smith et al., 1983) and the ability to conduct trains impulses at high frequencies is impaired (Honmou et al., 1996). In the most severely affected demyelinated fibers, conduction block occurs and action potentials fail to propagate beyond the region of myelin damage. These conduction abnormalities are known to be associated with clinical deficits (Waxman et al., 1995).

Injury to the myelin sheath represents an important component of the pathology in inflammatory (demyelinating) disorders including multiple sclerosis, genetic (dysmyelinating) disorders including the leukodystrophies, and traumatic disorders including con-

Cell Biology and Pathology of Myelin, edited by Juurlink *et al.*
Plenum Press, New York, 1997

tusive spinal cord injury and spinal cord compression (Gledhill et al., 1973; Blight, 1985; Byrne and Waxman, 1990). Decreased conduction velocity, increased refractory period, reduced ability to conduct high frequency impulse trains, and total conduction block occur as a result of damage to the myelin in these disorders. Functional repair of demyelinated and dysmyelinated axons, so that they can conduct action potentials in a normal manner, might thus be expected to result in clinical improvement, and is a major focus of current research.

Previous studies have demonstrated that glial cells can survive and form myelin with relatively normal morphological characteristics and compact periodicity following transplantation into the CNS (Blakemore and Franklin, 1991; Duncan et al., 1988a; Gout and Dubois-Dalcq, 1993; Lachapelle et al., 1994; Rosenbluth et al., 1990). However, most previous studies have focused on the morphology of the newly formed myelin, and only a few studies (Utzschneider et al. 1994; Honmou et al. 1996) have examined the *functional* effects of myelination by transplanted cells, i.e., the effects on axonal conduction (Utzschneider et al., 1994; Honmou et al., 1996). This has represented a significant gap in our knowledge. Functional studies of this type are essential, because the formation of myelin *per se* does not necessarily insure the reliable or rapid conduction of action potentials; secure and rapid conduction will only occur if myelin segments of the appropriate length and thickness have been formed (Huxley and Stämpfli, 1949; Waxman and Brill, 1978), and depends, in addition, on the clustering of sodium channels in high densities at the newly formed nodes of Ranvier (Ritchie and Rogart, 1977; Waxman, 1977; Moore et al., 1978; Hines and Shrager, 1991). Formation of myelin by endogenous oligodendrocytes or following invasion of endogenous Schwann cells from the PNS into the CNS has been demonstrated in certain experimental models of demyelination, and electrophysiological studies have demonstrated relatively normal action potential conduction following myelin formation by endogenous oligodendrocytes or Schwann cells in some of these models (Blight and Young, 1989; Felts and Smith, 1992; Smith et al., 1981; Smith et al., 1983).

In our laboratories we are exploring the possibility of functional repair of demyelinated axons within the spinal cord. As part of this research program, we are studying action potential conduction in experimental models which lack CNS myelin, into which we have transplanted myelin-forming glial cells. In order to assess the functional consequences of myelin formation following cell transplantation, we have used electrophysiological recording methods (Kocsis and Waxman, 1980; Kocsis et al., 1983) including intra-axonal recording and field potential analysis to examine conduction and to measure conduction velocity, refractory period for transmission, and the ability to conduct high-frequency trains of action potentials in spinal cord axons before and after glial cell transplantation. This chapter summarizes results from our laboratory which demonstrate that myelination by *exogenous, transplanted* glial cells can enhance the conduction of action potentials of dysmyelinated and demyelinated axons, with restoration of close-to-normal conduction velocity, refractory period, and ability to carry high-frequency trains of action potentials.

MYELINATION BY TRANSPLANTED CELLS ENHANCES CONDUCTION IN DYSMYELINATED SPINAL CORD AXONS

Our first studies utilized the myelin-deficient (*md*) rat as a model of dysmyelination (Utzschneider et al., 1994). This experimental model, which results from a point mutation in the proteolipid protein gene, lacks CNS myelin (Boison and Stoffel, 1989; Simons and

Riordan, 1990; Zeller et al., 1989; Hudson et al., 1989). The lack of endogenous myelin within the CNS permits definitive confirmation that physiological changes are due to myelination by exogenous, transplanted glial cells, and are not due to background host myelination (Duncan et al., 1988b). Earlier electrophysiological studies in the *md* rat demonstrated that the dysmyelinated *md* axons are capable of conducting action potentials although their conduction velocities are reduced (Waxman et al., 1990) they are capable of secure action potential conduction at relatively normal frequencies (Utzschneider et al., 1992). For the present studies, female litter mates, at 4–5 days postnatal, of the animals to be transplanted were sacrificed, and their spinal cords were removed for preparation of a glial cell suspension. Cells were maintained in culture overnight, and were concentrated at 50,000 per µl. Prior to transplantation, recipient rats were anesthetized with halothane and a dorsal laminectomy was performed at the thoracolumbar junction. Using a glass micropipette, injections of 1.0 µl of cell suspension were made into two or three sites along the dorsal columns of the spinal cord. Transplantation sites were marked for subsequent identification.

In order to study axonal conduction, the transplanted rats were sacrificed on postnatal day 20, 21, or 22, i.e., 15–17 days following cell transplantation. Following deep pentobarbital anesthesia, animals were perfused with chilled, oxygenated high sucrose solution (containing 124 mM sucrose, 26 mM $NaHCO_3$, 3 mM KCl, 1.3 mM NaH_2PO_4, 2 mM $MgCl_2$, 10 mM dextrose) and the spinal cord was then removed *en bloc* and allowed to come to room temperature over 30–45 minutes prior to positioning in a standard recording chamber. Field potential and single cell recordings were obtained as previously described (Utzschneider et al., 1994).

Within each of the transplanted spinal cords studied in this series (n = 10) oligodendrocytes migrated away from the injection site and formed myelin sheaths. The region of myelination could be visualized as an opaque white patch which extended for several mm longitudinally along the dorsal surface of the spinal cord (Utzschneider et al., 1994). Histological examination revealed that numerous oligodendrocytes were present within these patches and most axons had become myelinated. Rostral and caudal to these areas of myelination, the spinal cords remained pale and histological study showed that there were few or no myelinated axons within the dorsal columns, similar to non-transplanted *md* spinal cords.

Field potential recordings were obtained along the trajectories of dorsal column axons, both within the transplant region where they had become myelinated, and outside of the transplant region, as shown in Figure 1A. These records provide a measure of the pooled behavior of dorsal column fibers, and show a conduction velocity of 0.9 ± 0.03 m/sec outside of the transplant region (Figures 1B and 1C). Within the transplant region (Figure 1D), field potentials showed two negativities, which represent conduction in two populations of axons (those that had become myelinated and those that had not). The change in latency per increment of conduction distance is consistently smaller for the first negativity (N_1) than for the second (N_2) negativity. For the slower N_2 response, the change in latency, per increment of conduction distance, is similar to that recorded in the adjacent, non-transplanted dorsal columns. In contrast, for the N_1 response, which reflects conduction in axons that have been myelinated following transplantation, the latency per increment per conduction distance is consistently smaller (Figure 1D). Analysis of the field potentials (Fig. 1B) indicates, in fact, that conduction velocity of the most rapidly conducting fibers within the transplant region is 3–4 times faster (3.2 ± 0.2 m/sec) than the conduction velocity outside the transplant region (0.9 ± 0.03 m/sec).

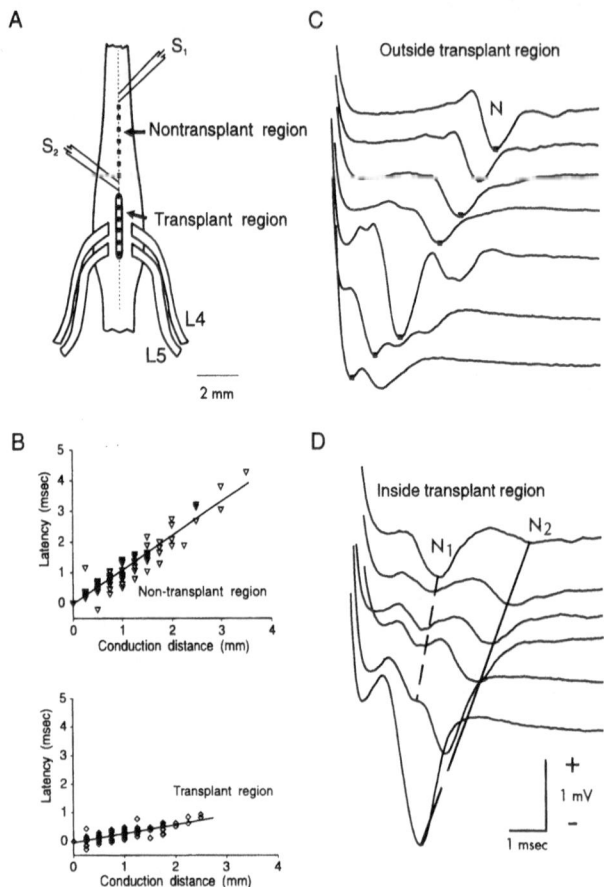

Figure 1. Conduction velocity is increased in dorsal column axons of the myelin-deficient *(md)* rat following transplantation of myelin-forming cells. (A) Field potentials were recorded from transplant region (~ 3 mm in length) and non-transplant region 16 days following glial cell transplantation. Stimulation at two sites (S₁ and S₂) provided recording tracks within and outside the transplant region. Recordings were made at 0.5 mm intervals for both tracks. (B) Aggregate conduction latencies in non-transplant (upper graph) and transplant regions (lower graph) in the *md* dorsal columns. The upper graph displays the latency of the main N negativity (from 100 recording sites from 17 recording tracks) outside the transplant region. The slope of the linear regression indicates an average conduction velocity of 0.9 ± 0.03 m/s. In the lower graph a significantly smaller increase in latency with increasing conduction distance can be seen within the transplant region, where there is an average conduction velocity of 3.2 ± 0.23 m/s. (C) Field potentials outside the transplant region typically display a single main negativity with occasional early or late components. (D) Field potentials from transplant region of the same animal show two negativities (N₁ and N₂) with increasingly distinct latencies as the conduction distance in increased. The N₁ component displays an increased conduction velocity. In this experiment, the stimulus site was outside the transplant region, demonstrating conduction of the impulse across the dysmyelinated-myelinated junction. (Modified from Utzschneider et al, 1994)

Field potential analysis also permitted us to study other aspects of axonal conduction in the transplanted spinal cords. In all six of the spinal cords that were studied, field potentials could be observed to propagate into, or out of, the transplant area. This is an important finding because abrupt changes in the pattern of myelination can cause impedance mismatch which prevents conduction (Waxman and Brill, 1978). In recordings in which the conduction track encompassed both myelinated and non-myelinated regions, we observed secure propagation of the action potential into the myelinated region and conduction velocity increased approximately two-fold in the region of myelination.

Frequency-following properties of the dorsal columns were also assessed using field potentials within the transplant region. Dorsal column fibers that had been myelinated displayed frequency-following characteristics that were essentially the same as those of control spinal cord axons (Figure 2). The ability of axons within the transplant region to carry high-frequency impulse trains, following stimulation at frequencies as high as 100 Hz, was similar to that of axons outside of the transplant region, which are known to have high-frequency transmission capabilities similar to those of spinal cord axons in control rats.

To assess the physiological consequences of myelination by transplanted cells at the single-cell level, intracellular recordings were obtained from dorsal root ganglion (DRG) neurons which project axons into the dorsal columns, and the axons of these cells were antidromically stimulated at the dorsal root entry zone and within the dorsal columns (Figure 3). Analysis of conduction in 67 axons which conducted through the transplant region demonstrated an average conduction velocity that was ~ 3 times faster than for axons from non-transplanted *md* spinal cords. Conduction velocities in these myelinated axons were comparable to conduction veclocities from age-matched controls (transplanted rats, 2.3 ± 0.3 m/sec, n=67; *md* rats, 0.76 ± 0.02 m/sec, n=258; age-matched controls 1.9 ± 0.13 m/sec, n=95). Figure 3C is a graph showing dorsal column conduction velocity plotted as a function of dorsal root conduction velocity for axons from each of the three experimental groups. Dorsal column axons in the transplant region display conduction velocities that are significantly faster than those for axons with comparable dorsal root conduction velocities from non-transplanted *md* rats. These results thus show, for single axons, that myelination by transplanted cells is associated with an increase in conduction velocity. These experiments also demonstrate, at the single axon level, the conduction of action potentials from the transplant region into non-transplanted zones, providing additional evidence for secure conduction through the critical zone of potential impedence mismatch at the junction between myelinated and non-myelinated parts of the nervous system (Figure 3B).

We conclude, from these observations in the dysmyelinated spinal cord of the *md* rat, that as a result of myelination by exogenous, transplanted glial cells in this model, axonal conduction is enhanced, with restoration of conduction velocity to essentially normal values. Notably, myelination by transplanted cells is not confined to the injection site; our experiments demonstrate that in this model system, transplanted glial cells can migrate away from their site of implantation, and successfully myelinate axons at distances as far as 5–10 mm rostrally and caudally.

The absence of endogenous CNS myelin in the *md* model is important because it facilitates confirmation that the functional changes are the results of myelination by exogenous, transplanted cells and are not due to background myelination by endogenous glial cells. However, this model has some disadvantages. Since affected *md* rats rarely survive more than 3–4 weeks postnatal, experiments using this model system are confined, by necessity, to an examination of the effects of cell transplantation into the CNS of immature hosts. Moreover, this system cannot be used to assess the long-term effects of glial cell transplantation. In order to address these issues, we have more recently begun to study the effects of glial cell transplantation into the acutely demyelinated spinal cord of adult rats.

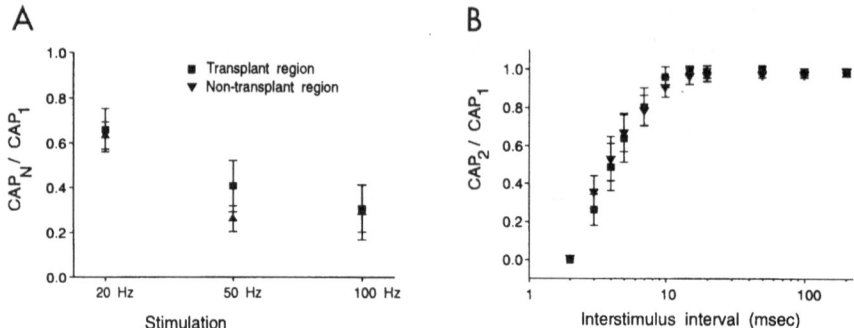

Figure 2. (A) High frequency conduction in *md* axons in response to tetanic stimuli is similar inside and outside the transplant region. The ratio of the amplitudes of the first (CAP_1) and last (CAP_N) compound action potentials (CAPs) for repetitive stimuli at 20 Hz (10 s), 50 Hz (10 s), and 100 Hz (2 s) are shown. (B) Double-shock experiments showing the ratio of test CAP (CAP_2) to control CAP (CAP_1) for interstimulus intervals of 2-200 ms. The time-course of recovery for impulse conduction is similar for axons inside and axons outside the transplant region, which are known to conduct high-frequency trains as well as spinal cord axons in control rats. (Modified from Utzschneider et al, 1994)

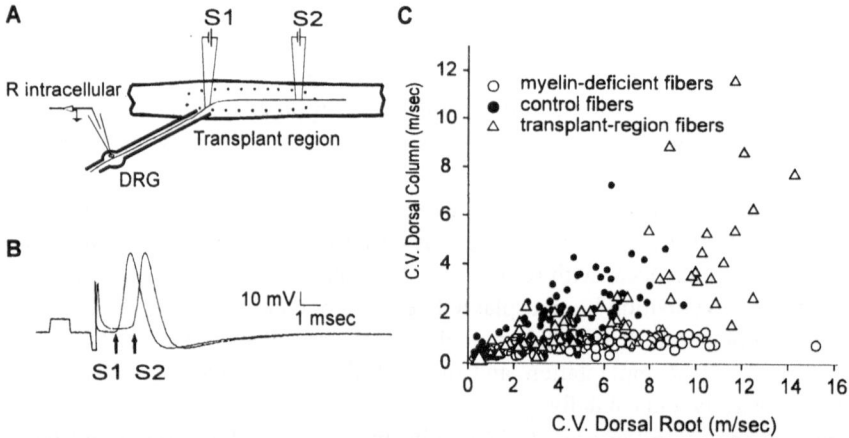

Figure 3. (A) Single-cell recording demonstrating conduction of action potentials through the transplant region. Placement of stimulating electrodes in the transplant zone, and an intracellular recording electrode in a dorsal root ganglion cell, are shown in the schematic. (B) Action potentials recorded from dorsal root ganglion cell in response to stimulation at two sites (S_1 and S_2) in the transplant region. Propagation of the action potential from stimulating electrode S_2 to the dorsal root ganglion cell indicates that conduction was not blocked at the zone of potential impedance mismatch between the transplant zone and non-transplanted parts of the host nervous system. From the latency shift and interstimulus distance, a conduction velocity of 2.6 m/s within the transplant zone can be calculated for this axon. (C) Aggregate data conduction velocity for axons within *md* transplant regions, *md* non-transplant region, and control spinal cord; note the increase in conduction velocity in axons within the transplant zone. (Modified from Utzschneider et al, 1994)

MYELINATION BY TRANSPLANTED CELLS ENHANCES CONDUCTION IN DEMYELINATED DORSAL COLUMN AXONS IN THE ADULT RAT

To examine the consequences of myelination by exogenous glial cells within the adult central nervous system, we have transplanted cultured Schwann cells, in some cases transfected with *lacZ* reporter gene, together with astrocytes derived from immature rats, into the acutely demyelinated dorsal columns in the adult rat spinal cord (Honmou et al., 1996). In these studies, an acute demylinating lesion with minimal endogenous remyelination was produced by irradiation using a method derived from that of Blakemore and Patterson (1978). In these experiments, a 40 Grays surface dose of X-irradiation was delivered to the spinal cord caudal to the T-10 level and, at three days post-X-irradiation, a demyelinating lesion was induced within the dorsal columns by direct injection of ethidium bromide, a nucleic acid chelating agent which induces oligodendrocyte death. The ethidium bromide-X radiation lesion remains glial cell-free for more than 5–6 weeks in contrast to most other models, in which endogenous remyelination can commence within days of demyelination.

Histological and electron microscopic examination (Figure 4, 5) confirmed that axons within the lesion were totally demyelinated and the lesion site was largely free of glial cells (see figure 4D and 5B). Demyelinated lesions were large and well-circumscribed, encompassing 70–80% of the transverse extent of the dorsal columns and, using our demyelination protocol, measured 7–8 mm along the longitudinal axis of the spinal cord. In the absence of glial cell transplantation, no axonal ensheathment was present within these lesions and, at intervals ranging from 1–6 weeks following induction of the lesion, there was no evidence of remyelination by endogenous oligodendrocytes or Schwann cells.

For these experiments we prepared Schwann cells for transplantation following culturing from the sciatic nerve of neonatal (P1-P3) rats using the method of Brockes et al. (1979). Primary astrocyte cultures were established from neonatal rat optic nerve using the method of McCarthy and de Vellis (1980). A replication-defective BAG retroviral vector (Price et al., 1987) containing the φ2 packaging line (Mann et al., 1983) was utilized to transfect the bacterial ß-gal gene into Schwann cell primary cultures so that they could be identified after transplantation. For construction of the BAG vector, the ß-gal gene was cloned into a pDOL construct derived from the Maloney murine leukemia virus (Mo-MuLV) which provided a promoter for the ß-gal gene. Insertion of the simian virus 40 early promoter and the Tn5 neomycin-resistant gene downstream from the ß-gal gene permitted selection of infected colonies. Supernatants from packaging cells were used to infect Schwann cells cultured in the presence of polybrene (8µg/ml), which were rapidly proliferating under the influence of forskolin (2µM) and glial growth factor (Brockes et al., 1979). Treatment with cytosine arabinoside (10µM) and antibody-complement mediated cell lysis with anti-Thy 1.1 antibody and rabbit complement were used to eliminate contaminating fibroblasts from cultures prior to transfection (Porter et al., 1986). Transfected Schwann cells were selected following incubation in the neomycin analog G-418. Suspensions of 5×10^4 Schwann cells and astrocytes (in a ratio of approximately 3:2) in 1 µl were injected, under Ketamine/xylazine anesthesia, into the middle of the ethidium bromide-x-irradiation induced lesion 3 days following ethidium bromide injection.

As in the *md* spinal cord, histological and ultrastructural examination confirmed that the transplanted Schwann cells migrated away from the injection site. In fact, in the demyelinated dorsal columns following this transplantation procedure, exogenous Schwann

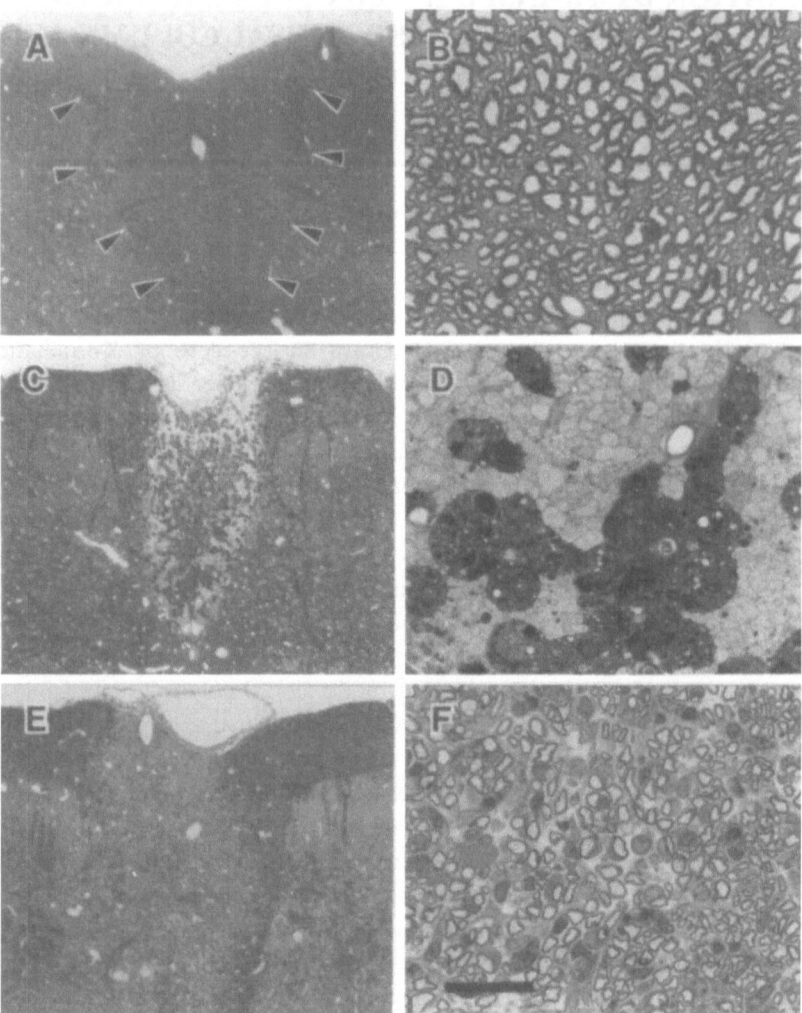

Figure 4. Light micrographs at low (A, C, and E) and high (B, D, and F) magnification showing transverse sections of control (A and B), demyelinated but not transplanted (C and D, 25 days following injection of ethidium bromide), and transplanted (E and F; 24 days post ethidium bromide injection) rats. For orientation, the dorsal columns have been outlined with arrow heads in A. The demyelinated lesion (C, D) in the absence of a transplant, is composed of naked, demyelinated axons and phagocytes which are filled with debris (D). Following transplantation of Schwann cells and astrocytes (E, F), there is a large region of remyelination (F). Scale bar = 500 μm for A, C and E; 50 μm for B, D, and F (modified with permission from Honmou et al, 1996)

cells could be seen to form myelin around axons throughout the lesion. Thus, virtually all of the axons within the lesion zone, with the exception of the smallest diameter axons which are normally nonmyelinated, exhibited remyelination by three weeks following Schwann cell transplantation (Figure 5C). In experiments where Schwann cells were transfected with the *lacZ* reporter gene, the presence of ß-gal gene product confirmed that the new myelin had been formed by transplanted, and not endogenous, Schwann cells (Figure 5D).

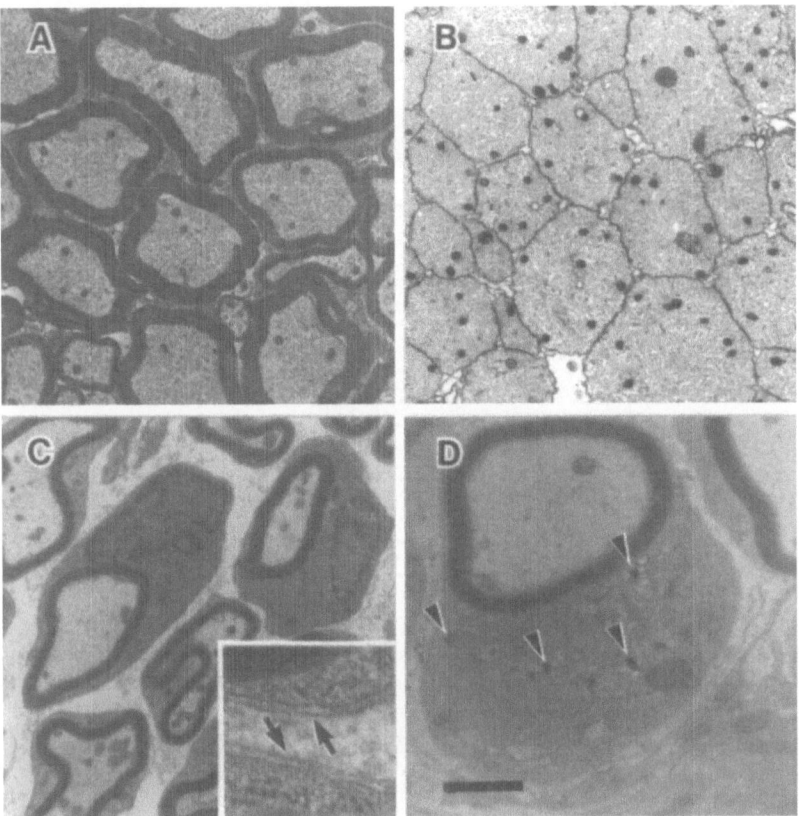

Figure 5. Electron micrographs showing transverse sections of normal (A) and demyelinated (B) axons in the dorsal columns of adult rat. Following cell transplantation all of the dorsal columns showed clear evidence of re-myelination (C). Examination at higher magnification (inset at C) demonstrated basal lamina (arrows) and extracellular collagen fibrils surrounding the individual axons. (D) Schwann cells carrying the ß-gal reporter gene (reaction product indicated by arrow heads) could be detected in the lesion following treatment of the tissue with the substrate X-Gal. Scale bar = 4 µm for A, B and C; 2 µm for D; 0.6 µm for inset in C. (Modified from Honmou et al, 1996)

Our electrophysiological experiments demonstrated that conduction within the de-myelinated dorsal columns was significantly impaired, compared to control dorsal columns; this impairment was reversed in transplanted spinal cords. Within control dorsal columns, in which there were no demyelinated lesions, after the compound action potential had been conducted for 5 mm, amplitude of the response remained at 13.4 ± 4.4% (mean ± SEM; n=5) of the amplitude at 2 mm (Figure 6B1). In contrast, within the de-myelinated dorsal columns, there was a precipitous fall-off in the compound action potential and its amplitude decreased much more rapidly as conduction distance increased; at a conduction distance of 5 mm virtually no response could be recorded, demonstrating conduction block in the demyelinated axons (Figure 6B2). This pathological fall-off in the compound action potential was reversed by glial cell transplantation; the amplitude decrement with distance was restored to normal values following myelination by transplanted cells and was indistinguishable from controls (Figures 6B3 and 6C). Thus, in transplanted

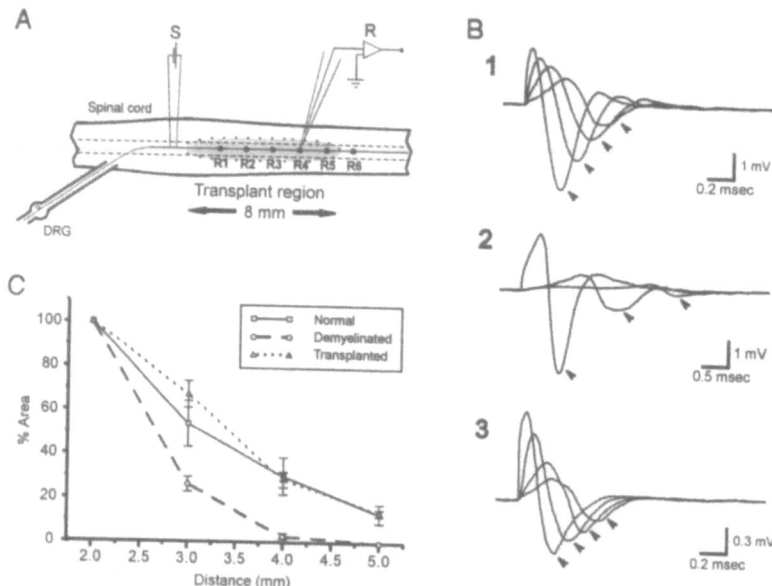

Figure 6. (A) Schematic illustrating the dorsal surface of spinal cord, showing the positions of the stimulating (S) and recording (R) electrodes. The area of demyelination or remyelination is shaded. (B) Compound action potentials recorded at 1 mm increments along the dorsal columns in control (1), ethidium bromide-X-irradiation demyelinated (2) and transplant-induced remyelinated (3) axons. (C) A plot of compound action potential area (% Area) vs. conduction distance for normal, demyelinated, and transplant-induced remyelinated dorsal columns (n=5). (Modified from Honmou et al, 1996)

spinal cords, action potentials propagated for a substantially greater distance into the lesion without block.

Conduction velocities in the demyelinated spinal cord, similar to those in the *md* model, were substantially reduced, but were restored to close-to-normal values following transplantation. Thus, conduction velocity for the compound action potential in the demyelinated dorsal column (0.9 ± 0.1 m/sec) was much lower than control values (10.2 ± 0.5 m/sec) at 36° C. Following remyelination by transplanted Schwann cells, essentially normal conduction velocities were restored (11.4 ± 0.7 m/sec).

To study conduction in single axons that traversed the lesion site we used intra-axonal recording methods. For these studies, arrays of stimulating electrodes were positioned on the dorsal surface of the spinal cord within the lesion and at distances up to several mm rostral and caudal to the lesion. Axons were impaled at a site between the arrays of stimulating electrodes within the non-demyelinated region, so as to permit examination of conduction along segments of the same axon running through, and excluding, the lesion (Figure 7A). Figure 7B confirms that, in demyelinated spinal cords that had not received a transplant, conduction velocity was considerably reduced within the region of demyelination. In contrast, in axons that had been myelinated by transplanted Schwann cells, conduction velocities were increased to essentially normal levels, and were similar within the lesion (where remyelination by exogenous Schwann cells had occurred) as compared to conduction outside of the demyelinated zone (Figure 7C, D). Thus, conduction velocity was restored to essentially normal levels as a result of cell transplantation and myelination. These experiments also showed that action potentials, evoked by stimu-

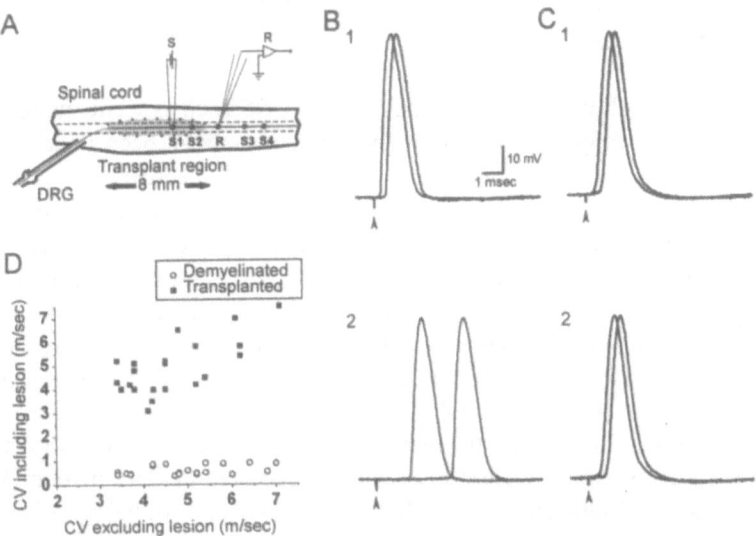

Figure 7. (A) Schematic showing arrangement of intra-axonal recording and extracellular stimulation sites. Intra-axonal impalements permitted recordings to be obtained from dorsal column axons outside of the lesion, where the axons were normally myelinated (R). Stimulating electrodes were positioned within and outside (S1-S4) of the lesion zone, permitting conduction velocity to be examined in the demyelinated or remyelinated parts of the axons, and from a normally myelinated segment of the same axon. (B) Pairs of action potentials, from a spinal cord that did not receive a transplant; the two records show conduction (for comparable conduction distances) along trajectories that included the demyelinated lesion (2) and excluded (1) the lesion. Conduction latency is increased due to slowed conduction through the demyelinated region. (C) Similar stimulation-recording protocol as B, but for transplant-induced remyelinated axons. The conduction latencies through the remyelinated part of the axon trajectory are similar to those of the axon segment outside of the lesion zone; this is due to an increase in conduction velocity as a result of myelin formation. (D) Conduction velocities of axon segments within the lesion are plotted vs. conduction velocities of the axon segment outside of the lesion, for demyelinated and transplant-induced remyelinated groups. Conduction velocities are increased in the transplanted axons. (Modified from Honmou et al. 1996)

lation within the transplant region, could reliably propagate into the non-demyelinated portion of the spinal cord (Figure 7A, C).

Refractory periods for transmission were prolonged in the demyelinated dorsal columns, as measured by paired pulse stimulation experiments. As shown in Figure 8A, recovery of the compound action potential, following a suprathreshold conditioning stimulus, was delayed in demyelinated axons. In axons remyelinated following transplantation, this functional abnormality was reversed, and the relative refractory period was reduced so that it approached, or was even less than, control values. Thus the onset of recovery occurred sooner, and the slope of recovery was greater in axons remyelinated following transplantation.

The ability of dorsal column axons to carry high-frequency trains was also impaired following demyelination, as shown in Figure 8B; amplitude of the field potential of the demyelinated axons was reduced compared to controls at stimulation frequencies of 50 Hz and higher. This conduction abnormality was reversed following transplantation of Schwann cells, and remyelinated axons were able to follow high-frequency stimulation as well as or better than controls, with smaller amplitude decrements at high stimulus frequencies than in control dorsal columns.

FUNCTIONAL REPAIR OF THE CNS BY CELL TRANSPLANTATION

Although the possibility of repairing parts of the CNS such as white matter by transplantation of myelin-forming glial cells, or their precursors, has been under study as a therapeutic strategy for dysmyelinating and demyelinating disorders for a number of years, only limited laboratory work has examined the electrophysiological or functional consequences of myelin formation within the CNS as a result of introduction of exogenous cells. Cultured oligodendrocytes, progenitor cells, and Schwann cells are capable of myelinating CNS axons that are devoid of myelin (Blakemore and Crang, 1985; Duncan et al., 1988a; Groves et al., 1993). Schwann cells provide a potential advantage since they are derived from peripheral nerve and, in theory, could be derived from the host (e.g. by sural nerve biopsy), thus permitting autologous transplantation and obviating the necessity for immunosuppression. In addition, the immune attack in inflammatory CNS demyelinating diseases such as multiple sclerosis is essentially confined to CNS myelin (Waxman, 1993), and Schwann cell remyelination might have the advantage of being resistant to further immunological attack.

If transplantation strategies are to be clinically applicable to dys- and demyelinating disorders, it will be necessary for transplanted cells to migrate to axons that lack myelin so that they can form myelin around them. There is, in fact, evidence for migration of transplanted glial cells within the CNS (Vignais et al., 1993); we observed migration of transplanted myelin-forming cells for millimeters within immature (Utzschneider et al., 1994) and mature (Honmou et al., 1996) spinal cord.

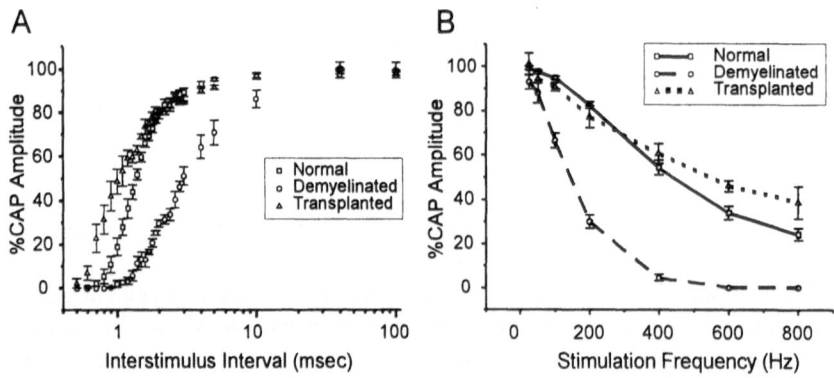

Figure 8. Refractory period and ability to follow high frequency stimuli are restored to close-to-normal values in the demyelinated dorsal columns as a result of cell transplantation. (A) Paired pulse stimuli at varying interstimulus intervals were used to measure the refractory period for transmission in normal, demyelinated, and transplant-induced remyelinated axons. Compound action potentials evoked by the second of two paired stimuli were measured at increasing interstimulus intervals. Amplitude recovery was impaired in the demyelinated axons as compared to normal axons, but transplant-induced remyelinated axons exhibited faster recovery properties than control axons. (B) To study high frequency conduction, the ratio of the compound action potential amplitudes of the last response of a train (0.5 sec, 36° C) over the first response was examined at various frequencies for normal, demyelinated and transplant-induced remyelinated axons. Demyelinated axons showed reduced frequency-response properties. Transplantation-induced remyelinated axons showed improved high-frequency conduction capabilities compared to those of the demyelinates axons; at high frequencies the transplanted axons, which had been remyelinated by Schwann cells, displayed less amplitude decrement than controls. (Modified from Honmou et al, 1996)

In the ethidium bromide-X irradiation model, we found that introduction of Schwann cells alone resulted in myelination that was limited to a region close to the injection site. As suggested by the work of Blakemore and Crang (1985), however, we also found that transplantation of a mixture of Schwann cells and astrocytes resulted in extensive migration of the transplanted Schwann cells, with remyelination extending throughout the entire 8 mm extent of the demyelinated zone within the dorsal column. This observation suggests that an astrocytic influence or factor confers, or amplifies, the migratory potential of Schwann cells. Astrocytic influences on transplanted Schwann cells appear to be complex. Within the *md* rat there is no endogenous CNS myelin but the CNS contains numerous astocytes. In the *md* model, introduction of Schwann cells and combinations of Schwann cells and astrocytes results in the formation of myelin around myelin deficient axons, but Schwann cell migration is less robust than in the astrocyte-free ethidium bromide-X irradiation lesion (Duncan et al., 1988a) suggesting that resident astrocytes may impede Schwann cell migration. Consistent with this, the astrocytic layer at the glia limitans, which delineates the interface between the spinal cord and CNS, impedes Schwann cell migration (Sims et al., 1985). Delineation of the molecular characteristics of astrocytic factors that influence Schwann cell migration may be useful in the development of transplantation procedures that will utilize the migratory potential of Schwann cells.

It is likely that, following transplantation of myelin forming glia, there are axo-glial interactions that trigger or facilitate the molecular remodeling of the newly myelinated axon membrane. The axon membrane in normal myelinated fibers contains a non-uniform distribution of Na^+ channels which are clustered at nodes of Ranvier at a density of about $1,000/\mu m^2$, with a much lower density (less than $25/\mu m^2$) in the internodal axon membrane under the myelin; the focal distribution of Na^+ channels is necessary to promote the high inward current density required for saltatory conduction of action potentials (Ritchie and Rogart, 1977; Waxman, 1977; Shrager, 1989). During normal development, clustering of nodal Na^+ channels is dependent on association of the axon with glial cells prior to myelination (Waxman and Foster, 1980; Wiley-Livingston and Ellisman, 1980). Although prior anatomical studies demonstrated the formation of myelin following transplantation of myelin-forming precursor cells into ethidium bromide-X irradiation lesions within the spinal cord, it could not be assumed, *a priori*, that axon-glial cell interactions in this pathological environment would permit the formation, along the newly myelinated fibers, of the high nodal Na^+ channel densities that are required for secure conduction. Moreover, morphological studies have demonstrated that internode distances, myelin sheath thickness, and axon diameter following endogenous remyelination are altered compared to normal in some experimental models (Harrison and McDonald, 1977). Since these changes can produce impedance mismatch which results in conduction failure (Waxman and Brill, 1978), it was not possible to predict, *a priori*, whether myelination by transplanted cells would enhance conduction in previously demyelinated axons. In the studies described above we observed that conduction extended for much longer distances in transplanted zones within the demyelinated dorsal columns as compared to non-transplanted demyelinated regions. This observation suggests that clustering of Na^+ channels at nodes of Ranvier had occurred in axons remyelinated by exogenous cells, permitting conduction block to be overcome. Our results also show that impedance mismatch did not block conduction along axons that had been remyelinated by transplanted cells.

Conduction velocities in previously demyelinated dorsal column axons, following myelination by transplanted glial cells, were restored to close-to-normal values. This result is consistent with the hyperbolic relationship, with a broad maximum, that has been demonstrated between nodal spacing and conduction velocity (Huxley and Stämpfli,

1949), and with the observation that reductions of 3-fold or less in internode distance should result in conduction of velocities close to normal (Brill et al., 1977). Interestingly, our experiments using the ethidium bromide-x irradiation model indicate that Schwann cell remyelination is associated with a relative refractory period shorter than in controls, and with enhanced frequency-following capability. This may reflect differences in nodal organization or in the geometry of the extracellular space surrounding nodes along Schwann cell-remyelinated axons, as compared to oligodendrocyte-myelinated axons (Peters, 1966; Berthold and Rydmark, 1995), or could be due to altered ionic regulation at nodes of Ranvier along Schwann cell-remyelinated axons in the CNS following cell transplantation. It is well-established that extracellular K^+ concentration, which can be modulated by glial cells, affects axonal conduction within the CNS (Kocsis et al., 1983).

On the basis of these findings, we conclude that repair of demyelinated white matter, via transplantation of exogenous myelin-forming cells, can elicit the formation of functionally appropriate myelinated internodes and associated specializations such as nodes of Ranvier, restoring secure impulse conduction, and a relatively normal conduction velocity, within the host CNS. A number of important issues, including the long-term effects of glial cell transplantation, and the selection and/or design of the optimal cell types for transplantation, remain to be explored. The functional consequences of shorter-than-normal refractory periods and higher-than-normal frequency-following capabilities, in axons myelinated by transplanted Schwann cells, require further study. Nevertheless, our results encourage us to predict that it may be possible to improve clinical status, in human disorders characterized by dysmyelination or demyelination, via the transplantation of myelin-forming cells.

ACKNOWLEDGMENTS

We thank the colleagues within our research group who participated in this work. This research was supported in part by grants from the Medical Research Service, Veterans Association, the NINDS, the National Multiple Sclerosis Society, and The Myelin Project, and the Nancy Davis foundation for Multiple Sclerosis.

REFERENCES

Berthold, C and Rydmark, M (1995): Morphology of normal peripheral axons. *The Axon; structure, function and pathophysiology*. New York, Oxford University Press S.G. Waxman, J.D. Kocsis and P.K. Stys, pp. 13–50.

Blakemore, WF and Crang, AJ (1985): The use of cultured autologous Schwann cells to remyelinate areas of demyelination in the central nervous system. J. Neurol. Sci. 70:207–223.

Blakemore, WF and Franklin, RJ (1991): Transplantation of glial cells into the CNS. Trends in Neurosci(14):323–327.

Blakemore, WF and Patterson, RC (1978): Suppression of remyelination in the CNS by X-irradiation. Acta. Neuropath. 42:105–113.

Blight, AR (1985): Delayed demyelination and macrophage invasion: A candidate for secondary cell damage in spinal cord injury. CNS Trauma 2:299–314.

Blight, AR and Young, W (1989): Central axons in injured cat spinal cord recover electrophysiological function following remyelination by Schwann cells. J. Neurol. Sci. 91:15–34.

Boison, D and Stoffel, W (1989): Myelin-deficient rat: a point mutation in exon III (A—C, Thr75—Pro) of the myelin proteolipid protein causes dysmyelination and oligodendrocyte death. The EMBO Journal 8:3295–3302.

Brill, MH, Waxman, SG, Moore, JW and Joyner, RW (1977): Conduction velocity and spike configuration in myelinated fibers: computed dependence on internode distance. J. Neurol. Neurosurg. Psychiat. 40:769–774.

Brockes, JP, Fields, KL and Raff, MC (1979): Studies on cultured rat Schwann cells. I. Establishment of purified populations from cultures of peripheral nerve. Brain Res. 165:105–118.

Byrne, TN and Waxman, SG (1990). *Spinal Cord Compression*. Philadelphia, F. A. Davis Co.

Duncan, ID, Hammang, JP and Gilmore, SA (1988b): Schwann cell myelination of the myelin deficient rat spinal cord following X-irradiation. Glia 1:233–239.

Duncan, ID, Hammang, JP, Jackson, KF, Wood, PM, Bunge, RP and Langford, L (1988a): Transplantation of oligodendrocytes and Schwann cells into the spinal cord of the myelin-deficient rat. J. of Neurocytology 17:351–360.

Felts, PA and Smith, KJ (1992): Conduction properties of central nerve fibers remyelinated by Schwann cells. Brain Res. 574:178–192.

Gledhill, RF, Harrison, BM and McDonald, WI (1973): Demyelination and remyelination after acute spinal cord compression. Exper. Neurol. 38:472–487.

Gout, O and Dubois-Dalcq, M (1993): Directed migration of transplanted glial cells toward a spinal cord demyelinating lesion. Intl. J. Dev. Neurosci. 11:613–623.

Groves, AK, Barnett, SC, Franklin, RJM, Crang, AJ, Mayer, M, Blakemore, WF and Noble, M (1993): Repair of demyelinated lesions by transplantation of purified O–2A progenitor cells. Nature 362:453–456.

Harrison, BM and McDonald, WI (1977): Remyelination after transient experimental compression of the spinal cord. Ann. Neurol. 1:542–551.

Hines, M and Shrager, P (1991): A computational test of the requirements for conduction in demyelinated axons. Restor. Neurol. and Neurosci. 3:81–93.

Honmou, O, Felts, PA, Waxman, SG and Kocsis, JD (1996): Restoration of normal conduction properties in demyelinated spinal cord axons in the adult rat by transplantation of exogenous Schwann cells. J. Neuroscience, in press.

Hudson, LD, Puckett, C, Berndt, J, Chan, J and Gencic, S (1989): Mutation of the proteolipid protein gene PLP in a human X chromosome-linked myelin disorder. Proc. Natl. Acad. Sci. 86:8128–8131.

Huxley, AF and Stämpfli, R (1949): Evidence for saltatory conduction in peripheral myelinated nerve fibres. J. Physiol. (Lond.) 108:315–339.

Kocsis, JD, Malenka, RC and Waxman, SG (1983): Effects of extracellular potassium concentration on the excitability of the parallel fibres of the rat cerebellum. J. Physiol. (Lond.) 334:225–244.

Kocsis, JD and Waxman, SG (1980): Absence of potassium conductance in central myelinated axons. Nature 287:348–349.

Lachapelle, F, Duhamel-Clerin, E, Gansmuller, A, Baron-Van Evercooren, A, Villarroya, H and Gumpel, M (1994): Transplanted transgenically marked oligodendrocytes survive, migrate and myelinate in the normal mouse brain as they do in the shiverer mouse brain. Eur. J. Neurosci. 6:814–824.

Mann, R, Mulligan, RC and Baltimore, D (1983): Construction of a retrovirus packaging mutant and its use to produce helper-free defective retrovirus. Cell 33:153–159.

McCarthy, KD and de Vellis, J (1980): Preparation of separate astroglial and oligodendroglial cell cultures from rat cerebral tissue. J. Cell Biol. 85:890–902.

Moore, JW, Joyner, RW, Brill, MH, Waxman, SG and Najar-Joa, M (1978): Simulations of conduction in uniform myelinated fibres: relative sensitivity to changes in nodal and internodal parameters. Biophys. J. 21:147–161.

Peters, A (1966): The node of Ranvier in the central nervous system. Quarterly J. Exper. Physiol. 51:229–36.

Porter, S, Blairclark, M, Glaser, L and Bunge, RP (1986): Schwann cells stimulated to proliferate in the absence of neurons retain full functional capability. J. Neurol. Sci. 6:3070–3078.

Price, J, Turner, D and Cepko, C (1987): Lineage analysis in the vertebrate nervous system by retrovirus-mediated gene transfer. Proc. Natl. Acad. Sci., USA 84:156–160.

Ritchie, JM and Rogart, RB (1977): The density of sodium channels in mammalian myelinated nerve fibers and the nature of the axonal membrane under the myelin sheath. Proc. Natl. Acad. Sci. USA 74:211–215.

Rosenbluth, J, Hasegawa, M, Shirasaki, N, Rosen, CL and Liu, Z (1990): Myelin formation following transplantation of normal fetal glia into myelin-deficient rat spinal cord. J. Neurocytol. 19:718–730.

Shrager, P (1989): Sodium channels in single demyelinated mammalian axons. Brain Res. 483:149–154.

Simons, R and Riordan, JR (1990): Single base substitution in codon 74 of the *md* rat myelin proteolipid gene. Ann. New York Acad. Sci. 605:146–154.

Sims, TJ, Gilmore, SA, Waxman, SG and Klinge, E (1985): Dorsal-ventral differences in the glia limitans of the spinal cord: An ultrastructural study in developing normal and irradiated rats. J. Neuropath. Exper. Neurol. 44:415–430.

Smith, KJ, Blakemore, WF and McDonald, WI (1981): The restoration of conduction by central remyelination. Brain 104:383–404.

Smith, KJ, Blakemore, WF and McDonald, WI (1983): Central remyelination restores secure conduction. Nature 280:395–396.

Utzschneider, D, Black, JA and Kocsis, JD (1992): Conduction properties of spinal cord axons in the myelin-deficient rat mutant. Neurosci. 49:221–228.

Utzschneider, DA, Archer, DR, Duncan, IR, Waxman, SG and Kocsis, JD (1994): Transplantation of myelin-forming cells enhances impulse conduction in amyelinated spinal cord axons in the myelin-deficient rat. Proc. Natl. Acad. Sci. 91:53–57.

Vignais, L, Nait Oumesmar, B, Mellouk, F, Gout, O, Labourdette, G, Baron-Van Evercooren, A and Gumpel, M (1993): Transplantation of oligodendrocyte precursors in the adult demyelinated spinal cord: migration and remyelination. Intl. J. Dev. Neurosci. 11:603–612.

Waxman, SG (1977): Conduction in myelinated, unmyelinated, and demyelinated fibers. Arch. Neurol. 34:585–590.

Waxman, SG (1993): Peripheral nerve abnormalities in multiple sclerosis. Muscle & Nerve 16:1–5.

Waxman, SG and Bennett, MVL (1972): Relative conduction velocities of small myelinated and non-myelinated fibers in the central nervous system. Nature New Biology 238:217–219.

Waxman, SG, Black, JA, Duncan, ID and Ransom, BR (1990): Macromolecular structure of axon membrane and action potential conduction in myelin deficient and myelin deficient heterozygote rat optic nerves. J. Neurocytology 19:11–27.

Waxman, SG and Brill, MH (1978): Conduction through demyelinated plaques in multiple sclerosis: computer simulations of facilitation by short internodes. J. Neurol. Neurosurg. Psychiat. 41:408–417.

Waxman, SG and Foster, RE (1980): Development of the axon membrane during differentiation of myelinated fibres in spinal nerve roots. Proc. Roy. Soc. (Lond.) B 209:441–446.

Waxman, SG, Kocsis, JD and Black, JA (1995): Pathophysiology of demyelinated axons. In: Waxman, S. G., Kocsis, J. D. and Stys, P. K. (ed), The Axon. Oxford University PressWaxman, S. G., Kocsis, J. D. and Stys, P. K., pp. 438–461.

Wiley-Livingston, CA and Ellisman, MH (1980): Development of axonal membrane specializations defines nodes of Ranvier and precedes Schwann cell myelin elaboration. Dev. Biol. 79:334–355.

Zeller, NK, Dubois-Dalcq, M and Lazzarini, RA (1989): Myelin protein expression in the myelin-deficient rat brain and cultured oligodendrocytes. J. Molec. Neurosci. 1:139–149.

29

AXONAL REGENERATION IN THE FISH AND AMPHIBIAN CNS

Myelin-Associated Neurite Growth Inhibitors and Adaptive Plasticity of Glial Cells

D. M. Lang, R. Ankerhold, and C. A. O. Stuermer

Faculty of Biology
University of Konstanz
D-78434 Konstanz
Germany

INTRODUCTION

For regeneration of severed nerve fibers to be successful, a number of basic requirements have to be met. First, the injured neuron must be able to reinitiate the cellular machinery required for axonal regrowth. Second, the environment of the nerve cells must be conducive to neurite growth, and allow axons to regenerate back to their targets.

There are marked differences in the regenerative capacity of nerve fibers in the central nervous system (CNS) of anamniotic vertebrates (fish and amphibians) on the one hand, and mammals and birds on the other. Axonal regeneration fails in the CNS of mammals and birds, but occurs throughout the CNS of fish (Gaze, 1970; Sharma et al., 1993). In amphibians, the situation is more complex: urodeles (newts and salamanders) are capable of axonal regeneration in both the optic nerve (ON; Turner and Singer, 1974; Stensaas and Feringa, 1977) and spinal cord (SC; Piatt, 1955; Clarke et al., 1988). Anurans (frogs and toads), however, regenerate axons in the transected ON (Gaze, 1970), but regeneration fails to occur in the frog SC (Forehand and Farel, 1982; Beattie et al., 1990). While this is so in adult frogs, axonal regeneration takes place in the tadpole SC (Forehand and Farel, 1982; Beattie et al., 1990). In this review, we consider the properties of the glial cell environment, in particular of oligodendrocytes and myelin in fish and frogs, and discuss their role in success or failure of axonal regeneration.

It has been demonstrated that mammalian oligodendrocytes and CNS myelin contain proteins that inhibit neurite outgrowth (Caroni and Schwab, 1988a). These myelin-associated neurite growth inhibitors (NI) are two low abundance glycoproteins of MW 35 and 250 kDa, which cause growth cone collapse in a wide variety of neurons (e.g. Fawcett et

Cell Biology and Pathology of Myelin, edited by Juurlink *et al.*
Plenum Press, New York, 1997

al., 1989; Bandtlow et al., 1990; Bastmeyer et al., 1991; Moorman and Hume, 1993). They render mammalian and avian oligodendrocytes and myelin non-permissive substrates for axon growth, and interfere with attachment and spreading of fibroblasts in vitro (Caroni and Schwab, 1988a). A monoclonal antibody, MAb IN-1, against NI proteins of the rat (Caroni and Schwab, 1988b), partially neutralizes these inhibitors. In the presence of MAb IN-1 mammalian oligodendrocytes and CNS myelin become permissive for neurite growth (Caroni and Schwab, 1988b; Bandtlow et al., 1991; Bastmeyer et al., 1991), and regeneration of some axons in the transected rat spinal cord axons can occur (Schnell and Schwab, 1990). As the inhibitory substrate properties of oligodendrocytes and myelin appear to play a central role in the failure of axonal regeneration in the mammalian CNS, we asked whether there is a correlation between the substrate properties of oligodendrocytes and myelin and the varying capacity for axonal regeneration in the CNS of frogs and fish. In fact, there is evidence to support this view. In addition, however, fish and frog CNS glial cells appear to possess other properties not found in their mammalian counterparts, in that they undergo changes in response to injury, and thus assist neurite regrowth.

TESTS FOR THE PRESENCE OF NEURITE GROWTH INHIBITORS IN AMPHIBIANS AND FISH

Substrate Properties of Amphibian CNS Myelin and Oligodendrocytes

Earlier results, obtained in fibroblast spreading assays, indicated that non-permissive substrate properties are only detectable in CNS myelin of mammals and birds (Caroni and Schwab, 1988a). Recently, however, the substrate properties of Xenopus myelin, derived from the optic nerve/tectum (OT), where axonal regeneration occurs, and the SC, where regeneration does not take place, were compared. In a series of in vitro experiments the numbers of axons emerging from small retinal explants of standardized size were counted and evaluated in dependence of the myelin substrate (Lang et al., 1995). Axons from the goldfish retina were used because they have been shown to be sensitive to the inhibitory substrate properties of rat CNS myelin (Bastmeyer et al., 1991). When myelin from the frog OT was offered as the sole substrate to retina explants, it allowed considerable outgrowth of retinal axons (Fig. 1). In contrast, myelin derived from the frog SC did not support axon growth, much like rat CNS myelin which inhibits neurite growth (Fig. 1; see also Bastmeyer et al., 1991). These results indicate that myelin from the Xenopus ON differs in its substrate properties from myelin of the SC, and this coincides with success or failure of axonal regeneration in these CNS regions.

This correlation between success of axon regrowth and permissive substrate properties of CNS myelin extends to the SC of the axolotl, a urodele capable of axonal regeneration in this CNS region (Piatt, 1955; Clarke et al., 1988). There is outgrowth of retinal axons, and the number of axons is at least as large as on goldfish CNS- or Xenopus OT-derived myelin (Fig. 1). Concerning the ability of axons to regenerate in the SC of Xenopus tadpoles, it was found that the SC of the larval frogs was largely devoid of myelination up to metamorphic climax (Lang et al., 1995). This may serve as an explanation for the success of axonal regeneration in the anuran SC at this developmental stage.

Whether the non-permissive substrate properties of Xenopus SC myelin are indeed caused by the presence of mammalian-like NI proteins was tested by using MAb IN-1 in a series of coculture experiments. Oligodendrocytes were isolated from the Xenopus ON or SC, and the reactions of growth cones from retinal or dorsal root ganglion explants upon

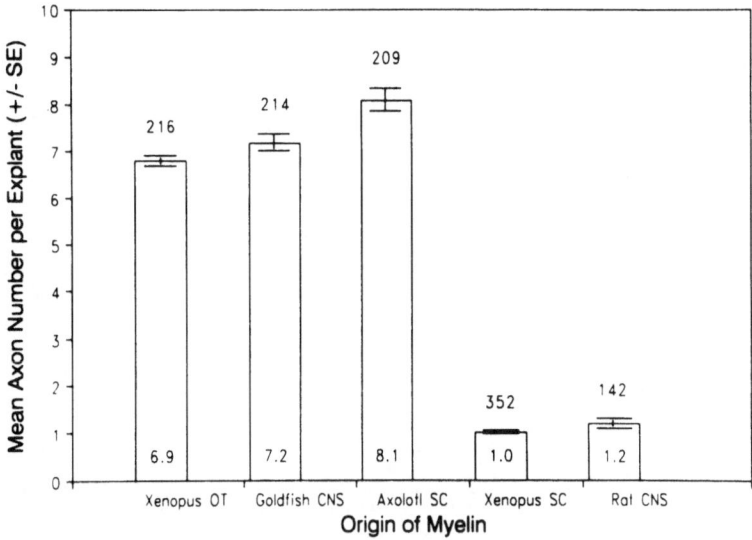

Figure 1. Comparison of the number of regenerating goldfish retinal axons extending on a substrate consisting of CNS myelin. The height of each column represents the mean axon number. The standard error (SEM) is indicated at the top of each column. Numbers of microexplants are given above each column. Figure taken from Lang et al., 1995.

encountering the glial cells was monitored. Differentiated Xenopus oligodendrocytes in coculture with retinal axons from both frog and goldfish (Lang et al., 1995) elicited growth cone collapse or avoidance reactions of the axons, when derived from the SC. This is reflected by neurite-free areas on and around the oligodendrocytes (Fig. 2). Similar neurite-free areas form around mammalian oligodendrocytes, known to express NI proteins (Schwab and Caroni, 1988; Fawcett et al., 1989). In contrast, when the oligodendrocytes were derived from the frog ON, they were mostly overgrown by axons (Fig. 2). This demonstrates that oligodendrocytes from the frog ON and SC possess differential substrate properties, in keeping with the differences in CNS myelin and the regenerative capacity of these CNS regions.

That the non-permissive substrate properties of Xenopus SC oligodendrocytes are largely due to the presence of mammalian-like NI proteins was demonstrated using MAb IN-1 (Lang et al., 1995). In the presence of MAb IN-1, SC-derived oligodendrocytes were overgrown by retinal axons significantly more often than those in control cocultures. With oligodendrocytes from the ON, no improvement of substrate properties was noted in the presence of IN-1, suggesting that NI proteins are absent from the ON. Still, even in the presence of MAb IN-1 neither ON- nor SC-derived differentiated Xenopus oligodendrocytes represented particularly good substrates for axon growth. Therefore, these cells probably express other molecules potentially inhibitory for growing axons, such as myelin-associated glycoprotein (MAG; Mukhopadhyay et al., 1994; McKerracher et al., 1994) or janusin (Taylor et al., 1993).

That NI proteins are present in the Xenopus SC, but not its ON is also supported by immunostaining experiments with MAb IN-1 (Lang et al., 1995). IN-1-immunoreactivity was present in the SC and hindbrain, but not the ON and optic tectum.

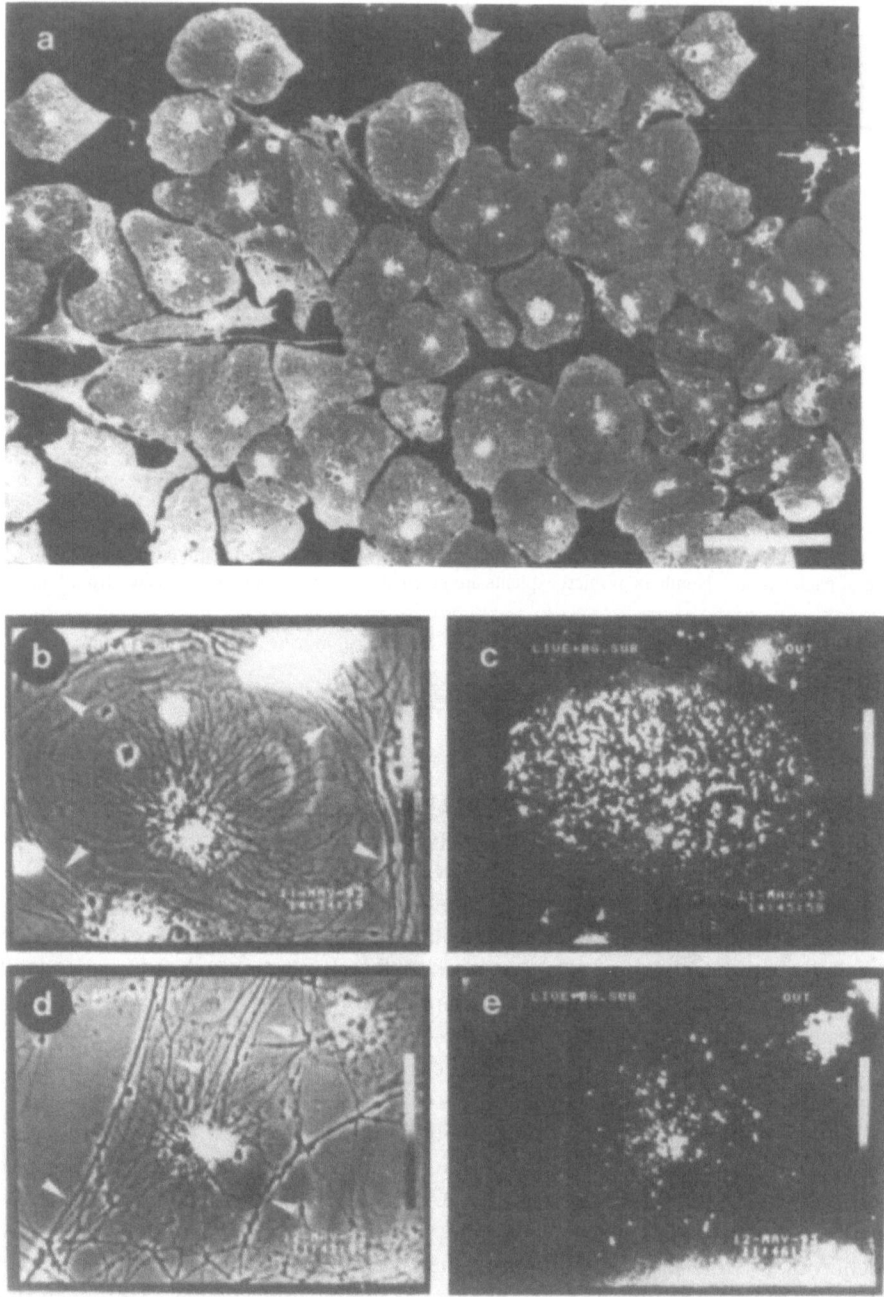

Figure 2. Growth cone/oligodendrocyte encounters. a) Differentiated oligodendrocytes derived from the Xenopus SC after exposure to forskolin. Oligodendrocytes exhibit a highly branched morphology and are immunostained with GalC-antibodies. b,c) The majority of growth cones avoid crossing O1-positive (c) SC-derived oligodendro-cytes (b), and instead grow around the perimeter of the glial cell (white arrowheads). d,e) The majority of growth cones can grow over O1-positive oligodendrocytes (e) derived from the Xenopus ON. Axons (d) are marked by white arrowheads. Scale bar in (a): 100 μm. Bars to the right of each frame represent 100 μm. Figure taken from Lang et al., 1995.

These findings are indicative of the presence of mammalian-like NI proteins in the SC of the frog Xenopus. Thus, these molecules seem to be highly conserved in vertebrate evolution.

Are There Neurite Growth Inhibitors in the Fish CNS?

Fishes can regenerate severed nerve fibers throughout their CNS. They regain normal vision after transection of the ON. Earlier studies (Vanselow et al., 1990; Bastmeyer et al., 1991; Lang et al., 1995; see also Fig. 1) reported that this regenerative success in the fish ON might be due to the growth-permissive substrate properties of myelin and the absence of NI proteins.

A recent study (Sivron and Schwartz, 1994) indicated that the substrate properties of goldfish ON myelin for retinal axons are much improved after incubation with MAb IN-1, thus arguing that NI proteins are present in this system. Comparable experiments performed in our lab, however, yielded no evidence for any neutralizing activity of MAb IN-1 when applied to goldfish ON myelin (Wanner et al., 1995). Surprisingly, though, MAb O4 with specificity for sulfatide (Sommer and Schachner, 1981), an antibody commonly used as a control in such in vitro assays, strongly reduced axon outgrowth on goldfish ON myelin. Thus, the selection of appropriate controls in these experiments is critical.

Moreover, growth cone collapse assays of goldfish retinal as well as rat dorsal root ganglion axons (Wanner et al., 1995) were performed. Myelin proteins were extracted from mammalian and goldfish ON myelin, reconstituted into liposomes, and applied to advancing growth cones. The results showed a clear qualitative difference between the mammalian- and the fish-derived myelin proteins: the mammalian myelin proteins elicited growth cone collapse and axon retraction in the majority of cases. Fish myelin proteins did not cause such dramatic responses. They caused no or sometimes at most a transient growth cone collapse, from which the growth cone recovered, and no axon retraction (Fig. 3). IN-1-treatment did not alter the effect of fish myelin proteins on growth cones. However, after treatment of the mammalian CNS-derived myelin proteins with MAb IN-1, the frequency of growth cone collapse was significantly reduced.

These results are consistent with the notion that fish oligodendrocytes and CNS myelin do not possess NI proteins - at least not to an extent that would cause growth cone arrest or collapse.

ADAPTIVE PLASTICITY OF FISH AND AMPHIBIAN GLIAL CELLS

The findings discussed above have one major shortcoming: they describe the properties of oligodendrocytes and myelin of the intact CNS. Substrate properties of glial cells may, however, change in response to injury of the nervous system. This is known to happen in peripheral nerves. Schwann cells dedifferentiate following a lesion, proliferate and express cell adhesion molecules, like L1 (Bixby et al., 1988), as well as neurotrophic factors, e.g. nerve growth factor (NGF; Heumann et al., 1987). Thus, Schwann cells provide a microenvironment conducive to axonal regrowth.

Similar adaptive processes are likely to occur in the CNS of fish and amphibians. This was first indicated by morphological studies of the amphibian ON (Reier and Webster, 1974; Bohn and Reier, 1985). Following nerve transection, glial cells changed their morphology and extended processes, which appeared to guide the regrowing axons.

304 D. M. Lang *et al.*

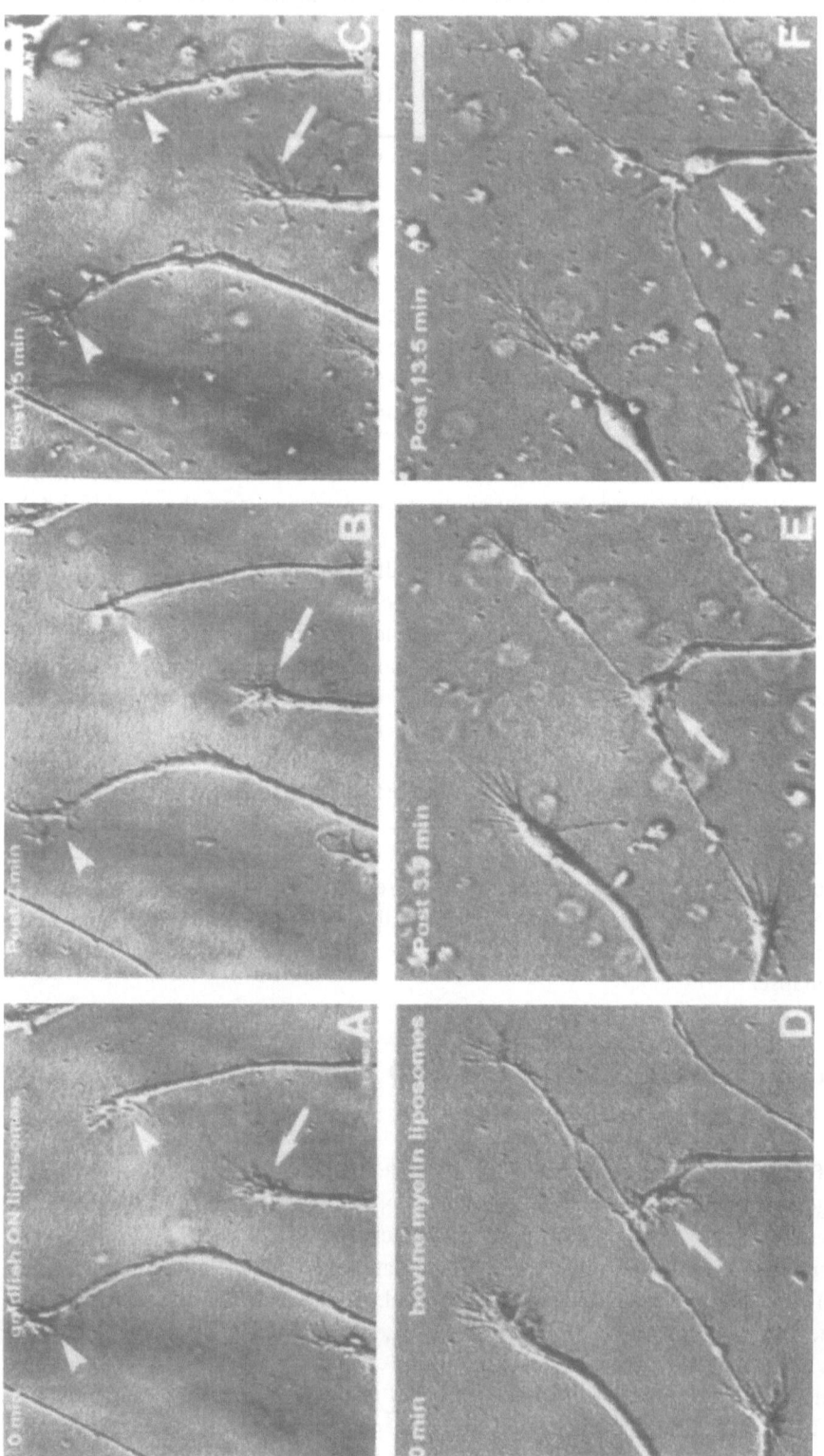

Figure 3. Selected images of sequences showing fish retinal growth cones A) prior to the addition of liposomes. B) at the time that liposomes containing proteins of goldfish optic nerve myelin contact the growth cones, C) 15 min after liposome addition. One growth cone (arrow in A, B, C) retracts its lamellipodia and filopodia when contacted by liposomes (B) but resumes its normal shape and motility within 15 min (C). Arrowheads (A, B, C) show growth cones which are unaffected by the addition of the liposomes. D-F) Growth cones collapse and retract their axons when contacted by liposomes with bovine myelin proteins (arrow). Scale bar: 25 μm. Figure taken from Wanner et al., 1995.

Recently, glial cells have been isolated from explants of Xenopus ON (Lang and Stuermer, 1996, in press). The majority of cells emigrating from such explants display a bipolar, elongated phenotype, are highly motile and are labeled by MAb O4, a marker of immature oligodendrocytes in mammals, as well as by antibodies to glial fibrillary acidic protein (GFAP). Under standard culture conditions, these frog glial cells proliferate and fail to differentiate - as opposed to mammalian oligodendrocyte precursors, which differentiate spontaneously (Raff et al., 1983). However, stimulation of these frog oligodendrocyte precursor-like cells with forskolin (an agent that elevates intracellular cAMP levels) leads to differentiation, indicated by morphlgical changes as well as expression of GalC (see Fig. 2) and myelin marker proteins (Lang and Stuermer, 1996, in press). This is reminiscent of the differentiation behaviour of Schwann cells (Morgan et al., 1991). Moreover, like Schwann cells, these oligodendrocyte precursors were found to promote outgrowth of retinal axons in vitro (Fig. 4), a property which is lost upon differentiation (Lang and Stuermer, 1996, in press).

There are indications that Xenopus oligodendrocytes apparently also undergo plastic changes after injury in vivo (Lang and Stuermer, 1996, in press). Upon ON transection, immunoreactivity for myelin marker proteins is rapidly lost from the nerve; however, GFAP- and O4-immunoreactivity appears in cells with elongated processes, while axons

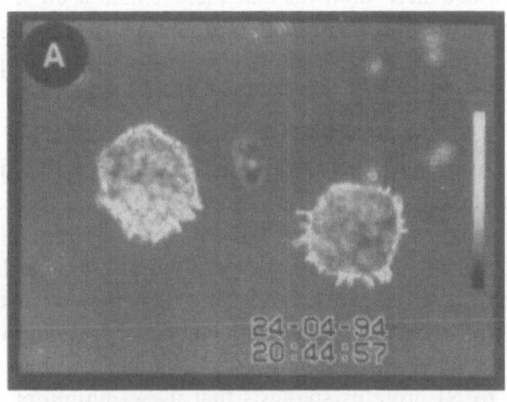

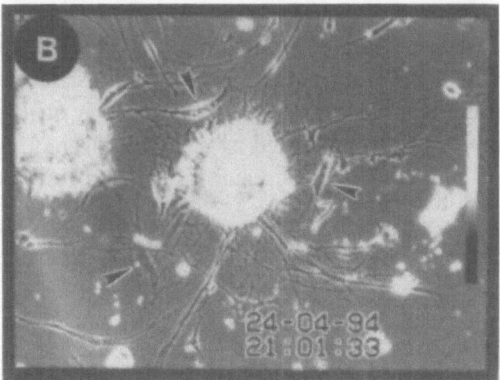

Figure 4. Miniexplants of Xenopus retina after 2 d in vitro. a) Explants on polylysine. b) Explants on Xenopus glial cells. No axonal outgrowth occurs on polylysine as the sole substrate. In the presence of glial cells, however, retinal axons grow profusely. Bars to the right: 100 μm. Figure taken from Lang and Stuermer, 1996, in press.

regenerate through the ON. After regrowth of retinal axons, remyelination and reexpression of myelin marker proteins is observed in those regions of the ON where axons are in contact with glial cell processes.

Immature oligodendrocytes derived from the goldfish ON appear remarkably similar to those of Xenopus. While they cannot be induced to differentiate by forskolin, they express myelin markers when cocultured over prolonged periods with retinal axons (Bastmeyer et al., 1994). Moreover, these cells express the L1-like cell adhesion molecule ε587-antigen (Bastmeyer et al., 1994) and have recently been shown to produce a neurite outgrowth promoting factor, AF-1 (Schwalb et al., 1995; Schwalb et al., 1996). They thus provide a substrate highly conducive to neurite growth. In culture, immature oligodendrocytes from the goldfish ON even promote growth of axons from the adult rat retina (Bastmeyer et al., 1993). These properties of oligodendrocytes derived from the goldfish and Xenopus ON are likely to contribute to the success of axonal regeneration in these systems.

CONCLUDING REMARKS

The findings discussed here indicate that there exists a close correlation between success of axonal regeneration and the substrate properties of oligodendrocytes and myelin. In the presence of oligodendrocytes and CNS myelin with non-permissive substrate properties, axon regrowth fails, and this is, at least in part, due to the presence of NI proteins.

The evolutive origin of these neurite growth inhibitory proteins remains yet to be elucidated. The outcome of the experiments described above is indicative of the absence of mammalian-like NI proteins from fish CNS myelin and from oligodendrocytes and myelin of the frog ON. In order to be able to perform a more thorough analysis of the distribution and phylogenetic origin of these inhibitors, their molecular characterization will be a prerequisite. Moreover, neither the receptor nor the signaling pathways activated by NI proteins are known to date. However, the wide variety of neuronal as well as non-neuronal cells sensitive to NI proteins suggests that these inhibitors act through a highly conserved mechanism.

In addition, the presence of growth promoting molecules in the visual pathway of fish and frogs may actively support axonal regeneration. Thus, successful regeneration of axons in the CNS of amphibians and fish not only depends on the absence of neurite growth inhibitors, but appears to be enhanced by a growth promoting microenvironment. The adaptive plasticity of glial cells seems crucial for the post-lesion changes accompanying axonal regeneration.

ACKNOWLEDGMENTS

The research work referred to in this paper was supported by grants from the Deutsche Forschungsgemeinschaft and from the Gemeinnützige Hertie-Stiftung to C.A.O. Stuermer. D.M. Lang and R. Ankerhold were supported by grants from the Boehringer Ingelheim Fonds.

REFERENCES

Bandtlow C, Zachleder T, Schwab ME (1990): Oligodendrocytes arrest neurite growth by contact inhibition. J. Neurosci. 10:3837–3848.

Bastmeyer M, Schlosshauer B, Stuermer CAO (1990): The spatiotemporal distribution of N-CAM in the retinotectal pathway of adult goldfish detected by the monoclonal antibody D3. Development 108:299–311.

Bastmeyer M, Jeserich G, Stuermer CAO (1994): Similarities and differences between fish oligodendrocytes and Schwann cells in vitro. Glia 11:300–314.

Bastmeyer M, Bähr M, Stuermer CAO (1993): Fish optic nerve oligodendrocytes support axonal regeneration of fish and mammalian retinal ganglion cells. Glia 8:1–12.

Bastmeyer M, Beckmann M, Schwab ME, Stuermer CAO (1991): Growth of regenerating goldfish axons is inhibited by rat oligodendrocytes and CNS myelin but not by goldfish ON tract oligodendrocytelike cells and fish CNS myelin. J. Neurosci. 11:626–640.

Beattie MS, Bresnahan JC, Lopate G (1990): Metamorphosis alters the response to spinal cord transection in Xenopus laevis frogs. J. Neurobiol 21:1108–1122.

Bixby MA, Pollock B (1983): Increased regeneration rate in peripheral axons following double lesions: enhancement of the conditioning lesion effect. J. Neurobiol.14:467–472.

Bohn RC, Reier PJ (1985): Retrograde degeneration of myelinated axons and reorganization in the optic nerves of adult frogs (Xenopus laevis) following nerve injury or tectal ablation. J. Neurocytol. 14:221–244.

Caroni P, Schwab ME (1988a): Two membrane protein fractions from rat central myelin with inhibitory properties for neurite growth and fibroblast spreading. J. Cell Biol. 106:1281–1288.

Caroni P, Schwab ME (1988b): Antibody against myelin-associated inhibitor of neurite growth neutralizes nonpermissive substrate properties of CNS white matter. Neuron 1:85–96.

Clarke JDW, Alexander R, Holder N (1988): Regeneration of descending'axons in the spinal cord of the axolotl. Neurosci. Lett. 89:1–6.

Fawcett JW, Rokos J, Bakst I (1989): Oligodendrocytes repel axons and cause axonal growth cone collapse. J. Cell Sci. 92:93–100.

Forehand CJ, Farel PB (1982): Anatomical and behavioral recovery from the effects of spinal cord transection: dependence on metamorphosis in anuran larvae. J. Neurosci. 2:654–662.

Gaze RM (1970): The formation of nerve connections. London:Academic Press.

Heumann R, Lindholm D, Bandtlow C, Meyer M, Radeke MJ, Misko TP, Shooter E, Thoenen H (1987): Differential regulation of mRNA encoding nerve growth factor and its receptor in rat sciatic nerve during development, degeneration, and regeneration: role of macrophages. Proc. Natl. Acad. Sci. USA 84:8735–8739.

Lang DM, Stuermer CAO (1996): Properties of Xenopus glial cells in vitro and after CNS fiber tract lesions in vivo. Glia, in press.

Lang DM, Rubin BP, Schwab ME, Stuermer CAO (1995): CNS Myelin and oligodendrocytes of the Xenopus spinal cord - but not optic nerve - are nonpermissive for axon growth. J.Neurosci. 15:99–109.

McKerracher L, David S, Jackson DL, Kottis V, Dunn RJ, Braun PE (1994): Identification of myelin-associated glycoprotein as a major myelin-derived inhibitor of neurite growth. Neuron 13:805–811.

Moorman SJ, Hume RI (1993): α-Conotoxin prevents myelin-evoked growth cone collapse in neonatal rat locus coeruleus neurons in vitro. J. Neurosci. 13:4727–4736.

Morgan L, Jessen KR, Mirsky R (1991): The effects of cAMP on differentiation of cultured Schwann cells: progression from an early phenotype (O4+) to a myelin phenotype (P0+, GFAP-, N-CAM-, NGF-receptor) depends on growth inhibition. J. Cell Biol. 112:457–467.

Mukhopadhyay G, Doherty P, Walsh FS, Crocker PR, Filbin MT (1994): A novel role for myelin-associated glycoprotein as an inhibitor of axonal regeneration. Neuron 13:757–767.

Piatt J. (1955): Regeneration of the spinal cord in the salamander. J. Exp. Zool. 129:177–207.

Raff MC, Miller RH, Noble M (1983): A glial progenitor cell that develops in vitro into an astrocyte or an oligodendrocyte depending on culture medium. Nature 303:390–396.

Reier PJ, Webster H deF (1974): Regeneration and remyelination of Xenopus tadpole optic nerve fibres following transection or crush. J. Neurocytol. 3:591–618.

Schnell L, Schwab ME (1990): Axonal regeneration in the rat spinal cord produced by an antibody against myelin-associated neurite growth inhibitors. Nature 343:269–272.

Schwab ME, Caroni P. (1988): Oligodendrocytes and CNS myelin are non-permissive substrates for neurite growth and fibroblast spreading in vitro. J. Neurosci. 8:2381–2393.

Schwalb JM, Gu M, Stuermer C, Bastmeyer M, Hu G, Boulis N, Irwin N, Benowitz LI (1996): Goldfish optic nerve glia secrete AF-1, a peptide that stimulates retinal ganglion cells to regenerate their axons. Neurosci., in press.

Schwalb JM, Boulis NM, Gu M, Winickoff J, Jackson PS, Irwin N, Benowitz LI (1995): Two factors secreted by the goldfish optic nerve induce retinal ganglion cells to regenerate axons in culture. J.Neurosci. 15:5514–5525.

Sharma SC, Jadhao AG, Prasada Rao PD (1993): Regeneration of supraspinal projection neurons in the adult goldfish. Brain Res. 620:221–228.

Sivron T, Schwartz M (1994): Nonpermissive nature of fish optic nerves to axonal growth is due to the presence of myelin-associated growth inhibitors. Exp. Neurol. 130:411–413.

Sommer I, Schachner M (1981): Monoclonal antibodies (O1 to O4) to oligodendrocyte cell surfaces: an immunocytological study in the central nervous system. Dev. Biol. 83:311–327.

Stensaas LJ, Feringa ER (1977): Axon regeneration across the site of injury in the optic nerve of the newt Triturus pyrrhogaster. Cell Tissue Res. 179:501–516.

Taylor J, Pesheva P, Schachner M (1993): Influence of janusin and tenascin on growth cone behavior in vitro. J. Neurosci. Res. 35:347–362.

Turner JE, Singer ME (1974): The ultrastructure of regeneration in the severed newt optic nerve. J.Exp. Zool. 190:249–268.

Vanselow J, Schwab ME, Thanos S (1990): Responses of regenerating rat retinal ganglion cell axons to contacts with central nervous myelin in vitro. Eur. J. Neurosci. 2:121–125.

Wanner M, Lang DM, Bandtlow CE, Schwab ME, Bastmeyer M, Stuermer CAO (1995): Reevaluation of the growth permissive substrate properties of goldfish optic nerve myelin. J. Neurosci. 15:7500–7508.

30

ROLE OF MAG AS AN AXON GROWTH INHIBITORY PROTEIN FOR REGENERATION OF INJURED NEURONS IN THE CNS

Lisa McKerracher[1] and Sam David[2]

[1]Département de pathologie
Université de Montréal
CP 6128, Centre-ville
Montréal, Québec H3C 3J7
[2]Centre for Research in Neuroscience
Montreal General Hospital Research Institute
1650 Cedar Ave., Montréal, Québec H3G 1A4

INTRODUCTION

Following trauma in the adult central nervous system (CNS) of mammals, injured neurons do not regenerate their transected axons. An important barrier to regeneration is an axon growth inhibitory activity that is present in CNS myelin (Schwab et al, 1993). The growth inhibitory properties of CNS myelin have been demonstrated by a wide variety of techniques in a number of different laboratories, yet until recently the identity of specific proteins important for the inhibitory activity remained elusive. It is now clear that multiple proteins in myelin have growth inhibitory activity (Caroni and Schwab, 1988a; 1988b; McKerracher et al, 1994) and the search for the identity of the myelin-derived growth inhibitory proteins continues to be an important domain of research in this field. One growth inhibitory protein that has been identified recently is a known component of myelin, the myelin-associated glycoprotein (MAG) (McKerracher et al, 1994; Mukhopadhay et al, 1994). The discovery of MAG as an inhibitor of neurite outgrowth was surprising because, previously, the homology of MAG with other members of the immunoglobulin family of adhesion molecules suggested an adhesive function (Arquint et al, 1987; Salzer et al, 1987). Indeed, for certain embryonic neurons it can promote adhesion (Sadoul et al, 1990; Johnson et al, 1989; Mukhopadhay et al 1994) although for many neurons, regardless of age, it acts as a neurite growth inhibitor (De Bellard et al, 1996). As an inhibitor of neurite outgrowth MAG is an important component of myelin-derived inhibitory activity, and acts on a wide variety of neuronal types (Mukhopadhay et al, 1994; Li et al. 1996; Schafer et al, 1996; DeBellard et al, 1996). We review here the evidence that growth inhibitory molecules are an important barrier to regeneration in the CNS, and describe the charac-

Cell Biology and Pathology of Myelin, edited by Juurlink *et al.*
Plenum Press, New York, 1997

terization of MAG as a growth inhibitory molecule. The ability of MAG to influence axon regeneration *in vivo* is also reviewed.

REGENERATION OF AXONS IN THE CNS AND THE RELATIONSHIP TO MYELINATION

Neurons that do not regenerate in the environment of the CNS will regrow their damaged axons when provided with an appropriate environment (David and Aguayo, 1981; Kromer et al., 1981). One experimental strategy to induce regeneration in the CNS is to provide a peripheral nerve (PN) graft as the favorable environment (reviewed by Aguayo et al, 1991). In such grafts, as in injured peripheral nerves, the myelin debris is cleared away, and the Schwann cells proliferate and provide a supportive environment for axon regeneration. Studies of peripheral nerves grafted to transected optic nerves have established that these PN grafts secrete factors that have both trophic and tropic influences (Aguayo et al 1991; Cho and So, 1992). However less than 10% of the transected axons successfully enter the PN grafts. Also, growth-associated changes in the tubulin cytoskeleton are observed only for those retinal gangliou cells (RGC) with an axon extension in the graft (Fournier and McKerracher 1995). Therefore, diffusible molecules from the graft are not responsible all changes in tubulin expression, and these findings suggest that the transected axon stubs must have access to an appropriate substrate for successful regeneration. Moreover, an impediment to regeneration may be growth inhibitory molecules at the optic nerve/PN graft interface, especially because early after injury many axons retract back into the optic nerve (Richardson et al, 1982). Therefore, even in the presence of a growth-permissive environment, the milieu of the cut axon, if unfavorable for regrowth, may have a significant impact on the success of axonal growth.

It has now been well documented that removing myelin *in vivo* improves the success of regenerative growth over the native terrain of the CNS, at least when the lesion is made in young animals. Regeneration occurs after irradiation of newborn rats, a procedure that kills oligodendrocytes and prevents the appearance of myelin proteins (Savio and Schwab, 1990). After such procedure in rats and following a corticospinal lesion made at 12 to 17 days of age, some corticospinal axons regrow around the lesion and extend long distances. Also, in a chick model of spinal cord repair, the onset of myelination correlates with a loss of regenerative ability of cut axons (Keirstead et al, 1992). The removal of myelin with anti-galactocerebroside and complement extends the permissive period for axonal regeneration. These experiments demonstrate a good correlation between myelination and the failure of axons to regenerate in the CNS. Moreover, it is well known that adult neurons cannot extend over CNS myelin *in vitro* (Schwab et al, 1993; Vanselow et al, 1990). However, in some transplantation studies where embryonic neurons were transplanted into the adult CNS, long distance axon growth from the transplants occurred along recipient, white matter regions (Davies et al, 1994; Wictorian et al, 1992). It has been suggested that in such cases, the embryonic neurons used for transplantation are less susceptible to myelin-associated growth inhibitory proteins, than neurons from fully adult animals (Bjorkland, 1994). This interpretation is in accord with the observations that some neurons are less susceptible to inhibition by MAG early in development (Mukhopadhay et al, 1994), and such may also prove to be the case with the other myelin-derived growth inhibitory molecules.

The existence of growth inhibitory molecules was suspected long before any inhibitors were identified in myelin. When PN and optic nerve explants were compared in tissue culture for their ability to support neurite extension it was observed that even in the presence of neurotrophic factors, axons did not grow into the optic nerve (Schwab and Thoenen, 1985). Later, two antigenically related proteins, recognized by a monoclonal antibody designated IN-1, were found to block the inhibitory activity of differentiated oligodendrocytes and CNS myelin *in vitro* (Caroni and Schwab, 1988a; 1988b). The application of this antibody *in vivo* allows some corticospinal axons to elongate long distances after CNS injury (Schnell and Schwab, 1990). Antibody treatment with IN-1 also allows raphespinal serotonergic neurons to regenerate, resulting in improvement in some aspects of locomotor function (Bregman et al, 1995). Therefore, the evidence to date suggests that removing or blocking the myelin-derived inhibitors of neurite outgrowth will be an important component of any strategy to improve regeneration in the adult CNS.

Recently, it has become clear that many growth inhibitory molecules are abundant in the adult CNS. Inhibitory molecules may be part of the growth substrate, or diffusible chemorepulsive molecules (Pini, 1993). Some diffusible neurite growth inhibitors include thrombin, a protease expressed in the CNS (Suidan et al, 1992), lysophosphatydic acid, a bioactive phospholipid (Jalinck et al, 1994), nitrous oxide (Hesse et al, 1993), and collapsin (Luo et al, 1993). The molecular identification of collapsin, a protein expressed both early and late in chick brain development, demonstrated that both membrane bound and diffusible forms of a similar growth inhibitory protein may exist (Luo and Raper, 1994; Kolodkin et al, 1993). In addition, components of the extracellular matrix can act as growth inhibitory molecules, such as tenascin (Lochter et al, 1991), chondroitin sulfate proteoglycan (Friedlander et al, 1994), and S-laminin (Hunter et al, 1991). Of these diverse inhibitors it is known that components of the tenascin family (Lochler et al, 1991) and members of the proteoglycan family (McKeon et al, 1991; Bansal et al, 1995) are expressed by oligodendrocytes. However, the extent to which these proteins may also be present in myelin as additional myelin-derived inhibitors is not known. Although myelin is an important barrier to axonal regeneration in the spinal cord, the inhibitory proteins that give rise to this activity have been elusive. To better understand the growth inhibition by CNS myelin it will be critical to identify all of the proteins that give rise to myelin-derived inhibitory activity.

MAG IS A NEURITE GROWTH INHIBITORY PROTEIN

To characterize the growth inhibitory protein present in purified myelin, we set out to purify the growth inhibitory molecules present in CNS myelin, and we recently reported the identification of MAG as one of several growth inhibitory proteins present in myelin (McKerracher et al, 1994). For these experiments non-denaturing extraction with 1% octylglycoside was used to solubilize myelin proteins, followed by chromatography on a DEAE anion exchange column to separate inhibitory activity. NG108–15 cells were used in a 24 hour bioassay to test substrates for neurite growth inhibition. Two major peaks of inhibitory activity were eluted with a 0.2 to 2 M salt gradient (Fig. 1). Through immunodepletion with anti-MAG antibody we determined that growth inhibition correlated with MAG in the first peak (McKerracher et al, 1994). More recently, we have demonstrated that the inhibitory activity in the peak that contains the MAG is significantly reduced in MAG null mutant mice (Li et al, 1996). The second peak, a peak of non-MAG inhibitory activity observed in the original experiments, remained detectable in the MAG knock-out mice, demonstrating that the second inhibitory peak in our preparations is not MAG. Moreover, some non-MAG inhibitory activity is also present in the MAG-enriched peak

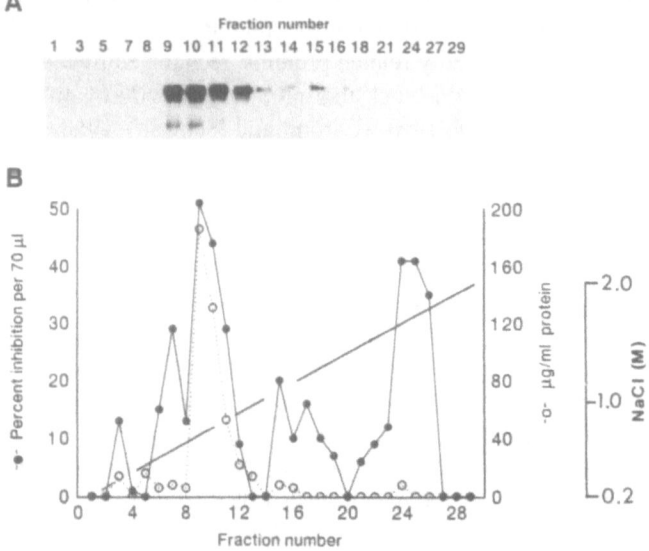

Figure 1. Analysis of neurite growth inhibition after separation of octlyglucoside extracts of myelin by DEAE anion exchange chromatography. A). Western blots of column fractions with anti-MAG antibody demonstrate that MAG is enriched in the first peak of inhibitory activity. B) Neurite growth inhibiton and protein profile present in the column fractions shown in (A). Two peaks of inhibitory activiy are detected when the fractions are tested. Reproduced from McKerracher et al. (1994) Neuron:13:805–811, with permission.

as revealed by the immunodepletion experiments (McKerracher et al., 1994) and by the inhibitory activity profile of the MAG knock-out mice (Li et al., 1996). These findings suggest that a third inhibitory protein might exist in myelin. Together, these experiments indicate that multiple inhibitory proteins are present in CNS myelin, and that the non-MAG inhibitors remain to be identified.

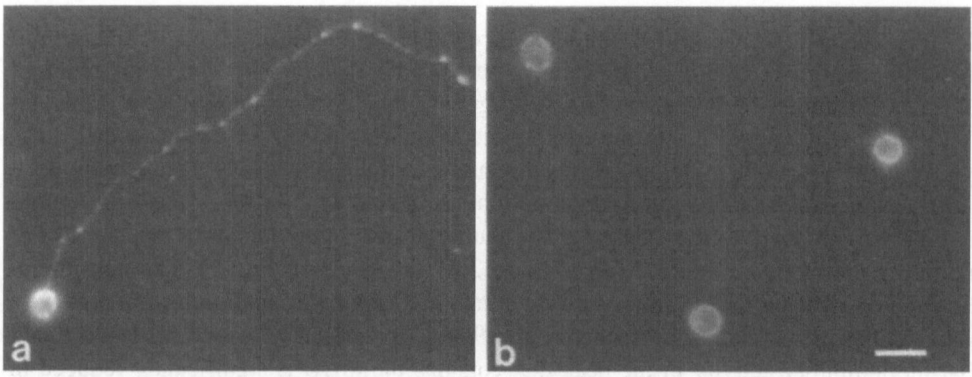

Figure 2. MAG inhibits neurite outgrowth from cerebellar granule cells. Cerebellar granule neurons were plated on laminin-coated coverslips (A) or on recombinant MAG (B). Note the lack of neurite growth on the MAG substrate.

Direct evidence for the inhibitory effect of MAG on neurite outgrowth was obtained by testing recombinant MAG produced in insect cells (McKerracher et al, 1994). When used as a growth substrate, recombinant MAG inhibited neurite growth from NG108 cells. MAG treated at 80°C to denature the protein did not inhibit growth, a finding that suggested that the conformation of the protein is important for the inhibitory activity. Recombinant MAG used as a substrate has now been found to inhibit the growth of neurites from cerebellar granule cells (Fig. 2), and hippocampal neurons (Li et al, 1996). Moreover, Mukhopadhay et al. (1994) have documented that cells that express recombinant MAG on their cell surface are an inhibitory substrate for neurite extension when compared to mock transfected cells, which provide a favorable growth substrate. More recent studies indicate that neurite length is significantly reduced when a wide variety of neuronal types are plated on MAG-expressing CHO cells compared to controls (DeBellard et al, 1996). Interestingly, of the neurons tested, only dorsal root ganglion neurons have been found to exhibit a clear age-dependent reaction to MAG, with the negative response developing only after postnatal day 3. All other neurons that have been tested, regardless of age, have significantly shorter neurites when plated on MAG. Moreover, it is now clear that both myelinated and unmyelinated classes of neurons can be inhibited by MAG. Such observations suggest that molecules that act as receptors for MAG may be widely expressed in the nervous system. It has been reported that adhesive interactions with MAG can be mediated by members of the sialoadhesion family of molecules (Kelm et al, 1994). It will be important to determine where these molecules are expressed in the nervous system and if they are involved in mediating neurite growth inhibition by MAG.

Experiments that use MAG as a substrate for growth do not address the relationship between growth cone motility and contact with MAG, such as might occur *in vivo*. Many neurite growth inhibitory proteins cause growth cone collapse, which is a rapid loss of elaborate growth cone structure that results in a cessation of axon extension (Luo and Raper, 1994). Alternatively, growth inhibitory molecules may cause growth cone turning and a redirection of growth. In a recent series of experiments we have coated beads with recombinant MAG or native MAG and observed encounters between the beads and growth cones from postnatal day 1 hippocampal neurons. Most growth cones that con-

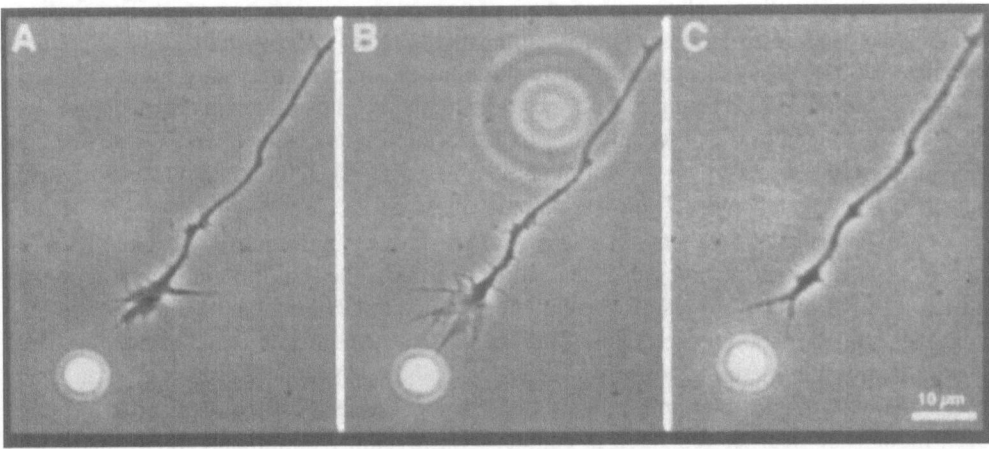

Figure 3. Growth cone collapse follows contact with MAG-coated beads. Hippocampal axon growth cones were followed before contact with the MAG-coated bead (A), during contact with the bead (B), and approximately 17 minutes after contact (C). Reproduced, with permission from Li et al (1996) J. Neurosci Res. In press.

tacted MAG-coated beads collapsed back from the bead (Fig.3). Growth cones coated with denatured MAG continued to grow past the bead (Li et al, 1996). These results demonstrate that MAG, like a growing number of molecules that influence growth cone motility, can exert a negative influence on the growth cone to prevent axon extension.

GROWTH INHIBITORY ACTIVITY OF MAG IN PERIPHERAL NERVE

Unlike the peripheral nervous system, myelin debris is not quickly cleared in the injured CNS (Stoll et al, 1989), and thus the myelin-associated proteins may have a large impact on the success or failure of axonal regrowth after injury. Although myelin in peripheral nerves is rapidly cleared to permit regeneration, it was reported that unlike CNS myelin, peripheral nerve myelin lacked inhibitory activity (Caroni and Schwab, 1988b). This observation was at odds with the finding that MAG is a growth inhibitory protein because MAG is also a component of peripheral nerve myelin (Trapp, 1990). Therefore, we have re-investigated the inhibitory properties of peripheral nerve myelin prepared from bovine sciatic nerve (David et al, 1995). We found that, in our hands, peripheral nerve myelin had a very potent inhibitory activity, similar to growth inhibition by CNS myelin when tested by bioassay with NG108 cells and cerebellar granule cells. To investigate possible differences between methods of preparation of peripheral nerve myelin, we tested myelin prepared after homogenization with a Polytron rather than homogenization with a Dounce homogenizer, as we have used in the past (McKerracher et al, 1994; David et al, 1995). In agreement with previous reports (Caroni and Schwab, 1988b), myelin prepared with a Polytron does not show inhibitory activity. Because of the high shear forces of the Polytron method, we reasoned that components of the basal lamina may have contaminated the myelin to provide a suitable growth substrate. Moreover, laminin is a protein in Schwann cell basal lamina that is well known to stimulate neurite growth (see McKerracher et al, 1996 for review). Immunoblotting experiments revealed that laminin was present in the peripheral myelin preparations that supported neurite growth, but greatly diminished in the myelin that retained its growth inhibitory properties (David et al, 1995). Moreover, when the growth-permissive myelin was pre-incubated with an antibody to laminin, the myelin regained much of its growth inhibitory properties. This finding suggests that laminin acts by stimulating neurite growth, rather than by masking the inhibitory domain of MAG. We have also tested directly the ability of laminin to override growth inhibition by MAG by testing mixed laminin/MAG substrates for their ability to support neurite growth by NG108 cells (unpublished observations). Similar to the results obtained with whole myelin, laminin was able to overcome growth inhibition by MAG. Therefore, laminin, in sufficient amounts, can override completely growth inhibition by MAG.

The observation that laminin could override growth inhibition in peripheral nerve myelin suggested that it might also be used to overcome the multiple inhibitory proteins present in CNS myelin. To test if laminin, when added to CNS myelin, could override the multiple growth inhibitory proteins that are present, we mixed together myelin and laminin and tested these mixed substrates for their ability to support neurite growth. This CNS myelin with added laminin now provided a supportive substrate for neurite growth, even though the neurites remained in contact with myelin. By immunocytochemistry we showed that the laminin was present in the myelin as fine specks, not as channels or patches, and the neurites were able to extend over myelin in these mixed substrates (David et al, 1995). These results indicate that neurite growth can occur directly on myelin if the appropriate balance of growth promoting molecules is present.

AXON REGENERATION IN MAG NULL-MUTANT MICE

The inhibitory activity of peripheral nerve myelin has also been documented *in vivo*. A strain of mice that is impaired in its ability to clear peripheral myelin after injury, (C57BL/wlds, previously called OLA mice) is unable to regenerate myelinated peripheral nerves (Chen and Bisby, 1993; Brown et al, 1994). To investigate the role of MAG in this inhibition Schafer et al (1996) cross-bred MAG null-mutant mice with C57BL/wlds mice. These mice failed to clear their peripheral nerve myelin after nerve injury, but the myelin that remained lacked MAG. After transection of axons in peripheral nerves, these mice showed significantly improved regeneration, with double the number of regenerated fibres compared to the C57BL/wlds parental strain (Schafer et al, 1996). These results, therefore, provide strong evidence that MAG acts as an important inhibitor of axon growth *in vivo*.

Because MAG comprises 1% of total CNS myelin protein, compared to 0.1% in peripheral nerve myelin (Trapp, 1990), the axon growth inhibitory properties of MAG may be just as important in the CNS. To test this possibility ,we examined the success of axon regeneration after spinal cord injury in 4–5 week old MAG-deficient mice (Li et al., 1996). The corticospinal tracts which lie in the deep portion of the dorsal columns were transected bilaterally by a dorsal hemisection in the lower thoracic region. The regeneration of the corticospinal tract fibers was assessed about 3 weeks later by anterogradely labeling these axons with wheat germ agglutinin conjugated horseradish peroxidase (WGA-HRP) injected into the sensory-motor cortex. Unlike the wild-type mice, WGA-HRP-labeled axons were found extending up to 13.2 mm distal to the site of injury in MAG-deficient mice. However, the number of axons that extend such distances were extremely small. The detection of some difference in these mice differ from the results reported for a similar study of a different line of MAG-deficient mice (Bartsch et al, 1995). In both cases, however, the results from the studies of MAG knock out mice injured in the CNS are strikingly different from the marked enhancement of peripheral nerve regeneration in the C57BL/wlds that are MAG-deficient (Schafer et al, 1996). The difference between the results in the CNS and PNS in the MAG knock out mice suggest that the lack of MAG in the CNS is masked by the presence of other axon growth inhibitory molecules. Indeed, the fractionation of octlyglucoside extracts of myelin from the knock out mice clearly show that multiple inhibitors of axon growth remain present in CNS myelin, although the inhibition associated with MAG is reduced (Li et al, 1996). Blocking the inhibitory actions of MAG *in vivo* may be better achieved by the use of function-blocking antibodies, as has been done with IN-1 antibodies (Schnell and Schwab, 1990). Such a strategy may permit axon regeneration to occur before the onset of compensatory changes in other molecules subserving a similar function during development of the knock out mice.

At present it is not possible to compare the relative contributions *in vivo* of MAG and the inhibitory proteins recognized by the IN-1 antibody. Appropriate comparisons of this nature will only be possible with the identification of the proteins that are recognized by the IN-1 antibody and the generation of corresponding knock-out mice. Moreover, all of the different inhibitory components of CNS myelin will need to be fully characterized before the contributions of the different proteins can be fully assessed. At present, the work on peripheral nerve regeneration in the C57BL/wlds mice that are MAG-deficient point to the importance of MAG in limiting axon regeneration *in vivo*. The new lines of research that are developing in the field of the myelin-derived inhibitors should yield additional insight to the problems of axon regeneration in the adult CNS of mammals.

ACKNOWLEDGMENTS

We gratefully acknowledge financial support from the Canadian NeuroScience Network (S.D. and L.M.), NSERC (L.M) and the American Paralysis Association (L.M.).

REFERENCES

Aguayo AJ, Rasminsky M, Bray GM, Carbonetto S, McKerracher L, Villegas-Pérez MP, Vidal-Sanz M, Carter DA (1991): Degenerative and regenerative responses of injured neurons in the central nervous system of adult mammals. Phil Trans R Soc Lond 331:337–343.

Arquint M, Roder J, Chia L-S, Down J, Wilkinson D, Bayley H, Braun P Dunn R (1987): Molecular cloning and primary structure of myelin-associated glycoprotein. Proc Natl Acad Sci USA 84:600–604.

Bansal R, Kumar M, Murray K, Pfeiffer SE (1995): Syndecans (1–4) and glypican are selectively expressed within a developmental lineage (oligodendrocytes) and are regulated by FGF-2. Molec Biol Cell 6:162a.

Bartsch U, Bandtlow L, Schnell L, Bartsch S, Spillmann AA, Rubin BP, Hillenbrand R, Montag D, Schwab ME, Schachner M (1995) Lack of evidence that myelin-associated glycoprotein is a major inhibitor of axonal regeneration in the CNS. Neuron 15:1375–1381.

Bregman BS, Kunkel-Bagden E, Schnell L, Dai HN, Gao D, Schwab ME (1995): Recovery from spinal cord injury mediated by antibodies to neurite growth inhibitors. Nature 378:498–501.

Björklund A (1994): Long distance axonal growth in the adult central nervous system. J Neurol 241:S33-S35.

Brown MC, Perry VH, Hunt SP, Lapper, SR (1994): Further studies on motor and sensory nerve regeneration in mice with delayed Wallerian degeneration. Eur J Neurosci 6:420–428.

Caroni P, Schwab ME (1988a): Antibody against myelin-associated inhibitor of neurite growth neutralizes nonpermissive substrate properties of CNS white matter. Neuron 1:85–96.

Caroni P, Schwab ME (1988b): Two membrane protein fractions from rat central myelin with inhibitory properties for neurite growth and fibroblast spreading. J Cell Biol 106:1281–1288.

Chen S, Bisby MA (1993): Impaired motor axon regeneration in the C57BL/Ola mouse. J. Comp. Neurol. 333:449–454.

Cho EYP, So KF (1992): Characterization of the sprouting response of axon-like processes from retinal ganglion cells after axotomy in adult hamsters: a model using intravitreal implantation of a peripheral nerve. J Neurocytol 21:589–603.

David S, Aguayo AJ (1981): Axonal elongation into peripheral nervous system "bridges" after central nervous system injury in adult rats. Science 214:931–933.

David S, Braun PE, Jackson DL, Kottis V, McKerracher L (1995): Laminin overrides the inhibitory effects of peripheral nervous system and central nervous system myelin-derived inhibitors of neurite growth. J Neurosci Res 42:594–602.

Davies SJA, Field PM, Raisman G (1994): Long interfascicular axon growth from embryonic neurons transplanted into adult myelinated tracts. J Neurosci 14:1596–1612.

DeBellard ME, Tang S, Mukhopadhyay G, Shen YJ, Filbin MT (1996): Myelin-associated glycoprotein inhibits axonal regeneration from a variety of neurons via interaction with a sialoglycoprotein. Molec Cell Neurosci 7:89–101.

Fournier AE, McKerracher L (1995): Tubulin expression and axonal transport in injured and regenerating neurons in the adult mammalian central nervous system. Biochem Cell Biol 73:659–664.

Friedlander DR, Milev P, Karthikeyan L, Margolis RK, Margolis RU, Grumet M (1994): The neuronal chondroitin sulfate proteoglycan neurocan binds to the neural cell adhesion molecules Ng-CAM/L1/NILE and N-CAM, and inhibits neuronal adhesion and neurite outgrowth. J Cell Biol 125:669–680.

Hess DT, Patterson SI, Smith DS, Pate Skene JH (1993): Neuronal growth cone collapse and inhibition of protein fatty acylation by nitric oxide. Nature 366:562–565.

Hunter DD, Cashman N, Morris-Valero R, Bulock JW, Adams SP, Sanes, JR (1991): An LRE (Leucine-arginine-glutamate)-dependent mechanism for adhesion of neurons to S-laminin. J Neurosci 11:3960–3971.

Jalinck K, van Corven E J, Hengeveld T, Mori N, Narumiya S, Moolenaar WH (1994): Inhibition of lysophosphatidate- and thrombin-induced neurite retraction and neuronal cell rounding by ADP ribosylation of the small GTP-binding protein Rho. J Cell Biol 126:801–810.

Johnson PW, Abramow-Newerly W, Seilheimer B, Sadoul R, Tropak MB, Arquint M, Dunn RJ, Schachner M, Roder JC (1989): Recombinant myelin-associated glycoprotein confers neural adhesion and neurite outgrowth function. Neuron 3:377–385.

Keirstead HS, Hasan SJ, Muir GD, Steeves JD (1992): Suppression of the onset of myelination extends the permissive period for the functional repair of embryonic spinal cord. Proc Natl Acad Sci 89:11664–11668.

Kelm S, Pelz A, Schauer R, Filbin MT, Tang S, de Bellard ME, Schnaar RL, Mahoney JA, Hartnell A, Bradfield P, Crocker PR (1994): Sialoadhesin, myelin-associated glycoprotein and CD22 define a new family of sialic acid-dependent adhesion molecules of the immunoglobulin superfamily. Curr Biol 4:965–972.

Kolodkin AL, Matthes DJ, Goodman CS (1993): The semaphorin genes encode a family of transmembrane and secreted growth cone guidance molecules. Cell 75:1389–1399.

Kromer LF, Bjorklund A, Stenevi U. (1981): Regeneration of the septohippocampal pathways in adult rats is promoted by utilizing embryonic hippocampal implants as bridges. Brain Res 210:173–200.

Li M, Shibata A, Li C, Braun PE, McKerracher L, Roder J, Kater SB, David S (1996) Myelin-associated glycoprotein inhibits neurite/axon growth and causes growth cone collapse. J Neurosci Res: 46: 404–414.

Lochter A, Vaughan L, Kaplony A, Prochiantz A, Schachner M, Faissner A (1991): J1/tenascin in substrate-bound and soluble form displays contrary effects on neurite outgrowth. J Cell Biol 113:1159–1171.

Luo Y, Raible D, Raper JA (1993): Collapsin: a protein in brain that induces the collapse and paralysis of neuronal growth cones. Cell 75:217–227.

Luo Y, Raper JA (1994): Inhibitory factors controlling growth cone motility and guidance. Curr Opin Neurobiol 4:648–654.

McKeon RJ, Schreiber RC, Rudge JS, Silver J (1991): Reduction of neurite outgrwoth in a model of glial scarring following CNS injury is correlated with the expression of inhibitory molecules on reactive astrocytes. J Neurosci 11:3398–3411.

McKerracher L, Chamoux M, Arregui CO (1996): Role of laminin and integrin interactions in growth cone guidance. Mol Neurobiol: 12: 95–116.

McKerracher L, David S, Jackson DL, Kottis V, Dunn RJ, Braun PE (1994): Identification of myelin-associated glycoprotein as a major myelin-derived inhibitor of neurite growth. Neuron 13:805–811.

Mukhopadhyay G, Doherty P, Walsh FS, Crocker PR, Filbin MT (1994): A novel role for myelin-associated glycoprotein, MAG, as an inhibitor of axonal regeneration. Neuron 13:757–767.

Pini A (1993): Chemorepulsion of axons in the developing mammalian central nervous system. Science 261:95–98.

Richardson PM, Issa VMK, Shemie S (1982): Regeneration and retrograde degeneration of axons in the rat optic nerve. J Neurocytol 11:949–966.

Sadoul R, Fahrig T, Bartsch U, Schachner M (1990): Binding properties of liposomes containing the myelin-associated glycoprotein MAG to neural cell cultures. J Neurosci Res 25:1–13.

Salzer J, Holmes WP, Colman DR (1987): The amino acid sequence of the myelin-associated glycoproteins: Homology to the immunoglobulin gene superfamily. J Cell Biol 104:57–66.

Savio T, Schwab ME (1990): Lesioned corticospinal tract axons regenerate in myelin-free rat spinal cord. Neurobiology 87:4130–4133.

Schafer M, Fruttiger M, Montag D, Schachner M, Martini R (1996): Disruption of the gene for the myelin-associated glycoprotein leads to improved axonal regrowth along non-degenerated myelin in C57BL/Wlds mice. Neuron: 16:1107–1113.

Schnell L, Schwab ME (1990): Axonal regeneration in the rat spinal cord produced by an antibody against myelin-associated neurite growth inhibitors. Nature 343:269–272.

Schwab ME, Kapfhammer JP, Bandtlow CE (1993): Inhibitors of neurite growth. Annu Rev Neurosci 16:565–595.

Schwab ME, Thoenen H (1985): Dissociated neurons regenerate into sciatic but not optic nerve explants in culture irrespective of neurotrophic factors. J Neurosci 5:2415–2423.

Stoll G, Trapp BD, Griffin JW (1989): Macrophage function during Wallerian degeneration of rat optic nerve: Clearance of degenerating myelin and Ia expression. J Neurosci 9:2327–2335.

Suidan H S, Stone SR, Hemmings B A, Monard D (1992): Thrombin causes neurite retraction in neuronal cells through activation of cell surface receptors. Neuron 8:363–375.

Trapp. BD (1990): Myelin-associated glycoprotein. Location and potential functions. Annals N Acad Sci 605:29–43.

Wictorin K, Brundin P, Sauer H, Lindvall O, Bjorklund A (1992): Long distance directed axonal growth from human dopaminergic mesencephalic neuroblasts implanted along the nigrostriatal pathway in 6-hydroxy-dopamine lesioned adult rats. J Comp Neurol 323:475–494.

Vanselow J, Schwab ME, Thanos S (1990): Response of regenerating rat retinal ganglion cell axons to contacts with central nervous system myelin in vitro. Europ J Neurosci 2:121–125.

31

SCHWANN CELLS AS FACILITATORS OF AXONAL REGENERATION IN CNS FIBER TRACTS

Mary Bartlett Bunge[1,2] and Naomi Kleitman[2]

The Chambers Family Electron Microscopy Laboratory, The Miami Project
  to Cure Paralysis, and
[1]Department of Cell Biology and Anatomy
[2]Department of Neurological Surgery
University of Miami School of Medicine
Miami, Florida 33136

INTRODUCTION

Studies of human spinal cord injury by Richard Bunge and colleagues in The Miami Project to Cure Paralysis emphasize that, in most cases, gray matter loss is localized to one or only a few segments. Despite the limited loss of spinal cord neurons in those cases in which damage is localized linearly, extensive injury across the diameter of the cord abolishes much of the signalling required for motor functioning below the level of injury. In 35% of the 46 injured human cords studied to date, injury extends across the entire width of the cord (Bunge et al., 1993, 1996). Thus, the most debilitating aspect of these injuries derives from interruption of nerve fibers in the long ascending and descending tracts in the white matter, rather than from damage to the neuronal population. We, therefore, have attempted to create bridges for nerve fiber regrowth across areas of spinal cord injury to re-establish the requisite signalling. It is now well known that after injury the central neuron is able to regenerate its axon if the local environment is suitable. Requirements include an appropriate substratum for growth, sufficient quantities of specific neurotrophic factors, and the absence of neurite growth inhibitors such as myelin- and astrocyte-related proteins.

Because Schwann cells have been shown to foster central nerve fiber regeneration and large numbers of Schwann cells derived from peripheral nerve segments can be generated in culture, we have created bridges of Schwann cells to test their ability to promote regeneration of fibers across injured areas of midthoracic spinal cord. As in the case of the grafting of peripheral nerve segments into lower regions of the spinal cord (Richardson et al., 1984; Houle, 1991), grafts of purified populations of Schwann cells in thoracic regions generate a substantial regenerative response from propriospinal neurons but not from supraspinally situated nerve cells. Our hypothesis is that Schwann cells in combination with

Cell Biology and Pathology of Myelin, edited by Juurlink et al.
Plenum Press, New York, 1997

neurotrophic, neurotropic, and neuroprotective substances will enable appropriate axonal regeneration across the site of spinal cord injury to enable meaningful functional recovery. That this hypothesis is valid will be shown in the description of our results and of recent work by other investigators. This chapter describes our experiments designed to test the efficacy of Schwann cells alone and in combination with neurotrophic and neuroprotective agents in promoting axonal regeneration across sites of spinal cord injury.

The use of peripheral nerve and Schwann cell grafts has been much reviewed (e.g., Aguayo, 1985; Kromer and Cornbrooks, 1987; Bray et al., 1991; Guénard et al., 1993; Houle et al., 1994) and, thus, will not be described in detail here. Briefly, it is the presence of Schwann cells in peripheral nerve grafts that has been considered to be critical for the effectiveness of peripheral nerve grafts in supporting axonal regeneration of central fibers. Schwann cells secrete neurotrophic factors, express cell adhesion molecules, and generate numerous extracellular matrix molecules, all known to influence nerve fiber growth. An additional consideration is that a piece of peripheral nerve could be obtained from a spinal cord injured person, the Schwann cells extricated from that piece of nerve and caused to proliferate extensively in culture. The cells could then be autologously transplanted into the site of injury in the spinal cord injured person (Bunge, 1975). Methods are now available for obtaining substantially large and purified populations of Schwann cells not only from adult rat peripheral nerve (Morrissey et al., 1991) but also from adult human nerve (Levi et al., 1995; Morrissey et al., 1995; Rutkowski et al., 1995; Casella et al., 1996). It is now possible to obtain adequate numbers of human Schwann cells for transplantation into either adult rat peripheral nerve (Levi and Bunge, 1994; Levi et al., 1994) or spinal cord (Guest and Bunge, 1994, 1995; Guest et al., 1996). Both rat and human Schwann cells taken from adult sources survive and function (ensheathe and myelinate axons) in the adult rodent peripheral nerve and spinal cord locations.

We have asked what bridge content will promote regeneration of long tract fibers from one transected stump to another in the adult rat. The model system we have chosen is transection of the adult Fischer rat thoracic spinal cord with removal of neighboring segments to create a substantial gap across which regeneration can be unequivocally evaluated. We elected to use this model to ensure complete spinal cord discontinuity, to test all fiber types in the cord for their ability to respond to the grafts, and to represent human spinal cord injury that spans the entire diameter of the spinal cord. This model presents a greater challenge for regrowth than do hemisection models and precludes consideration of sprouting from uninterrupted fibers. In the work reported by Schnell and Schwab (1993), for example, a variety of implanted materials (including embryonic spinal cord, newborn pons, placental extracellular matrix, collagen, laminin-coated nitrocellulose, and carbon filaments) did not attract corticospinal tract regenerative growth across the hemisected area. Growth occurred only into the tissue remaining in continuity beneath the implant (in the presence of antibody that neutralized myelin-associated neurite growth inhibitory protein). It is not known whether the surviving tissue abutting sites of human spinal cord injury will be supportive for axonal regrowth.

In our model, the gap in the spinal cord is then filled with a linear array, or cable, of Schwann cells and Matrigel suspended within a polymer guidance channel (Fig. 1; see also Xu et al., 1995a). The channels are made of a mixture of polyacrylonitrile and polyvinylchloride; polycarbonate tubes have also been used to transplant Schwann cells into the cord (Montgomery et al., 1996). The transected spinal cord stump is inserted 1 mm into the end of the guidance channel, occluding the opening and creating a protected cellular environment within the channel. Studies have been conducted to examine growth into

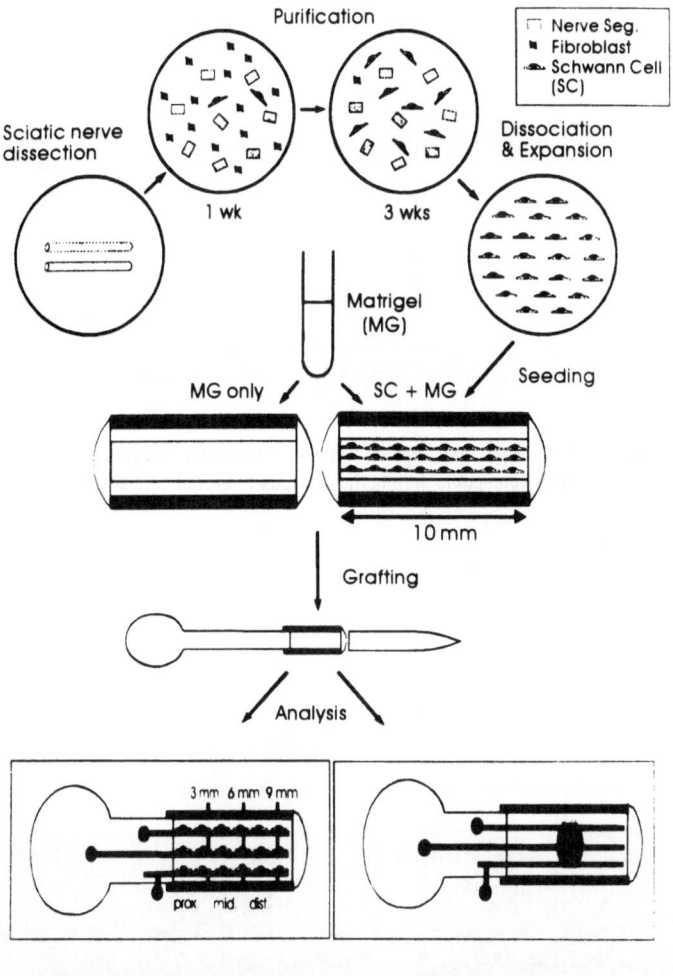

Figure 1. Overall schema of transplantation paradigm. SCs were purified from pieces of peripheral nerve and then combined with Matrigel to form a cellular cable inside a polymeric guidance channel. In initial experiments, the channel distal end was capped but, in others, the channel received a cut cord stump in each end. One month later, a tracer was injected into some cables to identify the nerve cells that extended axons into the graft. Other grafts were analyzed histologically at the 3, 6 or 9 mm level and immunocytochemically in intervening and host-graft interface regions.

and along the graft, by employing distally capped channels (Fig. 1) or channels open at both ends to receive both transected stumps to assess growth from the graft into the host cord distal to the graft (Fig. 2). Care is taken to appose the cord stumps and the graft as closely as possible. Animals are perfused one month after grafting and the graft, spinal cord, DRG, and brain are carefully extirpated for study in 1 μm plastic sections to assess cellular content and count myelinated axons, and in frozen sections to immunostain for axonal (Fig. 3) and cellular content. Both retrograde and anterograde tracing may be performed to identify the neurons that extend axons into the graft and to detect regenerative growth from the graft into the cord. An injection of a retrograde tracer, Fast Blue, into the cable midpoint remains well localized and does not spread to the cable-cord interface; its localization is carefully monitored after tissue processing (Xu et al., 1995b).

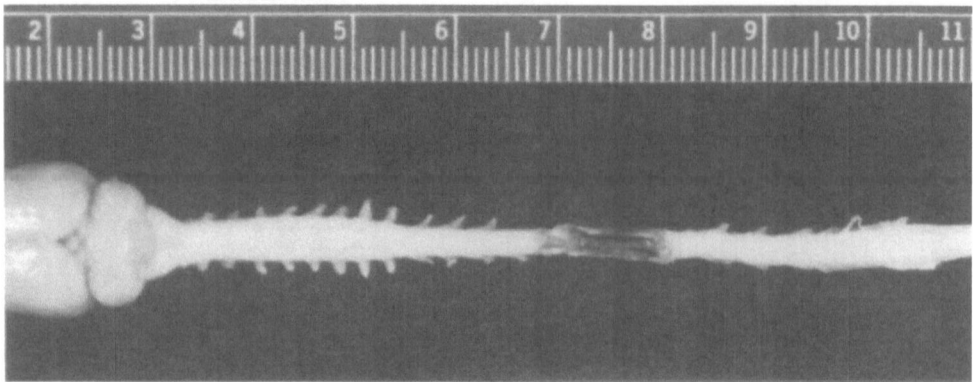

Figure 2. Appearance of a graft inside an open-ended channel a month after transplantation. The top of the channel has been removed to show the cellular cable uniting the previously severed cord stumps.

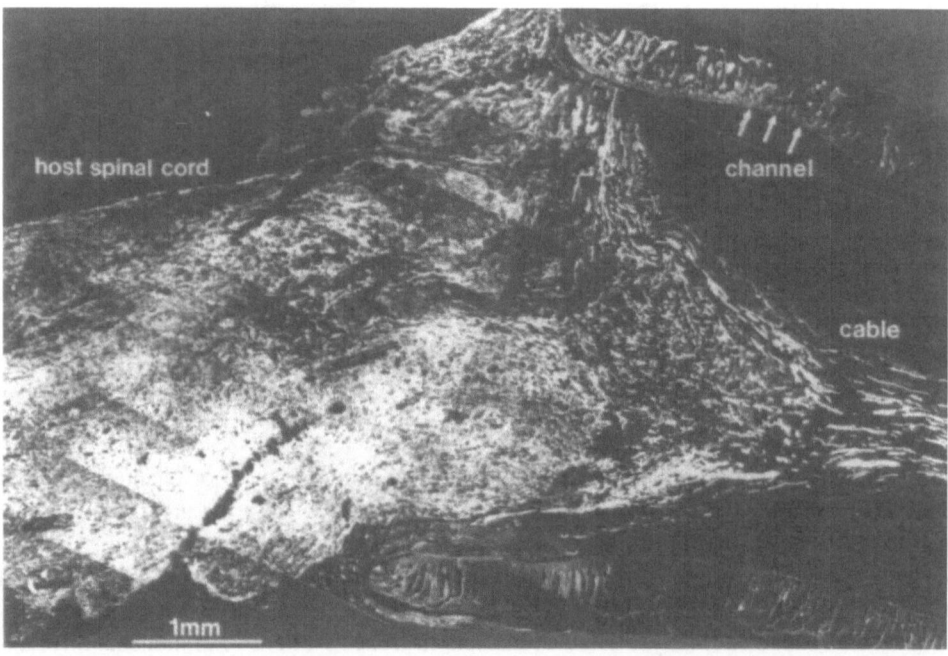

Figure 3. Sagittal section of a rostral cord-graft interface area immunostained for neurofilaments. The upper channel wall is indicated by arrows. The immunostained structures entering the cable appear thicker than those in the cord because the axons have been fasciculated by SCs. SC/Matrigel case. (From Xu et al., 1995a by copyright permission of Wiley-Liss, Inc.)

RESULTS

Schwann Cell Grafts in Distally Capped Channels

In our initial transplantation study to determine the effectiveness of purified populations of Schwann cells (suspended in a basal lamina matrix, Matrigel) to elicit axonal regrowth, the transected stump was inserted into the rostral end of the guidance channel with the distal end capped (Fig. 1) (Xu et al., 1995a). That the transplanted Schwann cells survived for at least a month in the graft was detected by prelabelling the cells with Hoechst dye (Xu et al., 1995a,b). Myelinated axons were present within the cable (mean, 501; Table 1) and many of these extended the length of the 10 mm cable. Electron microscopy revealed that the population of regenerating axons in the cable contained four times more unmyelinated than myelinated axons and that Schwann cell ensheathment and myelination were typical of peripheral nerve. Fewer axons were observed in the control channels filled with Matrigel only (mean, 71; Table 1). These axons were myelinated, undoubtedly by Schwann cells that migrated from the host into the Matrigel graft. Axons were not found unaccompanied by Schwann cells. Thus, the presence of Schwann cells was necessary for axonal ingrowth. A similar finding was particularly striking in earlier studies in which Schwann cells transplanted inside rolled collagen promoted substantial axonal ingrowth in contrast to collagen alone, in which axonal ingrowth did not occur (Paino and Bunge, 1991; Paino et al., 1994).

In our study utilizing Schwann cells in distally capped channels, lack of immunostaining for serotonin (5-HT) and dopamine ß hydroxylase (DBH) showed that supraspinal projections from raphe nuclei or locus coeruleus, respectively, did not regenerate into the cable; lack of response of brainstem nuclei was confirmed by retrograde tracing. The injection of the tracer, Fast Blue, into the midpoint of the cable demonstrated that most neurons (mean, 306; Table 2) whose axons regenerated into the graft were located in spinal cord gray matter as far rostral as nine segments (to C7, 25 mm from the rostral interface). A substantial but lesser number of dorsal root ganglion (DRG) neurons were retrogradely labelled as well. In this study, as in all the subsequent ones, a moderate enhancement of glial fibrillary acidic protein staining was observed at the host-graft interface. In conclusion, these SC grafts inside distally

Table 1. Number of myelinated axons in differing transplantation procedures

Graft level	Graft composition	Capped channel (mean ± SEM)	Open channel (mean ± SEM)
3 mm	Matrigel only	71 ± 25	
	Matrigel/SC	501 ± 83	
3 mm	Matrigel/SC/vehicle	882 ± 287	
	Matrigel/SC/NT	1523 ± 292	
4-5 mm	Matrigel/SC	355 ± 108	
	Matrigel/SC/MP	1159 ± 308	
3 mm	Matrigel/SC	504 ± 78	
	Matrigel/SC/IGF-1/PDGF	316 ± 42	
4-5 mm	Matrigel only		3 ± 0.9
	Matrigel/SC		1990 ± 594
4-5 mm	Matrigel/SC/vehicle		1324 ± 342
	Matrigel/SC/MP		3237 ± 2478

SC, Schwann cell; NT, neurotrophin-treated; MP, methylprednisolone

Table 2. Number of Fast Blue-labelled neurons in different types of transplantation procedures

Location of labelled neurons	Capped channel (mean ± SEM)			Open channel (mean ± SEM)	
	SC	SC/NT	SC/MP	SC	SC/MP
Spinal cord	306±69	967±104	1116±113	1064±145	2083±321
Brainstem	0	92±20	46±14	0	57±22

SC, Schwann cell; NT, neurotrophin-treated; MP, methylprednisolone

capped channels induced propriospinal and sensory but not brainstem or cortical neurons to regenerate axons into the graft. This is in concurrence with peripheral nerve grafting work (Richardson et al., 1984; Houle, 1991).

Schwann Cell Grafts Infused with Neurotrophins in Distally Capped Channels

Recent advances in our understanding of neurotrophins and their availability now enable us to test the regenerative response of specific neuronal populations to the presence

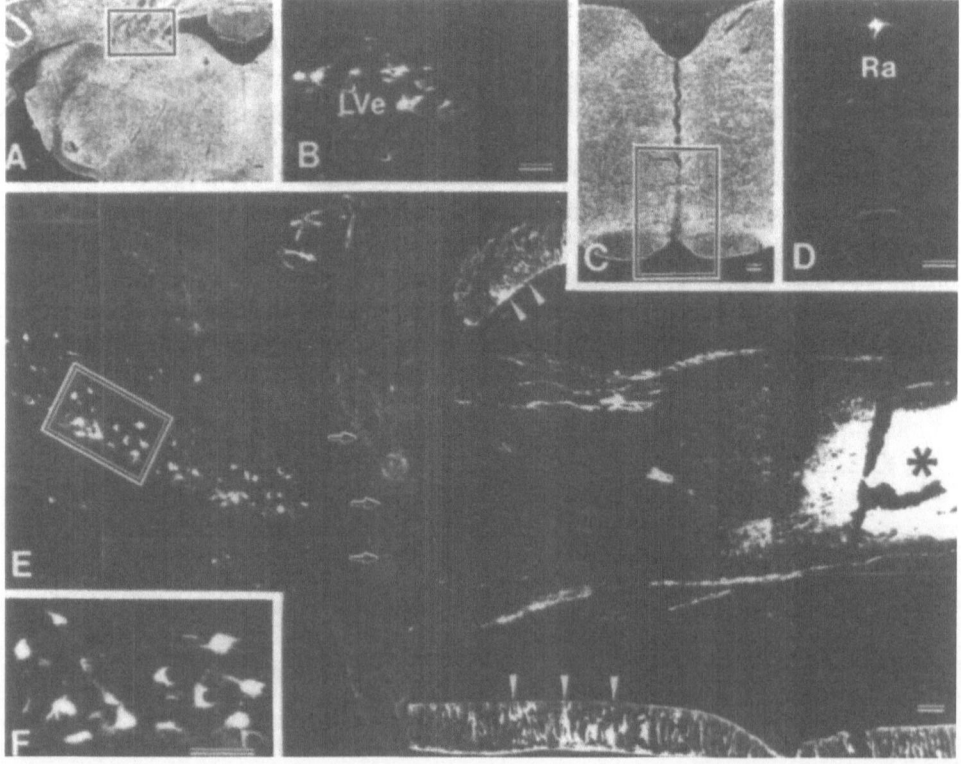

Figure 4. Sagittal section of a rostral cord-graft interface region in a SC/NT case, illustrating labelled neurons (boxes E,F) that picked up Fast Blue injected into the midpoint of the cable (asterisk). Note the position of labelled neurons closest to the graft. The cord-graft boundary is indicated by arrows; no tracer is visible at this border. Labelled cells were also found in brainstem nuclei; e.g. in the lateral vestibular (LVe, B) and raphe (Ra, D) nuclei. These labelled neurons are from areas indicated in the pons (A) and medulla (C), respectively. All bars, 200 μm. (From Xu et al., 1995b by copyright permission of Academic Press, Inc.)

of these factors. Although Schwann cells secrete numerous neurotrophic factors, these substances may not be made in sufficient quantity to engender a regenerative response. We, therefore, initiated a study testing the neurotrophins brain-derived neurotrophic factor (BDNF) and neurotrophin-3 (NT-3) (Xu et al., 1995b). In these experiments, the Schwann cell cable/guidance channel was constructed with tubing connected to the distal cap of the channel leading to an Alzet minipump containing the neurotrophins BDNF and NT-3. Over the first 14 days of the 30 day survival period, 12 μg/day of each neurotrophin was infused simultaneously into the distal end of the graft. Controls received vehicle solution. One month after grafting, a mean of 1,523 myelinated axons (Table 1) was present in Schwann cell/neurotrophin grafts, approximately twice as many as were present in Schwann cell/vehicle grafts.

Along with this increase in the number of myelinated axons in the graft, other differences were noted as well. Serotonergic axons regenerated into the Schwann cell/neurotrophin grafts for at least 5 mm, but did not enter control grafts. The regeneration of serotonergic axons of brainstem origin was confirmed by retrograde labelling of some raphe neurons with Fast Blue. When Fast Blue was injected at the Schwann cell/neurotrophin cable midpoint (Fig. 4), a mean of 92 retrogradely labelled neurons (Table 2) was found in ten nuclei in the brainstem, with the highest labelling (67%) consistently seen in vestibular nuclei. Thus, when neurotrophins were infused into Schwann cell grafts in the thoracic region, certain brainstem neurons were able to respond by regenerating axons into the graft. Thus, the combination of cells and factors improved CNS regeneration compared with the outcome of peripheral nerve grafting, which does not engender a response from brainstem neurons when grafts are placed in the thoracic region (Richardson et al., 1984; Houle, 1991). Labelled cells were also present throughout the cord rostral to the channel at an increased level compared with that seen in uninfused SC-containing close-ended channels (mean, 967; Table 2). In sum, regeneration of axons into the grafts from both spinal cord and some brainstem neuronal populations distant from the thoracic neurotrophin-containing Schwann cell implant was promoted by a combination of a permissive cellular substratum and the presence of neurotrophic factors. The efficacy of such a combination approach is in agreement with work by Schnell et al. (1994) who observed that a substratum made permissive by neutralization of neurite growth inhibiting proteins in combination with provision of NT-3 led to improved corticospinal axonal regrowth in the spinal cord.

Schwann Cell Grafts in Distally Capped Channels in Combination with Methylprednisolone

Clinically, the steroid methylprednisolone (MP) has been shown to be neuroprotective if administered within eight hours after spinal cord injury (Bracken et al., 1990, 1992). Using our experimental model, we investigated whether axonal regrowth into Schwann cell grafts is enhanced when MP is administered at the time of spinal cord transection and Schwann cell implantation (Chen et al., 1996). MP (30 mg/kg) or vehicle (control) was administered intravenously at 5 min, 2 h, and 4 h after transection. Thirty to 45 days later, the Schwann cell/MP group showed large diameter cables in the channels. A very striking finding was that there was more spinal cord tissue present at the rostral host cord-graft interface in Schwann cell/MP grafts than in the control grafts. This interface was consistently found *inside* the channel instead of at the rostral end of the channel as seen in our previous studies (Fig. 5). Significantly more myelinated axons (mean, 1159; Table 1) were present at the 5 mm level in Schwann cell/MP grafts than in Schwann cell/vehicle cables (mean, 355; Table 1). More unmyelinated than myelinated axons (4:1) were resolved by electron microscopy under both conditions. In the Schwann cell/MP group, serotonergic and noradrenergic fibers were detected immunocytochemically up to 2.5 and 2.0 mm, respectively, into the graft. This was in contrast to the Schwann cell/vehicle group and also to the Schwann cell graft group in the initial study in which monoaminergic axons were entirely absent from the cables. Fast Blue retrograde tracing showed that more propriospinal neurons (mean, 1116; Table 2) were present in Schwann cell/MP grafts than in Schwann cell/vehicle grafts (mean, 284±88). Also, spinal cord neurons 2–3 segments more distant from the graft (to C5) responded by regenerating axons into the graft when MP had been administered. Perhaps the most significant difference was that brainstem neurons extended axons into the graft when MP was administered (means, 46 vs. 0 in controls), even without provision of additional growth factors. Thus, MP improved axonal regeneration from both cord and brainstem neurons into Schwann cell grafts, possibly by reducing secondary injury of the cord adjacent to the graft.

Schwann Cell Grafts in Open-Ended Channels

Schwann cell grafts were placed between transected stumps of adult rat cord inside open-ended channels to test their efficacy to serve as bridges across the injured area (Xu et al., 1996). One month later, the gross and histological appearance and the demonstration of ingrowth of propriospinal axons from both stumps indicated that the Schwann cell bridge had effectively united the severed stumps (Fig. 2). At the bridge midpoint, there was a mean of 1990 myelinated axons (Table 1) and eight times as many nonmyelinated, ensheathed axons. Essentially no myelinated axons were observed in control (Matrigel-only) grafts. Propriospinal neurons (mean, 1064; Table 2) as far away as C3 and S4 were labelled by retrograde tracing with Fast Blue following injection into the bridge. Brainstem neurons were not retrogradely labelled from the graft. This was consistent with ob-

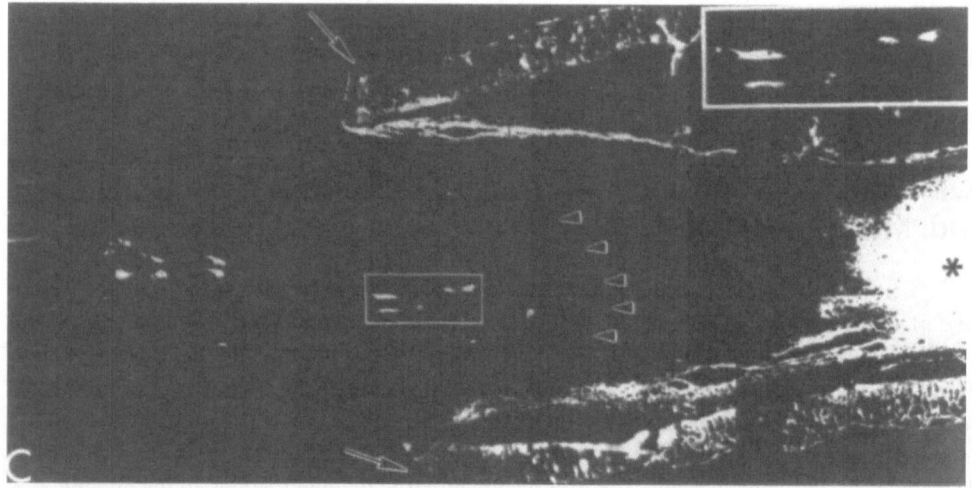

Figure 5. Host-cord and attached graft from a SC/MP case following sagittal sectioning. The tracer-containing neurons closest to the cord-graft interface (arrowheads) are in tissue inside the guidance channel (arrows), in contrast to their position in Fig. 4. An asterisk marks the injection site of the Fast Blue. Bar, 300 μm. (From Chen et al., 1996 by copyright permission of Academic Press, Inc.)

servations of growth of immunoreactive serotonergic and noradrenergic axons only a short distance into the rostral end of the graft and, therefore, not far enough to reach the area where the tracer had been injected at the graft midpoint. Anterograde tracing following injection of Phaseolus vulgaris leucoagglutinin (PHA-L) a few mm rostral to the graft was also performed to evaluate growth from the graft into the host cord distal to the graft. This tracing demonstrated that axons grew the length of the graft but did not leave the graft to grow into the host cord. In sum, this study demonstrated that Schwann cell grafts were able to serve as substrates that enabled regrowth of both ascending and descending axons across a gap in the adult rat thoracic spinal cord. There also was limited regrowth of serotonergic and noradrenergic fibers from the rostral stump only. Although this latter finding was different from our initial study in which Schwann cell grafts in distally capped channels were not found to contain any serotonergic or noradrenergic fibers, this investigation largely confirmed the initial study that found that the provision of Schwann cells alone was insufficient to promote axonal regrowth from brainstem neurons.

Schwann Cell Grafts in Open-Ended Channels in Combination with Methylprednisolone

This MP study was done in a manner similar to the MP study described above, except that the Schwann cells were placed inside open-ended rather than distally capped channels (Chen et al., 1994). As in the study with distally capped channels, the spinal cord-graft interfaces were consistently observed well inside the guidance channel one month after transplantation. This was true for the interfaces at both the rostral and distal ends of the graft. Thus, the reduction in secondary tissue loss after injury was observed in this study as well. The mean of myelinated axons in the graft was the highest observed in our studies, 3237, whereas in the Schwann cell graft in animals receiving vehicle injections, the mean was 1324 (Table 1). Retrograde tracing led to labelling of both propriospinal (mean, 2083; rostrally extending to C4 and caudally, to S4) and brainstem neurons (mean, 57) (Table 2). Serotonin and DBH-immunostained fibers were seen extending 2.0 and 2.5 mm, respectively, from the rostral interface into the graft. In sum, improvement in axonal regrowth by MP was observed in this study, as in the earlier study conducted with distally capped guidance channels. Moreover, in the animals receiving MP treatment, we observed and a modest axonal growth exiting the open-ended Schwann cell channels and into the distal cord (up to 2 mm) by PHA-L tracing.

Schwann Cell Grafts Containing Insulin-like Growth Factor-1 and Platelet-Derived Growth Factor in Distally Capped Channels

Because insulin-like growth factor-1 (IGF-1) promotes axonal regeneration in the peripheral nervous system and this effect is enhanced by platelet-derived growth factor (PDGF), we decided to study the effects of these factors on axonal regeneration into Schwann cell grafts (Oudega et al., 1996). During preparation of the guidance channels, the two factors were mixed with the Schwann cells and Matrigel, then drawn into the channel. Control animals received implants of only Matrigel and Schwann cells or only Matrigel and the factors. One month after transplantation, electron microscopic analysis showed that the addition of factors resulted in a decreased unmyelinated:myelinated fiber ratio in the Schwann cell grafts; fewer unmyelinated fibers were present and, consequently, fascicles were smaller. We also observed that myelin sheath thickness was increased two-fold in IGF-1/PDGF-containing channels. In contrast to the apparent promotion of myelination of regenerated axons, there was a 36% decrease in the total number of myelinated axons in the Schwann cell implants in the factor-containing cables (means, 316 vs. 504; Table 1). Calculations derived from these findings indicated that the overall fiber regeneration into Schwann cell implants was diminished up to 63% by the factors. More larger cavities were seen at the proximal spinal cord-graft interface in animals receiving Schwann cells plus the factors. This higher incidence of cavitation may have contributed to the diminished axonal ingrowth. The effect on axonal growth was mixed, however. In the presence of the factors, despite the overall decrease in axon numbers, a small number of serotonergic and noradrenergic fibers grew up to 2 mm into the Schwann cell graft. In sum, the results of this study indicated that the combination of the factors enhanced myelination of regenerating axons that entered Schwann cell grafts, but the overall regeneration of axons into these grafts was diminished.

BDNF-Secreting Schwann Cells Grafted to Transected Spinal Cord

The infusion of BDNF and NT-3 into Schwann cell-containing guidance channels promotes regeneration of brainstem neurons into grafts placed in transected adult rat spinal cord in the thoracic region (see above; Xu et al., 1995b). To individually assess the effects of these factors, and to shift experimentally toward a cellular delivery system for neurotrophin delivery, we initiated a project to genetically modify Schwann cells to secrete BDNF (Menei et al., 1996). One of the reasons to use Schwann cells in the grafts, as pointed out in the Introduction, is that they secrete neurotrophins. It appeared, however, that the amount of BDNF that they secrete in the grafts is not optimal because the infusion of exogenous BDNF (plus NT-3) led to improved axonal regeneration into the Schwann cell grafts. We started to investigate the efficacy of BDNF-secreting Schwann cells by transecting the cord and then grafting the Schwann cells in a trail 5 mm long beyond the cut edge of the distal stump and also into the transection site itself. When the distal stump was raised to receive the injection of Schwann cells, it was clear that the transection was complete. We compared transected animals receiving 1) normal Schwann cells, 2) BDNF-secreting Schwann cells, and 3) no Schwann cells. Schwann cells were infected with a replication-deficient retroviral vector pL(hBDNF)RNL encoding the human preproBDNF cDNA. The amounts of BDNF secreted by BDNF-secreting and normal Schwann cells, as detected by ELISA, were 23 and 5 ng/24 hours/10^6 cells, respectively. The secreted BDNF was biologically active, as detected by a retinal ganglion cell bioassay. The spinal cord was transected at level T8. Hoechst prelabelling of transplanted Schwann cells demonstrated that cohesive trails were maintained for at least a month. One month after grafting, neurofilament immunostained axons were present in the Schwann cell trails. More serotonergic and dopaminergic fibers were observed along BDNF-secreting Schwann cell transplants than when normal Schwann cells were used. When Fast Blue was injected 5 mm below the transection site, at the end of the trail, an average of 100 retrogradely labelled neurons was found in the brainstem, mostly in the reticular and raphe nuclei. This labelling was greater than when normal Schwann cells were transplanted. In the latter condition, only one-fifth as many neurons were labelled and the labelling was primarily in vestibular nuclei. Little augmentation of the number of labelled regenerating vestibular neurons was seen in the presence of BDNF. The mean of retrogradely labelled cells in rostral dorsal root ganglia was increased in the presence of BDNF-secreting Schwann cells, but the labelling of propriospinal neurons was similar. No labelled neurons were observed rostral to the transection site when no Schwann cells were transplanted. In sum, the implantation of Schwann cells secreting increased amounts of BDNF improved the regenerative response across a transection site and along the cord below. It is possible that the promotion of regeneration that we observed in this study may be specific because, in the presence of BDNF-secreting Schwann cells, the largest response was from neurons known to express the receptor for BDNF, *trk*B.

Human Schwann Cell Grafts in Distally Capped Channels Implanted into Nude Rats

An ongoing goal of The Miami Project has been the study of the biology of the human Schwann cell, in anticipation of clinical application of Schwann cell transplantation. This productive line of research has utilized both in vivo and in vitro techniques to study the behavior of adult-derived human Schwann cells in association with rat neurons. Accordingly, human Schwann cells have been transplanted into not only peripheral nerve (Levi and Bunge, 1994; Levi et al., 1995) but also into nude rat spinal cord (Guest and Bunge, 1994, 1995; Guest et al., 1996). The experimental parameters were similar to those described above except for the species of donor Schwann cells and rat hosts. The overall morphology of the cables one month after implantation was similar to that in the Fischer rat model in that connective tissue cells formed an epineurium-like structure around the perimeter of the cable, vascularization was substantial, and the typical peripheral nerve-like fascicles were located between the periphery of the graft and the Matrigel-rich interior. One difference was that the rostral cord-graft interface was observed within the rostral end of the guidance channel, signifying improved tissue survival after transection in the nude rats; this interface position was similar to that with MP in the Fischer rat even though MP was not administered in these experiments. The mean number of myelinated axons was 2795, contrasting with a mean of 146 in Matrigel-only grafts. Most serotonergic and noradrenergic fibers stopped at the host-graft interface, but some could be detected 3 and 5 mm, respectively, beyond the interface in the graft. Following injection of Fast Blue into the graft midpoint, 803 labelled neurons were detected in the spinal cord from T6 to C3 in a representative animal. In addition to propriospinal neuronal labelling, Fast Blue-labelled neurons were detected in the brainstem, predominantly in reticular, raphe, subcoeruleus, and vestibular nuclei. A mean of 113 brainstem neurons was found in these nuclei. In conclusion, human Schwann cells not only survived in this transplantation paradigm, but also supported the regeneration of both injured spinal cord and brainstem neurons. Similar observations were made when nude rat rather than human Schwann cells were transplanted into nude rat spinal cord. Brainstem labelling following transplantation of human or nude rat Schwann cells was different from the Fischer rat model in which there was no brainstem labelling without neurotrophin or MP administration. The amount of surviving spinal cord tissue near the interface with the graft of human Schwann cells may account, in part, for this improvement and also for the higher numbers of myelinated axons observed in the grafts and of propriospinal neurons that regrew axons into the grafts indicated by retrograde labelling.

DISCUSSION

The combination of bridging Schwann cell-containing grafts with neurotrophic factors and neuroprotective agents appears highly promising to achieve axonal regrowth across areas of mammalian spinal cord injury and also into areas beyond the injury. The combination of Schwann cell grafting with one or two neurotrophins or a neuroprotective compound improved regrowth into the prosthesis that spanned a gap in the cord, as described above. Our hypothesis that a combination of cells and factors will eventually lead to adequate regeneration to achieve functional recovery is being validated by work in other laboratories. When NGF is infused into peripheral nerve grafts inserted into adult rat spinal cord dorsal columns, more spinal cord neurons regenerate axons into the grafts and some of these neurons are farther from the graft site than when NGF is omitted (Fernandez et al., 1990). Ye and Houle (1995) observed that combining neurotrophic factor addition with cervically positioned peripheral nerve transplants increases the regenerative response of specific injured supraspinal neurons. The combination of peripheral nerve grafts and infusions of neurotrophins into the cord beyond the graft may be particularly useful to induce regenerating fibers to leave the graft to extend beyond the injury. Oudega and Hagg (1996) observed that infusing NGF into spinal cord beyond a peripheral nerve graft increased the number of regenerating sensory fibers that entered the cord beyond the graft.

The most successful combination of steps tested in a bridging paradigm so far was very recently reported by Cheng et al. (1996). They combined thin, multiple peripheral nerve implants and αFGF in fibrin glue to achieve not only fiber growth across a 5 mm thoracic cord gap following complete transection but also reported an improvement in hindlimb functional deficits observable 1–3 months after transplantation. These investigators also placed multiple grafts precisely to appose certain tracts, and the thoracic spinal column was rigidly fixed in dorsiflexion. Retrograde tracing studies showed transport from lumbar levels 25 mm below the graft to neurons at many levels of the neuraxis above the transection, including pyramidal nerve cell bodies. This result is novel in a number of ways. For example, corticospinal neurons have not been shown to respond to peripheral nerve (Richardson et al., 1984; Houle, 1991) or SC (Xu et al., 1995a) grafts used singly and placed in thoracic cord regions. Regenerative growth occurred not only across the graft-distal cord interface but also for a considerable length beyond (25 mm) in host spinal cord. Whereas axonal regeneration can be substantial, previously observed growth from grafts into the CNS has been limited to a few millimeters (Aguayo, 1985). It will be of interest to detect molecular changes in the interface area and beyond in the Cheng et al. (1996) paradigm to gain clues about permissive vs. inhibitory environments. Because of these recent observations, the goal of achieving return of function after spinal cord injury by utilizing bridging approaches appears very promising.

SUMMARY

We have reviewed in this chapter the results of transplantation experiments using a transection paradigm. When Schwann cells inside polymer guidance channels were positioned in a gap in the adult rat spinal cord, propriospinal and sensory neurons but neither brainstem nor cortical neurons responded by extending axons into the transplant. When BDNF and NT-3 were added to SC grafts or methylprednisolone was administered at the time of transplantation, myelinated axons in grafts increased in number, propriospinal neuron response improved, monoaminergic fibers grew into the grafts, and brainstem neurons responded by extending axons into the graft. Other factors, IGF-1 and PDGF, added to Schwann cell grafts led to a diminution in axonal ingrowth though myelin formation was enhanced. When Schwann cell grafts were inside open-ended channels, they effectively bridged the gap; axons regenerated into the graft from both stumps. Schwann cells genetically modified to secrete BDNF improved axonal regrowth across a transection site and growth along the cord distal to the injury. Finally, human Schwann cells were found to be as effective as rat Schwann cells in the guidance channel/transection paradigm. The approach of combining Schwann cell-containing grafts with neurotrophic and neuroprotective agents appears highly promising in the quest to improve axonal regeneration following spinal cord injury.

ACKNOWLEDGMENTS

Experimentation in the Bunge laboratory reviewed here was performed by Drs. X.M. Xu, A. Chen, V. Guénard, M. Oudega, P. Menei, C. Montero-Menei, and J. Guest. Mentorship was also provided by Drs. R.P. Bunge and S.R. Whittemore. Valuable laboratory support was provided by A. Gomez, M. Bates, J.-P. Brunschwig, E. Cuervo, A. Weber, C. Vargas, A. Rao., and C. Akong. These studies would not have been possible without the generosity of Dr. P. Aebischer (Univérsité de Lausanne) who provided the guidance channels, and the Transplant Procurement Team, University of Miami School of Medicine, which provided human peripheral nerve. We also thank D. Santiago for help with animal care, R. Camarena for photographic assistance, Dr. J. Klose for help with statistical analysis, and C. Rowlette for word processing. Drs. R. Lindsay and A. Acheson (Regeneron Pharmaceuticals, Inc.) gave us the neurotrophins and advice about dosage; the Institute for Molecular Biology donated IGF-1 and PDGF; Dr. I. Dickerson (Department of Physiology and Biophysics) provided CGRP antibody, and Dr. X. Breakefield (Harvard Medical School) sent us the Y2BDNF 1B producer line. This work was supported by NIH grants NS28059 and NS09923, The Miami Project, and the Hollfelder, Rudin and Heumann foundations. Drs. Xu and Oudega were Daniel Heumann International Scholars. Dr. Guest was a Fellow of the American Association of Neurological Surgeons. Drs. Menei and Montero-Menei were funded by IRME to work in the Bunge laboratory.

REFERENCES

Aguayo AJ (1985): Axonal regeneration from injured neurons in the adult mammalian central nervous system. In C.W. Cotman (ed) *Synaptic Plasticity*, The Guilford Press: New York. pp. 457–484.

Bracken MB, Shepard MJ, Collins WF, Holford TR, Young W, Baskin DS, Eisenberg HM, Flamm E, Leo-Summers L, Maroon J, Marshall LF, Perot PL, Piepmeier J, Sonntag VKH, Wagner FC, Wilberger JE, Winn HR (1990): A randomized, controlled trial of methylprednisolone or naloxone in the treatment of acute spinal cord injury. Results of the second National Acute Spinal Cord Injury Study. *N Engl J Med* 322:1405–1411.

Bracken MB, Shepard MJ, Collins WF, Jr., Holford TR, Baskin DS, Eisenberg HM, Flamm E, Leo-Summers L, Maroon JC, Marshall LF, Perot PL, Piepmeier J, Sonntag VKH, Wagner FC, Wilberger JL, Winn HR, Young W (1992): Methylprednisolone or naxolone treatment after acute spinal cord injury: 1-year follow-up data: Results of the second National Acute Spinal Cord Injury Study. *J Neurosurg* 76:23–31.

Bray GM, Villegas-Peréz MP, Vidal-Sanz M, Carter DA, Aguayo AJ (1991): Neuronal and nonneuronal influences on retinal ganglion cell survival, axonal regrowth, and connectivity after axotomy. *Ann New York Acad Sci* 633:214–228.

Bunge RP (1975): Changing uses of nerve tissue culture 1950-1975. In: Tower DB (ed) *The Nervous System, The Basic Neurosciences*. New York: Raven Press, Ltd, Vol 1, pp. 31–42.

Bunge RP, Puckett WR, Becerra JL, Marcillo A, Quencer RM (1993): Observations on the pathology of human spinal cord injury. A review and classification of 22 new cases with details from a case of chronic cord compression with extensive focal demyelination. IN: Seil FJ (ed) *Neural Regeneration. Advances in Neurology, Vol 59*. Raven Press, Ltd:New York, pp 75–89.

Bunge RP, Puckett WR, Hiester ED (1997): Observations on the pathology of several types of human spinal cord injury, with emphasis on the astrocyte response in penetrating injuries. IN: Seil FJ (ed) *Neuronal Regeneration, Reorganization and Repair. Advances in Neurology, Vol 72.* Raven Press, Ltd:New York pp. 305–315.

Casella G, Bunge RP, Wood PM (1996): Improved method for harvesting human Schwann cells from mature peripheral nerve and expansion in vitro. *Glia* 17: 327–338.

Chen A, Xu XM, Kleitman N, Bunge MB (1994): Methylprednisolone administration improves axonal regeneration into Schwann cell (SC) grafts in thoracic rat spinal cord. *Soc Neurosci Abstr* **20**:1111.

Chen A, Xu XM, Kleitman N, Bunge MB (1996): Methylprednisolone administration improves axonal regeneration into Schwann cell grafts in transected adult rat thoracic spinal cord. *Exp Neurol* **138**:261–276.

Cheng H, Cao Y, Olson L (1996): Spinal cord repair in adult paraplegic rats: Partial restoration of hind limb function. *Science* **273**:510–513.

Fernandez E, Pallini R, Mercanti D (1990): Effects of topically administered nerve growth factor on axonal regeneration in peripheral nerve autografts implanted in the spinal cord of rats. *Neurosurg* **26**:37–42.

Guénard V, Xu XM, Bunge MB (1993): The use of Schwann cell transplantation to foster central nervous system repair. *Sem Neurosci* **5**:401–11.

Guest JD, Bunge RP (1994): Human Schwann cells can enhance axonal regeneration and myelination in the nude rat spinal cord. *Soc Neurosci Abstr* **20**:1111.

Guest JD, Bunge RP (1995): Functional studies of human Schwann cells transplanted to the nude rat spinal cord. *J Neurotrauma* **12**:427.

Guest JD, Kleitman N, Aebischer P, Bunge MB, Bunge RP (1996): Axonal regeneration into human Schwann cell grafts placed to span the transected spinal cord of the nude rat. *Soc Neurosci Abstr* **22**: 1231.

Houle JD (1991): Demonstration of the potential for chronically injured neurons to regenerate axons into intraspinal peripheral nerve grafts. *Exp Neurol* **113**:1–9.

Houle JD, Wright JW, Ziegler MK (1994): After spinal cord injury, chronically injured neurons retain the potential for axonal regeneration. IN: Marwah, J., Teitelbaum H, Prasad KN (eds) *Neural Transplantation, CNS Neuronal Injury, and Regeneration.* Boca Raton:CRC Press, pp 103–118 .

Kromer LF, Cornbrooks CJ (1987): Identification of trophic factors and transplanted cellular environments that promote CNS axonal regeneration. *Ann New York Acad Sci* **495**:207–223.

Levi ADO, Bunge RP (1994): Studies of myelin formation after transplantation of human Schwann cells into the severe combined immunodeficient mouse. *Exp Neurol* **130**:41–52.

Levi ADO, Guénard V, Aebischer P, Bunge RP (1994): The functional characteristics of Schwann cells cultured from human peripheral nerve after transplantation into a gap within the rat sciatic nerve. *J Neurosci* **14**:1309–1319.

Levi ADO, Bunge RP, Lofgren JA, Meima L, Hefti F, Nikolics K, Sliwkowski MX (1995): The influence of heregulins on human Schwann cell proliferation. *J Neurosci* **15**:1329–1340.

Menei P, Montero-Menei C, Whittemore SR, Bunge MB, Bunge RP (1996): Schwann cells genetically engineered to produce BDNF promote axonal regeneration of brainstem neurons across transected adult rat spinal cord. *Soc Neurosci Abstr* **22**:1022.

Montgomery CT, Tenaglia EA, Robson JA (1996): Axonal growth into tubes implanted within lesions in the spinal cords of adult rats. *Exp Neurol* **137**:277–290.

Morrissey TK, Kleitman N, Bunge RP (1991): Isolation and functional characterization of Schwann cells derived from adult peripheral nerve. *J Neurosci* **11**:2433–42.

Morrissey TK, Levi ADO, Neuijens AA, Sliwkowski MX, Bunge RP (1995): Axon-Induced mitogenesis of human Schwann cells involves heregulin and p185^{erbB2}. *Proc Nat'l Acad Sci USA* **92**:1431–1435.

Oudega M and Hagg T (1996): Nerve growth factor promotes regeneration of sensory axons into adult rat spinal cord. *Exp Neurol* **140**: 218–229

Oudega M, Xu XM, Guénard V, Kleitman N, Bunge MB (1997): A combination of insulin-like growth factor-1 and platelet-derived growth factor enhances myelination but diminishes axonal regeneration into Schwann cell grafts in the adult rat spinal cord. *Glia* **19**:247–258.

Paíno CL, Bunge MB (1991): Induction of axon growth into Schwann cell implants grafted into lesioned adult rat spinal cord. *Exp Neurol* **114**:254–257.

Paíno CL, Fernandez-Valle C, Bates ML, Bunge MB (1994): Regrowth of axons in lesioned adult rat spinal cord: promotion by implants of cultured Schwann cells. *J Neurocytol* **23**:433–52.

Richardson PM, Issa VMK, Aguayo AJ (1984) Regeneration of long spinal axons in the rat. *J Neurocytol* **13**:165–182.

Rutkowski JL, Kirk CJ, Lerner MA, Tennekoon GI (1995): Purification and expansion of human Schwann cells *in vitro. Nature Med* **1**:80–83.

Schnell L, Schwab ME (1993): Sprouting and regeneration of lesioned corticospinal tract fibres in the adult rat spinal cord. *Eur J Neurosci* **5**:1156–71.

Schnell L, Schneider R, Kolbeck R, Barde Y-A, Schwab M (1994) Neurotrophin-3 enhances sprouting of corti-cospinal tract during development and after adult spinal cord lesion. *Nature* **367**:170–173.

Xu XM, Guénard V, Kleitman N, Bunge MB (1995a): Axonal regeneration into Schwann cell-seeded guidance channels grafted into transected adult rat spinal cord. *J Comp Neurol* **351**:145–60.

Xu XM, Guénard V, Kleitman N, Aebischer P, Bunge MB (1995b): A combination of BDNF and NT-3 promotes supraspinal axonal regeneration into Schwann cell grafts in adult rat thoracic spinal cord. *Exp Neurol* **134**:261–72.

Xu XM, Chen A, Guénard V, Kleitman N, Bunge MB (1997): Bridging Schwann cell transplants promote axonal regeneration from both the rostral and caudal stumps of transected adult rat spinal cord. *J Neurocytol*. In press.

Ye J-H, Houle JD (1995): Trophic factor enhanced regeneration by chronically injured supraspinal neurons following cervical spinal cord injury. *Soc Neurosci Abstr* **21**:1056.

32

OLFACTORY ENSHEATHING CELLS CAN BE INDUCED TO EXPRESS A MYELINATING PHENOTYPE

R. Doucette

Department of Anatomy and Cell Biology
College of Medicine
University of Saskatchewan
107 Wiggins Road
Saskatoon, Saskatchewan
Canada S7N 5E5

INTRODUCTION

Considerable progress has been made in developing therapeutic approaches to promoting axonal growth and remyelination in the adult mammalian CNS. It has proved possible, for example, to manipulate the cellular environment of lesion cavities or demyelinated areas of the CNS so as to promote axon regeneration and/or remyelination. Schwann cells are good promoters of axonal growth, and under certain conditions even astrocytes have been shown to possess growth-promoting properties (Ard and Bunge 1988; Ard et al. 1987; Bahr and Bunge 1989; Hatten et al. 1991; Kawaja and Gage 1991; Neugebauer et al. 1988; Silver 1988; Smith et al. 1986). Both Schwann cells and oligodendrocyte progenitor cells remyelinate axons when transplanted into the adult mammalian CNS (Duncan 1996). However, Schwann cells fail to become integrated into the CNS and astrocytes somehow perturb the neuron-Schwann cell interactions that lead to myelination (Franklin et al. 1992; Guenard et al. 1994).

More recently another glial cell type has attracted interest as a possible candidate for CNS transplantation to promote axon regeneration and remyelination. This cell type is the olfactory ensheathing cell, which supports the growth of regenerating dorsal root (Ramon-Cueto and Nieto-Sampedro 1994) and fimbria/fornix (Smale et al. 1996) axons, and can assemble a myelin sheath around axons of an appropriate diameter (Devon and Doucette 1992, 1995; Franklin et al. 1996). It was hypothesized that these cells are a more attractive candidate than either Schwann cells or oligodendrocytes because ensheathing cells can readily switch their phenotype from that resembling an astrocyte to one more like that of a myelinating Schwann cell (Devon and Doucette 1992; Devon and Doucette 1995).

Cell Biology and Pathology of Myelin, edited by Juurlink *et al.*
Plenum Press, New York, 1997

Ensheathing cells provide ensheathment for primary olfactory axons along both the PNS and CNS portions of the primary olfactory pathway (Doucette 1990, 1993a; Raisman 1985; Ramon-Cueto and Valverde 1995). They possess a highly malleable phenotype, express a mixture of astrocyte-specific and Schwann cell-specific phenotypic features, and perform the roles of both astrocytes and Schwann cells (Doucette 1990, 1993a; Doucette and Devon 1993). It is the malleable phenotype of ensheathing cells that enables them to adopt several different roles as the need arises (Doucette 1995; Doucette and Devon 1993). There has been much debate as to whether ensheathing cells belong to the astrocyte or the Schwann cell family (Barber and Lindsay 1982; Doucette 1993b; Norgren et al. 1992; Pixley 1992; Ramon-Cueto and Nieto-Sampedro 1992). It has even been suggested that they are a novel glial cell type (Barnett et al. 1993; Doucette and Devon 1993; Ramon-Cueto and Valverde 1995; Wozniak 1993), which possess the ability to become more astrocyte-like or more Schwann cell-like as the need arises. It is highly likely that ensheathing cells could perform the roles of astrocytes and Schwann cells when transplanted into a CNS lesion cavity, due to their possessing such a highly malleable phenotype (Doucette and Devon 1993). In their experiments dealing with the x-irradiated/ethidium bromide-treated dorsal funiculus of the adult rat spinal cord, Franklin et al. (1996) found that a clonal olfactory ensheathing cell line differentiated not only into cells resembling myelinating Schwann cells but also into ones resembling astrocytes.

EXPRESSION OF A NON-MYELINATING PHENOTYPE *IN VIVO*

In vivo, ensheathing cells express a non-myelinating phenotype, providing ensheathment for small C fibers (i.e., olfactory axons). In addition to providing ensheathment for olfactory axons, ensheathing cells also contribute to the formation of the glia limitans of the olfactory bulb (Barber and Lindsay 1982; Berger 1971; Doucette 1984, 1990, 1993a; Valverde and Lopez-Mascaraque 1991), a job that elsewhere in the mammalian CNS is the exclusive domain of astrocytes (Peters et al. 1990). In fact, during embryonic and fetal stages of development, ensheathing cells are solely responsible for forming a glia limitans along the periphery of the developing bulb, with astrocytes of the olfactory bulb not contributing to this function until sometime later in development (Doucette 1989, 1993a).

An additional phenotypic feature ensheathing cells share in common with astrocytes is the expression of central type glial fibrillary acidic protein (GFAP) (Barber and Dahl 1987) and both glial cell types also express the connexin-43 (Cx43) gap junctional protein (Miragall et al. 1992). Within the nerve fiber layer (NFL) of the olfactory bulb, the plasma membranes of ensheathing cells and astrocytes frequently come into close contact with no intervening basal lamina (Doucette 1984, 1990, 1993a; Valverde and Lopez-Mascaraque 1991). At the electron microscopic level, Scotti et al. (1994) noted the presence of gap junctions between astrocytes and ensheathing cells in the olfactory bulb of adult gerbils. Cx43 is most likely the connexin protein used to form these gap junctions, which allow ensheathing cells and astrocytes to communicate with one another along the CNS portion of the primary olfactory pathway.

The full range of phenotypic features expressed by non-myelinating ensheathing cells during development, as well as in adult animals, is only now becoming apparent. Several studies have revealed the phenotype of ensheathing cells at different stages of development, providing an initial glimpse of the changes that occur in these cells as they differentiate into mature non-myelinating cells. For example, it has been learned that ensheathing cells are heterogeneous in terms of the cellular phenotype they express, even as early as late fetal stages of development (Franceschini and Barnett 1996; Gong et al. 1994; Miragall et al. 1994).

Differentiation into Mature Non-Myelinating Cells: Prenatal Development

The progenitor cells that differentiate into ensheathing cells are derived from the olfactory placode, eventually giving rise to glial cells in both the PNS and CNS portions of the primary olfactory pathway. Soon after the progenitor cells enter the mesenchyme separating the placode from the telencephalic vesicle they express the cell adhesion molecules N-CAM and L1/Ng-CAM (Miragall et al. 1989). From the very beginning, these progenitor cells co-express both the embryonic and adult isoforms of N-CAM, which are known to differ in their polysialic acid content. The p75 neurotrophin receptor is a third molecule that is known to be expressed by ensheathing cell progenitors shortly after they migrate out of the placode to accompany the olfactory axons (Gong et al. 1994).

These earliest appearing ensheathing cell progenitors also express a nervous system specific fatty acid binding protein (Kurtz et al. 1994), which has been appropriately named brain-fatty acid binding protein (B-FABP). FABPs belong to a multigene family of small intracellular proteins, which function to bind hydrophobic ligands such as fatty acids, eicosanoids and retinoids. With respect to the retinoids, it has been suggested that the protein-ligand complexes are a direct substrate in retinoid conversion. Kurtz et al. (1994) suggested ensheathing cells may utilize B-FABP in the metabolism of retinoids, which LaMantia and colleagues (LaMantia et al. 1993) have implicated in the development of the primary olfactory system.

Upon reaching the rostral wall of the cerebral vesicle, the ensheathing cell progenitors gradually become incorporated into the CNS, thus contributing to the formation of the NFL of the olfactory bulb (Doucette 1989; Marin-Padilla and Amieva 1989). During these stages of embryonic development, ensheathing cell progenitors continue to express B-FABP (Kurtz et al. 1994), p75 neurotrophin receptor (Gong et al. 1994; Yan and Johnson 1988), and the cell adhesion molecules N-CAM and L1/Ng-CAM (Miragall et al. 1989). An additional molecule known to be expressed by ensheathing cells at this time is the L14 lectin (Mahanthappa et al. 1994), which binds to molecules containing polylactosamine chains (Cooper et al. 1991; Zhou and Cummings 1993). There are at least two ligands for L14 in the primary olfactory system, namely a lactosamine-containing glycolipid that is present on immature olfactory axons and the extracellular matrix molecule laminin (Mahanthappa et al. 1994). It is not known whether the progenitor cells express L14 at earlier stages of development.

Sometime after they become incorporated into the CNS one sees the first evidence of heterogeneity in the cellular phenotype of ensheathing cells. Gong et al. (1994) described a distinct p75-negative band of cells that occupied the deepest portion of the NFL. These p75-negative cells presumably are ensheathing cells, since no other glial elements are present in the NFL prenatally (Doucette 1989, 1993a). Ensheathing cells also begin to express the tight junctional protein ZO-1 as early as embryonic day 18 (E18) in mouse embryos (Miragall et al. 1994). However, ZO-1 is only expressed by ensheathing cells residing within the PNS portion of the primary olfactory pathway, never by those within the NFL of the olfactory bulb.

Differentiation into Mature Non-Myelinating Cells: Postnatal Development

After birth (in rats and mice) astroblasts migrate from deeper layers of the developing bulb, eventually differentiating into the interfascicular astrocytes of the NFL (Doucette 1989; Marin-Padilla and Amieva 1989). The glial progeny of these astroblasts participate in the formation of the glia limitans of the olfactory bulb, but do not assume either the role of ensheathing olfactory axons within the CNS or of forming the glia limitans at the PNS-CNS transitional zone. These latter roles continue to be performed by the progeny of ensheathing cell progenitors (Doucette 1989; Marin-Padilla and Amieva 1989).

Postnatally (in rats and mice), and continuing into adulthood, ensheathing cells continue to express both L1/Ng-CAM and the adult form of N-CAM, but eventually no longer express the embryonic form of N-CAM (Miragall et al. 1989). This is not to say that the polysialic acid rich isoform of N-CAM disappears totally from the NFL. Rather, its expression becomes restricted to some of the glial cells that form the glia limitans (Miragall et al. 1988, 1989). It remains to be determined whether the embryonic N-CAM expressing cells that contribute to the glia limitans of the bulb in adult mice are ensheathing cells or astrocytes. The GFAP intermediate filament protein also begins to be expressed by ensheathing cells during postnatal development.

During early postnatal development, the L14 lectin (Mahanthappa et al. 1994) and B-FABP (Kurtz et al. 1994) continue to be expressed by ensheathing cells (Mahanthappa et al. 1994). Using immunofluorescence, Mahanthappa et al. (1994) found that L14 expression colocalized with laminin immunostaining, as well as with 1B2 immunostaining, in the primary olfactory pathway of P2 and P14 rat pups. 1B2 is a monoclonal antibody that recognizes terminal lactosamine epitopes on molecules. There continues to be a heterogeneous expression of the ZO-1 tight junctional protein (Miragall et al. 1994) and of the p75 neurotrophin receptor (Gong et al. 1994; Vickland et al. 1991) by non-myelinating ensheathing cells during early postnatal development. All ensheathing cells eventually downregulate p75 expression (Gong et al. 1994; Vickland et al. 1991), most likely as a result of glial-neuronal interactions, as has been found for Schwann cells in peripheral nerves. Ramon-Cueto et al. (1993) co-cultured olfactory ensheathing cells from the olfactory bulb of adult rats with explants of olfactory mucosa from embryonic rats. Those glial cells that ensheathed olfactory neurites down-regulated their expression of p75. Furthermore, this down-regulated expression occurred only along the part of the cell membrane that was directly apposed to the neurite (Ramon-Cueto et al. 1993).

Miragall et al. (1992) studied the expression of connexin (Cx) 26, 32 and 43 in the primary olfactory pathway of embryonic, neonatal and adult mice. They used both immunofluorescence and immunoelectron microscopy. At the electron microscopic level, gap junctions were detected between adjacent ensheathing cells in tissues harvested from adult mice. Neither Cx26 nor Cx 32 were expressed by ensheathing cells in any of the animals. However, ensheathing cells began to express Cx43 as early as P0, and continued to do so thereafter into adulthood.

Griffiths et al. (1995a) used in situ hybridization to detect proteolipid protein (PLP)/DM-20 mRNAs and immunofluorescence to detect the respective proteins in the NFL of the olfactory bulb of 5–40 day old mice. The PLP protein was not expressed by any of the glial cells of the NFL. However, the DM-20 protein was, and Griffiths et al. (1995a) suggested it was ensheathing cells that were responsible for its synthesis. The immunostaining was very strong in the P5 and P10 mice, but was considerably reduced in older animals. DM-20 is an isoform of PLP, which itself is a CNS-specific myelin-associated molecule. The DM-20 isoform is encoded for by the PLP gene and is produced by alternative splicing of exon 3B of the PLP gene. It is the predominant isoform of the gene that is expressed by non-myelin forming glial cells of the PNS at all ages and by oligodendrocyte progenitor cells in the CNS during embryonic development (Griffiths et al. 1995b). Griffiths et al. (1995a) concluded that since ensheathing cells are non-myelinating glial cells, the putative role of DM-20 must not be related to myelination per se but rather be concerned with some aspect of glial cell development.

Scotti et al. (1994) used immunoelectron microscopy to examine the cellular localization of protease nexin-1 (PN-1; also referred to as glia-derived nexin; see (Knauer et al. 1987)) in the nerve fiber layer of the olfactory bulb of adult gerbils. This protease is a 43 Kd cell-secreted slow-binding inhibitor of trypsin, urokinase, thrombin, and plasminogen activator (Guenther et al. 1985; Stone et al. 1987). In addition to being a serine protease inhibitor, thus modulating the degradation of some components of the extracellular matrix, glia-derived nexin is also a neurite-promoting factor for mouse neuroblastoma cells (Guenther et al. 1985; Monard et al. 1973) and for chick sympathetic neurons growing in primary culture (Zurn et al. 1988). It had previously been shown that PN-1 was synthesized by cells (presumably ensheathing cells) along the length of the primary olfactory pathway (Reinhard et al. 1988). Doucette (1990) suggested it was ensheathing cells that were responsible for its synthesis, and that it was an ideal molecular candidate for a chemotropic factor in the primary olfactory pathway. Subsequently, Scotti et al. (1994) confirmed that the PN-1 protein was expressed by glial cells of the nerve fiber layer of the bulb, but they found its expression was not restricted to ensheathing cells. The astrocytes in this layer of the bulb, as well as in the adjacent part of the glomerular layer, also expressed the protein. It was suggested that this restricted expression of PN-1 by astrocytes was due to the well-documented neuronal turnover and continual neurogenesis that is occurring within the olfactory epithelium. In other words, PN-1 continues to be synthesized by ensheathing cells and astrocytes in the adult olfactory bulb because this protease is needed to promote the ingrowth of olfactory axons into and within the CNS.

EXPRESSION OF A MYELINATING PHENOTYPE *IN VITRO*

Using tissue culture (Devon and Doucette 1992, 1995) and transplantation (Franklin et al. 1996) paradigms, it has been shown that ensheathing cells can be induced to express a myelinating phenotype, which is very different than the phenotype they normally express in vivo. It appears that the expression of a myelinating phenotype by ensheathing cells is heavily dependent on axonal contact (Barnett et al. 1993; Devon and Doucette, 1995; Doucette and Devon 1995), a requirement that also applies to Schwann cells (Brunden and Brown 1990; Brunden et al. 1990). The connexin 32 (Cx32) gap junctional protein is believed to play an important role in the biology of myelin-forming cells (Scherer et al. 1995), but Miragall et al. (1992) found that ensheathing cells in vivo expressed Cx43 but not Cx32. According to Scherer et al. (1995), Cx32 is the only connexin expressed by Schwann cells, which may explain the latter cell's susceptibility to mutations in the Cx32 gene. The X-linked form of Charcot-Marie-Tooth disease is caused by a mutation in the CX32 gene (Bergoffen et al., 1993), which presumably impedes the passage of small molecules through gap junctions in the Schmidt-Lanterman incisures and paranodal regions of a Schwann cell myelin sheath, ultimately resulting in a demyelinating neuropathy. Whether ensheathing cells switch to expressing Cx32 instead of, or in addition to, Cx43 as they begin to assemble a myelin sheath remains to be determined.

DRG-Glial Cell Co-Cultures

When co-cultured with dorsal root ganglion (DRG) neurons ensheathing cells form contiguous series of cells along the neurites, providing ensheathment for either a myelinated neurite or for several unmyelinated neurites but not both (Devon and Doucette 1992, 1995). These myelinating ensheathing cells assemble peripheral type myelin around those neurites large enough to be myelinated. Franklin et al. (1996) have shown that ensheathing cells will also myelinate axons in vivo after transplantation into a demyelinated area of the adult rat spinal cord, assembling peripheral type myelin in the process. Ensheathment of DRG neurites stimulates Schwann cells to assemble a basal lamina along their abaxonal surfaces (Bunge et al. 1989; Clark and Bunge 1989) and the presence of such a basal lamina covering is a necessary prerequisite for Schwann cells to form a myelin sheath (Eldridge et al. 1989). Ensheathment and myelination of neurites by ensheathing cells also appear to stimulate these glial cells to assemble a basal lamina (Devon and Doucette 1992, 1995; Franklin et al. 1996).

Must ensheathing cells, like Schwann cells, be apposed to a basal lamina in order to myelinate neurites? In a recent experiment that attempted to address this question, ensheathing cells were co-cultured with DRG neurons in the presence or absence of ascorbic acid (Devon and Doucette 1995). Ascorbic acid functions as a cofactor for both prolyl and lysyl hydroxylase, two enzymes that hydroxylate procollagen molecules so they can be assembled into a triple helical collagen molecule (Prockop et al. 1979). Collagen type IV is a major component of the basal lamina and may form a scaffolding onto which other molecules are added during basal lamina assembly. Ensheathing cells, unlike Schwann cells, can myelinate DRG neurites regardless of whether ascorbic acid is included in the growth medium and these glial cells can assemble a basal lamina in the absence of added ascorbic acid (Devon and Doucette 1995). It appears that Schwann cells and ensheathing cells have different growth media requirements for the assembly of a basal lamina and a myelin sheath.

Neuron-Free Cultures

Ensheathing cells express a nonmyelinating phenotype when grown in neuron-free cultures, even when examined within the first few days after plating (Barnett et al. 1993; Doucette and Devon 1994, 1995). This is the expected phenotype because ensheathing cells provide ensheathment for unmyelinated axons in vivo and thus would not have expressed any myelin associated molecules at the time of plating. Like Schwann cells, the full expression of a myelinating phenotype by ensheathing cells is heavily dependent on their making contact with appropriately sized axons. In contrast to these findings, Ramon-Cueto and Nieto-Sampedro (1992) isolated a population of MBP+ve/GFAP+ve cells from the adult rat olfactory bulb that they believed to be ensheathing cells. What is most puzzling about these cells is why they would be MBP+ve in vitro if they had not been myelinating axons in vivo.

Axons also modulate the expression of a myelinating phenotype by oligodendrocytes. However, these glial cells can synthesize significant levels of myelin associated molecules in the absence of axonal contact, provided the growth medium contains insulin, hydrocortisone or triiodothyronine (Lopes-Cardozo et al. 1989; McMorris et al. 1986,

1990; Mozell and McMorris 1991; Poduslo et al. 1990; Ved et al. 1989). Hydrocortisone and triiodothyronine most likely regulate the oligodendrocyte expression of myelin basic protein (MBP) at the level of gene transcription (Farsetti et al. 1991; Vielkind et al. 1990). In a recent study using neuron-free cultures of ensheathing cells, growth media that supported constitutive expression of myelin molecules by oligodendrocytes failed to induce ensheathing cells to express either galactocerebroside (GAL-C) or MBP (Doucette and Devon 1994). Barnett et al. (1993) obtained similar results when they fed cell cultures of O4+ve ensheathing cells with a serum-free medium containing insulin and triiodothyronine, as well as thyroxine.

These results point towards a different regulatory control over the promoter region of the MBP gene in oligodendrocytes and ensheathing cells. Unlike oligodendrocytes, the promoter region of the MBP gene in ensheathing cells does not appear to contain a functional thyroid hormone response element. Furthermore, if ensheathing cells do express glucocorticoid receptors then they must play a different role in these glial cells than they do in oligodendrocytes. Studies with transgenic mice have shown that the promoter region of the MBP gene is also controlled differently in oligodendrocytes and Schwann cells. Gow et al. (1992) used a 1.9 Kb sequence of the MBP promoter to generate transgenic mice in which the Lac Z reporter gene was expressed correctly by oligodendrocytes, but in which the gene was not expressed by Schwann cells. Foran and Peterson (1992) obtained a similar result using a 3.2 Kb sequence of the MBP promoter. Gow et al. (1992) suggested that this promoter may contain unidentified oligodendrocyte-specific enhancer elements, as well as additional components that either expand the specificity to include Schwann cells or that function independently as Schwann cell-specific enhancer elements. Doucette and Devon (1994) hypothesized that these promoter constructs would also fail to direct ensheathing cells taken from these transgenic mice to express the Lac Z gene even if the cells were co-cultured with DRG neurons.

A more effective inducer of GAL-C expression by ensheathing cells in neuron-free cultures is to increase the intracellular level of cAMP (Doucette and Devon 1994, 1995). However, treatment with cAMP analogues failed to induce ensheathing cells to express MBP (Doucette and Devon 1995). In contrast to our findings, Barnett et al. (1993) were unable to induce O4+ve ensheathing cells to express GAL-C in cultures that were treated with dBcAMP; these cells also failed to express the Po myelin protein. In both oligodendrocytes and Schwann cells, elevated cAMP levels only indirectly affect the expression of the MBP gene. This gene belongs to the group of genes that respond very slowly to elevated levels of cAMP (Roesler et al. 1988), with increased expression of this gene not occurring for at least several hours (Monuki et al. 1989; Zhang and Miskimins 1993). The reason for the delayed response to cAMP elevation is most likely due to the absence of a cAMP response element (i.e., CRE) in the promoter region of the MBP gene (Zhang and Miskimins 1993).

CONCLUDING STATEMENTS

What factors control the expression of a myelinating phenotype by ensheathing cells? The main one seems to be ensheathment of axons of an appropriate diameter, which is thus far the only identified requirement that ensheathing cells have in common with Schwann cells. Thus, even though in assembling a myelin sheath ensheathing cells assemble peripheral type myelin (Devon and Doucette 1992, 1995), this assembly is under a different regulatory control than it is in either Schwann cells or oligodendrocytes. Unlike

Schwann cells, cAMP analogues did not induce ensheathing cells to express MBP (Doucette and Devon 1995), nor did ensheathing cells require the addition of L-ascorbic acid to the medium for the cells to assemble a myelin sheath or a basal lamina covering (Devon and Doucette 1995). Even growth media that are known to promote the expression of a myelinating phenotype by oligodendrocytes (Doucette and Devon 1994) or the partial re-expression of such a phenotype by Schwann cells (Doucette and Devon 1995) were ineffective on ensheathing cells. Clearly, this is an area that is ripe for further investigation.

ACKNOWLEDGMENTS

The research referred to in this article was supported (in part) by grants from the Medical Research Council of Canada and by research grant number 5R01 DC 02370–02 from the National Institute on Deafness and Other Communicative Disorders, National Institutes of Health. I thank Mr. Lyndon Osman, Ms. Kelly Gratto and Ms. Cindy Farrar for their technical assistance.

REFERENCES

Ard, MD and Bunge, RP (1988): Heparin sulfate proteoglycan and laminin immunoreactivity on cultured astrocytes: Relationship to differentiation and neurite growth. J Neurosci 8: 2844–2858.

Ard, MD, Bunge, RP and Bunge, MB (1987): Comparison of the Schwann cell surface and Schwann cell extracellular matrix as promoters of neurite growth. J Neurocyt 16: 539–555.

Bahr, M and Bunge, RP (1989): Functional status influences the ability of Schwann cells to support adult rat retinal ganglion cell survival and axonal regrowth. Exper Neurol 106: 27–40.

Barber, PC and Dahl, D (1987): Glial fibrillary acidic protein (GFAP)-like immunoreactivity in normal and transected rat olfactory nerve. Exper Brain Res 65: 681–685.

Barber, PC and Lindsay, RM (1982): Schwann cells of the olfactory nerves contain glial fibrillary acidic protein and resemble astrocytes. Neurosci 7: 3077–3090.

Barnett, SC, Hutchins, AM and Noble, M (1993): Purification of olfactory nerve ensheathing cells from the olfactory bulb. Dev Biol 155(2): 337–350.

Berger, B (1971): Etude ultrastructurale de la degenerescence Wallerienne experimentale d'un nerf entierement amyelinique: II. Reactions cellulairs. J Ultrastruct Res 37: 479–494.

Brunden, KR and Brown, DT (1990): Po mRNA expression in cultures of Schwann cells and neurons that lack basal lamina and myelin. J Neurosci Res 27(2): 159–168.

Brunden, KR, Windebank, AJ and Poduslo, JF (1990): Role of axons in the regulation of Po biosynthesis by Schwann cells. J Neurosci Res 26(2): 135–143.

Bunge, MB, Bunge, RA, Kleitman, N and Dean, AC (1989): Role of peripheral nerve extracellular matrix in Schwann cell function and in neurite regeneration. Devel Neurosci 11: 348–360.

Clark, MB and Bunge, MB (1989): Cultured Schwann cells assemble normal-appearing basal lamina only when they ensheath axons. Devel Biol 133: 393–404.

Cooper, DN, Massa, SM and Barondes, SH (1991): Endogenous muscle lectin inhibits myoblast adhesion to laminin. J Cell Biol 115: 1437–1448.

Devon, R and Doucette, R (1992): Olfactory ensheathing cells myelinate dorsal root ganglion neurites. Brain Res 589: 175–179.

Devon, R and Doucette, R (1995): Olfactory ensheathing cells do not require L-ascorbic acid in vitro to assemble a basal lamina or to myelinate dorsal root ganglion neurites. Brain Res 688(1–2): 223–229.

Doucette, R (1984): The glial cells in the nerve fiber layer of the rat olfactory bulb. Anat Record 210: 385–391.

Doucette, R (1989): Development of the nerve fiber layer in the olfactory bulb of mouse embryos. J Comp Neurol 285: 514–527.

Doucette, R (1990): Glial influences on axonal growth in the primary olfactory system. Glia 3: 433–449.

Doucette, R (1993a): Glial cells in the nerve fiber layer of the main olfactory bulb of embryonic and adult mammals. Micr Res Techn 24: 113–130.

Doucette, R (1993b): Glial progenitor cells of the nerve fiber layer of the olfactory bulb: Effect of astrocyte growth media. J Neurosci Res 35(3): 274–287.

Doucette, R (1995): Olfactory ensheathing cells: Potential for glial cell transplantation into areas of CNS injury. Histol Histopath 10(2): 503–507.

Doucette, R and Devon, R (1993): Olfactory ensheathing cells: Factors influencing the phenotype of these glial cells. "Biology and Pathology of Astrocyte-Neuron Interactions." S. Fedoroff, B. H. J. Juurlink and R. Doucette. New York, Plenum Press. 3: 117–124.

Doucette, R and Devon, R (1994): Media that support the growth and differentiation of oligodendrocytes do not induce olfactory ensheathing cells to express a myelinating phenotype. Glia 10(4): 296–310.

Doucette, R and Devon, R (1995): Elevated intracellular levels of cAMP induce olfactory ensheathing cells to express GAL-C and GFAP but not MBP. Glia 13(2): 130–140.

Duncan, ID (1996): Glial cell transplantation and remyelination of the central nervous system. Neuropathol Appl Neurobiol 22(2): 87–100.

Eldridge, CF, Bunge, MB and Bunge, RP (1989): Differentiation of axon-related Schwann cells in vitro: II. Control of myelin formation by basal lamina. J Neurosci 9: 625–638.

Farsetti, A, Mitsuhashi, T, Desvergne, B, Robbins, J and Nikodem, VM (1991): Molecular basis of thyroid hormone regulation of myelin basic protein gene expression in rodent brain. J Biol Chem 266(34): 23226–23232.

Foran, DR and Peterson, AC (1992): Myelin acquisition in the central nervous system of the mouse revealed by an MBP-Lac-Z transgene. J Neurosci 12(12): 4890–4897.

Franceschini, IA and Barnett, SC (1996): Low-affinity NGF-receptor and E-N-CAM expression define two types of olfactory nerve ensheathing cells that share a common lineage. Devel Biol 173(1): 327–343.

Franklin, RJ, Crang, AJ and Blakemore, WF (1992): Type 1 astrocytes fail to inhibit Schwann cell remyelination of CNS axons in the absence of cells of the 0–2A lineage. Devel Neurosci 14: 85–92.

Franklin, RJM, Gilson, JM, Franceschini, IA and Barnett, SC (1996): Schwann cell-like myelination following transplantation of an olfactory bulb-ensheathing cell line into areas of demyelination in the adult CNS. Glia 17(3): 217–224.

Gong, QZ, Bailey, MS, Pixley, SK, Ennis, M, Liu, W and Shipley, MT (1994): Localization and regulation of low affinity nerve growth factor receptor expression in the rat olfactory system during development and regeneration. J Comp Neurol 344(3): 336–348.

Gow, A, Friedrich, V and Lazzarini, R (1992): Myelin basic protein gene contains separate enhancers for oligodendrocyte and Schwann cell expression. J Cell Biol 119(3): 605–616.

Griffiths, IR, Dickinson, P and Montague, P (1995a): Expression of the proteolipid protein gene in glial cells of the postnatal peripheral nervous system of rodents. Neuropathol Appl Neurobiol 21(2): 97–110.

Griffiths, IR, Montague, P and Dickinson, P (1995b): The proteolipid protein gene. Neuropathol Appl Neurobiol 21(2): 85–96.

Guenard, V, Gwynn, LA and Wood, PM (1994): Astrocytes inhibit Schwann cell proliferation and myelination of dorsal root ganglion neurons in vitro. J Neurosci 14(5 Part 2): 2980–2992.

Guenther, J, Nick, H and Monard, D (1985): A glia-derived neurite-promoting factor with protease inhibitory activity. EMBO J 4: 1963–1966.

Hatten, ME, Liem, RK, Shelanski, ML and Mason, CA (1991): Astroglia in CNS injury. Glia 4: 233–243.

Kawaja, MD and Gage, FH (1991): Reactive astrocytes are substrates for the growth of adult CNS axons in the presence of elevated levels of nerve growth factor. Neuron 7: 1019–1030.

Knauer, DJ, Orlando, RA and Rosenblatt, D (1987): The glioma cell-derived neurite promoting activity protein is functionally and immunologically related to human protease nexin-1. J Cell Physiol 132: 318–324.

Kurtz, A, Zimmer, A, Schnutgen, F, Bruning, G, Spener, F and Muller, T (1994): The expression pattern of a novel gene encoding brain fatty acid binding protein correlates with neuronal and glial cell development. Develop 120(9): 2637–2649.

LaMantia, AS, Colbert, MC and Linney, E (1993): Retinoic acid induction and regional differentiation prefigure olfactory pathway formation in the mammalian forebrain. Neuron 10(6): 1035–1048.

Lopes-Cardozo, M, Sykes, JE, Van der Pal, RH and Van Golde, LM (1989): Development of oligodendrocytes. Studies of rat glial cells cultured in chemically-defined medium. J Devel Physiol 12: 117–127.

Mahanthappa, NK, Cooper, DNW, Barondes, SH and Schwarting, GA (1994): Rat olfactory neurons can utilize the endogenous lectin, I-14, in a novel adhesion mechanism. Develop 120(6): 1373–1384.

Marin-Padilla, M and Amieva, MR (1989): Early neurogenesis of the mouse olfactory nerve: Golgi and electron microscopic studies. J Comp Neurol 288: 339–352.

McMorris, FA, Furlanetto, RW, Mozell, RL, Carson, MJ and Raible, DW (1990): Regulation of oligodendrocyte development by insulin-like growth factors and cyclic nucleotides. Ann NY Acad Sci 605: 101–109.

McMorris, FA, Smith, TM, DeSalvo, S and Furlanetto, RW (1986): Insulin-like growth factor 1/somatomedin C: A potent inducer of oligodendrocyte development. PNAS 83: 822–826.

Miragall, F, Hwang, TK, Traub, O, Hertzberg, EL and Dermietzel, R (1992): Expression of connexins in the developing olfactory system of the mouse. J Comp Neurol 325: 359–378.

Miragall, F, Kadmon, G, Husmann, M and Schachner, M (1988): Expression of cell adhesion molecules in the olfactory system of the adult mouse: Presence of the embryonic form of N-CAM. Devel Biol 129: 516–531.

Miragall, F, Kadmon, G and Schachner, M (1989): Expression of L1 and N-CAM cell adhesion molecules during development of the mouse olfactory system. Devel Biol 135: 272–286.

Miragall, F, Krause, D, Devries, U and Dermietzel, R (1994): Expression of the tight junction protein zo-1 in the olfactory system - presence of zo-1 on olfactory sensory neurons and glial cells. J Comp Neurol 341(4): 433–448.

Monard, D, Solomon, F, Rentsch, M and Gysin, R (1973): Glia-induced morphological differentiation in neuroblastoma cells. PNAS 70: 1894–1897.

Monuki, ES, Weinmaster, G, Kuhn, R and Lemke, G (1989): SCIP: A glial POU domain gene regulated by cyclic AMP. Neuron 3: 783–793.

Mozell, RL and McMorris, FA (1991): Insulin-like growth factor 1 stimulates oligodendrocyte development and myelination in rat brain aggregate cultures. Journal of Neuroscience Research 30: 382–390. (330)

Neugebauer, KM, Tomaselli, KJ, Lilien, J and Reichardt, LF (1988): N-cadherin, N-CAM, and integrins promote retinal neurite outgrowth on astrocytes in vitro. J Cell Biol 107: 1177–1187.

Norgren, RB, Ratner, N and Brackenbury, R (1992): Development of olfactory nerve glia defined by a monoclonal antibody specific for Schwann cells. Devel Dyn 194: 231–238.

Peters, A, Palay, SL and Webster, H (1990): "The Fine Structure of the Nervous System." Philadelphia, W.B. Saunders Co.

Pixley, SK (1992): The olfactory nerve contains two populations of glia, identified both in vivo and in vitro. Glia 5: 269–284.

Poduslo, SE, Pak, CH and Miller, K (1990): Hydrocortisone induction during oligodendroglial differentiation. Neurosci Lett 113: 84–88.

Prockop, DJ, Kivirikko, KI, Tuderman, L and Guzman, NA (1979): The biosynthesis of collagen and its disorders. New Eng J Med 301(1): 13–23.

Raisman, G (1985): Specialized neuroglial arrangement may explain the capacity of vomeronasal axons to reinnervate central neurons. Neurosci 14: 237–254.

Ramon-Cueto, A and Nieto-Sampedro, M (1992): Glial cells from adult rat olfactory bulb: Immunocytochemical properties of pure cultures of ensheathing cells. Neurosci 47(1): 213–220.

Ramon-Cueto, A and Nieto-Sampedro, M (1994): Regeneration into the spinal cord of transected dorsal root axons is promoted by ensheathing glia transplants. Exper Neurol 127(2): 232–244.

Ramon-Cueto, A, Perez, J and Nieto-Sampedro, M (1993): In vitro enfolding of olfactory neurites by p75 NGF receptor positive ensheathing cells from adult rat olfactory bulb. Europ J Neurosci 5(9): 1172–1180.

Ramon-Cueto, A and Valverde, F (1995): Olfactory bulb ensheathing glia: A unique cell type with axonal growth-promoting properties. Glia 14(3): 163–173.

Reinhard, E, Meier, R, Halfter, W, Rovelli, G and Monard, D (1988): Detection of glia-derived nexin in the olfactory system of the rat. Neuron 1: 387–394.

Roesler, WJ, Vandenbark, GR and Hanson, RW (1988): Cyclic AMP and the induction of eukaryotic gene transcription. J Biol Chem 263(19): 9063–9066.

Scherer, SS, Deschenes, SM, Xu, YT, Grinspan, JB, Fischbeck, KH and Paul, DL (1995): Connexin32 is a myelin-related protein in the PNS and CNS. J Neurosci 15(12): 8281–8294.

Scotti, AL, Hoffmann, MC and Nitsch, C (1994): The neurite growth promoting protease nexin 1 in glial cells of the olfactory bulb of the gerbil: an ultrastructural study. Cell Tiss Res 278(2): 409–413.

Silver, J (1988): Transplantation strategies using embryonic astroglial cells to promote CNS axon regeneration in neonatal and adult mammals. Clin Res 36: 196–199.

Smale, KA, Doucette, R and Kawaja, MD (1996): Implantation of olfactory ensheathing cells in the adult rat brain following fimbria-fornix transection. Exper Neurol 137(225–233).

Smith, GM, Miller, RH and Silver, J (1986): Changing role of forebrain astrocytes during development, regenerative failure, and induced regeneration upon transplantation. J Comp Neurol 251: 23–43.

Stone, SR, Nick, H, Hofsteenge, J and Monard, D (1987): Glia-derived neurite-promoting factor is a slow-binding inhibitor of trypsin, thrombin, and urokinase. Arch Biochem Biophy 252: 237–244.

Valverde, F and Lopez-Mascaraque, L (1991): Neuroglial arrangements in the olfactory glomeruli of the hedgehog. J Comp Neurol 307: 658–674.

Ved, HS, Gustow, E and Pieringer, RA (1989): Effect of hydrocortisone on myelin basic protein in developing primary brain cultures. Neurosci Lett 99: 203–207.

Vickland, H, Westrum, L, Kott, J, Patterson, S and Bothwell, M (1991): Nerve growth factor receptor expression in the young and adult rat olfactory system. Brain Res 565: 269–279.

Vielkind, U, Walencewicz, A, Levine, JM and Churchill Bohn, M (1990): Type II glucocorticoid receptors are expressed in oligodendrocytes and astrocytes. J Neurosci Res 27: 360–373.

Wozniak, W (1993): Ensheathing cells in the nerve fiber layer of the olfactory bulb: A novel glial cell type. Folia Morphol Warsz 52(3): 121–127.

Yan, Q and Johnson, EM (1988): An immunohistochemical study of the nerve growth factor receptor in developing rats. J Neurosci 8: 3481–3498.

Zhang, XP and Miskimins, R (1993): Binding at an NFI site is modulated by cyclic AMP-dependent activation of myelin basic protein gene expression. J Neurochem 60(6): 2010–2017.

Zhou, Q and Cummings, RD (1993): L-14 lectin recognition of laminin and its promotion of in vitro cell adhesion. Arch Biochem Biophy 300: 6–17.

Zurn, AD, Nick, H and Monard, D (1988): A glia-derived nexin promotes neurite outgrowth in cultured chick sympathetic neurons. Devel Neurosci 10: 17–24.

33

OVERCOMING MYELIN-ASSOCIATED INHIBITION OF AXONAL REGENERATION AFTER CNS INJURY

Jason K. Dyer, John McGraw, Jason Bourque, and John D. Steeves

Collaboration on Repair Discoveries (CORD)
Departments of Zoology, Anatomy and Surgery
University of British Columbia
Biological Science Building
6270 University Boulevard
Vancouver, BC
V6T 1Z4, Canada

1. INTRODUCTION

After damage to the peripheral nervous system (PNS), a cascade of events unfold that results in the regeneration of some injured axons towards peripheral targets with an accompanying restoration of function [Pollock, 1995; Terenghi, 1995; Brecknall and Fawcett, 1996]. Our understanding of the mechanisms supporting PNS regeneration has been characterized over the past 5 decades, with the specific roles of glial cells (e.g. Schwann cells) [Bunge, 1994], invading macrophages [Perry and Brown, 1992], extracellular matrix [Martini, 1994] and growth factors [Bunge, 1993] being well documented.

In contrast, injuries to the adult central nervous system (CNS), as in spinal cord or head trauma, most often results in limited (~1mm) axonal regeneration [Bjorklund et al., 1971; Li et al., 1995] followed by axonal degeneration [Ramon-y-Cajal, 1928; Bjorklund et al., 1971] and a subsequent slower degeneration of the surrounding myelin [Waller, 1850]. The factors contributing to this lack of CNS regeneration have remained elusive, although the questions have become more directed. For example, to what degree does inflammation and secondary neural cell death, at the site of injury, alter the downstream outcome? Is it partially an inability of CNS neurons to re-activate intrinsic programs for axonal growth? Which essential growth promoting factors and/or extracellular matrix molecules are missing from the injured adult CNS? And finally, has CNS development contributed to the formation of a mature extraneuronal environment that is inhibitory to axonal regeneration?

Cell Biology and Pathology of Myelin, edited by Juurlink *et al.*
Plenum Press, New York, 1997

2. IDENTIFICATION OF THE INHIBITORY FACTOR(S) IN THE CNS

Some of the earliest experiments documenting the ability of CNS axons to grow within the supportive environment of the PNS are those of Tello [1911]. More recently, Aguayo and co-workers [Richardson et al., 1980; David and Aguayo, 1981; Richardson et al., 1984] re-examined these early observations using more refined approaches. By grafting a PNS bridge around a localized spinal cord injury, they observed CNS axons growing in either direction through the graft, but abruptly halting upon re-entry into the CNS. The permissive nature of PNS grafts for CNS axonal regeneration has also been documented in several other regions of the CNS, including the injured optic tract [Bray et al., 1987]. These findings demonstrated that at least some CNS neurons have the intrinsic ability to re-grow axons within a permissive environment, and CNS neurons have differential responses to the different extrinsic signals of the PNS and CNS. These observations served to dispel previous suggestions that the lack of CNS axonal regeneration in higher adult vertebrates (i.e., birds and mammals) was solely due to the irreversible suppression of intrinsic neuronal growth programs. Nevertheless, distinct neuronal phenotypes may possess differing intrinsic abilities with regards to axonal injury, cell survival and any subsequent axonal outgrowth [Tetzlaff et al., 1991; Fawcett, 1992].

The presence of a distinct type of myelin within the adult CNS of higher vertebrates has been suggested as one factor that might limit functional repair after a neurotraumatic injury. Part of the evidence supporting the idea of myelin-associated inhibition comes from observations on the robust regenerative ability exhibited by 'lower' adult vertebrates that lack compact myelin and from the developing CNS of 'higher' vertebrates, prior to normal myelination. For example, the lamprey possesses little compact CNS myelin and generates both anatomical repair and physiological synaptogenesis after either a larval or adult lesion of the spinal cord, resulting in substantial functional recovery [Cohen et al., 1988; McClellan, 1990; Lurie and Selzer, 1991]. After a spinal cord transection close to the head of a lamprey (10% of body length, BL), animals displayed a gradual recovery of function over a period of 32 weeks. Retrograde tracing indicated regeneration of brain stem-spinal neurons to levels approximating 40% of BL. More interesting is the observation of significant functional recovery to 60% BL (~60mm), without an accompanying large number (<50 cells) of regenerated brain stem-spinal axons to that level of the cord [Davis and McClellan, 1994]. This suggests that only a small percentage of regenerated axons may be necessary for substantial functional recovery and also that the re-organization (i.e. plasticity) of spinal cord circuits, caudal to the injury, may also contribute to adequate sensorimotor recovery. Teleost fish also display a robust ability to regenerate axons after CNS lesions [Bernhardt, 1989; Bastmeyer et al., 1991; Schwartz, 1993; Sharma et al., 1993].

A phylogenic transition in the regenerative ability of the injured CNS is observed in *Xenopus*: a lesion of the tadpole spinal cord results in functional recovery associated with axonal regeneration [Beattie et al., 1990]. This spinal cord regenerative potential is lost upon transition to adult life stages, i.e. during metamorphosis [Forehand and Farel, 1982; Beattie et al., 1990; Lang et al., 1995]. In contrast, the optic nerve of adult *Xenopus* will continue to regenerate after injury [Lang et al., 1995]. CNS myelin of fish and amphibians may differ from that of the rat, especially within the optic nerve [Bastmeyer et al., 1991; Schwartz, 1993; Sivo et al., 1994; Wanner et al., 1995]. It should also be remembered that many CNS neurons of fish and amphibians retain the ability for continuous proliferation and differentiation (i.e. axonal outgrowth) throughout adult life and these intrinsic abilities may also contribute to promoting repair after injury.

In birds and mammals, myelin formation in the CNS is either a late embryonic (birds) or early post-natal event (mammals). This has provided an opportunity for the study of CNS axonal regeneration both before and after CNS myelination. As most mammals are viviparous, it is difficult to manipulate of spinal cord development and/or axonal repair in utero without compromising the survival of the developing embryo and/or mother. Since chickens are oviparous, the embryo is readily accessible and will survive extensive experimental manipulations (e.g. surgery) that cannot be successfully achieved using a mammalian fetus. Another avian advantage is the relatively accelerated rate of CNS neurodevelopment when compared to most mammals. Hatchling chicks are precocial and their CNS (especially those brain stem and spinal systems concerned with locomotion) can be considered mature and adult-like in function at the time of hatching [Okado and Oppenheim, 1985; Shimizu et al., 1990; Shiga et al., 1991; Glover and Pettursdottir, 1991; Hasan et al., 1991, 1993; Glover, 1993]. Since it is well established that the overall structure and function of the vertebrate brain stem and spinal cord has remained relatively unchanged throughout evolutionary history [Sarnat and Netsky, 1981; McClellan, 1990], studies of the development and repair of avian brain stem and spinal locomotor mechanisms can be illustrative of the fundamental processes in most, if not all, vertebrates [Bekoff, 1976; Okado and Oppenheim, 1985; Sholomenko and Steeves, 1987; Steeves et al., 1987; Webster and Steeves, 1988; Valenzuela et al., 1990; Hasan et al., 1991; Sholomenko et al., 1991a,b,c; O'Donovan et al., 1992]. For example, previous work in our laboratory showed that the anatomical, physiological, and pharmacological organization of avian brain stem-spinal locomotor circuits is analogous to that of all other vertebrates, including mammals [Sholomenko and Steeves, 1987; Steeves et al., 1987; Webster and Steeves, 1988, 1991; Valenzuela et al., 1990; Sholomenko et al., 1991a,b,c]. Moreover, after a complete or incomplete spinal cord injury, adult birds suffer the same motor deficits as adult mammals [Sholomenko and Steeves, 1987; Webster and Steeves, 1991].

We and others have previously determined the duration of the 21 day in ovo development period over which an embryonic chick is capable of axonal repair after a complete transection of the thoracic spinal cord [Shimizu et al., 1990; Hasan et al., 1991, 1993; Keirstead et al., 1992; Steeves et al., 1993]. Briefly, within a week of an embryonic spinal cord transection at the low cervical or high thoracic level, neuroanatomical repair can be evaluated with a restricted injection into the lumbar cord (caudal to the transection site) of fluorescent retrograde tracing dyes [Hasan et al., 1991, 1993]. Functional (physiological) repair of descending supraspinal pathways can be evaluated with behavioral observations of locomotor functions by hatching chicks and also directly tested by focal electrical stimulation of brain stem locomotor regions known to have direct projections to the lumbar cord [Valenzuela et al., 1990; Hasan et al., 1991, 1993]. Severing the embryonic chick spinal cord prior to embryonic day 13 (E13) results in relatively complete neuroanatomical and physiological repair leading to total functional recovery upon hatching [Shimizu et al., 1990; Hasan et al., 1991, 1993]. We have called the period prior to E13 the *permissive period for functional spinal cord repair*. Conversely, if the spinal cord is transected on E13 or E14, the repair of descending supraspinal pathways progressively diminishes, resulting in little or no functional recovery [Shimizu et al., 1990; Hasan et al., 1991, 1993]. By E15, a thoracic transection results in no axonal repair/regeneration or functional recovery. The chick, upon hatching, is as paralyzed as an adult bird or mammal with a spinal cord transection [Sholomenko and Steeves, 1987; Shimizu et al., 1990; Hasan et al., 1991, 1993; Webster and Steeves, 1991; Keirstead et al., 1992]. For these reasons we have

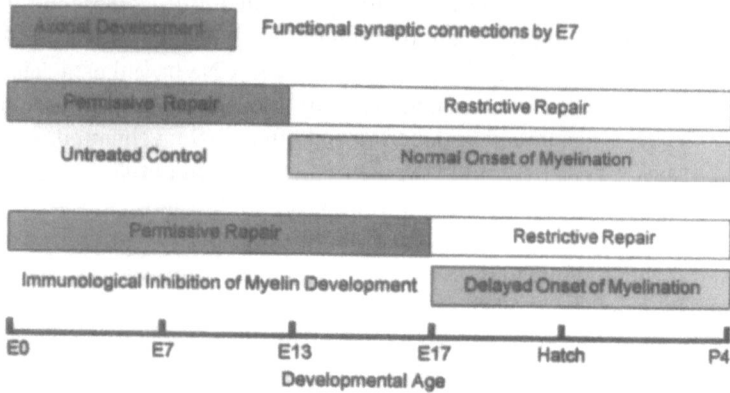

Figure 1. Schematic representation for the extension of the permissive period for functional axonal regeneration in the embryonic chick spinal cord. Brain stem-spinal neurons complete their differentiation, including axonal out-growth to target spinal neurons, early in development. Fully functional axonal regeneration is evident after a transection of the low cervical or high thoracic spinal cord prior to embryonic day (E)13 (the permissive period for repair). Transection on or after E13 results in limited or no axonal regeneration with a corresponding lack of functional recovery (the restrictive period for repair). The developmental transition from the permissive to restrictive period for repair coincides with the onset of spinal cord myelination. Intraspinal infusion of serum complement proteins with an oligodendrocyte-specific antibody that binds complement delays the onset of myelination and extends the permissive period for functional axonal regeneration to later stages of development [Keirstead et al., 1992; Hasan et al., 1993].

called the developmental period after E13 the *restrictive period for spinal cord repair* (Figure 1).

Several neuroanatomical responses could underlie this repair process, including: 1) neurogenesis of new descending brain stem-spinal neurons, 2) subsequent projections from late developing brain stem-spinal neurons, and 3) axonal regeneration of previously axotomized brain stem-spinal projections. Brain stem-spinal projections and synaptogenesis with spinal target neurons progress over a period extending from approximately E3 to E11 [Okado and Oppenheim, 1985; Shimizu et al., 1990; Glover and Pettursdottir, 1991; Hasan et al., 1991; Shiga et al., 1991; Glover, 1993] with the first functional synaptic connections evident on E6-E7 [Bekoff, 1976; Shiga et al., 1991; Sholomenko and O'Donovan, 1995]. By E11 the distribution and number of retrograde labeled brain stem-spinal neurons is equivalent to those labeled in a hatchling chick [Okado and Oppenheim, 1985; Shimizu et al., 1990; Hasan et al., 1991; Shiga et al., 1991]. In short, axonal growth is completed before the end of the permissive regenerative period and before the onset of spinal cord myelination at E13 [Benstead et al., 1957; Hartman et al., 1979; Macklin and Weill, 1985]. In addition, brain stem-spinal neurons become post-mitotic very early in development, usually prior to E5 [McConnell and Sechrist, 1980] and well before E12. Therefore, it is unlikely that our findings of functional repair of an E7-E12 transected cord is dependent on the neurogenesis of additional brain stem-spinal neurons. If spinal repair was not in part due to 'true' axonal regeneration, but only due to subsequent axonal projections from late developing neurons, a reduced number of brain stem-spinal neurons should be retrogradely labeled after an E11-E12 transection. The similar number and distribution of retrograde labeled brain stem-spinal neurons in both E11-transected embryos and sham-operated control embryos argues in favor of axonal regeneration contributing to the repair process [Hasan et al., 1991; 1993]. Nevertheless, these potential repair mechanisms are not mutually exclusive and were investigated with the use of double retrograde tracing techniques where the first fluorescent dye was injected before the transection and the second fluorescent tracer was injected approximately 7 days after injury. The presence of double labeled brain stem-spinal neurons indicated that 'true' axonal regeneration did contribute to the observed functional regeneration (Hasan et al., 1993).

As mentioned above, one notable developmental event correlating with the transition from the permissive to restrictive repair period at E13 is the onset of myelination within the developing chick spinal cord [Benstead et al., 1957; Hartman et al., 1979; Macklin and Weill, 1985]. Using immunocytochemical protocols for either myelin basic protein (MBP), myelin associated glycoprotein (MAG) or 2',3'-cyclic nucleotide 3'-phosphodiesterase (CNPase), we confirmed the previous histological findings, noting the coincidental onset of myelination with the transition from the permissive to restrictive period for spinal cord repair [Keirstead et al., 1992; Keirstead et al, unpublished observations]. These results extended the proposition that the presence of CNS myelin is one factor that contributes to the inhibition of neuronal repair after an adult CNS injury [Keirstead et al., 1992]. The onset of spinal cord myelination was then suppressed with an intraspinal injection at E9-E11 of a combination of serum complement proteins and a myelin specific antibody that binds complement (see below). This immunological approach to suppress the development of myelin extended the permissive period for axonal regeneration and functional recovery to later stages of development (e.g. E15), a time when normally there is no evidence for repair after spinal injury [Keirstead et al., 1992].

A similar correlation between the extent of axonal regeneration and the presence of myelin is evident from mammalian studies. For example, marsupials, such as the opossum, are born in a highly undeveloped state with subsequent development being ex-utero. This animal model also provides increased access for experimental examination during CNS development [Martin et al., 1994]. A time-frame has been established for the onset of spinal cord myelination (P7–12) in the Brazilian opossum [Varga et al., 1995], and this developmental stage correlates with the loss of axonal regeneration after injury [Treherne et al., 1992; reviewed in Nicholls and Saunders, 1996]. Likewise, the embryonic rat spinal cord can be maintained in culture [Saunders et al., 1992] and after a crush lesion, the growth of axons and recovery of electrical conductivity can be demonstrated through the injured region. Importantly, the embryonic rat CNS has yet to develop CNS myelin. For some precocial mammals, such as horses, myelin develops prior to birth [Sweasey et al., 1982]. However in most other mammals, CNS myelin first appears around the time of birth. In rats and mice, myelin appears throughout the spinal cord over the first two weeks of neonatal life [Foran and Peterson, 1992]. Carefully severing the rat spinal cord during the first two weeks of postnatal life, prior to establishing an extensive network of compact CNS myelin, can be followed by substantial anatomical and functional repair [Commissiong and Toffano, 1989; Iwashita et al., 1994]. Finally cutting a dorsal root results in increased axonal sprouting within the neonatal cord by adjacent intact dorsal root ganglion neurons, but only when spinal cord myelination is suppressed by exposure of the cord to X-rays [Schwegler et al., 1995].

In summary, all these avian and mammalian developmental paradigms indicate that the presence of CNS myelin is deleterious to axonal re-growth after injury. Clearly there is no evolutionary drive for CNS myelin being an inhibitor of axonal re-growth. The inhibition of regeneration may be a consequence of a developmental function of myelin. The above evidence supports the suggestion that the late developmental appearance of myelin acts as a brake on the continued development of axonal projections within the differentiated CNS (i.e. limits 'spontaneous' axon collateral formation). Thus, this myelin stop or stabilizer signal would also inhibit any axonal regeneration attempts within the injured adult CNS.

Recently a number of myelin molecules have been characterized that actively *inhibit* the growth of axons. In vitro assay experiments have noted that explants of CNS tissue, co-cultured with peripheral neurons, did not support the ingrowth of axons, as compared to a peripheral nerve explant substrate [Schwab and Thoenen, 1985]. Frozen tissue sections of adult rat CNS tissue (optic nerve, spinal cord and brain) are also a non-permissive substrate for neurite outgrowth [Carbonetto et al., 1987; Crutcher, 1989; Savio and Schwab, 1989; Watanabe and Murakami, 1989; Tuttle and Matthew, 1991]. More specifically, CNS white matter environments are more inhibitory than CNS gray matter for axonal outgrowth [Crutcher, 1989; Savio and Schwab, 1989; Watanabe and Murakami, 1989; Siegal et al., 1990]. Likewise, results from oligodendrocyte culture experiments indicated that these cells were a relatively non-permissive substrate for neurite outgrowth and contact often induced the collapse of axonal growth cones [Schwab and Caroni, 1988; Fawcett et al., 1989; Bandtlow et al., 1990]. Finally, purified myelin substrates, from a number of vertebrate species, are inhibitory for cell adhesion in vitro [Schwab and Caroni, 1988; Bandtlow et al, 1991; Keller et al., 1993; Ng et al., 1996] and will induce active collapse of any outgrowing neurite growth cones [Bandtlow et al., 1993].

The exact molecular nature of this myelin inhibitory activity has been actively pursued. The specificity of myelin inhibitory activity has perhaps best been demonstrated by its reversal. For example, antibodies have been raised against incompletely characterized myelin proteins isolated from bovine CNS white matter. The first myelin inhibitory protein was termed NI35/250, on the basis of the apparent molecular weights, and the best monoclonal antibody to NI35/250 was termed IN-1 [Caroni and Schwab, 1988a]. Both in vitro and in vivo [Caroni and Schwab, 1988b; Schnell and Schwab, 1990; Schnell et al., 1994] application of IN-1 suppressed the inhibitory activity of myelin on axonal growth. Interestingly, some lower vertebrates (e.g. teleost fish) appear to lack the NI35/250 myelin protein [Bastmeyer et al., 1991; Sivon et al., 1994]. Similarly, the *Xenopus* tadpole CNS does not react with the IN-1 antibody until after metamorphosis when the CNS becomes a less permissive environment for axonal outgrowth or repair [Lang et al., 1995].

Another myelin protein, myelin associated glycoprotein (MAG) [McKerracher et al., 1994; Mukhopadhyay et al., 1994; Martini, 1994; Filbin, 1995] has recently been cited as a potential inhibitor of neurite outgrowth. Myelin isolated from both the CNS and PNS has been shown to contain inhibitory activity that was related to MAG [Bähr and Przyrembel, 1995]. *In vitro*, the inhibitory activity of MAG substrates can be overcome by the addition of specific anti-MAG antibodies, suggesting the possibility of a MAG specific receptor/ligand interaction mediating the inhibitory effect of MAG on neurite outgrowth. Another question is whether the MAG inhibition is developmentally dependent and only observed in neurons when they would normally be exposed to myelin. This possibility is supported by the observation that immature DRG-derived neurite outgrowth fails to be inhibited by MAG, whereas adult DRG- and cerebellar-derived neurite outgrowth is inhibited on MAG substrates [DeBellard et al., 1996]. The receptor binding and growth inhibitory domain of MAG has recently been postulated to be a sialoglycogen moiety [DeBellard et al., 1996], desialyation of the cultured neurons, or inclusion of small sialic-acid-bearing sugars in the culture media abolishes the inhibitory activity of MAG.

Of course, there are several other axonal growth inhibitors not associated with myelin and brief reference to some of these molecules is warranted. Molecules, such as collapsin [Luo et al., 1993], a secreted glycoprotein, and other inhibitory members of the semaphorin [Kolodkin et al., 1993] family, have been shown to induce changes in growth cone motility, including growth cone collapse. Tenascin (cytotactin in the chick) is a member if the Ig superfamily of molecules, secreted by astrocytes into the basement membrane [Grierson et al., 1990; Bartsch et al., 1992]. Tenascin has anti-adhesive properties for a number of cell types including neurons [Spring et al., 1989; Wehrle-Haller and Chiquet, 1993]. Produced throughout normal CNS development, but subsequently down-regulated within the adult CNS, tenascin may act to 'sculpt' the developing nervous system. After an adult CNS injury, however, tenascin is up-regulated in astrocytes associated with the glial scar [Laywell et al., 1992]. Oligodendrocytes also produce a related molecule, Janusin/J1–160/180 [Pesheva et al., 1989; Bartsch et al., 1993], that actively repulses growth cones [Taylor et al., 1993]. Proteoglycans (PGs) also have growth regulating functions [Herndon and Lander, 1990]. Some PGs, such as heparin sulphate-proteoglycan (HSPG) have neurite promoting activity, however two members of the group, chondroitin sulphate- proteoglycan (CSPG) and keratin sulphate-proteoglycan (KSPG) exert inhibitory influences on neurite outgrowth [Snow et al, 1990; Oohira et al., 1991]. The ratio of HSPG to CSPG expression during CNS development in the chick drops significantly at the transition between the permissive and restrictive periods for axonal regeneration [Dow et al., 1994]. CSPG and KSPG are expressed by cells associated with scar formation after injury [McKeon et al., 1991; Dow et al., 1994; Levine, 1994]. It should be remembered that the 'scar' formed at the site of a CNS injury by astrocytes and fibroblasts may be both a physical and a chemical barrier to any subsequent axonal regeneration attempt.

3. OVERCOMING MYELIN INHIBITORS TO FACILITATE FUNCTIONAL RECOVERY

A number of approaches are currently being examined that may overcome the inhibitory influence of myelin to axonal regeneration by CNS neurons after injury. One approach is to 'bridge' the damaged region with a transplant graft that establishes a permissive environment for axonal outgrowth. Several studies have demonstrated the potential for peripheral nerve grafts to facilitate axonal re-growth across a lesion in the spinal cord [e.g. Richardson et al., 1980; David and Aguayo, 1981; Houle, 1991], resulting in some cases of limited anatomical and functional recovery [Cheng et al., 1996]. A developmental extension of this concept is the grafting of embryonic tissue into the lesion site [for review see Reier et al., 1992]. This not only provides a structural scaffold that is similar to that seen during normal development by the growing axon, it may also be a source of essential developmental growth promoting molecules. In some cases, 'downstream' outgrowth by immature neurons, of the embryonic graft, back into the adult CNS may act as a functional relay for 'upstream' injured axons that can only successfully enter the graft, but not grow through the graft. A possible reason for the robust outgrowth by immature neurons into the supposedly hostile environment of the adult CNS is their naïveté to the inhibitory molecules surrounding them. In short, they have not yet completed their differentiation, and thus may have not yet expressed receptors for the inhibitory 'stop' molecules associated with myelin.

Block grafts of fetal tissue [Iwashita et al., 1994], dissociated neurons [Yakovleff et al., 1995] and Schwann cell suspensions [Morrisey et al., 1991; Guénard et al., 1992; Montgomery and Robson, 1993] have been employed to promote repair. For example, Kawaguchi and coworkers have observed that the transplant of E16 spinal cord tissue into the cord of a neonatal rat resulted in brain stem-spinal axonal regeneration and functional recovery within 4 months [Iwashita et al., 1994]. The graft became integrated with the host tissue and showed little evidence of a glial scar at the injury site. The absence of myelin within the host neonatal spinal cord at the time of transplantation may have contributed to the observed repair. A previous study [Bregman et al., 1993] demonstrated that the degree of recovery after fetal tissue transplantation at the site of a spinal cord injury is greater in younger animals. The major limitation of grafting strategies in CNS repair has been the failure of severed axons to re-enter the CNS after coursing through the grafted 'bridge'. One recently reported success is that of Olsen and co-workers who deliberately directed white matter tracts to grow through an intercostal nerve transplant that was attached to a spinal gray matter region on the distal side of the lesion [Cheng et al., 1996]. Gray matter has previously been shown to be far less inhibitory than white matter as a supportive environment for axonal growth [Crutcher, 1989; Savio and Schwab, 1989; Watanabe and Murakami, 1989; Siegal et al., 1990].

As there is more than one myelin-associated inhibitory molecule in the CNS, more than one method of transiently removing myelin, or transiently blocking these inhibitory factors may be required. A number of experimental methods have been developed that will *remove* myelin from the CNS [Bradl and Linnington, 1996; also see chapter in this volume by Blakemore et al.]. These include chemical methods, such as cuprizone [Ludwin, 1978], ethidium bromide [Yajima and Suzuki, 1979], dietholamine [Melnick et al., 1994], lysolecithin [Dousset et al., 1995] application and viral factors such as JC virus, polyoma virus [Stoner, 1993], Theiler's virus [Rodriguez, 1992; Tsunoda and Fujina, 1996], and canine distemper virus [Vandevelde and Zurbriggen, 1995]. However, many of these methods are often irreversible or alter other neural cell types including astrocytes, microglia and neurons [Mitchell and Caren, 1982]. On another active research front and ignoring the obvious social dilemmas, gene therapy is only likely to become an effective clinical reality when the in vivo technologies for targeted gene transfer and specific gene induction are improved. Finally, multiple antibodies and/or small peptide fragments may be required to block the various functional epitopes of these antigens.

One strategy we have been investigating as a potential therapeutic intervention is the transient removal or disruption of CNS myelin near the injury site. The protocol involves an immunological approach which directs and restricts the in vivo lytic capabilities of serum complement proteins to myelin through the simultaneous infusion of a myelin-specific antibody (e.g. an IgG antibody to galactocerebroside, GalC) that has a high affinity for binding serum complement proteins. Such immunological treatment of the developing chick spinal cord transiently suppresses the onset of myelination and facilitates functional axonal regeneration after injury of the chick spinal cord in ovo [Keirstead et al., 1992]. In the mature, myelinated spinal cord of the hatchling chick, a similar treatment *after a spinal cord injury* causes transient demyelination and/or myelin disruption and also facilitates axonal regeneration (with the formation of functional synapses), albeit to a lesser degree than that observed after an embryonic injury [Keirstead et al., 1995]. In adult rats and mice, similar intraspinal injections of serum complement proteins with a complement-fixing, myelin-specific antibody causes focal demyelination with a larger surrounding zone of myelin disruption [Dyer et al., 1995; Dyer and Steeves, 1996].

Antisera against GalC have been shown to inhibit myelination and reversibly demyelinate CNS tissue in vitro [Dubois-Dalcq et al., 1970; Fry et al., 1974; Hruby et al., 1977; Dorfman et al., 1979; Dyer and Benjamins, 1990]; as well as the optic nerve [Sergott et al., 1984; Ozawa et al., 1989] and spinal cord [Mastiglia et al., 1989] *in vivo*. It is not known whether this effect requires the presence of and action of endogenous serum complement proteins. GalC is the major sphingolipid produced by oligodendrocytes and is primarily localized to myelin membranes of differentiated oligodendrocytes. The GalC structure is highly conserved across all vertebrate species [Ranscht et al., 1982]. Anti-GalC-induced demyelination in vitro may involve microtubule disassembly and be mediated by an influx of extracellular calcium [Dyer and Benjamins, 1990]. Although, the cellular mechanism for in vivo myelin disruption has not been fully characterized, we have found that a single injection of a monoclonal, or polyclonal, GalC antibody, with heterologous (guinea-pig) serum complement proteins, into the thoracic spinal cord of an E9-E12 chick embryo (i.e. around the time of oligodendrocyte differentiation) results in a delay in the onset of spinal cord myelination as determined by electron microscopic examination [Keirstead and Steeves, unpublished observations] and MBP, CNP and MAG immunoreactivity (i.e. myelination) until E17 [Keirstead et al., 1992]. Equally effective suppression of myelin development can also be produced by simultaneously injecting serum complement along with the complement-fixing, oligodendrocyte-specific O4 antibody of Melitta Schachner and colleagues [Sommer and Schachner, 1981]. No effect on CNS myelination was observed if antibody only, serum complement proteins only, vehicle only, or GalC antibody plus heat-inactivated serum were injected into the spinal cord. Also a non-myelin, complement binding antibody (anti-glial fibrillary acidic protein, GFAP) had no effect on the myelin development [Keirstead et al., 1992]. This indicates that both the GalC antibody and serum complement proteins are necessary to inhibit the normal development of myelin.

A thoracic spinal cord transection as late as E15 (i.e. during the normal restrictive period for repair) in a myelin-suppressed embryo resulted in substantial neuroanatomical repair and functional recovery [Keirstead et al., 1992]. This is in contrast to (normally myelinated) control embryos transected on E15 which showed no axonal repair/regeneration and, upon hatching, were paralyzed. In brief, the number and distribution of retrograde labeled brain stem-spinal neuronal cell bodies in myelination-inhibited E15-transected animals was similar from non-transected control embryos or embryos transected during the permissive repair period (e.g. E12, when myelin has yet to appear). The locomotor functions of myelin-suppressed E15-transected embryos were similar to untransected control animals or embryos transected during the permissive period [Keirstead et al., 1992]. In short, these findings demonstrated that, after a spinal transection, the suppression of myelination extends the permissive period for neuroanatomical repair/regeneration and functional recovery.

More critical questions were then asked of this procedure: 1) will the immunological therapy, outlined above, transiently remove myelin-associated inhibitory influences from the adult avian and mammalian CNS, 2) will myelin-disruption of the CNS *after* a traumatic injury facilitate regeneration of severed axonal pathways, and 3) does myelin-disruption alone facilitate neuroanatomical regeneration to a sufficient degree to promote functional recovery after a CNS injury? Current published and unpublished data suggest the first two questions can be answered with a qualified yes, whereas the third question appears to be a qualified no [Keirstead et al., 1995; Dyer and Steeves, unpublished observations].

Recently we have shown that direct spinal cord infusion (via an osmotic pump) of serum complement proteins with an oligodendrocyte-specific antibody (e.g., GalC or O4) that binds complement will transiently disrupt the myelin in several segments of the spinal cord in hatchling chicks [Keirstead et al., 1995]. The effects can be observed within 12 hours after the start of infusion and can be maintained for as long as the complement and antibody are infused. Based on MBP immunofluorescence and ultrastructural analysis, remyelination begins within days after terminating the infusion (Bourque, Dyer and Steeves, unpublished observations). We do not know whether the remyelination is due to the re-ex-

tension of myelin processes from mature oligodendrocytes that have survived the immunological treatment or due to the extension of processes from oligodendrocytes newly differentiated from progenitor cells (Keirstead & Blakemore, personal communication).

Intraspinal infusion of complement and GalC antibodies after complete transection
of the hatchling chick spinal cord resulted in axonal regeneration of approximately 5–20%
of brain stem-spinal neurons [Keirstead et al., 1995]. These results are based on single retrograde labeling of brain stem-spinal neurons from an injection within the lumbar cord,
2–4 weeks after spinal cord injury. Double retrograde labeling protocols also indicated
that 'true' axonal regeneration contributed to the observed repair (Figure 2). The percentage figure for axonal regeneration is based on comparisons with the number of labeled

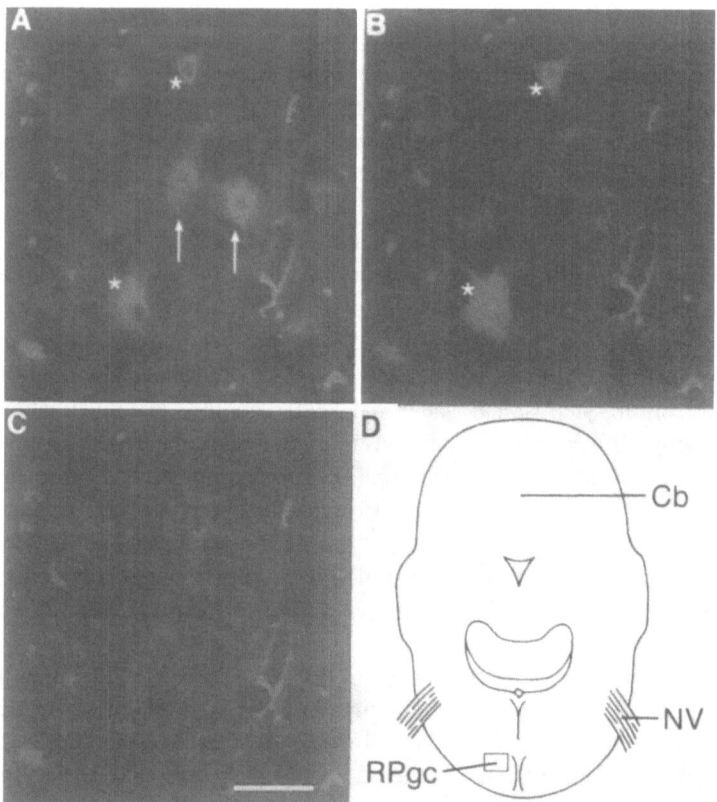

Figure 2. Photomicrographs of double-labeled brain stem-spinal neurons within the hatchling chick brain stem indicating myelin suppression after a spinal cord transection facilitates regeneration. A) Fluorescent rhodamine dextran amine (RDA) tracer, injected into the lumbar cord of a hatchling chick prior to a thoracic spinal cord
transection, labeled reticulospinal neurons within the ventromedial pontomeduallry reticular formation. B) Approximately 28 days following transection and subsequent immunological myelin suppression [Keirstead et al.,
1995], a second fluorescent tracer (Cascade blue dextran amine, CBDA) was injected into the lumbar cord. The
CBDA tracer retrogradely-labeled some of the same reticulospinal neurons labeled with the first fluorescent RDA
tracer (denoted by *). The double-labeled neurons indicate 'true' axonal regeneration contributed to the observed
repair [Keirstead et al., 1995]. There was a notable absence of neurons singly labeled with the CBDA (second)
tracer indicating there was no de novo development of brain stem-spinal projections after the injury. C) The use of
an FITC-filter indicates there was no non-specific fluorescent labeling of reticulospinal neurons. D) Schematic diagram of the chick pontine brain stem indicating the location of the photomicrographs in A-C. Cb, cerebellum; NV,
nervus trigeminus; RPgc, N. reticularis pontis caudalis, pars gigantocellularis. Scale bar for A-C = 50 um.

brain stem-spinal neurons observed in untransected control animals. Although focal electrical stimulation of brain stem locomotor regions can evoke stepping in these experimental animals [Keirstead et al., 1995], we did not observe any voluntary functional recovery. Nevertheless, the physiological data indicated that regenerating axons were able to make functional synapses with target cells of the caudal cord. One factor that may have contributed to the poor behavioral recovery was that all experimental transected animals showed significant muscular atrophy. Spinal transected hatchling chicks that did not receive the 14 day immunological protocol showed no evidence for axonal regeneration or functional recovery as assessed by retrograde labeling and focal electrical stimulation.

Although it seems possible that more than 20% of the available brain stem-spinal projections need to regenerate for outward signs of functional recovery to occur, the recent data of others suggests that large numbers of regenerating fibers may not be required for limited functional recovery [Davis and McClellan, 1994; Bregman et al., 1995; Cheng et al., 1996]. Finally, the lower degree of observed recovery after injury of the mature spinal cord may also be due to the lack of other, as yet unidentified, trophic or tropic molecule(s) and/or the presence of additional unidentified inhibitory factor(s).

More recent work has demonstrated the ability of the immunological treatment to disrupt the myelin in both murine and rat models, and the potential of this approach to facilitate regeneration of the injured adult mammalian spinal cord. Both single injections and continuous infusion of the immunological reagents results in modification to the structure and biochemistry of myelin of the rodent spinal cord. Shortly after initiation of treatment by continuous infusion, via mini-osmotic pumps, there is a rapid focal demyelination (within the same segment), a distal down-regulation of the structural proteins of the compact myelin sheath (MBP, PLP, MAG) and a profound unraveling of the compact lamella of the myelin sheathes at greater distances from the injections site (Figure 3). This unraveling does not appear to lead to a loss of myelin.

4. CONCLUDING REMARKS

The therapeutic use of trophic factors to promote the survival of neurons [Rich, 1992] and axonal re-growth [Richardson, 1991; Mocchetti and Wrathall, 1995] after injury has long been suggested. However, trophic factor application alone does not appear to be sufficient to overcome the inhibitory environment of the adult CNS. Nevertheless, a combinatorial approach may be more successful [Schnell et al., 1994; Cheng et al., 1996]. Thus it is not unreasonable to suggest that combination therapies, addressing both the intrinsic growth potential of neurons and the permissibility of their environment, may provide a more effective approach to facilitate axonal regeneration and functional recovery after injury.

ACKNOWLEDGMENTS

We are grateful for the constructive suggestions of Wolfram Tetzlaff and David Pataky. Some of the work outlined in this article was supported by the following Canadian agencies: Medical Research Council, Natural Sciences and Engineering Research Council, and the Canadian NeuroScience Network. JKD is supported by a Postdoctoral Fellowship from the Rick Hansen Man in Motion Foundation and a training award from the NeuroScience Network; JM is supported by a NeuroScience Network Studentship. The Steeves

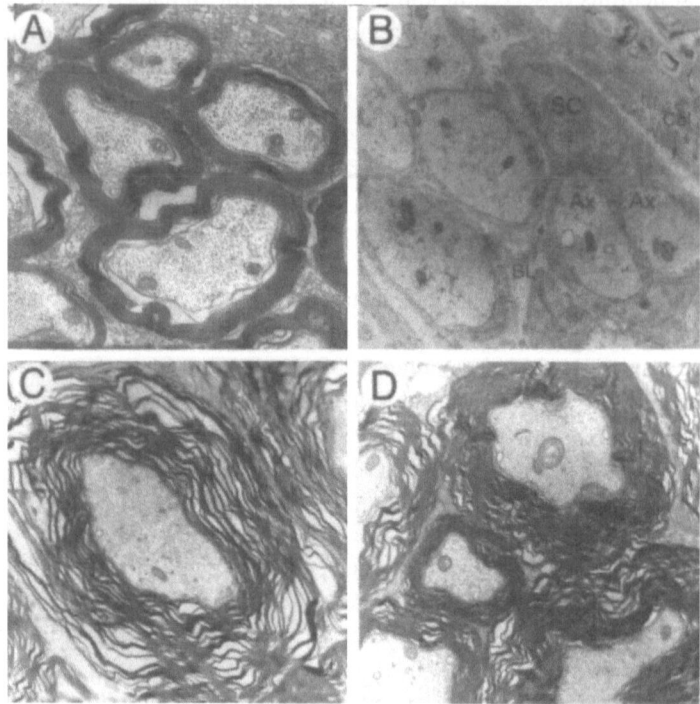

Figure 3. Myelin structure within the adult rat thoracic spinal cord following 14 days intraspinal infusion of a Galactocerebroside (GalC) antibody along with serum complement proteins. A) control treated section (33% complement). Compact nature of myelin was preserved in all control animals, including GalC only, PBS vehicle only. B) section from the site of the intraspinal infusion, of GalC antibody along with complement, illustrating the demyelination that occurs within 2mm of the cannula. There is evidence of invading peripheral Schwann cells initiating some remyelination of central axons. Schwann cells (SC) were identified by their close association with axons (Ax), the ability to produce a basal lamina (BL) and collagen deposits (Col). At greater distances from the site of infusion only myelin disruption is evident, C) section taken 8mm (2 spinal segments) from the infusion site indicating myelin disruption. In other areas (not shown), oligodendrocytes are visible, appearing normal, with no significant astroglial or mononuclear phagocytotic response being observed. D) section taken 16mm (4 spinal segments) from infusion site. Original magnification was x10 000.

laboratory is a member of CORD (Collaboration On Repair Discoveries). Our Internet website at UBC can be found at http://cord.ubc.ca.

REFERENCES

Bähr M and Przyrembel C (1995): Myelin from peripheral and central nervous system is a non-permissive substrate for retinal ganglion cell axons. Exp. Neurol. 134: 87–93.

Bandtlow CE Schmidt MF Hassinger TD Schwab ME and Kater SB (1993): Role of intercellular calcium in NI-35-evoked collapse of neuronal growth cones. Science 259:80–83.

Bandtlow CE and Schwab ME (1991): Purification and biochemical characterization of rat and bovine CNS myelin associated neurite growth inhibitors NI-35 and NI-250. Soc. Neurosci. Abstr. 17: 1495.

Bandtlow CE Zachleder T and Schwab ME (1990): Oligodendrocytes arrest neurite growth by contact inhibition. J. Neurosci. 10: 3837–3848.

Bartsch U Bartsch S Dörries U and Schachner M (1992): Immunohistological localization of tenascin in the developing and lesioned adult mouse optic nerve. Eur. J. Neurosci. 4:338–352.

Bartsch U Pesheva M Raff M and Schachner M (1993): Expression of Janusin (J1–160/180): in the retina and optic nerve of the developing and adult mouse. Glia 9:57–69.

Bastmeyer M Beckmann M Schwab ME and Stuermer CAO (1991): Growth of regenerating goldfish axons is inhibited by rat oligodendrocytes and CNS myelin but not by goldfish optic nerve tract oligodendrocyte-like cells and fish CNS myelin. J. Neurosci. 11:626–650.

Beattie MS Bresnahan JC and Lopate G (1990): Metamorphosis alters the response to spinal cord transection in Xenopus laevis frogs. J. Neurobiol. 21:1108–1122.

Bekoff A (1976): Ontogeny of leg motor output in the chick embryo: a neural analysis. Brain Res. 106: 271–291.

Benstead JPM Dobbing J Morgan RS Reid RTW and Payling-Wright G (1957): Neuroglial development and myelination in the spinal cord of the chick embryo. J. Embryol. Exp. Morph. 5: 428–437.

Bernhardt R (1989): Axonal pathfinding during the regeneration of the goldfish optic pathway. J. Comp. Neurol. 284:119–134.

Bjorklund A Katzman R Stenevi U and West KA (1971): Development and growth of axonal sprouts from noradrenaline and 5-hydroxytryptamine neurons in the rat spinal cord. Brain Res. 31: 21–33.

Bradl M and Linnington C (1996): Animal models of demyelination. Brain Pathol. 6:303–311.

Bray GM Villegas-Perez MP Vidal-Sanz M and Aguayo AJ (1987): The use of peripheral nerve grafts to enhance neuronal survival promote growth and permit terminal reconnections in the central nervous system. J. Exp. Biol. 132:5–19.

Brecknall JE and Fawcett JW (1996): Axonal regeneration. Biol. Rev. Cam. Phil. Soc. 71:227–355.

Bregman BS Kunkel-Bagden E Reier PJ Ning Dai H McAtee M and Gao D (1993): Recovery of function after spinal cord injury: mechanisms underlying transplant-mediated recovery of function differ after spinal cord injury in newborn and adult rats. Exp. Neurol. 123:3–16.

Bregman BS Kunkel-Bagden E Schnell L Dal HN Gao D and Schwab ME (1995): Recovery from spinal cord injury mediated by antibodies to neurite growth inhibitors Nature 378: 498–501.

Bunge RP (1993): Expanding roles for the Schwann cell: ensheathment myelination tropism and regeneration. Curr. Op. Neurobiol. 3:805–809.

Bunge RP (1994): The role of the Schwann cell in trophic support and regeneration. J. Neurol. 242:S19–21.

Carbonetto S Evans D and Cochard P (1987): Nerve fiber growth in culture on tissue substrata from central and peripheral nervous systems. J. Neurosci. 7:610–620.

Caroni P and Schwab ME (1988a): Antibody against myelin-associated inhibitor of neurite outgrowth neutralizes non-permissive substrate properties of CNS white matter. Neuron 1: 85–96.

Caroni P and Schwab ME (1988b): Two membrane protein fractions from rat central myelin with inhibitory properties for neurite growth and fibroblast spreading. J. Cell Biol. 106:1281–1288.

Cheng H Cao Y and Olsen L (1996): Spinal cord repair in adult paraplegic rats: partial restoration of hindlimb function. Science 273:510–513.

Cohen AH Mackler SA and Selzer ME (1988): Behavioral recovery following spinal transection: functional regeneration in the lamprey CNS. Trends Neurosci. 11:227–231.

Commissiong JW and Toffano G (1989): Complete spinal cord transection at different postnatal ages: recovery of motor coordination correlated with spinal cord catecholamines. Exp. Brain Res. 78:597–603.

Crutcher KA (1989): Tissue sections from the mature rat brain and spinal cord as substrates for neurite outgrowth in vitro: extensive outgrowth on gray matter but little growth on white matter. Exp. Neurol. 104:39–54.

David S and Aguayo AJ (1981): Axonal elongation into peripheral nervous system "bridges" after central nervous system injury in adult rats. Science 214: 931–933.

Davis GR and McClellan AD (1994): Extent and time course of restoration of descending brain stem projections in spinal-cord transected lamprey. J. Comp. Neurol. 344:65–82.

DeBallard M-E Tang S Mukhopadhyay G Shen Y-J and Filbin M (1996): Myelin-associated glycoprotein inhibits axonal regeneration from a variety of neurons via interaction with a sialoglycoprotein. Mol. Cell. Neurosci. 7: 89–101.

Dorfman SH Fry J.M and Silberberg DH (1979): Antiserum induced demyelination inhibition in vitro without complement. Brain Res. 177: 105–114.

Dousset V Brochet B Vital A Gross C Benazzouz A Boullerne A Bidabe AM Gin AM and Caille JM (1995): Lysolecithin-induced demyelination in primates: preliminary in vivo study with MR and magnetization transfer. Am. J. Neuroradiol. 16:225–231.

Dow KE Ethell DW Steeves JD and Riopelle RJ (1994): Molecular correlates of spinal cord repair in the embryonic chick: heparan sulphate and chondroitin sulphate proteoglycans. Exp. Neurol. 128:233–238.

Dubios-Dalcq M Niedieck B and Buyse M (1970): Action of anti-cerebroside sera on myelinated tissue cultures. Pathol. Eur. 5: 331–347.

Dyer CA and Benjamins JA (1990): Glycolipids and transmembrane signaling: antibodies to galactocerebroside cause an influx of calcium in oligodendrocytes. J. Cell Biol. 111: 625–633.

Dyer JK Keirstead HS and Steeves JD (1995): Immunological and ultrastructural studies of adult chick and mouse myelin after intraspinal injection of serum complement proteins and myelin specific antibodies. Soc. Neurosci. Abstr. 21(1):313.

Dyer JK and Steeves JD (1996): Regeneration of descending brain stem-spinal and corticospinal axons after immunological myelin disruption of the adult rat spinal cord. Soc. Neurosci. Abstr. 22(1):764.

Fawcett JW (1992): Intrinsic neuronal determinants of regeneration. Trends Neurosci. 15:5–8.

Fawcett JW Rokos J and Bakst I (1989): Oligodendrocytes repel axons and cause axonal growth cone collapse. J. Cell Sci. 92:93–100.

Filbin M (1995): Myelin-associated glycoprotein: a role in myelination and in the inhibition of axonal regeneration? Curr. Op. Neurobiol. 5:588–595.

Foran DR and Peterson A.C (1992): Myelin acquisition in the central nervous system of the mouse revealed by an MBP-Lac-Z transgene. J. Neurosci. 12:4890–4897.

Forehand CJ and Farel P.B (1982): Anatomical and behavioral recovery from the effects of spinal cord transection: dependence on metamorphosis in anuran larvae. J. Neurosci. 2:654–662.

Fry JM Weissbarth S Lehrer GM and Burnstein MB (1974): Cerebroside antibody inhibits sulfatide synthesis and myelination and demyelination in cord tissue cultures. Science 183: 540–542.

Glover J (1993): The development of brain stem projections to the spinal cord in the chicken embryo. Brain Res. Bull. 30: 265–272.

Glover J and Pettursdottir G (1991): Regional specificity of developing reticulospinal vestibulospinal and vestibulo-ocular projections in the chicken embryo. J. Neurobiol. 22: 353–376.

Grierson JP Petroski RE Ling SF and Geller HM (1990): Astrocyte topography and tenascin/cytotactin expression: correlation with the ability to support neuritic outgrowth. Dev. Brain Res. 55:11–19.

Guénard V Kleitman N Morrissey TK Bunge RP and Aebischer P (1992): Syngeneic Schwann cells derived from adult nerves seeded in semipermeable guidance channels enhance peripheral nerve regeneration. J. Neurosci. 12:3310–3320.

Hartman BK Agrawal HC Kalmbach S and Shearer WTJ (1979): A comparative study of the immunohistochemical localization of basic protein to myelin and oligodendrocytes in rat and chicken brain. J. Comp. Neurol. 188: 273–290.

Hasan SJ Nelson BH Valenzuela JI Keirstead HS Schull SE Ethell DW and Steeves JD (1991): Functional repair of transected spinal cord in embryonic chick. Restor. Neurol. Neurosci. 2: 137–154.

Hasan SJ Keirstead HS Muir GD and Steeves JD (1993): Axonal regeneration contributes to repair of injured brain stem-spinal neurons in embryonic chick. J. Neurosci. 13: 492–507.

Herndon ME and Lander AD (1990): A diverse set of developmentally regulated proteoglycans is expressed in the rat central nervous system. Neuron 4:949–961.

Houle JD (1991): Demonstration of the potential for chronically injured neurons to regenerate axons into intraspinal peripheral nerve grafts. Exp. Neurol. 113:1–9.

Hruby S Alroul EC-Jr and Seil FJ (1977): Synthetic galactocerebroside evoke myelination-inhibiting antibodies. Science 195: 173–175.

Iwashita Y Kawaguchi S and Murata M (1994): Restoration of function by replacement of spinal cord segments in the rat. Nature 367:167–170.

Keirstead HS Hasan SJ Muir GD and Steeves JD (1992): Suppression of the onset of myelination extends the permissive period for the functional repair of embryonic spinal cord. Proc. Natl. Acad. Sci. (USA): 89: 11664–11668.

Keirstead HS Dyer JK Sholomenko GN McGraw J Delaney KR and Steeves JD (1995): Axonal regeneration and physiological activity following transection and immunological disruption of myelin within the hatchling chick spinal cord. J. Neurosci. 15(10):6963–6974.

Keller F Rubin B Oesch B and Schwab ME (1993): Purification and characterization of a neurite growth inhibitor from CNS myelin. Soc. Neurosci. Abstr. 19:877.

Kolodkin AL Matthes DJ and Goodman CS (1993): The semaphorin genes encode a family of transmembrane and secreted growth cone guidance molecules. Cell 75:1389–1399.

Lang DM Rubin BP Schwab ME and Stuermer CAO (1995): CNS myelin and oligodendrocytes of the Xenopus spinal cord - but not optic nerve - are nonpermissive for axon growth. J. Neurosci. 15:99–109.

Laywell ED Dörries U Bartsch U Faissner A Schachner M and Steindler DA (1992): Enhanced expression of the developmentally regulated extracellular matrix molecule tenascin following adult brain injury. Proc. Natl. Acad. Sci. (USA): 89:2634–2638.

Levine JM (1994): Increased expression of the NG2 chondroitin-sulphate proteoglycan after brain injury. J. Neurosci. 14:4716–4730.

Li D Field PM and Raisman G (1995): Failure of axon regeneration in postnatal rat. Eur. J. Neurosci. 7:1164–1171.

Ludwin SK (1978): Central nervous system demyelination and remyelination in the mouse: and ultrastructural study of cuprizone toxicity. Lab. Invest. 39:597–612.

Luo Y Raible D and Raper JA (1993): Collapsin: a protein in the brain that induces the collapse and paralysis of neuronal growth cones. Cell 75:217–227.

Lurie DI and Selzer ME (1991): Axonal regeneration in the adult lamprey spinal cord. J. Comp. Neurol. 306:409–416.

Macklin WB and Weill CL (1985): Appearance of myelin proteins during development in the chick central nervous system. Dev. Neurosci. 7: 170–178.

Martin GF Ghooray GT Wang XM Xu XM and Zou XC (1994): Models of spinal cord regeneration. Prog. Brain Res. 103:175–202.

Martini R (1994): Expression and functional roles of neural cell surface molecules and extracellular matrix components during development and regeneration of peripheral nerves. J. Neurocytol. 23:1–28.

Mastiglia F Carroll W and Jennings A (1989): Spinal cord lesions induced by anti-galactocerebroside serum. Clin. Exp. Neurol. 26: 33–44.

McClellan AD (1990): Locomotor recovery in spinal-transected lamprey: role of functional regeneration of descending axons from brain stem locomotor command neurons. Neuroscience 37: 781–798.

McConnell J and Sechrist J (1980): Identification of early neurons in the brain stem and spinal cord. I. An autoradiographic study in the chick. J. Comp. Neurol. 192: 769–783.

McKeon RJ Schreiber RC Rudge JS and Silver J (1991): Reduction of neurite outgrowth in a model of glial scarring following CNS injury is correlated with the expression of inhibitory molecules on astrocytes. J. Neurosci. 11:3398–3411.

McKerracher L David S Jackson DL Kottis V Dunn RJ and Braun PE (1994): Identification of myelin-associated glycoprotein as a major myelin-derived inhibitor of neurite outgrowth. Neuron 13: 805–811.

Melnick RL Mahler J Bucher JR Thompson M Hejmancik M Ryan MJ and Mezza LE (1994): Toxicity of dietholamine. 1. Drinking water and topical application exposures in F344 rats. J. Appl. Toxicol. 14:1–9.

Mitchell J and Caren CA (1982): Degeneration of non-myelinated axons in the rat sciatic nerve following lysolecithin injection. Acta Neuropath. 56:187–193.

Mocchetti I and Wrathall JR (1995): Neurotrophic factors in central nervous system trauma. J. Neurotrauma. 12:853–870.

Montgomery CT and Robson JA (1993): Implants of cultured Schwann cells support axonal growth in the central nervous system of adult rats. Exp. Neurol. 122:107–124.

Morrissey TK Kleitman N and Bunge RP (1991): Isolation and functional characterization of Schwann cells isolated from adult peripheral nerves. J. Neurosci. 11:2433–2442.

Mukhopadhyay G Doherty P Walsh FS Crocker PR and Filbin MT (1994): A novel role for myelin-associated glycoprotein as an inhibitor of axonal regeneration. Neuron 13: 757–767.

Ng WP Cartel N Roder J Roach A and Lozano A (1996): Human central nervous system myelin inhibits neurite outgrowth. Brain Res. 720: 17–24.

Nicholls J and Saunders N (1995): Regeneration of immature mammalian spinal cord after injury. Trends Neurosci. 19:229–234.

O'Donovan M Sernagor E Sholomenko G Ho S Antal M and Yee W (1992): Development of spinal motor networks in the chick embryo. J. Exp. Zool. 261: 261–273.

Okado N and Oppenheim RW (1985): The onset and development of descending pathways to the spinal cord in the chick embryo. J. Comp. Neurol. 232: 143–161.

Oohira A Matsui F and Katho-Semba R (1991): Inhibitory effects of brain chondroitin sulphate proteoglycans on neurite outgrowth from PC12D cells. J. Neurosci. 11:822–827.

Ozawa K Saida T Saida K Nishitani H and Kameyama M (1989): In vivo CNS demyelination mediated by anti-galactocerebroside antibody. Acta Neuropathol. 77: 621–628.

Paino CL Fernandez-Valle C Bates ML and Bunge MB (1994): Regrowth of axons in lesioned adult spinal cord: promotion by implants of cultured Schwann cells. J. Neurocytol. 23:433–452.

Perry VH and Brown MC (1992): Role of macrophages in peripheral nerve degeneration and repair. Bioessays 14:401–406.

Pesheva P Spiess E and Schachner M (1989): J1-160 and J1-180 are oligodendrocyte-secreted nonpermissive substrates for cell adhesion. J. Cell Biol. 109:1765–1778.

Pollock M (1995): Nerve regeneration. Curr. Op. Neurol. 8:354–358.

Ramon-y-Cajal S (1928): Degeneration and Regeneration of the Nervous System; (English translation by Raoul M. May): Hafner New York 1959.

Rancht B Clapshaw PA Price J Noble M and Seifert W (1982): Development of oligodendrocytes and Schwann cells studied with a monoclonal antibody against galactocerebroside. Proc. Natl. Acad. Sci. (USA): 79: 2709–2713.

Reier RJ Anderson DK Thompson FJ and Stokes BT (1992): Neural tissue transplantation and CNS trauma: anatomical and functional repair of the injured spinal cord. J. Neurotrauma. 9:S223–242.

Rich KM (1992): Neuronal death after trophic factor deprivation. J. Neurotrauma. 9:S61–69.

Richardson PM (1991): Neurotrophic factors in regeneration. Curr. Op. Neurobiol. 1:401–406.

Richardson PM McGuinness UM and Aguayo AJ (1980): Axons from CNS neurons regenerate into PNS grafts. Nature 284:264–265.

Richardson PM Issa VM and Aguayo AJ (1984): Regeneration of long spinal axons in the rat. J. Neurocytol. 13:165–182.

Rodriguez M (1992): Central nervous system demyelination and remyelination in multiple sclerosis and viral models of disease. J. Neuroimmunol. 40:255–263.

Saunders NR Balkwill P Knott G Habgood MD Mollgard K Treherne JM and Nicholls JG (1992): Growth of axons through a lesion in the intact CNS fetal rat maintained in long-term culture. Proc. R. Soc. Lond. B. Biol. 250:171–180.

Savio T and Schwab ME (1989): Rat CNS white matter but not gray matter is non-permissive for neuronal cell adhesion and fiber outgrowth. J. Neurosci. 9:1126–1133.

Schnell L and Schwab ME (1990): Axonal regeneration in the rat spinal cord produced by an antibody against myelin-associated neurite growth inhibitors. Nature 343: 269–272.

Schnell L Schneider R Kolbeck R Barde Y-A and Schwab ME (1994): Neurotrophin-3 enhances sprouting of corticospinal tract during development and after adult spinal cord lesion. Nature 367: 170–173.

Schwab ME and Caroni P (1988): Oligodendrocytes and CNS myelin are nonpermissive substrates for neurite outgrowth and fibroblast spreading in vitro. J. Neurosci. 8:2381–2393.

Schwab ME and Thoenen H (1985): Dissociated neurons regenerate into sciatic but not optic nerve explants in culture irrespective of neurotrophic factors. J. Neurosci. 5:2415–2423.

Schwartz M (1993): Nwe light on nerve regeneration in the mammalian nervous system. Endevour. 17:38–40.

Schwegler G Schwab ME and Kapfhammer JP (1995): Increased collateral sprouting of primary afferents in the myelin-free spinal cord. J. Neurosci. 15:2756–2767.

Sergott RC Brown MJ Silberberg DH and Lisak RPJ (1984): Antigalactocerebroside serum demyelinates optic nerve in vivo. J. Neurol. Sci. 64: 297–303.

Sharma SC Jadhao AG and Rao PD (1993): Regeneration of supraspinal projection neurons in the adult goldfish. Brain Res. 620:221–228.

Shiga T Kunzi R and Oppenheim RW (1991): Axonal projections and synaptogenesis by supraspinal descending neurons in the spinal cord of the chick embryo. J. Comp. Neurol. 305: 83–95.

Shimizu I Oppenheim RW O'Brien M and Schneiderman A (1990): Anatomical and functional recovery following spinal cord transection in the chick embryo. J. Neurobiol. 21: 918–937.

Sholomenko GN and O'Donovan M (1995): Development and characterization of pathways descending to the spinal cord in the developing chick. J. Neurophys. 73:1223–1233.

Sholomenko GN and Steeves JD (1987): Effects of selective spinal cord lesions on hind limb locomotion in birds. Exp. Neurol. 95: 403–418.

Sholomenko GN Funk GD and Steeves JD (1991a): Locomotor activities in the decerebrate bird without phasic afferent input. Neuroscience 40: 257–266.

Sholomenko GN Funk GD and Steeves JD (1991b): Avian locomotion activated by brain stem infusion of neurotransmitter agonists and antagonists. I. Acetylcholine excitatory amino acids and substance P. Exp. Brain Res. 85: 659–673.

Sholomenko GN Funk GD and Steeves JD (1991c): Avian locomotion activated by brain stem infusion of neurotransmitter agonists and antagonists. II gama-Aminobutyric acid. Exp. Brain Res. 85: 674–681.

Sivon T Schwab ME and Schwartz M (1994): Presence of growth inhibitors in fish optic nerve myelin: post injury changes. J. Comp. Neurol. 343:237–246.

Siegal JD Kliot M Smith GM and Silver J (1990): A comparison of the regeneration potential of dorsal root fibers into gray or white matter of the adult rat spinal cord. Exp. Neurol. 109:90–97.

Snow D.M Lemmon V Carrino DA Caplan AI and Silver J (1990): Sulphated proteoglycans in astroglial barriers inhibit neurite outgrowth in vitro. Exp. Neurol. 109:111–130.

Sommer I and Schachner M (1981): Monoclonal antibodies (O1 and O4): to oligodendrocyte cell surfaces: an immunocytological study in the central nervous system. Dev. Biol. 83: 311–327.

Spring J Beck K and Chiquet-Ehrismann R (1989): Two contrary functions of tenascin: dissection of the active sites by recombinant tenascin fragments. Cell 59:325–334.

Steeves JD Hasan SJ Keirstead HS Muir GD Ethell DW Pataky DM McBride CB Rott ME and Wisniewska AB (1993): The embryonic chicken as a model for central nervous system injury and repair. Neuroprotocols 3: 35–43.

Steeves JD Sholomenko GN and Webster DMS (1987): Stimulation of the pontomedullary reticular formation initiates locomotion in decerebrate birds. Brain Res. 401: 205–212.

Stoner GL (1993): Polyoma virus models of brain infection and the pathogenesis of multiple sclerosis. Brain Pathol. 3:213–227.

Sweasey D Patterson DS and Leadon DP (1982): Chemical composition of the spinal cord in the normal developing fetus and in the premature foal. J. Reprod. Fert. Suppl. 32:563–567.

Taylor J Pesheva P and Schachner M (1993): Influence of janusin and tenascin on growth cone behavior in vitro. J. Neurosci. Res. 35:347–362.

Tello F (1911): La influencia del neurotropismo en la regeneracion de los centros nerviosos. Trabajos del Lab de Invest. Biol. book 9: 124–159.

Terenghi G (1995): Peripheral nerve injury and regeneration Histol. Histopath. 10:709–718.

Tetzlaff W Alexander S Miller F and Bisby M (1991): Response of facial and rubrospinal neurons to axotomy: changes in mRNA expression for cytoskeletal proteins and GAP-43. J. Neurosci. 11: 2528–2544.

Treherne JM Woodward SKA Varga ZM Ritchie JM and Nicholls JG (1992): Restoration of conduction and growth of axons through injured spinal cord of neonatal opossum in culture. Proc. Natl. Acad. Sci. (USA): 89: 431–434.

Tsunoda I and Fujina RS (1996): Two models for multiple sclerosis: experimental allergic encephalomyelitis and Theiler's murine encephalomyelitis virus. J. Neuropathol. Exp. Neurol. 55:673–683.

Tuttle R and Matthew WD (1991): An in vitro bioassay for neurite growth using cryostat sections of nervous tissue as substratum. J. Neurosci. Methods 39:193–202.

Valenzuela JI Hasan SJ and Steeves JD (1990): Stimulation of the brain stem reticular formation evokes locomotor activity in embryonic chicken in ovo. Dev. Brain Res. 56: 13–18.

Vandervelde M and Zurbriggen A (1995): The neurobiology of canine distemper virus infection. Vet. Microbiol. 44:271–280.

Varga ZM Bandtlow CE Erulkar SD Schwab ME and Nicholls JG (1995): The critical period of repair of CNS neonatal opossum (Monodelphis domestica): in culture: correlation with development of glial cells myelin and growth inhibitory molecules. Eur. J. Neurosci. 7:2119–2129.

Waller AV (1850): Experiments on the section of the glossopharyngeal and hypoglossal nerves in the frog and observation on the alterations produced thereby in the structure of their primitive fibers. Phil. Trans. R. Soc. London (Biol.). 140:423.

Wanner M Lang DM Bandtlow CE Schwab ME Bastmeyer M and Stuermer CA (1995): Reevaluation of the growth-permissive substrate properties of goldfish optic nerve myelin and myelin proteins. J. Neurosci. 15:7500–7508.

Watanabe E and Murakami F (1989): Preferential adhesion of chick central neurons to the gray matter of the central nervous system. Neurosci. Lett. 97:69–74.

Webster DMS and Steeves JD (1988): Origins of brain stem-spinal projections in the duck and goose. J. Comp. Neurol. 273: 573–583.

Webster DMS and Steeves JD (1991): Funicular organization of avian brain stem-spinal projections. J. Comp. Neurol. 312: 467–476.

Wehrle-Haller B and Chiquet M (1993): Dual function of tenascin: simultaneous promotion of neurite outgrowth and inhibition of glial migration. J. Cell Sci. 106:597–610.

Yakovleff A Cabelguen J-M Orsal D Gimenez y Robotta M Rajaofetra N Drian M-J Bussel B and Privat A (1995): Fictive activities in adult chronic spinal rats transplanted with embryonic brain stem neurons. Exp. Brain Res. 106:69–78.

Yajima K and Suzuki K (1979): Demyelination and remyelination in the rat central nervous system following ethidium bromide injection. Lab. Invest. 41:385–392.

ABSTRACTS

THE STRUCTURE OF MYELIN BASIC PROTEIN DETERMINED BY HIGH RESOLUTION ELECTRON MICROSCOPY AND MOLECULAR MODELLING

D.R. Beniac, R.A. Ridsdale, M.D. Luckevich, T.A. Tompkins[*] and G. Harauz. Department of Molecular Biology & Genetics, University of Guelph, Guelph, Ontario, Canada, N1G 2W1.

Knowledge of the tertiary structure of myelin basic protein (MBP) is essential to understanding the organisation of the myelin membrane and the mechanisms of development of autoimmunity in multiple sclerosis. We have initiated high-resolution electron microscopical (EM) analyses of MBP to determine its structure. When imaged as individual particles, MBP demonstrates a "C" shape in projection, of approximately 1.5 nm in thickness and 8 nm in length. Two-dimensional crystallisation experiments on MBP using monolayers of DHAA (dehydroabietylamine) or lipid mixtures (phosphatidylcholine, phosphatidylserine, dioleoyl phosphatidylethanolamine, galactocerebroside) yielded lamellar or occasionally crystalline structures with a repeating pattern of packed protein molecules with a unit cell size of approximately 2 nm by 7 nm, suggesting that the MBP extends itself upon interaction with the monolayer. Molecular modelling of MBP based on a proposal by Stoner (J. Neurochem. 43, 433–447, 1984), using coordinates of the B sheet backbone of bacteriochlorophyll A protein and energy minimisation, yielded a compact structure of size and appearance consistent with both our EM data and with biochemical data on accessibility of specific sites (e.g., Ser7, Thr98) to modification. The bend of the C shape appears due to a natural curvature in the B sheet. These results are fully congruous with MBP having a regular secondary structure and domains that are potentially flexible with respect to one another. This work was supported by the Multiple Sclerosis Society of Canada.

CRYSTALLIZATION OF MYELIN BASIC PROTEIN - SOME REFLECTIONS

J. Sedzik and S. Hjertén. Department of Biochemistry, BMC Box 576, Uppsala University, 751-23 Uppsala, Sweden.

* Present address: Lallemand Inc., Laboratoire R&D, 6100, Royalmount, Montreal. Quebec. Canada. H4P 2R2.

365

In spite of the tremendous effort of several laboratories, myelin basic protein (MBP; MW 18.5 kDa) resisted attempts at 2-D or 3-D crystallization. This failure totally afflicted determination of its atomic structure. In view of the simplicity of the crystallization techniques, it seems most likely that the persistent difficulties of MBP crystallization may occur from the complexity of purification protocols, clearly implying the importance of purification. The "bottleneck" is to find the conditions for 2-D or 3-D crystallization. In this report we present a novel method of MBP purification and crystallization. We have purified myelin basic proteins in the presence of 20 mM phosphate buffer, pH 7.4, 0.1–0.5 M NaCl and 30% ethylene glycol using immobilized Zn^{2+} metal ion chromatography followed by the cation exchange chromatography on a "continuous bed" column. The protein used for crystallization is "pure" (one single band, on SDS-PAGE or capillary electrophoresis) and it is enzymatically active against p-nitrophenyl acetate. Unfortunately, due to the lack of diffracting power, the obtained crystals cannot be used for x-ray determination. In conclusion, MBP - which by the overwhelming amount of data is classified as a flexible or random coil - surely can form crystals, *i.e., it is crystallizable*. The next, very difficult task will be to obtain such a form of crystals which will diffract.

VESICULATION - A NEW APPROACH TO STUDYING MYELIN MEMBRANE

J.Sedzik, E. Brekkan and P. Lundahl, Department of Biochemistry, BMC Box 576, Uppsala University, 751–23 Uppsala, Sweden.
The main purpose of vesiculating myelin membrane is to gain more information about its structure; to obtain more information on the topographical distribution of proteins in the membrane; to access and to characterize proteins that play key roles in maintaining the myelin sheath integrity; to search for a putative role of myelin enzymes; and to improve the understanding of myelinogenesis and myelin function. When myelin membrane is vesiculated it is possible to study the in/out transport of water, ions, drugs. Hitherto, the vesiculation of myelin (particularly CNS0 was a difficult task. In this work sonication was applied as a method of creating vesicles. Myelin membrane was purified from bovine spinal cord and dispersed in buffer, 50 mm Tris-HC1, pH 7.4, protein concentration 2.5 mg/ml. This suspension (4 ml) was sonicated 10 x 1 min at maximal power (VibraCell VC 600, 3 mm microtip). As judged by the EM (negative staining) myelin vesicles of diameter approximately 2,500 Å were reproducibly created. The internal volume of vesicles was determined by inclusion of radioactive D-glucose during dispersion of myelin. Free glucose and glucose entrapped in myelin vesicles were separated on a Sephadex G-50 column. The amount of radioactivity was quantified on-line with a flow scintillation detector (Radiomatic FLO-ONE Beta 525TR). The internal volume of myelin vesicles was 34 ml/mg protein (or 14 ml/mg lipids).

MODULATION OF MAG AND SMP BINDING BY SIALYLATION

Michael B. Tropak and John C. Roder. Samuel Lunenfeld Research Institute at Mt. Sinai Hospital, Toronto, Ontario, Canada.
Myelin associated glycoprotein (MAG), as well as SMP, CD22 and CD33, are members of a newly defined family of cell adheson molecules belonging to the immunoglobulin superfamily which recognize specific terminal sialic acid residues on N-and/or

O-linked oligosaccharides. The behaviour of mutant mice which no longer express MAG on myelinating glia suggests that the protein mediates interactions with counterreceptors on axons which are important for the maintenance of the periaxonal region of the myelin sheath. MAG is highly glycosylated. To determine the importance of the N-linked oligosccharides on MAG, carbohydrates were removed either enzymatically or by mutagenesis of the eight predicted carbohydrate addition sites. The results suggested that all eight N-liked glycosylation sites are utilized in vivo and that they do not appear to be required for MAG binding to its ligand. However, N-linked glycosylation of MAG is required for maximal expression of the folded protein. Previously it had been shown by others, that erythrocyte binding by CD22 and CD33, can be inhibited by sialylation. Here we show that MAG, SMP and especially the MAG mutant lacking all N-linked glycosylation sites are susceptible to the inhibitory effects of sialylation. The degree to which the various constructs are affected by sialylation depend in part on 1) the level at which MAG is expressed on the surface of the host cell and 2) the presence of appropriately sialylated ligands on the host cell. A model is proposed whereby MAG can interact with sialylated ligands on the host cell as well as on the target cell. This model may account for the alterations observed both in the myelinating glia and axons of older mice with a mutated MAG gene.

OXIDATION OF PROTEOLIPID PROTEIN DURING COLUMN SEPARATION

F. Giordano, B. Chang, A.M. La Ronde, A. Al-Sabbagh and M. Kretschmer. AutoImmune, Inc., 128 Spring St., Lexington, Massachusetts 02173.

Various protocols exist for the purification of proteolipid protein from brain tissue. In all of these purification schemes, sufficient delipidation is a key issue in achieving water solubility of the final product. One widely used procedure for removal of lower molecular weight lipid components is gel filtration on Sephadex LH60 with chloroform / methanol / acetic acid as eluent. Here we present evidence that bovine proteolipid protein delipidated by the above column method contains a significant amount of oxidized tryptophan at position 211. Lys-C peptide maps of proteolipid protein purified with and without the column step revealed that peptide $PLP_{192-217}$ elutes in two peaks from a C-4 column. The molecular masses of the peptides in these two peaks were 2744.0 ± 0.8 and 2727.2 ± 0.2 for the earlier and later eluting peaks, respectively. The later eluting peak corresponds to the non oxidized peptide and dominates in material that was not subjected to gel filtration in acidified chloroform / methanol. The oxidized (earlier eluting) peptide is more prevalent in digests of column purified material and might be due to acid catalyzed oxidation of tryptophan 211.

CELL-SPECIFIC FACTORS BIND TO REGULATORY ELEMENTS LOCATED DOWNSTREAM OF THE TATA BOX ELEMENT IN THE MOUSE MYELIN BASIC PROTEIN (MBP) PROMOTER

A. Asipu and G.E. Blair, Department of Biochemistry and Molecular Biology, University of Leeds, Leeds LS2 9JT. U.K.

Cell-specific transcription of the MBP gene in primary oligodendrocytes (OL) is regulated by cis-acting regulatory elements located downstream of the TATA-box region of the MBP promoter. To identify the cell-specific trans-acting factors that bind to these downstream regulatory elements (in the region -53 to + 1 05) and the factors that may be responsible for the control of MBP expression in non-neural cells, we utilised DNase I footprinting analysis and gel retardation assays using nuclear extracts from myelin form-ing OL, and non-myelin forming cells C6 glioma (C6) and baby rat kidney (BRK) cells. Since the DNase I protected region in C6, from +58 to +64, showed extensive sequence homology to the interferon (IFN) consensus sequence of major histocompatibility class I genes (MHC I-ICS), we also performed gel retardation assay using MHC 1- ICS oligonu-cleotide and transient transfection analysis using MBP-CAT constructs to study the effect of IFNs on the MBP promoter activity in OL and C6. The results provided evidence for the existence of a specific transcription factor(s) that appeared to bind the negative regula-tory element (NRE) at +48 to +56 of the MBP promoter and conferred cell-specific ex-pression of MBP in OL. In OL, IFN-α/β caused little induction of CAT activity, but IFN-γ resulted in a 2- to 3.5-fold decrease in CAT activity. In contrast, in C6 cells both IFN-α/β and IFN-γ induced a significant (1.5 to 2.5-fold) increase in activity. Although our results support previous observations that IFN-γ inhibits MBP expression in OL, it is difficult to explain why IFNs displayed an opposite effect in C6 cells by stimulating MBP expression. It is possible that in OL, NRE binding factors play a key role by modulating the sensitivity of the ICS-like sequence in the MBP promoter to IFN-γ. Trans-acting factors binding to the NRE may be responsible for the selective synthesis of OL-specific isoforms of MBP, while suppressing the synthesis of other isoforms in OL. (Supported by a grant from the Multiple Sclerosis Society of Great Britain and Northern Ireland).

EXPRESSION OF PROTEOLIPID PROTEIN GENE TRANSCRIPTS IS UNALTERED IN THE ND4 TRANSGENIC MOUSE

B. Mak and M.A. Moscarello, Department of Biochemistry Research, Hospital for Sick Children, Toronto, Ontario, Canada M5G 1X8.

The ND4 demyelinating mouse model containing 70 copies of a DM20 cDNA trans-gene has been partially characterized in our lab. With over-production of DM20 protein in these animals, proteolipid protein (PLP) was found to be reduced throughout development (F.Mastronardi and M.A.Moscarello, J.Neurocsci.Res., 1993). Early studies showing high levels of transgene RNA suggest that competition for transcription factors may be occur-ring between the transgene and the endogenous PLP gene (R.Simon-Johnson et al, J.Neurochem., 1995). Decreased PLP gene transcription would result in reduced protein synthesis. Using a mobility shift assay to measure transcription factors, two complexes binding to a transcriptionally active sequence of the PLP promoter were resolved from the brain nuclear extract of both normal and transgenic mice. One of these complexes was shown to be specific by competition with an unlabeled oligonucleotide. The amount of this complex was the same in both normal and transgenic samples showing that the factors specific to this sequence are not sequestered. The expression levels of PLP and endo-genous DM20 RNA were examined using RNA blots and were found to be unaltered in transgenic mice over development. Transgene RNA levels were found to increase rapidly leveling off at 2–3 times higher than endogenous transcripts. Despite this, the unaltered levels of endogenous PLP gene transcripts does not reflect the reduced protein amounts. Unchanged levels of transcription factors specific to the PLP promoter sequence suggests that the defect resulting in reversed protein expression patterns in the ND4 mice does not lie at the transcriptional level. Although the effect arises from the addition of DM20 cDNA, a post-transcriptional mechanism may be the underlying defect resulting in de-myelination in these mice.

OLFACTORY ENSHEATHING CELLS CONTINUE TO EXPRESS THE P75 NEUROTROPHIN RECEPTOR IN CULTURES FED WITH TRIIODOTHYRONINE, TGF-BETA OR RETINOIC ACID

R. Doucette, K. Gratto and V. Verge. Department of Anatomy & Cell Biology and the Multiple Sclerosis and Neuroscience Research Center, University of Saskatchewan, Saskatoon, Saskatchewan Canada S7N 5E5.

Olfactory ensheathing (En) cells are the glial cells that ensheath the primary olfac-tory axons in both the PNS and CNS portions of the primary olfactory pathway. In vivo, these cells express the p75 neurotrophin receptor, but only during fetal and early neonatal development. In rats made hypothyroid by including PTU in the drinking water, p75 con-tinues to be expressed in the olfactory nerve in adult animals. The objective of this study was to determine whether triiodothyronine (T3) might play a role in downregulating the expression of p75 by En cells during neonatal development. Cell cultures of En cells were initiated from the olfactory bulbs of E18 rat embryos. These neuron-free cultures were fed with serum-free medium containing 10 to 160 nM T3 for up to 14 days in vitro (DIV). The cultures were stained with the 192-IgG monoclonal antibody to p75, using the Vectastain ABC Elite kit. Additional cultures were fed with serum-free medium containing T3 (20–80 nM) as well as TGF-beta (2 ng/ml) and/or retinoic acid (1–20 ng/ml). En cells in all treatment groups continued to express the p75 receptor. The results do not support a role for T3 in acting directly on En cells to downregulate their expression of p75. These experiments need to be repeated using co-cultures of neurons and En cells. (Supported by grants from the NIDCD to R.D. and from the MRC of Canada to V.V.).

OLFACTORY ENSHEATHING CELLS CONTINUE TO EXPRESS THE HOX-A2, -B3 & -B4 TRANSCRIPTION FACTORS IN CULTURES FED WITH TRIIODOTHYRONINE, TGF-BETA, NGF OR RETINOIC ACID

R. Doucette[1], J. Yeung[2] and A. Nazarali[3]. [1]Department of Anatomy & Cell Biology, College of Medicine & [3]Laboratory of Molecular Biology, College of Pharmarmacy & Nutrition, University of Saskatchewan, Saskatoon, Saskatchewan Canada S7N 5E5.; [2]Health Protection Branch, Sir Frederick Banting Research Centre, Ottawa, ON, Canada.

Olfactory ensheathing (En) cells are the glial cells that ensheath the primary olfac-tory axons in both the PNS and CNS portions of the primary olfactory pathway. We are interested in comparing the repertoire of transcription factors expressed by nonmyelinat-ing and myelinating En cells. The objective of this study was to study the expression of three members of the Hox gene family (Hoxa2, Hoxb3 & Hoxb4) by nonmyelinating En cells in vitro. These neuron-free cell cultures were initiated from the olfactory bulbs of E18 rat embryos. Using rabbit polyclonal antibodies to identify expressing cells (Vectas-tain ABC Elite Kit), we found that each transcription factor was expressed by nonmyelinating En cells fed with either serum-free or serum-containing medium. We then initiated a series of experiments to identify the molecular mechanisms through which Hox gene expression could be downregulated in En cells in these cultures. The cells were fed with serum-free medium containing: TGF-beta (1–4 ng/ml), triiodothyronine (20–80 nM), NGF (1–25 ng/ml) or retinoic acid (1 ng/ml). Ensheathing cells in all treatment groups continued to express Hoxa2, Hoxb3 and Hoxb4, even when being fed with medium containing two or more of these factors at the same time. These experiments need to be repeated using co-cultures of neurons and En cells to determine the effect of axonal contact and myelination on the expression of these three Hox proteins. (Supported by grants from the NIDCD to R.D. and from the NSERC of Canada to A.N.)

THE REGULATION OF KROX-20 EXPRESSION REFLECTS IMPORTANT STEPS IN THE CONTROL OF PERIPHERAL GLIAL CELL DEVELOPMENT

P. Murphy, P. Topilko, S. Schneider-Maunoury, T. Seitanidou, A. Baron-Van Evercooren[†] and P. Charnay. Unité 368 de l'Institut National de la Santé et de la recherche Medicale, Ecole Normale Supérieure, 46, rue d'Ulm, 75230 Paris, France.

The zinc finger transcription factor gene Krox-20 is expressed in Schwann cells and is required for the myelination of peripheral nerves. We have shown that the regulation of Krox-20 expression in peripheral glial cells reflect three important processes in the development and differenciation of these cells: i) Expression of Krox-20 in Schwann cells requires continuous neuronal signalling via direct axonal contact. Therefore Krox-20 appears to be a key component of the transduction cascade linking axonal signalling to myelination. ii) Krox-20 inducibility is acquired by Schwann cells at the time these cells are formed from their precursors. Diffusible factor(s) synthesised by the neural tube can mediate this transition and this effect is mimicked by NDF-2 or a combination of CNTF and bFGF. iii) In sensory ganglia, the microenvironment is capable of negatively regulating Krox-20, presumably by preventing the conversion of satellite cells toward a Schwann cell-like phenotype.

THE MYELIN GENE P_0 IS CONSTITUTIVELY EXPRESSED BEFORE MYELINATION AND IS RESPECTIVELY DOWN- AND UP-REGULATED DURING THE DEVELOPMENT OF NON-MYELIN AND MYELIN-FORMING CELLS

M.J. Lee, A. Brennan, A. Tabernero, Z Dong, A. Blanchard, G. Zoidl, K.R. Jessen and R. Mirsky. Department of Anatomy and Developmental Biology, University College London, Gower Street, London WC1E 6BT, England.

The expression of the myelin protein P0 is examined in the rat Schwann cell lineage using in situ hybridisation and immunohistochemical methods that have been adjusted to higher sensitivity than that needed to detect the exceptionally high levels of P0 in myelinating cells in vivo. This reveals unambiguous P0 mRNA expression in migrating neural

† Unité 134 de l'Institut National de la Santé et de la recherche Medicale, Hôpital de la Salpêtriere, 75651 Paris.

crest cells, Schwann cell precursors from embryo day 14 (E14) nerves, and in embryonic Schwann cells. The Schwann cells in the E18 sympathetic trunk are P0 positive using this adjusted method, although 95% of cells in this nerve remain non–myelin–forming in the adult, and the level is very much the same as most pre-myelinating Schwann cells in E18 sciatic nerves. Transecting the sciatic nerve at birth does not alter this basal P0 expression significantly in nerves observed two days later (postnatal day 2). In contrast, it abolishes the strikingly high P0 mRNA and protein seen in some cells (those forming myelin sheaths) in the contralateral control nerves. Furthermore, non-myelin-forming cells in adult nerves fail to show the basal P0 levels, indicating the P0 expression has been suppressed in these cells. In this case, nerve transection results in up-regulation of P0 mRNA expression in the denervated non-myelin-forming cells. We conclude that the postnatal diversification of immature Schwann cells to form myelin-forming and non-myelin-forming cells involves axon-dependent amplification and suppression, respectively, of myelin-independent P0 expression. In the Schwann cell lineage, this basal expression is a constitutive early phenotype that is likely to appear as one of the first signals of glial lineage choice in neural crest development.

DYSMYELINATION AND MICROTUBULAR DEFECT IN OLIGODENDROCYTES OF THE MYELIN MUTANT *TAIEP* RAT

E. Couve, F. Cabello, J. Krsulovic and M. Roncagliolo. Facultad de Ciencias, Universidad de Valparaíso, Valparaíso, Chile.

Myelinogenesis is a complex process in which several genetic defects affecting oligodendrocytes have been identified. The *taiep* rat is a new myelin mutant in which there is a microtubular defect that affects only oligodendrocytes. This defect is inherited as an autosomal recessive trait. *Taiep* opens new opportunities for understanding the myelination process and the alterations produced by dysmyelinating diseases. In *taiep* animals severe neurological alterations become evident after the second postnatal week. These observations induced us to investigate: i) the time of appearance of the microtubular defect, ii) the topographical location of membranous components that become associated with microtubules in *taiep* oligodendrocytes, and iii) the characterization of the early developmental defects during myelination of axons. Optic nerve (ON) samples from 8 to 30 day old homozygote mutant male rats were processed for electron microscopy. Oligodendrocytes were morphologically characterized and microtubules (MTs), endoplasmic reticulum (ER), transitional elements and Golgi apparatus analyzed. Microtubules appear physically bound to smooth ER profiles of oligodendrocytes. The most prominent association of MTs to ER profiles was observed between days 12 and 15. The ER membranes associated with MTs were morphologically characterized as transitional elements that constitute the intermediate compartment according to their topographical location. At 15 days of age, redundant profiles of aberrant myelin sheaths become prominent in ON of the *taiep* rats. The number of myelinated axons was significantly lower than in control animals. The microtubular defect observed in *taiep* could explain the dysmyelinating condition observed in this mutant. Moreover, these observations cause us to propose that there is a blockage of protein trafficking at the intermediate compartment of *taiep* oligodendrocytes. We show morphological evidences indicating that MTs accumulating in *taiep* oligodendrocytes are physically bound to specific subdomains of the ER, characterized as transitional elements, a defect that constitutes the most conspicuous cytological alteration observed in this myelin mutant. Fondecyt 1960293-CHILE

LOCALIZATION OF CONNEXIN32 IN OLIGODENDROCYTES AND ALONG SUBPOPULATIONS OF MYELINATED FIBERS IN THE CNS

J. Li, E.L. Hertzberg and J.I. Nagy. Department of Physiology, University of Manitoba, Winnipeg, Manitoba, Canada R3E 0W3.

A series of monoclonal and polyclonal anti-connexin32 (Cx32) antibodies were used to determine immunohistochemically the cellular localization of Cx32 in rat and mouse central nervous system (CNS). Antibodies were raised against whole Cx32 protein or different peptide sequences of Cx32, and all produced similar immunostaining patterns. Double fluorescence staining with oligodendrocyte markers (CNPase and RIP) indicated the presence of punctate labelling for Cx32 in oligodendrocyte cell bodies and their processes. Punctate Cx32 staining was also found along myelinated fibers and Cx32-positive oligodendrocytic processes could were seen extending to these fibers. Not all myelinated fibers exhibited Cx32 labelling. In cerebellum, for example, labelling was associated with the myelinated axons of Purkinje cells, and not with many other CNPase-positive fibers. This raises the possibility of either differential expression of Cx32 in subpopulations of oligodendrocytes or differential transport of Cx32 in the processes of these cells. Our results also suggest a selective functional requirement for gap junctions along the myelin sheath of certain classes of myelinated fibers, where we have observed a prevalence of such junctions between astrocytic processes and myelinating oligodendrocytic processes.

HYPOMYELINATION IN ADULT MOUSE MUTANTS DEFICIENT IN THE CONNEXIN32 GENE

Dirk H.-H. Neuberg[1‡], Patrizia Anzini[2°], Eric Nelles[3], Klaus Willecke[3], Melitta Schachner[2], Rudolf Martini[2+], and Ueli Suter[1]* [1]Department of Cell Biology and [2]Neurobiology, ETH Hönggerberg, CH-8093 Zürich, Switzerland; [3]Institut für Genetik, Abteilung Molekulargenetik, Universität Bonn, D-53117 Bonn, Germany.

Charcot-Marie-Tooth disease (CMT) is a pathologically and genetically heterogeneous group of disorders which are characterized by progressive degeneration of peripheral nerves. Genetic studies have demonstrated that some forms of X-linked Charcot-Marie-Tooth (CMTX) are caused by mutations affecting the connexin32 (Cx32) gene. Cx32 belongs to a multigene family that consists of at least a dozen members which are expressed in a cell-type specific manner. Functionally, the connexins constitute gap junctions by associating apposed hexameric hemi-channels thereby allowing ionic and metabolic exchanges between adjacent cells. In the peripheral nervous system, Cx32 protein expression is mainly confined to paranodal myelin loops and Schmidt-Lanterman incisures of myelinating Schwann cells but the exact functional role of Cx32 in these structures remains to clarified. We have studied mice carrying a null mutation in the Cx32 gene which were generated by gene targeting techniques in embryonic stem (ES) cells in order to examine the functional role of Cx32 and to prove its causative role in hereditary peripheral neuropathies. Peripheral nerves of mice deficient in the Cx32 gene exhibit signs of myelin

‡ Equal Contribution
° Equal Contribution
+ Equal Contribution

degeneration and remyelination (onion bulb formation). These features of hypomyelination display a late onset by being very rare at the age of four weeks but becoming prominent at the age of four months and in older animals. In conclusion, Cx32-deficient animals
develop pathological changes in peripheral nerves which are comparable to CMTX patients. These results prove that Cx32 is the culprit gene in causing CMTX in humans and
identify Cx32-deficient mice as potential animal models for the evaluation of possible
treatment strategies.

PLASTICITY OF OLIGODENDROCYTES DURING RETINAL AXON DE- AND REGENERATION IN THE INJURED GOLDFISH OPTIC NERVE

R. Ankerhold and C.A.O. Stuermer, University of Konstanz, D-78434 Konstanz, Germany.

Retinal axons in fish are myelinated by oligodendrocytes which morphologically
resemble oligodendrocytes in other vertebrate species. Fish oligodendrocytes, however,
possess specific properties which seem advantageous for CNS fiber tract repair which
occurs so reliably after optic nerve transection (ONS) in fish. When deprived of axons by
culturing them in vitro, fish oligodendrocytes cease expression of advanced myelin
markers, dedifferentiate, divide and promote axon growth (Bastmeyer et al. 1993,
Schwalb et al. 1996) and thus resemble Schwann cells. Like Schwann cells they require
axon contact to re-express advanced myelin markers (Bastmeyer et al. 1994). Likewise,
after ONS in vivo, markers characteristic for myelinating oligodendrocytes are transiently lost as judged by immunocytochemistry. This correlates with a change in morphology of the oligodendrocytes as determined by dye injections into individual cells.
When retinal axons degenerate and subsequently regenerate, cells without any or with
1–2 short processes (perhaps representing dedifferentiated oligodendrocytes) were found.
But when the regenerating axons have reached their target (at about 3 weeks after ONS),
oligodendrocytes with many and profuse processes in contact with the axons re-appear.
By 4 weeks after ONS, oligodendrocytes have developed myelinating segments reminiscent of those in normal optic nerves. Along with the restoration of the oligodendrocyte
morphology, at 3 weeks after ONS markers typically associated with myelination in fish
(IP1, IP2, 36K, MBP) are re-expressed in the regenerating optic nerve and they increase
in density at 4 weeks after ONS. Thus, oligodendrocytes in the fish optic nerve seem to
adapt to the state of the retinal axons, and this adaptive plasticity apparently is useful or
prerequisitory for fiber tract repair. We thank G. Jeserich (Osnabrück) for providing antibodies against IP1/2 and 36K. RA is fellow of the Boehringer Ingelheim Fonds. This
work is supported by the DFG, SFB 156 to CAOS.

FGF-9 AND FGF-2 REGULATE FIBROBLAST GROWTH FACTOR RECEPTOR SUBTYPE EXPRESSION IN THE OLIGODENDROCYTE LINEAGE

Karen J. Chandross[1], W.T. Norton[2], L.D. Hudson[1], and R.I. Cohen[2], NINDS-LDN, National Institutes of Health[1], Bethesda, MD and Department of Neurology[2], Albert Ein-stein College Medicine, Bronx, New York.

In vitro fibroblast growth factor-2 (FGF-2 or basic FGF) stimulates proliferation in oligodendrocyte-type-2 astrocytes (O2A) and committed oligodendrocyte precursors and promotes the survival of their differentiated descendants. However, FGF-2 lacks the traditional hydrophobic signal sequence thought to be necessary for secretion and the protein is localized to the cell cytoplasm and nucleus making it unclear how FGF-2 exerts its effects on oligodendrocyte development in vivo. In contrast, the most recently cloned member of this family, FGF-9, is processed through the ER and Golgi, and is targeted for secretion. In this study, we compared the effects of FGF-2 and -9 on oligodendrocyte development and differentiation using both immunocytochemical and western analyses. Initially, we examined FGF receptor subtype expression in cultured rat oligodendrocytes. A 180 kDa form of FGFR-1 is expressed at all stages of oligodendrocyte development, however, this receptor is not present in highly purified CNS myelin. Three forms of FGFR-2, ranging in size from 90–150 kDa are detected in differentiated oligodendrocytes (O1$^+$) and myelin, but not in progenitor cells (A2B5$^+$). Moreover, FGFR-3 and FGFR-4 are both expressed in progenitor and differentiated cells, however, neither are detectable in myelin. Chronic exposure (up to 4 days) of oligodendrocyte progenitor cells to FGF-2 or -9 up-regulates expression of FGFR-1 while decreasing levels of FGFR-2. However, there is no effect on expression of either FGFR-3 or -4. In differentiated cells neither FGF-2 nor -9 has any effect on the expression of any FGFR subtype. In order to examine the extent to which FGF alters cellular differentiation the expression of three myelin proteins was examined. In progenitor cells FGF-2 or -9 down-regulates expression of myelin basic protein (MPB) and proteolipid protein (PLP), while 3′-5′-cyclic nucleotide phosphodiesterase (CNP) is not affected. In differentiated oligodendrocytes the levels of all myelin proteins examined remained similar to untreated cultures. To examine receptor activation, phosphorylation of MAPK was assayed. In both progenitor and differentiated cells either FGF-2 or -9 stimulates MAPK phosphorylation within 5 min, however, in differentiated cells the potency and efficacy of the FGF-9 effect is increased. These findings indicate that while FGF-2 and -9 effect signal transduction at all stages of oligodendrocyte development they regulate FGFR and myelin protein expression in differentiating cells. Moreover, subcellular receptor localization may be important for the responses of these cells to the FGFs at different stages of development.

A NEURONAL CELL LINE SECRETES BOTH STIMULATORY AND REGULATORY FACTORS FOR OLIGODENDROCYTE PRECURSOR PROLIFERATION

K. Asakura. S.F. Hunter. and M. Rodriguez, Departments of Neurology and Immunology, Mayo Clinic and Foundation, Rochester, Minnesota 55905.

Conditioned medium derived from rat central nervous system neuronal cell line B 104 (B 104 CM) was shown previously to contain uncharacterized potent mitogen(s) for oligodendrocyte/ type-2 astrocyte (0–2A) progenitor cells (Dev. Brain Res. 49:33, 1989). In this study we showed that B104 cells highly express platelet-derived growth factor (PDGF)-A chain but not PDGF-B chain mRNA by reverse transcription-polymerase chain reaction. PDGF-AA homodimer was detected in all B104 cells by immunostaining and in B104 CM by Western blotting. B104 cells did not express other potent mitogens for 0–2A progenitor cells including basic fibroblast growth factor and neurotrophin-3. Additionally, B104 cells expressed transcripts of transforming growth factor-β1 (TGF-β1) and -β2 (TGFb2) which are known to regulate and inhibit 0–2A progenitor cell differentiation and proliferation and secreted 25 kilodalton active forms of TGF-β1 and TGF-β2 into the culture medium. In contrast to the bipolar morphology of A2B5-positive progenitor cells cultured in control media, multipolar A2B5-positive progenitor cells were observed frequently when cultured with B104 CM, suggesting that this enhanced cellular differentiation may be attributed to TGF-β. Even though TGF-β has been reported to down-regulate the expression of PDGF-α receptor, B104 CM maintained high expression of PDGF-α receptor mRNA in oligodendrocytes. This suggested that B104 cells may secrete additional unidentified growth factor(s) by which oligodendrocytes preserve their sensitivity to PDGF via maintaining high PDGF-α receptor expression.

FGF-2 INDUCES MATURE OLIGODENDROCYTES IN CULTURE TO CHANGE TO A NOVEL PHENOTYPE

Rashmi Bansal, Susan Winkler and S. E. Pfeiffer, Departments of Pharmacology and Microbiology, and Program in Neurological Sciences, University of Connecticut Medical School, Farmington, CT 6030–3205, USA.

FGF-2 differentially regulates oligodendrocyte progenitor proliferation and differentiation in culture, and modulates gene expression of its own receptors, in a developmentally and receptor type specific manner (Bansal et al., 1996a,b). Three FGF receptors (FGFR-1, -2, -3) are expressed in post-mitotic, differentiated oligodendrocytes. These cells respond to FGF-2 by (a) down-regulating myelin-specific gene expression (*e.g.*, galactocerebroside transferase, cyclic nucleotide phosphohydrolase, myelin basic protein, proteolipid protein); (b) dramatically increasing process length in a time- and dose-dependent manner; (c) re-entering the cell cycle; (d) altering the expression of both low- and high-affinity FGF receptor expression. Although the treated cells have some characteristics of late oligodendrocyte progenitors, they clearly represent a different, apparently novel, phenotype both morphologically and biochemically. Thus they do not express gangliosides recognized by monoclonal antibody A2B5, they do not respond to PDGF, they incorporate BdUR at a slower rate that neonatal late oligodendrocyte progenitors, and their pattern of FGF high and low-affinity (syndecans) receptors is different. Whether this phenomenon occurs in vivo, where it could have significance for regeneration of oligodendrocytes and myelin sheaths following demyelination, is under investigation. Supported by a grant from the USPHS, NIH, NS10861.

ENHANCEMENT OF PROCESS OUTGROWTH OF ADULT HUMAN OLIGODENDROCYTES BY INTERACTION BETWEEN bFGF AND ASTROCYTE EXTRACELLULAR MATRIX

Y.S. Oh. and V.W. Yong, Montreal Neurological Institute, Department of Neurology and Neurosurgery, McGill University, Montreal, Canada H3A 2B4.

In purified cultures, the rate of process outgrowth by adult human oligodendrocytes (OLs) is very slow when compared to oligodendrocytes from adult rodent brains. However, in co-culture with human astrocytes, the extent of process formation by adult human OLs is significantly enhanced and is comparable to that induced by phorbol esters. To investigate the mechanism by which astrocytes promote oligodendroglial process outgrowth, we exposed adult human OLs to growth factors known to be produced by astrocytes (IGF, FGF, CNTF, PDGF, EGF, NGF, BDNF, NT-3 & -4). Of these, only bFGF exerted significant promoting effects on process extension by OLs; a neutralizing antibody to bFGF retarded the ability of astrocytes to promote oligodendroglial process outgrowth. However, the effects of bFGF was not as potent as that of live astrocytes, and bFGF did not synergize with other growth factors. Thus, we further investigated the role of astrocyte extracellular matrix (ECM) and found that the effects of astrocyte ECM alone was as potent as that of bFGF. Significantly, astrocyte ECM synergized with bFGF to match the effects of live astrocytes. Purified ECM molecules known to be produced by astrocytes (laminin, vitronectin, fibronectin, HSPG & CSPG) were tested. None of these by themselves promoted process outgrowth by OLs, but laminin and fibronectin augmented the effects of bFGF. We conclude that bFGF and astrocyte matrix, likely laminin/fibronectin, are important factors that promote process extension by adult human OLs in vitro.

CELL CULTURE MODEL FOR THE EXAMINATION OF THE FUNCTIONAL CAPABILITY OF ADULT HUMAN OLIGODENDROCYTES

B.H.J. Juurlink[1], R.W. Griebel[2] and R.M. Devon[3]. Departments of Anatomy & Cell Biology[1], Surgery (Neurosurgery)[2], Oral Biology[3], and the Multiple Sclerosis and Neuroscience Research Center, University of Saskatchewan, Saskatoon, Saskatchewan Canada S7N 5E5.

Adult human oligodendrocytes are obtainable from temporal lobectomies performed to treat intractable epilepsy. Our objective was to develop a culture system wherein we could examine: i) the ability of mature human oligodendrocytes obtained from the adult to remyelinate axons, ii) the effect of environmental perturbations on this remyelination activity. The culture model developed involved isolating highly enriched populations of projection neurons and various glial cell populations and then recombining specific cell populations. The cells were then grown in suspension culture as aggregates. This model gives us the advantage of controlling the cell types that interact within the aggregate. Adult human oligodendrocytes were isolated from temporal lobe resections, labelled with colloidal gold and combined with hippocampal neurons isolated from E15 mouse embryos. The cells were grown for up to 5 weeks in suspension as aggregates. The majority of cells in such aggregates are mouse neurons and human oligodendrocytes and less than 10% are either mouse astrocytes or human microglia. A few profiles of myelinated axons were already seen by two weeks, the majority of such profiles were close to astrocyte processes suggesting that interactions with astrocytes are necessary for successful myelination. Experiments are currently underway comparing cultures obtained by recombining mouse neurons, mouse astrocytes and human oligodendrocytes with cultures prepared by recombining just mouse neurons and human oligodendrocytes. *Supported by the Medical Research Council of Canada*

A-$_{1A}$ ADRENOCEPTORS REGULATION IN OLIGODENDROCYTES

A. Khorchid and G. Almazan, Department of Pharmacology, McGill University, Montreal, Quebec, Canada H3G 1Y6.

Oligodendroglial lineage cells express functional adrenoceptors. In this study we examined the homologous regulation of the $\alpha\text{-}_{1A}$ adrenoceptors, coupled to phosphoinositide metabolism, and the possible implication of protein kinase C (PKC) in the process of desensitization. Total inositol phosphates ($[^3H]IP$) levels were measured in $[^3H]$myo-inositol-labelled oligodendrocyte progenitor cultures following stimulation with 10 mM norepinepherine (NE) in the presence of 3 mM propanolol (to block β-adrenoceptor). In desensitization experiments, pretreatment with NE resulted in a time-dependent decrease in $[^3H]IP$ formation in response to a subsequent agonist challenge. This decrease was 50% of non-pretreated values by 60 min and reached basal level by 6 hrs. The time course of recovery for agonist stimulated $[^3H]IP$ formation was determined following 60 min desensitization. Agonist stimulated levels of inositol phosphates reached the non-pretreated levels in 8 hrs, indicating full recovery of the functional response. We also showed that activation of PKC with 1 mM phorbol-12-myristate 13-acetate (PMA) decreased significantly the agonist stimulated $[^3H]IP$ formation. This result suggested that $\alpha\text{-}_{1A}$ adrenoceptors might be regulated by a PKC-dependent feedback mechanism. To further assess such possibility, oligodendrocyte progenitor cultures were treated with H7, an inhibitor of PKC. This treatment blocked the effect of PMA without effecting NE-mediated desensitization. Similarly, down-regulation of PKC by overnight treatment with 1mM PMA had no effect on NE-mediated desensitization while completely blocking the acute effect of PMA. These results suggest that PKC is not implicated in the agonist induced desensitization mechanism, although it can negatively modulate $\alpha\text{-}_{1A}$ adrenoceptors activation. *(Supported by MRC of Canada)*

ACTIVATION OF PHOSPHOLIPASE C BY AMPA/KAINATE RECEPTORS IN OLIGODENDROCYTE PROGENITORS IS CALCIUM DEPENDENT

H.-N. Liu and G. Almazan, Department of Pharmacology and Therapeutics, McGill University, Montreal, Quebec, Canada, H3G 1Y6.

Oligodendrocytes are the myelin-forming cells in the central nervous system. They arise from progenitors which proliferate and differentiate into mature oligodendrocytes. In vitro studies have identified a number of neurotransmitters and growth factors affecting their growth and differentiation. The excitatory neurotransmitter glutamate is one of the molecules involved in their development. We have found that glutamate activates phospholipase C (PLC) causing accumulation of $[^3H]IP3$ in both a concentration- and time-dependent manner in oligodendrocyte progenitors. The two analogs AMPA and kainate were equally efficacious and exhibited higher potency. CNQX, an AMPA/kainate receptor antagonist, blocked the glutamate-induced $[^3H]IP3$ formation, in contrast, MK-801 and L-AP3, antagonists of NMDA and metabotropic receptors, were ineffective. These results suggest that glutamate-stimulated phosphoinositide hydrolysis is mediated through ionotropic AMPA/kainate receptors. Glutamate-stimulated $[^3H]IP3$ accumulation was prevented by EGTA indicating extracellular calcium requirement. This observation was further supported by the ability of the calcium ionophore A23187 to stimulate $[^3H]IP3$ formation. Pretreatment of progenitors with the voltage-gated calcium channel blockers, diltiazem and nifedipine, partially blocked glutamate-stimulated $[^3H]IP3$ formation suggesting the participation of the voltage-dependent calcium channels. By using cyclothiazide, an agent blocking AMPA receptor desensitization, we demonstrated that AMPA receptors are rapidly desensitized in the presence of agonist. These findings suggest that the activation of PLC by glutamate and downstream events may play important roles in the development of oligodendrocytes. *(Funded by a grant from the MRC and a student-ship from the Multiple Sclerosis Society of Canada.)*

SELENIUM REQUIREMENT FOR OLIGODENDROCYTE DIFFERENTIATION

G.W. Konat Gu Jin and R.C. Wiggins, Department of Anatomy, West Virginia University Medical School, P.O.Box 9128, Morgantown, West Virginia 26506–9128.

There are strong indications that selenium may be an essential microelement required for normal development and function of mammalian central nervous system (CNS). The purpose of this study was to characterize the effects of selenium on CNS myelinogenesis. In primary mixed glial cultures established from newborn rat brains, selenium deprivation profoundly inhibited temporal upregulation of myelin-specific genes, i.e., proteolipid protein (PLP), basic protein (BP), and myelin-associated glycoprotein (MAG) as assessed by Northern blot analysis. 30 nM selenium was found to be optimal for the gene upregulation. Selenium deprivation administered at early developmental stages irreversibly suppressed the gene upregulation indicating the existence of a critical period in oligodendrocyte differentiation. Also in purified oligodendrocyte cultures prepared by mechanical dislodging of progenitor (O2A) cells from mixed glial cultures, selenium deficiency strongly inhibited developmental upregulation of myelin genes, whereas the total cell number was virtually unaffected. In addition, selenium deficiency appears to inhibit the progression of preoligodendrocytes into mature oligodendrocytes as seen from the expression of specific surface antigens, i.e., A2B5, 04 and GC detected by immunocytochemical staining. The deleterious effects of selenium deficiency on oligodendrocyte differentiation may be related to the impaired defence of the oligodendrocyte lineage cells against oxidative stress.

SUSCEPTIBILITY OF OLIGODENDROGLIAL PRECURSOR CELLS TO OXIDATIVE STRESS IS RELATED TO LOW GLUTATHIONE AND HIGH IRON CONTENTS

S.K. Thorburne and B. H.J. Juurlink, Department of Anatomy and Cell Biology & the Saskatchewan Stroke Research Centre, University of Saskatchewan, Saskatoon, Saskatchewan, Canada S7N 5E5.

Previous work demonstrated that oligodendroglial precursors are more readily damaged by free radicals than are astrocytes (Husain and Juurlink. 1995. Brain Res. 698:86–94). Using the oxidation of dichlorofluorescin as a measure of oxidative stress we demonstrate that i) oligodendroglial precursors under normal culture conditions are under six times as much oxidative stress than astrocytes and ii) oxidative stress experienced by oligodendroglial precursors increases six-fold when exposed to 140 mW/m^2 of blue light, whereas, astrocytic oxidative stress only doubles. We also show that astrocytes have a 3 times higher concentration of GSH than oligodendroglial precursors and that oligodendroglial precursors have more than 20 times higher iron content than do astrocytes. GSH in conjunction with glutathione peroxidase converts hydrogen peroxide and lipid peroxides into water and lipid alcohols. Peroxides can be hazardous to cells since in the presence of free iron they are converted into potent oxidants such as the hydroxyl, peroxyl and alkoxyl radicals. We demonstrate that the photo-induced increase in oxidative stress in oligodendroglial precursors can be prevented either by chelating intracellular free iron or by raising intracellular GSH to astrocytic values. We conclude that GSH plays a central role in preventing free radical-mediated damage in glia. *Supported by the Heart and Stroke Foundation of Saskatchewan*

THE EFFECT OF A FREE RADICAL GENERATOR ON CELL VIABILITY AND MYELIN MORPHOLOGY IN MYELINATED RAT SPINAL CORD AGGREGATE CULTURES

B.L. Bartnik and R.M. Devon Department of Anatomy and Cell Biology, University of Saskatchewan, Saskatoon, SK. S7N 5E5 Canada.

Recent in vitro experiments suggest that, compared to astrocytes, oligodendrocytes and their precursors have reduced abilities to scavenge free radicals due to lower levels of glutathione and increased iron content. The following study was performed to determine the effect of free radicals on both myelin and oligodendrocyte morphology. Myelinating E15 rat spinal cord aggregate cultures were used as an in vitro model of CNS myelination. Aggregate cultures were maintained for 4 weeks in DMEM/F12 supplemented with 10% FBS and 1% glucose until adequately myelinated. Aggregates were then treated with 3-morpholinosyndnonamine (SIN-1), a O_2^- and NO^- generator, in concentrations ranging from 1.0 mM to 0.01mM and sampled at 24, 48 and 96 hours. Treating the cultures with 1mM, 100µM and 10 µM for 24 hours resulted in a dramatic loss of neuropil in the aggregate accompanied by widespread non-specific cell death. By 48 hours the aggregates were almost completely devoid of cellular detail and by 96 hours all the aggregates were destroyed. Subsequent experiments using concentrations of 1µM, 0.1µM and 0.01µM demonstrated that the SIN-1 had little effect on the aggregate at 24 hours. However when sampled at 48 hours the aggregates showed an increase in damage to both the cells and myelin. This damage was both time and concentration dependent as determined by cell viability assays using Mitotracker as a viability probe. Alterations to the myelin structure in the form of delamination and demyelination of axons were determined by electron microscopy. Stereological analysis of cells in the aggregate indicate that 1mM SIN-1 caused a decrease in the cell number and nuclear volume fraction of oligodendrocytes compared to the control aggregates. Taken together, these results suggest that the oligodendrocytes appear to sustain free radical induced damage in response to SIN-1 treatment. This work was supported by a Multiple Sclerosis of Canada Grant to RMD

GLUCOSE UPTAKE IN ASTROGLIA AND OLIGODENDROGLIA: EFFECT OF ANOXIA

H. Marrif, L. Hertz and B.H.J. Juurlink. Departments of Pharmacology and Anatomy & Cell Biology as well as the Saskatchewan Stroke Research Centre, University of Saskatchewan, Saskatoon, SK Canada.

We hypothesize that the ability of neural cells to survive an hypoxic-ischemic insult (HI) is, in part, dependent upon their ability to undergo anaerobic glycolysis and that there are differences in the ability of neural cells to modulate glycolytic activity. We have examined glucose uptake, using 2-deoxyglucose as a tracer, in two neural cell lineages with differing susceptibility to HI: i) oligodendroglia that are readily injured by HI, and ii) astroglia that are very resistant to HI. Under normoxic conditions astrocytes take up 17.4 ± 0.4 nmoles glucose/min/mg protein. This is increased by ~20% when astrocytes are directly exposed to anoxia. Oligodendroblasts and oligodendrocytes behave like astrocytes in that anoxia increases glucose uptake by ~20%. In contrast, when oligodendrocyte precursors are subjected to anoxia there is no increase in glucose uptake, suggesting that these cells do not generate much ATP via glycolysis. When CO_2 production from lactate was compared, the precursors produced 4.25 nmoles/min/mgprotein, while astrocytes produced only 2.28 nmoles/min/mg protein. These findings suggest that the precursors are much more dependent upon oxidative metabolism than the astrocytes. To examine whether prolonged exposure to anoxia alters the ability of the cells to use glucose, oligodendroblasts and astrocytes were exposed to anoxia for up to 8 hr followed by an examination of glucose uptake. These experiments showed that prolonged exposure to anoxia further increased glucose uptake in oligodendroblasts and astrocytes with maximum glucose uptake following 4 hr of anoxia in oligodendroblasts and 8 hr in astrocytes. *Supported by the Heart and Stroke Foundation of Saskatchewan*

THE PERIVENTRICULAR GERMINAL REGION AND DEVELOPING CORPUS CALLOSUM IS DAMAGED IN THE ONE-WEEK-OLD RAT PUP FOLLOWING A BRIEF SEVERE HYPOXIC-ISCHEMIC INSULT

S.E. Jelinski, J.Y. Yager†, and B.H.J. Juurlink. Departments of Anatomy & Cell Biology, Pediatrics† and the Saskatchewan Stroke Research Centre, University of Saskatchewan, Saskatoon, SK, Canada, S7N 5E5.

Periventricular leukomalacia, a predominant antecedent of cerebral palsy, is the most common hypoxic-ischemic (HI) lesion of developing white matter in premature infants. A suggested pathogenesis is that the glial cells in the periventricular region are undergoing active differentiation resulting in an intrinsic vulnerability to an HI injury (Volpe, 1989). In vitro work has shown that oligodendroglial precursors cells are more readily damaged by oxidative stress brought on by hypoxia than are astrocytes (Husain and Juurlink. 1995. Brain Res. 698:86–94). To examine the effects of an HI lesion on oligodendroglia in vivo, one-week-old rat pups, developmentally equivalent to premature infants born between 34 and 36 weeks of gestation, were subjected to a 10 min bilateral carotid ligation in the presence of 8% oxygen. Over the next 24 hr, during the reperfusion period, cells in the germinal zones and corpus callosum underwent pycnosis. Immunocytochemistry of living vibratome slices of the tissue demonstrated that the affected cells bound the O4 monoclonal antibody demonstrating the oligodendroglial nature of the cells. The cell death appears to have been caused by an inability to cope with oxidative stress since it was coincident with both lipid peroxidation, as demonstrated by the TBAR reaction, and induction of hsp32. We suggest that the increased vulnerability of oligodendroglial precursors to oxidative stress following an HI insult may contribute to the pathogenesis of periventricular leukomalacia. *Supported by the Heart and Stroke Foundation of Saskatchewan.*

MYELIN- AND OLIGODENDROCYTE-RELATED ENZYMES IN WHITE MATTER OF HYDROCEPHALIC RAT BRAINS

M.R. Del Bigio and J.N. Kanfer, Departments of Pathology and Biochemistry, University of Manitoba, Winnipeg, MB Canada R3E 0W3.

Hydrocephalus is a dilatation of the cerebral ventricles with consequent damage to surrounding brain structures. We induced hydrocephalus in 3 week old rats by injecting kaolin into the cisterna magna. Ventricular size was assessed by MR imaging. After 1, 2, or 4 weeks, rats were killed or treated by diversionary shunting of cerebrospinal fluid. Samples of corpus callosum/supraventricular white matter, fimbria, and medulla were rapidly dissected from each brain. The myelin-related enzymes 2',3'-cyclic nucleotide 3'-phosphodiesterase (CNP) and p-nitrophenylphosphorylcholine phosphodiesterase (PNP), and the oligodendrocyte-related enzyme UDP-galactose: ceramide galactosyltransferase (CGalT) were assayed. In the corpus callosum of hydrocephalic rats, CNP was reduced by 25% ($p<.0001$) and PNP was reduced by 47% ($p<.0001$) in comparison to controls. CGalT, which exhibited a significant maturation-related decline, was unaffected by hydrocephalus but showed 43% increase ($p<0.0019$) following early shunting. In the medulla, CGalT was reduced by 27% ($p<.0001$) in the 1 wk hydrocephalic rats. No changes were found in the fimbria. Although the corpus callosum was obviously thinned in hydrocephalic brains, immunohistochemical labeling intensity of myelin basic protein and 011 was unchanged. Damage to white matter by ventriculomegaly is reflected by reduced activity of myelin-related enzymes. CGalT activity in the corpus callosum, however, suggests that early reduction of ventricular size allows re-activation of oligodendrocytes. (Funded by the P.H.T. Thorlakson Fdn.)

LONGTERM ORAL ADMINISTRATION OF BOVINE MYELIN PREVENTS EXACERBATION AND SUPPRESSES CNS INFLAMMATION AND DEMYELINATION FOCI IN SJL/J MICE WITH CHRONIC RELAPSING EAE

A. AL-Sabbagh, *H.L. Weiner, and P.A. Nelson. AutoImmune Inc., Lexington MA 02173 and Center for Neurologic Diseases, Brigham and Women's Hospital, Harvard Medical School, Boston, Massachusetts 02115.

OBJECTIVE: This study was designed to investigate the effect of prolonged (six months) oral administration of bovine myelin in Chronic Relapsing Experimental Autoimmune Encephalomyelitis (CR-EAE) diseased mice, in terms of disease exacerbation and CNS histopathological lesions. *BACKGROUND:* Oral administration of bovine myelin is currently being tested in a multicenter double blinded clinical trial in multiple sclerosis patients. Murine CR-EAE has been shown to be a model of inflammatory demyelinating disease of the central nervous system (CNS) with great similarity to multiple sclerosis (MS) in humans. In this model we investigated the effect of prolonged oral administration of myelin to CR-EAE diseased mice during a six month period. *METHOD:* SJL/J mice were immunized for CR-EAE with mouse spinal cord homogenate in complete Freund's adjuvant. At the end of the first neurological attack, the EAE positive mice were distributed into different groups with a representative distribution of mice at various levels of EAE severity. One group was gavaged PBS as control and three groups were gavaged bovine myelin at doses of 1 mg, 10mg, and 20mg every other day three times a week for six months to investigate the effect of prolonged oral delivery of bovine myelin during ongo-ing disease. *RESULTS:* After six months of oral administration of myelin, no exacerbation of disease severity was found, in terms of cumulative clinical disease score (CCS), disease related death or CNS histopathological lesions (mean maximal inflammatory and demyeli-nation foci) as seen in the table below .

Group	CCS	Inflammatory FOCI		Demyelination FOCI	
		Brain	Spinal cord	Brain	Spinal cord
PBS 0.25ml	1519.0 ± 28	2.8 ± 0.5	0.9 ± 0.1	1.0 ± 0.7	4.2 ± 0.7
Myelin lmg	1251.0 ± 27	2.0 ± 0.7	0.4 ± 0.3	0.2 ± 0.2	3.2 ± 0.3
Myelin 10mg	1302.0 ± 29	0.3 ± 0.2(P<0.004)	0.0 ± 0.0(P<0.001)	0.2 ± 0.2	1.1 ± 0.3(P<0.007)
Myelin 20mg	1496.5 ± 75	0.5 ± 0.3(P<0.007)	0.2 ± 0.2(P<0.023)	0.2 ± 0.2	2.3 ± 0.4(P<0.039)

Unpaired 2-tailed Student's t-test: Bovine myelin vs PBS

CONCLUSION: It was found that prolonged oral delivery of neuroantigen sup-presses inflammatory and demyelination foci in the CNS of mice fed myelin (10mg and 20mg) with no exacerbation of clinical and neurological status as compared to the control group. These results have important implications for the use of neuroantigen(s) from the target organ for the treatment of human autoimmune disease such as multiple sclerosis.

IMMUNOLOGICAL MYELIN DISRUPTION AS A FACILITATOR OF AXONAL REGENERATION IN THE ADULT MAMMALIAN SPINAL CORD AFTER HEMISECTION INJURY

J.K. Dyer[1], J. Bourque[1] & J.D. Steeves[123]. Depts. [1]Zoology, [2]Anatomy & [3]Surgery, UBC, 6270 University Boulevard, Vancouver, BC, V6T 1Z4 CANADA.

The regenerative ability of the adult vertebrate nervous system is limited. During de-velopment of the CNS, the loss of regenerative ability, coincides with the onset of myeli-nation. This failure has recently be ascribed, in part, to the oligodendrocyte-derived, myelin-associated inhibitory molecules, N135/250 and MAG. Demyelination of the in-jured nervous system has been postulated as a possible way of overcoming these inhibi-tory influences. Most demyelination protocols (chemical, viral or irradiation-based) are irreversible, as well as having adverse effects on other non-myelin associated cells. We have previously described a transient immunological, myelin specific, antibody-mediated (anti-GalC), complement-dependent protocol that (I) suppresses the onset of myelination in the avian spinal cord, and facilitates axonal regeneration at "post-myelination" embry-onic stages; and (2) disrupts the myelin structure in the adult avian CNS thereby promot-ing limited anatomical regeneration and functional recovery after complete transection of the fully developed spinal cord. We now described the effects of injection/infusion of the immunological reagents on adult rodent CNS myelin and the possible application in facili-tating the regeneration of specific descending locomotor-associated tracts. Myelin disrup-tion is evident in the rodent spinal cord shortly (<1d) after administration of the reagents. There is a down regulation of myelin-associated proteins and a concomitant unraveling of the myelin lamellae. Long term infusion results in local demyelination, with distal myelin-disruption. Our previous studies in the hatching chick indicated that this change in the myelin is sufficient for axonal regeneration after injury. We are currently investigating whether this approach will facilitate regeneration of axons after injury to descending loco-motor-associated tracts (corticospinal. rubrospinal and vestibulospinal) in the adult rodent.

JKD is a Rick Hansen MIM Post-doctoral Fellow and a trainee of the NeuroScience Network. Funded by grants to JDS from the MRC & the NeuroScience Network. JKD & JDS are members of CORD (Collaboration On Repair Discoveries) at UBC.

A PROTEIN PHOSPHATASE 2C FAMILY MEMBER ("PETRIN") IS INVOLVED IN NEURITE GROWTH REGULATION ON MYELIN

Monika Labes*, John Roder* and Arthur Roach[#]. Samuel Lunenfeld Research Institute, Toronto, Ontario, MSG IX5* and Allelix Biopharmaceuticals Inc., Mississauga, Ontario, L4V IV7[#].

Mammalian CNS myelin contains molecules which exert inhibitory effects on neuronal growth. We have isolated and characterized a cDNA encoding a neuronal protein that is involved in the cellular response to myelin inhibitors. A rat brain cDNA expression library was screened with an antibody that had been raised against PC12 cell membrane proteins. For one of the positive clones isolated, introduction of antisense RNA or oligonucleotides which perturb the expression of the encoded gene product, conferred on NG108 cells the ability to grow neurites on an inhibitory CNS myelin substrate. The cDNA encodes a novel gene and contains an open reading frame (ORF) of 647 amino acids. The most closely related protein described to date is serine/threonine protein phosphatase 2C (PP2C) with amino acid sequence identities of up to 60% in distinct domains. A polyclonal antibody generated against the C-terminus of the ORF specifically precipitates a protein of ca. 62–65kD and a Mg^{2+}-dependent phosphatase activity from NG108 cell and brain lysates. The expression of the protein - designated as PETRIN - appears to be brain and neuron specific as determined by northern blot and *in situ* hybridization analyses, and is developmentally regulated, being first detectable after embryonic day E13 and increasing with age. Highly homologous sequences are detectable in mouse, hamster and human genomes. The presented data suggest that we have isolated a novel phosphatase which is part of a signalling pathway regulating the cellular growth response in the presence of myelin-associated inhibitors.

EXPRESSION OF GAP-43 BY NEURONS CULTURED IN THE PRESENCE OF CNS MYELIN FRAGMENTS

Parker L. Andersen and David J. Schreyer, Cameco MS Neuroscience Research Center and Dept. Anatomy and Cell Biology, University of Saskatchewan, Saskatoon, Canada S7N 5E5.

Central nervous system (CNS) myelin has been shown to inhibit axon growth from neurons in vivo and in vitro. An inverse pattern of myelin concentration and immunoreactivity for the growth-associated protein GAP-43 has been noted in the CNS, and this has led to the suggestion that CNS myelin represses GAP-43 expression. However dorsal root ganglion (DRG) neurons can regenerate their axons in the peripheral nervous system (PNS) (or within intraspinal PNS grafts) and display elevated expression of GAP-43 despite maintaining substantial contact with CNS myelin. We examined levels of GAP-43 expression in cultured adult DRG neurons using a quantitative ELISA for GAP-43. Cultured neurons underwent chronic exposure to CNS myelin membrane fragments (gift of Dr. L. McKerracher), applied either as a surface coating or as a particle suspension. GAP-

43 expression in adult DRG neurons remained unchanged in these experiments. MAP2 immunocytochemistry showed that neuron survival was also unchanged by CNS myelin exposure. The growth inhibitory effect of CNS myelin on DRG neurons thus appears not to involve downregulation of this growth-associated protein. Rather, CNS myelin growth inhibition may be due solely to local effects on growth cone motility.

PHOTOABLATION OF OLIGODENDROCYTES IN TRANSGENIC MICE FOR THE STUDY OF AXONAL SPROUTING AND REGENERATION IN THE DORSAL COLUMNS

J.L. Vanderluit, A. Peterson*, W. Tetzlaff Dept. of Zoology, University of British Columbia, Vancouver, B.C., Canada; *McGill University, Montreal, Canada.

We have previously demonstrated photoablation as a technique to target the destruction of oligodendrocytes in transgenic mice which express β-galactosidase from the myelin basic protein (MBP) promotor. Focal demyelination in the dorsal columns is produced by performing a laminectomy at T9–10 followed by application of a fluorescein-linked, β-gal substrate (fluorescein di-β-galactopyranoside, FDG) and illumination. Demyelination of axons in the dorsal columns has been demonstrated 12 days following photoablation by ultrastructural analysis and immunohistochemistry. In addition, we have shown successful repair of the demyelinated area by remyelinating oligodendrocytes and Schwann cells invading from the periphery (dorsal root entry zone). Currently, we are examining the ability of dorsal column axons to sprout and regenerate in this myelin/oligodendrocyte free zone following a dorsal hemisection of the spinal cord. Supported by the Neuroscience Network of Canada.

NORMAL NUMBER OF MYELINATING SCHWANN CELLS BUT ALTERED SCHWANN CELL PHENOTYPE IN YOUNG CMT1a PATIENTS

C.O. Hanemann, A. Gabreëls-Festen* and H.W. Mueller. Molecular Neurobiology Lab, Heinrich-Heine-University, Dusseldorf, Germany, *Institute of Neurology Nijmegen, Netherlands.

Since the myelin protein PMP22 is homologous to the growth arrest-specific gene gas3 and has a growth retarding effect on cultured Schwann cells we investigated Schwann cell phenotype and a number of myelinating Schwann cells in biopsies of morphometrically characterized CMT1a patients with duplication of the PMP22 gene. Previously we showed abnormal expression of LNGF-R in Schwann cells of CMT1a patients. LNGF-R was already abnormally expressed in young CMT1a patients with minimal nerve pathology. LNGF-R is also expressed in early onion bulbs. By immunohistochemistry with an monoclonal antibody to NCAM we now show expression of NCAM in all stages of CMT1a. Onion bulbs show expression of NCAM. In contrast onion bulbs in CIDP or Schwannomma show no NCAM expression. Thus we provide further evidence for an abnormal Schwann cell differentiation in CMT1a. Electronmicroscopic picutres, covering approx. 1/10 of the TTFA of 5 young CMT1a patients were used to look for Schwann cell death and proliferation. No evidence for apoptosis could be found. Counting myelinated axons, demyelinated axons and onion bulbs as an estimation of Schwann cells, which are in a 1:1 relation with the axon, indicated a normal number of Schwann cells in CMT1a compared to age matched controls. Analysis of maximum internodal length, as a measurement for decreased proliferation in the longitudinal axis using teased fiber preparations did not show increased internode length in CMT1a. Thus we found no evidence for decreased Schwann cell proliferation in CMT1a using the above mentioned experiments, but found further evidence for altered Schwann cell phenotype already in the early stages of CMT1a.

p75 DEFICIENT MICE EXHIBIT ALTERED RESPONSE TO NERVE INJURY

L.A. Karchewski and V.M.K. Verge. Department of Anatomy and Cell Biology & Cameco MS/Neuroscience Research Center, University of Saskatchewan, Saskatoon, Canada S7N 5E5.

Sciatic nerve injury invokes a number of cell body responses believed to be influenced by alterations in trophic support and/or altered expression of p75 and trk receptors. As a first step in elucidating the role of p75 in intact and injured sensory neurons, unilateral sciatic nerve transection time course studies were performed on mice carrying a null mutation in the p75 locus and compared to host strain mice. *In situ* hybridization to detect p75, trk receptors, various neuropeptides, cytoskeletal and immediate early gene mRNAs were performed on pairs of experimental and control DRG cryostat sections. Preliminary results show that, although these DRG contain approximately 50% fewer neurons (a lack of primarily small neurons), there is not a proportionate loss of trkA, trkB, trkC or CGRP expressing neurons. Time course analysis of injury responses indicate that at 4d & 7d, reduced expression of trkA, trkC, CGRP, SP and elevated expression of GAL, NPY and GAP43 are much more dramatic in the p75 deficient mice than in controls. However by 2 weeks, p75 deficient mice show enhanced levels of hybridization signal detecting trkA, SP, CGRP and BDNF mRNAs as compared to control mice. These results suggest that availability and/or response to NGF may be altered in these mice, perhaps through altered p75/trk ratios, different endoneurial trophic support and/or retrograde transport of NGF following injury. Whether axonal regeneration is altered in p75 knock-out mice is being determined. *Supported by the Neuroscience Network for the Centers of Excellence, Canada.*

STUDIES OF THE ADHESIVE PROPERTIES OF MYELIN P_0-PROTEIN DERIVED PEPTIDES

M. Foldvari and M.R. Jaafari. College of Pharmacy and Nutrition, University of Saskatchewan, 110 Science Place, Saskatoon, Sk. Canada S7N 5C9.

Using a novel liposomal system we have shown that P_0 protein, the major glycoprotein of peripheral nerve myelin, has the capability of mediating heterophilic interactions in cell culture. The main objective of the study was to characterize the binding domains of P_0 protein by conducting competition studies with synthetic P_0 peptides based on the chicken P_0 sequence and determine whether there is interaction between P_0 protein and ICAM-1. Human M21 melanoma cells were incubated with P_0-liposomes composed of ^{14}C-dipalmitoylphosphatidylcholine and cholesterol, 10:1 molar ratio at a concentration of 80 µg phospholipid/5×10^6 cells in the presence or absence of P_0 peptides (P_0-peptide-1: residues 90–96: Y-T-D-N-G-T-F, P_0-peptide-2: residues 140–146: V-A-L-L-V-A-V and P_0-peptide-3: residues 201–207:K-A-A-A-E-K-K; 90 µg/5X106 cells) at 37°C for 1h. After incubation, the medium was removed, the cells collected, washed and the association of liposomes with cells determined by scintillation counting. The presence of intact protein in the liposome bilayer (P_0-liposomes) increased the rate of interaction of liposomes with M21 cells 3–4 times in comparison with control liposomes. A control protein, Glycophorin A (transmembrane glycoprotein), when reconstituted in liposomes had no effect on the binding of liposomes with M21 cells. The increased binding of P_0-liposomes with M21 cells was inhibited by anti-chick P_0 Fab fragments. Therefore, P_0 protein plays a specific role in the binding of liposomes to M21 cells. Competition studies in the presence of synthetic P_0-peptides indicated that the association of P_0-liposomes with M21 cells decreased by 30% in the presence of P_0-peptide-1 and by 40% in the presence of P_0-peptide-3, while P_0-peptide-2 had no effect. The cell uptake of control liposomes in the presence of P_0-peptide-1 and P_0-peptide-2 increased by 240% and 290%, respectively. P_0-peptide-3 had no effect on control liposome interaction with cells. Both extracellular and intracellular P_0-peptides (1 and 3) showed competition and may be involved in the heterophilic interaction with M21 cells. The increasing effect of P_0-peptide-1 and 2 on control liposomes binding may be indicative of adhesion of these peptides to the liposome surface and mediating cellular uptake of liposomes. The expression of ICAM-1 on M21 cells was quantitatively determined and the effects of P_0 protein and P_0-peptides on the interaction of CD54 and FITC conjugated anti-CD54 was investigated by FACScan flow cytometry. The percentage of stained cells with FITC anti-human CD54 mAb was 95% for M21 cells and did not change in the presence of P_0 protein or P_0-peptides. It is concluded that although P_0-liposomes have a specific interaction with M21 cells, P_0 protein or P_0-peptides may interact with different epitopes on the CD54 molecule than anti-human CD54 mAb or with a different cell adhesion molecule.

CONTRIBUTORS

R. Aebersold, Department of Molecular Biotechnology, University of Washington, Seattle, WA 98195

R. Ankerhold, Faculty of Biology, University of Konstanz, Postfach 5560/M 625, D 78343 Konstanz, Germany

K. Asakura, Department of Neurology and Immunology, Mayo Clinic and Foundation, Rochester, MN 55905

V. Avellana-Adalid, INSERM U134, Laboratoire de Neurobiologie Cellulaire, Moleculaire, et Clinique Hôpital de la Salpêtriere, 75651 Paris cedex 13, France

R.J. Balice-Gordon, Department of Neuroscience, University of Pennsylvania Medical Center, 3400 Spruce St., Philadelphia, PA 19104

P.A. Ballenthin, Department of Pathology, Northwestern University Medical School, 303 East Chicago Avenue, Chicago, IL 60611

A. Baron-Van Evercooren, INSERM U134, Laboratoire de Neurobiologie Cellulaire, Moleculaire, et Clinique Hôpital de la Salpêtriere, 75651 Paris cedex 13, France

P. Beserman, Weizman Institute of Science, Department of Neurobiology, IL-76100, Rehovot, Israel

W.F. Blakemore, University of Cambridge, Cambridge Centre for Brain Repair, Madingley Rd, Cambridge CB0 0ES, England, United Kingdom

L.J. Bone, Department of Molecular and Cellular Biology, University of Pennsylvania Medical Center, 3400 Spruce St., Philadelphia, PA 19104

J. Bourque, Department of Zoology , University of British Columbia, Vancouver, BC V6T 1Z4, Canada

P.E. Braun, Department of Biochemistry, McGill University, 3655 Drummond St., Montreal, PQ H3G 1Y6, Canada

A. Brennan, Department of Anatomy and Developmental Biology, University College, London WC1E 6B2, England, United Kingdom

M.B. Bunge, Miami Project to Cure Paralysis, University of Miami School of Medicine, Rm 1044, 1600 NW 10th Ave. R-48, Miami, FL 33136

A.T. Campagnoni, Mental Retardation Research Center, and Brain Research Institute, UCLA Medical School, 760 Westwood Plaza, Los Angeles, CA 90024

R.E. Clark, Department of Biochemistry and Molecular Biology, University of South Dakota, School of Medicine, Vermillion, SD 57069

A. Compston, University of Cambridge, Addenbrookes Hospital, Neurology Unit, Hills Rd, Cambridge CB2 2QQ, England, United Kingdom

A.J. Crang, University of Cambridge, Cambridge Centre for Brain Repair, Madingley Rd, Cambridge CB0 0ES, England, United Kingdom

S. David, Centre for Research in Neuroscience, Montreal General Hospital Research Insitute, 1650 Cedar Ave., Montreal PQ H3G 1A4, Canada

S.M. Deschênes, Department Molecular and Cellular Biology, University of Pennsylvania Medical Center, 3400 Spruce St., Philadelphia, PA 19104

Z. Dong, Department of Anatomy and Developmental Biology, University College, London WC1E 6B2, England, United Kingdom

R. Doucette, Department of Anatomy and Cell Biology, College of Medicine, University of Saskatchewan, 107 Wiggins Rd., Saskatoon SK S7N 5E5, Canada

M. Dubois-Dalcq, Unité Neurovirologie et Régénération du Système Nerveux, Institute Pasteur, 28 Rue Dr. Roux, 75724 Paris Cedex 15, France

C. Duchala, Department of Neurosciences, Cleveland Clinic Foundation, NC-30, 9500 Euclid Ave., Cleveland, OH 44195

A. Ducret, Department of Biological Chemistry, UCLA School of Medicine, 33–257CHS, Los Angeles, CA 90024–1737

C.A. Dyer, Department of Biomedical Sciences, E.K. Shriver Center, 200 Trapelo Road, Waltham, MA 02554

J.K. Dyer, Department of Zoology , University of British Columbia, Vancouver, BC V6T 1Z4, Canada

S. Eitan, Weizmann Institute of Science, Department of Neurobiology, IL-76100, Rehovot, Israel

O. Eizenberg, Weizmann Institute of Science, Department of Neurobiology, IL-76100, Rehovot, Israel

R.G. Farrer, Myelin and Brain Development Section, Laboratory of Molecular and Cellular Neurobiology, NINDS, Bldg. 49, Rm. 2A10 NIH, Bethesda, MD 20892

M.T. Filbin, Department Biological Sciences, Hunter College of the City, University of New York, 695 Park Ave., New York, NY 10021

K. Fischback, Department Neurology, Univ. of Pennsylvania Medical Center, 3400 Spruce St., Philadelphia, PA 19104

R.J.M. Franklin, University of Cambridge, Cambridge Centre for Brain Repair, Madingley Rd, Cambridge CB0 0ES, England, United Kingdom

M.V. Gardinier, Department of Pathology W127, Northwestern University, Medical School, 303 E Chicago Ave., Chicago, IL 60611–3008

M. Gravel, Department Biochemistry, McGill University, 3655 Drummond St., Montreal, PQ H3G 1Y6, Canada

I. R. Griffiths, Applied Neurobiology Group, University of Glasgow, Scotland, United Kingdom

M. Hajihosseini, Unité Neurovirologie et Régenération du Système Nerveux, Institute Pasteur, 28 Rue Dr. Roux, 75724 Paris Cedex 15, France

P.L. Hirschberg, Weizmann Institute of Science, Department of Neurobiology, IL-76100, Rehovot, Israel

S.F. Hunter, Departments of Neurology and Immunology, Mayo Clinic and Foundation, 100 1ST St., Rochester, MN 55905

K. Jessen, Department of Anatomy and Developmental Biology, University College, London, England WC1E 6B2, United Kingdom

M. Jung, Zentrum für Molekulaire Biologie, University of Heidelberg, Im Neuenheimer Feld 282, D-69120 Heidelberg, Germany

H.S. Keirstead, University of Cambridge, Cambridge Centre for Brain Repair, Madingley Rd, Cambridge, CB0 0ES, England, United Kingdom

H.A. Kim, Department Cell Biology, Neurobiology & Anatomy, College of Medicine, University Cincinnati, P.O. Box 870521, 231 Bethesda Avenue, ML #521, Cincinnati, OH 45267–0521

N. Kleitman, Miami Project to Cure Paralysis, University of Miami School of Medicine, Rm 1044, 1600 NW 10th Ave. R-48, Miami, FL 33136

M. Klugman, Zentrum für Molekulaire Biologie, University of Heidelberg, Inf 282, D69120 Heidelberg, Germany

J.D. Kocsis, PVA/EPVA Center for Neuroscience, VA Medical Center, West Haven, CT 06516

J.F. Kroepfl, Department of Pathology, Northwestern University Medical School, 303 East Chicago Avenue, Chicago, IL 60611

F. Lachapelle, INSERM U134, Laboratoire de Neurobiologie Cellulaire, Moleculaire, et Clinique Hôpital de la Salpêtriere, 75651 Paris cedex 13, France

D.M. Lang, Faculty of Biology, University of Konstanz, Postfach 5560/M 625, D 78343 Konstanz, Germany

W.H. Li, Department Biological Science, Hunter College of the City University of New York, 695 Park Ave., New York, NY 10021, U.S.A.

W. Macklin, Department of Neuroscience, Cleveland Clinic and Foundation, NC-30, 9500 Euclid Ave., Cleveland, OH 44195

D.L. Madison, Department Microbiology, University of Connecticut Health Center , Program Neurological Science, Farmington, CT 06030–3205

J. McGraw, Department of Zoology, University of British Columbia, Vancouver, BC V6T 1Z4, Canada

L. McKerracher, Départment de pathologie, Université de Montréal, CP 6128, Centre-ville, Montréal, PQ H3C 3J7, Canada

J. Merrill, Department of Immunology, Berlex Biosciences, 15049 San Pablo Ave., P.O.B. 4099, Richmond, CA 94804–0099

D.J. Miller, Mayo Clinic and Foundation, Departments of Neurology and Immunology, 100 1st St., Rochester, MN 55905

R. Mirsky, Department of Anatomy and Developmental Biology, University College, London, England WC1E 6B2, United Kingdom

R. Miskimmins, Department of Biochemistry and Molecular Biology, University of South Dakota, School of Medicine, Vermillion, SD 57069

M.A. Moscarello, Hospital for Sick Children, Division of Biochemical Research, 555 University Avenue, Toronto, ON M5G 1XB, Canada

K. Murray, Unité Neurovirologie et Régenération du Système Nerveux, Institute Pasteur, 28 Rue Dr. Roux, 75724 Paris Cedex 15, France

B. Nait-Oumesmar, INSERM U134, Laboratoire de Neurobiologie Cellulaire, Moleculaire, et Clinique Hôpital de la Salpêtriere, 75651 Paris cedex 13, France

K.-A. Nave, Zentrum für Molekulaire Biologie, University of Heildelberg, Im Neuenheimer Feld 282, D-69120 Heidelberg, Germany

L.Y.S. Oh, Department Neurol & Neurosurgery, Montreal Neurological Institute, McGill University, 3801 University, Montreal, PQ H3A 2B4, Canada

J. Peterson, Cleveland Clinic Foundation, Department of Neuroscience, NC-30, 9500 Euclid Ave., Cleveland, OH 44195

S.E. Pfeiffer, Department of Microbiology, University of Connecticut Health Center, Program Neuroligical Science, Farmington, CT 06030–3205

J.F. Poduslo, Molecular Biology Laboratory, Department Neurology & Biochem/Molec Biology, Mayo Clinic and Foundation, 200 First Street SW, Rochester, MN 55905

A. Pühlhofer, Zentrum für Molekulaire Biologie, University of Heildelberg, Im Neuenhe-
    imer Feld 282, D-69120 Heidelberg, Germany

R.H. Quarles, Myelin and Brain Development Section, Laboratory of Molecular and Cel-
    lular Neurobiology, NINDS, Bldg. 49, Rm. 2A10 NIH, Bethesda, MD 20892

S.J. Quinlivan, Department Biological Chemistry, UCLA School of Medicine,
    33–257CHS, Los Angeles, CA 90024–1737

N. Ratner, Department Cell Biol., Neurobiology & Anatomy, College of Medicine, Uni-
    versity Cincinnati, P.O. Box 870521, 231 Bethesda Avenue, ML #521, Cincinnati,
    OH 45267–0521

S.D. Raval, Department of Biological Chemistry, UCLA School of Medicine,
    33–257CHS, Los Angeles, CA 90024–1737

M. Rodriguez, Departments of Neurology and Immunology, Mayo Clinic and Foundation,
    100 1st St., Rochester, MN 55905

L.H. Rome, Department Biological Chemistry, UCLA School of Medicine, 33–257CHS,
    Los Angeles, CA 90024–1737

L.A. Sawant, Department of Biological Chemistry, UCLA School of Medicine,
    33–257CHS, Los Angeles, CA 90024–1737

S.S. Scherer, Department of Neurology, University of Pennsylvania Medical Center, 3400
    Spruce St., Philadelphia, PA 19104

A. Schneider, Zentrum für Molekulaire Biologie, University of Heildelberg, Im Neuenhe-
    imer Feld 282, D-69120 Heidelberg, Germany

M. Schwab, Zentrum für Molekulaire Biologie, University of Heildelberg, Im Neuenhe-
    imer Feld 282, D-69120 Heidelberg, Germany

M. Schwartz, Weizmann Institute of Science, Department of Neurobiology, IL-76100, Re-
    hovot, Israel

N.J. Scolding, University of Cambridge, Addenbrookes Hospital, Neurology Unit, Hills
    Rd, Cambridge CB2 2QQ, England, United Kingdom

C. Shaw, Department of Neurology, Institute of Psychiatry, De Crespigny Parts, London,
    England, United Kingdom

B. St. Pierre, Department of Neurology, UCLA School of Medicine, Los Angeles, CA
    90024

J.D. Steeves, Department of Zoology, University of British Columbia, Vancouver, BC
    V6T 1Z4, Canada

C.A. Stuermer, Faculty of Biology, University of Konstanz, Postfach 5560/M 625, D
    78343 Konstanz, Germany

U. Suter, Institute of Cell Biology, Swiss Federal Institute of Technology, ETH-Hongger-
    berg, CH-8093 Zurich, Switzerland

S. Szuchet, Department of Neurology, University of Chicago, 5841 S Maryland Ave., MC
    2030, Chicago, IL 60637

G.I. Tennekoon, Division of Neurology Research, The Children's Hospital of Philadel-
    phia, 34th Street & Civic Center Boulevard, Philadelphia, PA 19104

C. Tornatore, NINDS, Laboratory of Molecular Medicine, Bethesda, MD 20892

B.D. Trapp, Cleveland Clinic Foundation, Department of Neuroscience, NC-30, 9500 Eu-
    clid Ave., Cleveland, OH 44195

L.R. Viise, Department of Pathology, Northwestern University Medical School, 303 East
    Chicago Avenue, Chicago, IL 60611

S. Vitry, INSERM U134, Laboratoire de Neurobiologie Cellulaire, Moleculaire, et
    Clinique Hôpital de la Salpêtriere, 75651 Paris cedex 13, France

R.S. Walikonis, Molecular Biology Laboratory, Department Neurology & Biochem/Molec Biology, Mayo Clinic and Foundation, 200 First Street SW, Rochester, MN 55905

S.G. Waxman, Department of Neurology, Yale University School of Medicine, LCI 708, PO Box 20708 , New Haven, CT 06510–8018

M.H. Wong, Department Biological Science, Hunter College of the City University of New York, 695 Park Ave., New York, NY 10021

J.L. Wong, Department of Neurology, UCLA School of Medicine, Los Angeles, CA 90024

S.H. Yim, Myelin and Brain Development Section, Laboratory of Molecular and Cellular Neurobiology, NINDS, Bldg. 49, Rm. 2A10 NIH, Bethesda, MD 20892

X. Yin, Cleveland Clinic and Foundation, Department of Neuroscience, NC-30, 9500 Euclid Ave., Cleveland, OH 44195

V. Wee Yong, Department Neurology and Neurosurgery, Montreal Neurological Institute, McGill University, 3801 University, Montreal, PQ H3A 2B4, Canada

Z. Zapp, Department of Biochemistry and Molecular Biology, University of South Dakota, School of Medicine, Vermillion, SD 57069

K. Zhang, Department Biological Science, Hunter College of the City University of New York, 695 Park Ave., New York, NY 10021, U.S.A.

F. Zimmermann, Zentrum für Molekulaire Biologie, University of Heildelberg, Im Neuenheimer Feld 282, D-69120 Heidelberg, Germany

INDEX